Service Life Prediction of Polymeric Materials

Global Perspectives

Jonathan W. Martin • Rose A. Ryntz
Joannie Chin • Ray A. Dickie
Editors

Service Life Prediction of Polymeric Materials
Global Perspectives

 Springer

Editors

Jonathan W. Martin
National Institute of Standards &
 Technology (NIST)
Materials & Construction Research
 Division
100 Bureau Drive
Gaithersburg, MD 20899 USA

Joannie Chin
National Institute of Standards &
 Technology (NIST)
Materials & Construction Research Division
100 Bureau Drive
Gaithersburg, MD 20899 USA

Rose A. Ryntz
IAC Group North America
5300 Auto Club Drive
Dearborn, MI 48126
USA

Ray A. Dickie
Federation of Societies for Coatings
 Technology (FSCT)
527 Plymouth Road, Suite 415
Plymouth Meeting, PA 19462
USA

ISBN 978-0-387-84875-4   ISBN 978-0-387-84876-1 (eBook)
DOI 10.1007/978-0-387-84876-1

Library of Congress Control Number: 2008935442

Mathematics Subject Classification (2000): 01-01, 04-01, 11Axx, 26-01

Printed on acid-free paper

springer.com

Preface

The global marketplace is a demanding master. To remain competitive, manufacturers must continually innovate and market advanced materials that satisfy their customers' needs and expectations. Advances in material performance and durability, however, have been hampered by our inability to measure and predict their long-term performance.

The mission of the 4th International Symposium on Service Life Prediction, which was held December 3–8, 2006, in Key Largo, Florida, was to provide an international forum for presenting and discussing the latest scientific and technical advances leading to more reliable and quantitative predictions of the weathering performance of polymeric materials. The objectives set forth included the following:

- To critically examine the existing methodology and alternatives to the existing methodology used in assessing the service life of polymeric materials;
- To present advances in accelerated and field exposure testing protocols leading to quantitative results that are both repeatable and reproducible;
- To introduce advanced methods including high throughput and combinatorial analyses, models, data collection and storage formats, and decision support systems having a strong scientific basis;
- To discuss strategies for implementing these advances; and
- To identify outstanding scientific issues that need to be addressed.

The symposium, which included 29 invited talks and a poster session, was attended by approximately 85 distinguished international scientists, including 19 participants from nine European and Pacific Rim countries.

The opening papers described efforts to link accelerated laboratory tests with outdoor field exposures by connecting degradation of materials to their thermal and irradiance resistance in an attempt to build predictive models. The findings of a ten-year weathering study on a variety of coil applied coatings were presented, as were the results from a reliability-based service life prediction methodology. The reliability-based results are noteworthy because, for the first time, it has been demonstrated that a single, scientifically-based model, derived from laboratory data and relying on a minimal set of assumptions, was able to successfully link field and laboratory exposure results. Furthermore, this model was used to predict field performance results for a series of independent exposures over a four-year period.

Several sessions presented the latest measurement techniques utilized to study mechanisms of failure of materials ranging from propylene fibers to coatings on metal or plastic to polyethylene pipelines. Risk analyses were discussed in relation to probability of failure, while reciprocity and scaling laws, like the additivity law, were utilized and validated in characterizing coating performance in accelerated weathering devices.

Advances in exposure to acid rain and plasma-generated radical showers, the effect of coating applied strain, the effect of pigment interactions with binders, as well as the efficiency and longevity of stabilizers in polymers were discussed and related to long-term performance.

Later in the week-long symposium, computer graphic techniques were described that were capable of measuring automotive paint color. Computer graphic techniques that could render the appearance of aging materials also were presented. The importance of the ability to characterize the optical properties of materials was correlated to another important topical durability concern in coatings, scratch resistance.

The significance of laboratory automation can never be over-emphasized. A variety of combinatorial and high-throughput testing regimes utilized to develop robust coating formulations were presented. Screening tests allow for rapid, accurate, precise, repeatable, and reproducible physical property measurements, which can be cross-referenced to coating degradation. These tools will allow the formulator to develop libraries of information useful in optimizing the short- and long-term properties of coatings.

As has been the standard in the previous three service life prediction conferences, the Key Largo conference concluded with a group discussion on research and measurement needs in service life prediction. The ability to shorten test time without degradation of the correlation to polymer structural breakdown, the increased use of high-throughput screening to enhance understanding of formulation variables in relation to durability, and the indoctrination of reciprocity and scaling laws all were discussed as advancements made in service life predictive capabilities in the past ten years. There is still much to accomplish.

The editors wish to thank the Key Largo organizing committee as well as the authors and participants for making the 4th International Symposium a success. The organizing committee, co-chaired by Jonathan Martin (NIST), Rose A. Ryntz (IAC), Ray A. Dickie (FSCT), and Joannie Chin (NIST), included:

Karlis Adamsons, DuPont
Masahiko Akahori, Nippon Paint Company
Dana Bres, Dept. of Housing and Urban Development (HUD)
Matthew Celina, Sandia National Laboratory
Jean Courter, Cytec
Ray Fernando, California Polytechnic State University
F. Louis Floyd, FLF Consulting
Grace Hsuan, Drexel University
Larry Kaetzel, K-Systems
Mark Nichols, Ford Motor Company
Tinh Nguyen, NIST
Thomas Reichert, Fraunhofer Institute
Austin Reid, DuPont
Agnes Rivaton, University of Blaise Pascal in Clermont-Ferrand
Edward Schmitt, Rohm and Haas Company
Marek Urban, University of Southern Mississippi
Sam Williams, Forest Products Laboratory (FPL)
Guy Wilson, The Sherwin-Williams Company
Kurt Wood, Arkema

Contents

Characterizing and Modeling Color and Appearance

Mechanistic Measurements

High Throughput Measurements, Combinatorial Methods, Informatics

Contributions of Pigments, Additives, and Fillers

Wrap-Up: Implementation

Linking Field and Laboratory
Exposure Results

Chapter 1

Linking Accelerated Laboratory Test with Outdoor Performance Results for a Model Epoxy Coating System

Xiaohong Gu,[1] Debbie Stanley,[1] Walter E. Byrd,[1] Brian Dickens,[1] Iliana Vaca-Trigo,[2] William Q. Meeker,[2] Tinh Nguyen,[1] Joannie W. Chin,[1] and Jonathan W. Martin[1]

[1] Materials and Construction Research Division, National Institute of Standards and Technology, Gaithersburg, MD 20899.
[2] Department of Statistics, Iowa State University, Ames, IA 50011.

Laboratory and field exposure results have been successfully linked for a model epoxy coating system. The mathematical model used in making this linkage only assumed that the total effective dosage, additivity law, and reciprocity laws were valid. In this study, accurate and time-based measurements on both exposure environments and degradation properties for polymer specimens exposed to accelerated laboratory weathering device and outdoor environments have been carried out. Laboratory weathering tests were conducted in the extremely well-controlled NIST SPHERE. A factorial design consisting of four temperatures, four relative humidities (RH), four ultraviolet (UV) spectral wavelengths, and four UV spectral intensities was selected for the SPHERE exposure conditions. Based on SPHERE exposures, effects of critical environmental conditions on chemical degradation of the UV-exposed epoxy materials have been assessed. The outdoor exposure experiments were carried out on the roof of a NIST laboratory located in Gaithersburg, MD. The temperature and RH of the outdoor exposure were continuously recorded every minute and the solar spectrum was recorded every 12 minutes. The chemical degradation for specimens exposed to the SPHERE and outdoor environments was quantified by transmission FTIR and UV-visible spectroscopies. It is found that the mechanism of chemical degradation for samples exposed to outdoor environments is similar to those exposed to SPHERE. Three approaches, chemical ratios as a metric, a model-free heuristic approach, and a mathematical predictive model, have been used to combine the chemical degradation data from the SPHERE and the outdoor exposures. Successful linkages have been made via all three approaches. It has been shown that the reliability-based methodology is capable of linking laboratory and field exposure data and predicting the service life of polymeric materials.

INTRODUCTION

Attempts at linking field and laboratory exposure results and at predicting the service life of a polymeric material exposed in its service environment have been a high priority research topic of the polymeric materials community for over a century.[1-3] The inability to generate accurate, precise, and timely service life estimates for polymeric systems exposed in their intended service environments hinders product innovation as well as the timely introduction of new polymeric materials into the market.

At present, the only accepted way of generating a performance history for a new product is to expose it for many years at one or more "standard" field exposure sites such as Florida or Arizona. Performance data generated from such experiments, however, are neither repeatable nor reproducible, since the weather never repeats itself over any time scale or at any location.[1] Laboratory weathering experiments are typically designed to simulate and accelerate outdoor degradation by exposing materials for extended periods of time to high UV irradiance, elevated temperature, and relative humidity environments. However, the lack of rigorous temporal and spatial experimental control over each of the weathering factors comprising an exposure environment within and among laboratory exposure devices has made it impossible to correlate results between laboratory devices and to link laboratory and field exposure data.[4-7] The sources of laboratory exposure device experimental errors have been well-documented in the literature and are largely associated with the emission physics of line light sources.[1,5] Deficiencies include aging of the light source,[8,9] non-uniform spatial irradiance of the specimens,[5] differences in the spectral emission distributions of the sun and laboratory light sources, and the inability to independently, temporally, and spatially monitor and control the temperature and moisture content of the exposed specimens.[8,9] These deficiencies have made it difficult to study the physiochemical mechanisms underlying the degradation of the exposed materials.

Reliability-based methodology has been applied successfully in the fields of electronics, medical, aeronautic, and nuclear industries.[10,11] It was introduced in the field of polymeric materials by Martin et al.[1,12,13] To date, a number of successful studies in polymer degradation have been published.[12-16] Unlike the current descriptive methodology,[17] the reliability-based methodology is a measurement intensive, predictive methodology. In this methodology, laboratory exposure experiments are experimentally designed and the data generated from these experiments are viewed as the standard of performance against which field exposure data are compared. As such, laboratory exposure equipment and experiments must be designed to be able to independently, precisely, and accurately control each and every exposure variable over extended periods of time. Field exposure experiments, on the other hand, are viewed like a laboratory experiment; albeit, one in which individual weathering factors cannot be controlled, but each variable can be monitored and characterized in the same manner and with the same degree of precision and accuracy as its equivalent laboratory exposure variable. Linkage between field and laboratory exposure results is achieved through scientifically based mathematical models. For weathering experiments, this predictive model includes parameters for total effective dosage, additivity law, and reciprocity law. The parameters of this predictive model are estimated solely from laboratory experimental results, while outdoor exposure results are used to verify the predictions of the parameterized model using the temporal field values for spectral irradiance, panel temperature, and panel moisture content as input.

In this chapter, a successful linkage between the laboratory and field exposure results for a polymeric coating will be demonstrated using reliability-based methodology. The chapter is divided into three major sections: (1) investigation of the effects of critical environmental conditions on the photodegradation of specimens exposed in the SPHERE; (2) characterization of chemical degradation for specimens exposed outdoors; (3) linking the laboratory results to the field data based on chemical degradation via a total effective dosage model.

EXPERIMENTAL*

Materials and Specimen Preparation

The amine-cured epoxy was a stoichiometric mixture of a pure diglycidyl ether of bisphenol A (DGEBA) with an epoxy equivalent of 172 g/equiv (DER 332, Dow Chemical) and 1,3-bis(aminomethyl)-cyclohexane (1,3 BAC, Aldrich). Appropriate amounts of toluene were added to the mixture, and then all components were mechanically mixed for 7 min. After degassing in a vacuum oven at ambient temperature to remove most of the bubbles, the epoxy/curing agent/solvent mixture was applied to the substrates. Thin films of approximately 6 µm thick used for FTIR and ultraviolet-visible (UV-visible) spectroscopy measurements were obtained by spin casting the solution onto calcium fluoride (CaF_2) substrates at 209 rad/s for 30 s. All samples were cured at room temperature for 24 h in a CO_2-free dry glove box, followed by heating at 130°C for 2 h in an air-circulated oven. The glass transition temperature, T_g, of the cured films was 123°C ± 2°C, as estimated by dynamic mechanical analysis. The thickness and cure state of the films were measured and only films meeting predetermined specifications were selected for laboratory or field exposure.

Outdoor Experiments

Outdoor UV exposures were conducted on the roof of a NIST laboratory located in Gaithersburg, MD. Specimens were first loaded into multiple-window exposure cells, which were then placed in an outdoor environmental chamber at 5° from the horizontal plane and facing south. The bottom of the chamber was made of black-anodized aluminum, the top was covered with borofloat glass, and all sides were enclosed with a breathable fabric material that allowed water vapor, but prevented dust from entering the chamber. UV-visible spectral results showed that the borofloat glass did not alter the solar spectrum over the exposure times in Gaithersburg, MD. The specimen exposure chamber was equipped with a thermocouple and a relative humidity sensor, and the temperature and relative humidity in the chamber were recorded once every minute, 24 hours per day and 365 days per year.

Sample exposures were performed in 20 different months over a four-year period starting in June 2002 and ending in April 2006. The exposed specimens were designated

---

*Instruments and materials are identified in this chapter to describe the experiments. In no case does such identification imply recommendation or endorsement by the National Institute of Standards and Technology (NIST).

G1 to G20, where G1 indicates the first exposure and G20 represents the last exposure. Since the weathering conditions, and, hence, the rates of degradation of the epoxy films, exposed at different starting dates differed, the exposure durations required to achieve the same amount of degradation for each were different. The starting months covered most months of the year. Every four months, a separate set of specimens was exposed outside the chamber to assess changes occurring in the real outdoor exposure. The exposure conditions for specimens outside the chamber were measured in the same manner as those in the covered chamber.

The chemical degradation of the specimens exposed to outdoor environments was monitored using FTIR and UV-visible spectroscopies at regular intervals (e.g., every three to four days). In estimating the solar dose on the roof, all radiation was assumed to come directly from the sun and none from the blue sky. The dose was estimated from STARS spectra of the sun supplied by the Smithsonian and was corrected for solar inclination and, where appropriate, for reflection at a cover glass, using the equations in Duffie and Beckman.[18] The dosage was corrected for the angle at which the rays traversed the specimen, since the UV-visible spectra were measured normal to the specimen.

High Radiant Laboratory Exposure

The NIST SPHERE (Simulated Photodegradation via High Energy Radiant Exposure) is an integrating sphere-based weathering device that effectively reduces or eliminates all known sources of experimental error occurring in laboratory weathering devices and does not appear to introduce any new source of experimental error.[19,20] The SPHERE is a uniform light source in which panel temperature, exposure relative humidity (RH), spectral irradiance, and spectral radiant intensity can be independently, precisely, and accurately controlled over a wide range. Radiant uniformity of the SPHERE is ensured by the optical physics underlying integrating sphere technology. Currently, the SPHERE is equipped with six high intensity light lamps; the maximum total intensity achievable is equivalent to 22 suns of ultraviolet radiation. Each lamp is equipped with a dichroic reflector that effectively transmits most of the visible and infrared radiation emitted by the lamp to a heat sink while preserving and reflecting into the SPHERE the ultraviolet radiation. Ultraviolet radiation below 290 nm is eliminated from the reflected beam by positioning a cut-off filter between the lamps and the interior of the integrating sphere. The spectra radiant flux within the SPHERE was monitored continuously from 290 nm to 800 nm using a UV/visible spectrometer. The temperature within each exposure chamber can be independently controlled over extended periods of time (i.e., thousands of hours) from 25°C to 75°C with a precision of ± 0.1°C, while RH can be independently controlled from 0% to 95% within ± 0.2%.

The NIST SPHERE is equipped with 32 exposure chambers. Within each chamber, multiple specimens can be exposed. In the current configuration, each chamber exposes 17 specimens. Effectively, therefore, the NIST SPHERE is capable of simultaneously exposing 544 specimens (32 chambers x 17 specimens per chamber). Each specimen can be exposed to its own unique, well-delineated and well-controlled exposure environment. The temperature and relative humidity for each exposure chamber are controlled using a custom network of microprocessor-based systems while spectral radiant flux and spectral irradiance for each specimen can be individually controlled through the insertion of interference and neutral density filters in front of each specimen. A factorial experiment was designed that included four temperatures, four relative humidity levels, four spectral wavelengths, and four spectral intensities. The four temperatures were 25°C, 35°C, 45°C, and 55°C, and the four relative humidities were 0% RH, 5% RH, 50% RH, and 75% RH. The four wavelengths were controlled by four bandpass (BP) filters that had nominal center wavelengths of 306 nm, 326 nm, 353 nm, and 450 nm, and full-width-half maximum values of ±3 nm, ±6 nm, ±21 nm, and ±79 nm, respectively. The spectral intensity for each BP filter could be modified by placing a neutral density filter in the beam path. One of four neutral density (ND) filters (10%, 40%, 60%, or 100%) was selected for each specimen.

In each treatment, four replicates were exposed. For each specimen, the chemical damage was monitored using FTIR and UV-visible spectroscopy at regular intervals (e.g., every three to four days) and the corresponding dosage was calculated.

Measurements

FTIR: Chemical degradation of the epoxy coatings was measured by FTIR transmission using a PIKE auto-sampling accessory (PIKE Technologies). This automated sampling device allowed efficient and rapid recording of the initial FTIR transmission spectra and the FTIR spectral after some specified exposure intervals for each specimen. Since the exposure cell was mounted precisely on the auto-sampler, errors due to variation of sampling at different exposure times were minimized. The auto-sampler accessory was placed in a FTIR spectrometer compartment equipped with a liquid nitrogen-cooled mercury cadmium telluride (MCT) detector. Spectra were recorded at a resolution of 4 cm^{-1} and were the average of 128 scans. The peak height was used to represent IR intensity, which was expressed in absorbance.

UV-VISIBLE SPECTROSCOPY: In this study, the field and laboratory chemical degradation results are linked using the total effective dosage model. In this model, the total effective dosage,[13] defined as the radiation that is incident upon and absorbed by the specimen, can be calculated via:

$$D_{total}(t) = \int_0^t \int_{\lambda_{min}}^{\lambda_{max}} E_o(\lambda, t)(1 - e^{-A(\lambda)})\phi(\lambda)d\lambda \; dt$$

Where $D_{total}(t)$ is the total effective dosage at time t; λ_{min} and λ_{max} are the minimum and maximum photolytically effective wavelengths; $E_0(\lambda, t)$ is the spectral UV irradiance of the light source at time t; $A(\lambda, t)$ is the spectral absorption of specimen at wavelength and time t, and $\phi(\lambda)$ is a spectral quantum efficiency. To obtain the dosage, the spectra irradiance of the light source, the spectral transmittance of the neutron density filters and the band pass filters, and the UV-visible absorbance spectra of the specimens coated on CaF$_2$ were recorded with an HP 8452a UV-visible spectrometer between 190 nm and 820 nm at a resolution of 2 nm. The UV-visible measurement was taken each time when an FTIR measurement was taken on the specimens. To convert the UV-visible spectrometer signal into irradiance units (W/m^2), the HP 8452a UV-visible spectrometer was calibrated in NIST's Spectra Irradiance and Radiance Calibrations with Uniform Sources Facility.

A custom software program was developed at NIST to analyze the IR and UV-visible spectra, and to calculate dose (energy incident upon the specimen), dosage (energy absorbed by the specimen), and the damage (changes in the absorbance of FTIR bands). Basically, for each time slice, the spectra irradiance of the SPHERE UV source was multiplied by the spectra transmittance of the filters (ND and BP) to yield the dose, and then the dose was convoluted with the UV-visible absorbance spectrum of the specimen to yield the dosage for that time slice. After integrating the time and the wavelength, the total effective dosage at selected time was obtained.

RESULTS AND DISCUSSION

Exposure to SPHERE

CHEMICAL CHANGES DURING LABORATORY EXPOSURE: The chemical degradation of amine-cured epoxy exposed to SPHERE at different environmental conditions was monitored by FTIR transmission spectroscopy. Typical FTIR absorbance spectra of the samples as a function of SPHERE exposure time are presented in *Figure* 1, along with the different spectra that were obtained by subtracting the spectrum of the unexposed sample from the exposed spectrum after baseline correction. In *Figure* 1, the bands at 1250 cm^{-1}, 1510 cm^{-1}, and 2925 cm^{-1} are typically observed for an amine-cured epoxy before and after UV exposure. The 1250 cm^{-1} band is attributed to C–O stretching of aryl ether, the 1510 cm^{-1} band to benzene ring stretching, and the 2925 cm^{-1} band to CH$_2$ anti-symmetric stretching.[21,22] Decreases in intensities of these bands indicate that chain scission and mass loss in the films have taken place during UV exposure. In addition to the intensity decreases of the existing bands, the spectra show the formation of new chemical species in the 1620 cm^{-1} to 1800 cm^{-1} region as a result of exposures. Two prominent bands at 1658 cm^{-1} and 1728 cm^{-1}, which could be attrib-

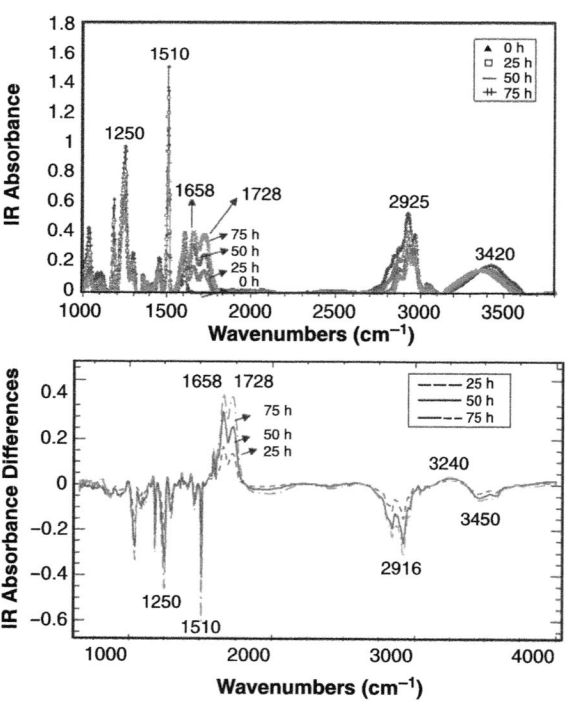

Figure 1—FTIR spectra of the amine-cured epoxy after exposure to SPHERE with 353 nm band pass (BP) filters and 100% neutral density (ND) filters at 35°C, 25% RH for different times (0 h, 25 h, 50 h, and 75 h): (upper) absorbance spectra; (bottom) difference spectra.

uted to C=O stretching of amide and ketone,[23,24] respectively, are due to formation of oxidation products. However, the band near 1728 cm⁻¹ is very broad, indicating the formation of a variety of carbonyl products. Aldehyde, ketone, acids, and esters have been reported as the residual degradation products left in the films.[21-25] The OH stretching bands near 3400 cm⁻¹ also shift to lower frequency and new bands appear around 3225 cm⁻¹. The above FTIR results are in good agreement with the photooxidative mechanisms proposed by Bellinger, et al.[21,22] and Rivaton, et al.[25] for epoxy cured with aliphatic amines.

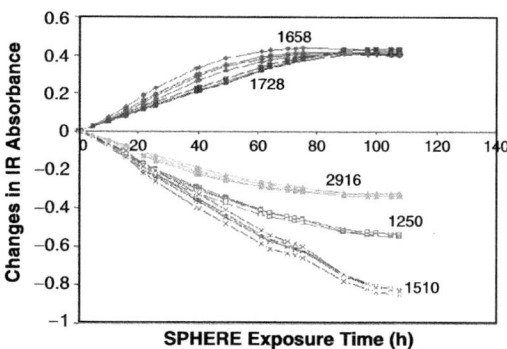

Figure 2—FTIR intensity changes of the bands at 1658 cm⁻¹, 1728 cm⁻¹, 2916 cm⁻¹, 1250 cm⁻¹, and 1510 cm⁻¹ as a function of exposure time for amine-cured epoxy exposed to SPHERE with 353 nm BP and 100% ND filters at 35°C, 25% RH. Results from four replicates are shown.

Figure 2 displays the changes in the absorbance for the bands of interest as a function of exposure time in SPHERE. Results from four replicates are presented. The superposition of these four curves indicates the accuracy and precision for the sample preparation, measurement, and environmental control. As shown, the intensity of photooxidation (1658 cm⁻¹, 1728 cm⁻¹) bands increases with exposure time until a plateau has reached, and then it starts to decrease. Continuous losses in the intensities of bands at 2916 cm⁻¹, 1250 cm⁻¹, and 1510 cm⁻¹ are attributed to chain scission. Since the high flux UV radiation is so aggressive, it is nearly impossible to keep the optical components of an exposure condition at the same level over extended periods of time. Therefore, dose or dosage is a better metric than time because variation of the lamp intensity is taken into account. Figure 3 shows the chemical changes as a function of dose for the same specimens shown in Figure 2. Compared to the plot using exposure time, Figure 3 reveals the effect of the actual UV irradiance upon the specimens on chemical degradation regardless of changes of light source or total radiation. The similarity between the plots in Figure 2 and in Figure 3 indicates that the light intensity of the SPHERE was mostly constant during the observed exposure period.

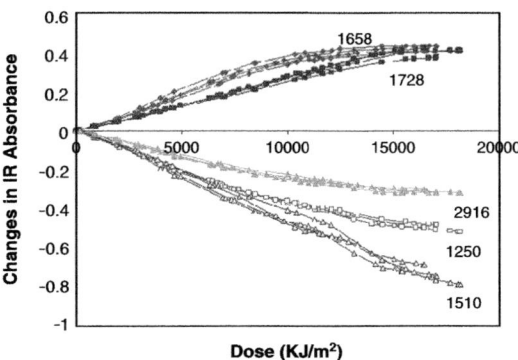

Figure 3—FTIR intensity changes of the bands at 1658 cm⁻¹, 1728 cm⁻¹, 2916 cm⁻¹, 1250 cm⁻¹, and 1510 cm⁻¹ as a function of dose for amine-cured epoxy exposed to SPHERE with 353 nm BP filters and 100% ND filters at 35°C, 25% RH. Results from four replicates are shown.

TOTAL EFFECTIVE DOSAGE: Only the incidence absorbed energy can cause a specimen to

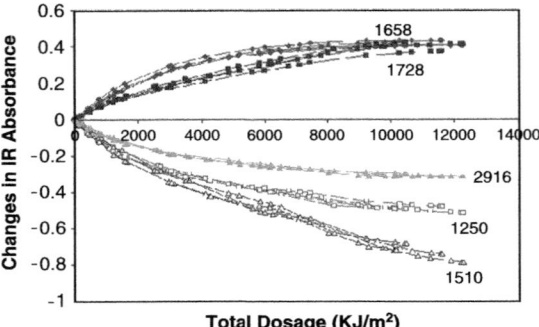

Figure 4—FTIR intensity changes of the bands at 1658 cm⁻¹, 1728 cm⁻¹, 2916 cm⁻¹, 1250 cm⁻¹, and 1510 cm⁻¹ as a function of dosage for amine-cured epoxy exposed to SPHERE with 353 nm BP filters and 100% ND filters at 35°C, 25% RH. Results from four replicates are shown.

degrade; thus, the total effective dosage is probably the most scientifically accurate metric for quantifying the severity of damage in the reliability-based methodology. As mentioned earlier, the dosage is calculated from the dose and the UV-visible spectra of the specimens [equation (1)]. By calculating the total effective dosage (hereafter referred to as dosage) using the custom written software, the chemical changes as a function of dosage were graphed, as shown in *Figure* 4. Compared to the near linear relationship between the chemical changes and the exposure time or dose, the slopes of these curves decrease gradually with the increase in the dosage. This phenomenon can be explained by the shielding effect of the degradation products. Note that the initial slope of the curve is generally defined as apparent spectral quantum yield and can be used to evaluate the initial efficiency of the radiation at different wavelengths.[26]

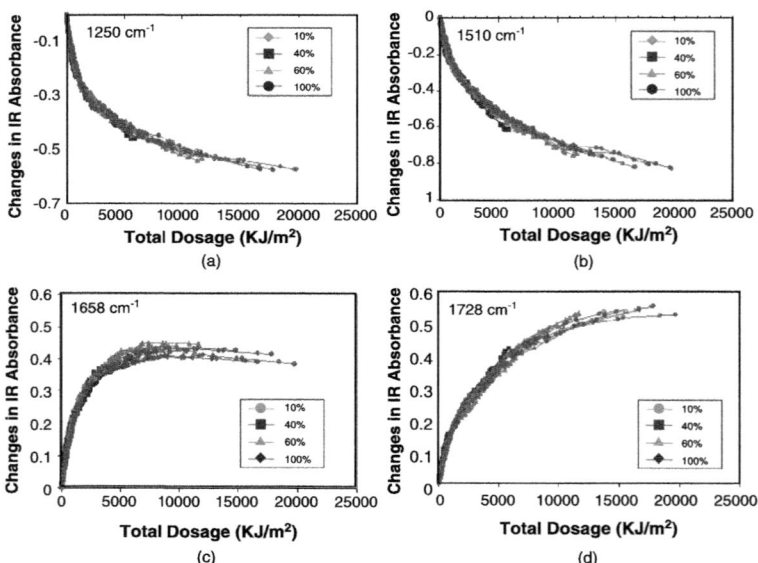

Figure 5—FTIR intensity changes of the bands at 1658 cm⁻¹, 1728 cm⁻¹, 1250 cm⁻¹, and 1510 cm⁻¹ as a function of dosage for amine-cured epoxy exposed to SPHERE with 326 nm filter and four different irradiance levels (10%, 40%, 60%, and 100%) at 25°C and 0% RH. Results from four replicates are shown.

EFFECT OF MAIN ENVIR-
ONMENTAL PARAMETERS ON
CHEMICAL DEGRADATION: Since
there were many combinations of
the environmental conditions used
in this study, the results given in
this section only show representative data to demonstrate the general effects of the environments on
this epoxy material. More complete results will be presented in
future publications. *Figure* 5 illustrates the intensity change of four
interested IR bands as a function of
dosage for specimens exposed to
SPHERE under 326 nm BP filters
and four ND filters with the target
transmittance values of 10%, 40%,
60%, and 100% at 25°C and 0%
RH. There are four replicates for

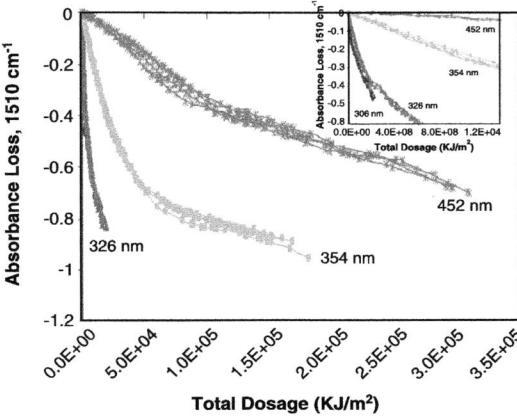

Figure 6—FTIR intensity changes of the bands at 1510 cm⁻¹ as a function of dosage for amine-cured epoxy exposed to the SPHERE with 100% ND filters at 25°C and 0% RH under the 306 nm, 326 nm, 354 nm, and 452 nm BP filters, respectively. The inset is for expanded early stage of degradation. The dosage is up to 1.6 × 10⁴ kJ/m². Results from four replicates are shown.

each exposure condition. The curves for all irradiance levels superimpose onto a single master curve, suggesting that the dosage needed to cause a given level of damage is inde-pendent of the incident intensity. This behavior is observed for other exposure conditions as well. These results indicate that reciprocity law is valid for this amine-cured epoxy sys-tem over the experiment ranges of temperature and relative humidity. Similar phenomena have been found for the UV degradation of the acrylic-melamine polymer.[27]

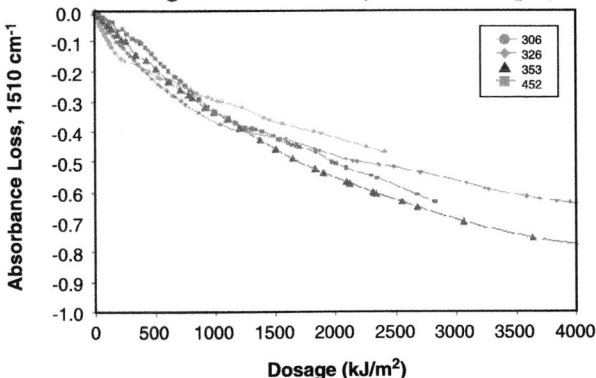

Figure 7—FTIR intensity changes of the bands at 1510 cm⁻¹ as a function of dosage for amine-cured epoxy exposed to the SPHERE with 100% ND filters at 25°C and 0% RH under the 306 nm, 326 nm, 354 nm, and 452 nm BP filters when the damage versus dosage curves are horizontally shifted. The shifting factors for the 306 nm, 326 nm, 354 nm, and 452 nm here are 100:50:6:1, respectively. Results from four replicates are shown.

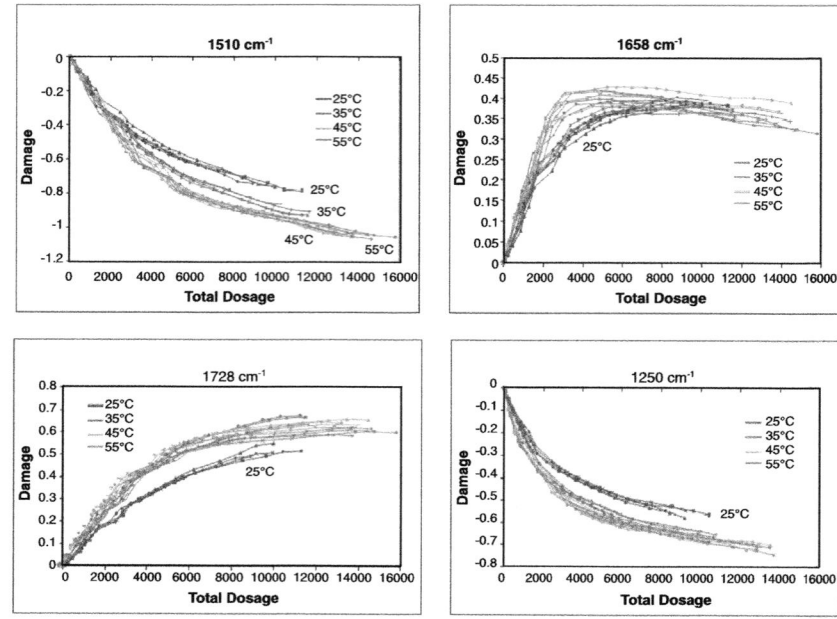

Figure 8—FTIR intensity changes of the bands at 1510 cm⁻¹, 1658 cm⁻¹, 1728 cm⁻¹, and 1250 cm⁻¹ as a function of dosage for amine-cured epoxy exposed to SPHERE with 326 nm BP filters and 100% ND filters under 0% RH at four temperatures (25°C, 35°C, 45°C, and 55°C). Results from four replicates are shown.

The effect of spectral wavelength on the efficiency of photodegradation of amine-cured epoxy can be seen in *Figure* 6, in which specimens were exposed to 100% ND filters at 25°C and 0% RH under the 306 nm, 326 nm, 354 nm, and 452 nm BP filters, respectively. The initial part (the dosage is up to 1.6 x 10⁴ kJ/m²) was expanded and shown in the inset. A clear sequence for the four wavelengths was observed with regard to the damage per unit dosage. The longer the wavelength, the lower is the damage efficiency. The curves for all wavelengths can be nearly superimposed when the damage versus dosage curve for 452 nm is used as a reference and the other curves are horizontally shifted. The shifting factors for the 306 nm, 326 nm, 354 nm, and 452 nm in *Figure* 7 are 100:50:6:1, respectively. These numbers can be used to evaluate the relative efficiencies of different wavelengths. The ratios obtained from other conditions for the four wavelengths are also similar, indicating that the relative photodegradation efficiencies of the wavelengths are independent of the studied UV irradiance intensity, temperature, and humidity for the same material; that is, the additivity law appears to be obeyed over the range of temperature, relative humidity, and spectral irradiance bands investigated in this study. Relative efficiencies obtained from other IR damage bands also follow the same trend.

The effect of temperature on photodegradation can be demonstrated by *Figure* 8, which shows the damage versus dosage for specimens exposed to SPHERE with 326 nm BP filters and 100% ND filters under 0% RH at different temperatures. Higher temperature

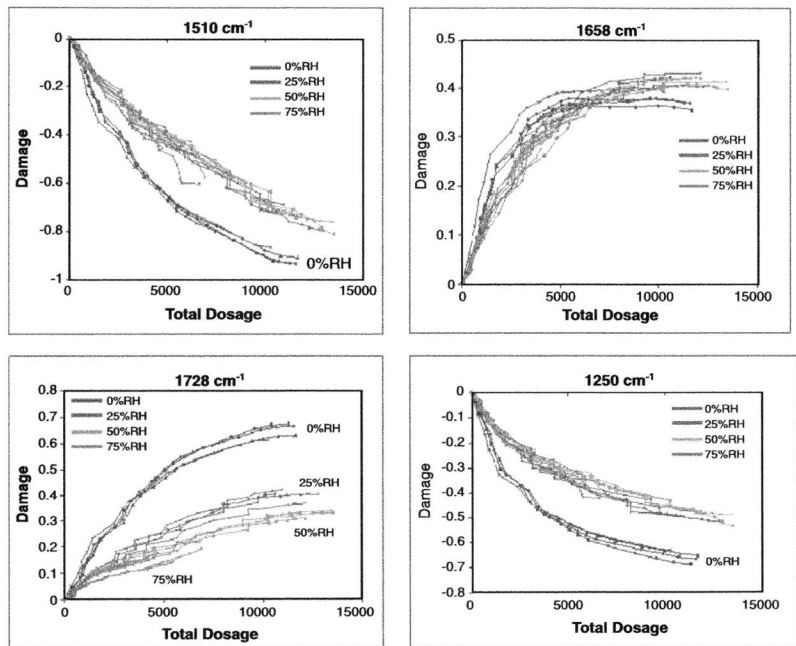

Figure 9—FTIR intensity changes of the bands at 1510 cm⁻¹, 1658 cm⁻¹, 1728 cm⁻¹, and 1250 cm⁻¹ as a function of dosage for amine-cured epoxy exposed to SPHERE with 326 nm BP filters and 100% ND filters at 35°C under different RH (0% RH, 25% RH, 50% RH, and 75% RH). Results from four replicates are shown.

exhibits higher damage per dosage for all major IR damage bands. However, the increase in the damage efficiency seems to slow down when the temperature is above 35°C. Compared to the exposure at 0% RH, a similar temperature effect is observed for specimens exposed to SPHERE at higher humidities. Details will be shown in future publications.

The effect of relative humidity on the photodegradation of amine-cured epoxy is displayed in *Figure* 9, where the specimens were exposed to SPHERE with 100% ND filters and 326 nm BP filters at 25°C under four different RHs. When RH is increased from 0% RH to 25% RH, the damage per dosage substantially decreases for the degradation modes corresponding to the bands at 1510 cm⁻¹, 1728 cm⁻¹, and 1250 cm⁻¹, but only slightly declines for 1658 cm⁻¹. Additionally, the humidity effect seems to slow down when the RH is higher than 25% RH. Among the four IR bands, the 1728 cm⁻¹ band shows the most sensitivity to the RH change.

The different responses of the different IR damage bands on the humidity changes result from the differences in their degradation mechanisms. The carbonyl absorption band at 1728 cm⁻¹ has been attributed to the formation of ketone products by Bellinger[21,22] or to aldehyde formation by Patterson-Jones.[28] In either case, the products are formed through the release of water molecules by dehydration of the hydroxyl groups (Schemes 1 and 2 in *Figure* 10). The process of water release could be hindered at high RH so that the formation of the products corresponding to 1728 cm⁻¹ band would be retarded with respect to the same amount of dosage at dry exposure condition. However,

14

Scheme 1. Formation of Ketone C=O (Bellinger[21,22])

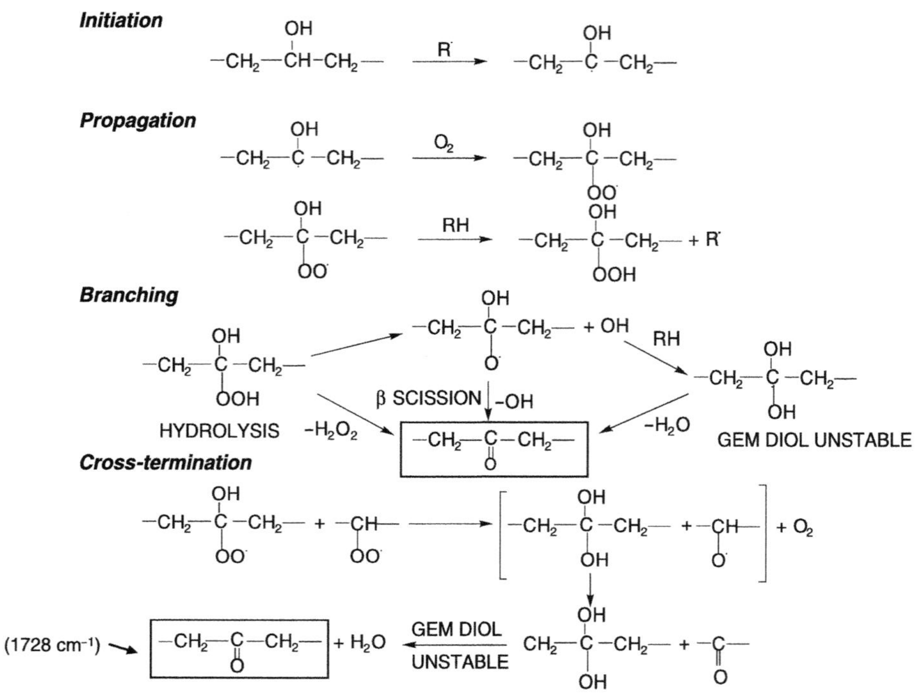

Initiation

$$-CH_2-\underset{OH}{\overset{}{CH}}-CH_2- \xrightarrow{R^{\cdot}} -CH_2-\underset{OH}{\overset{}{C}}-CH_2-$$

Propagation

$$-CH_2-\underset{OH}{\overset{}{C}}-CH_2- \xrightarrow{O_2} -CH_2-\underset{OO^{\cdot}}{\overset{OH}{C}}-CH_2-$$

$$-CH_2-\underset{OO^{\cdot}}{\overset{OH}{C}}-CH_2- \xrightarrow{RH} -CH_2-\underset{OOH}{\overset{OH}{C}}-CH_2- + R^{\cdot}$$

Branching

$$-CH_2-\underset{OOH}{\overset{OH}{C}}-CH_2-$$

$-CH_2-\underset{O^{\cdot}}{\overset{OH}{C}}-CH_2- + OH$ $\xrightarrow{RH}$ $-CH_2-\underset{OH}{\overset{OH}{C}}-CH_2-$

β SCISSION | −OH

HYDROLYSIS −H$_2$O$_2$ → $-CH_2-\underset{O}{\overset{}{C}}-CH_2-$ ← −H$_2$O GEM DIOL UNSTABLE

Cross-termination

$$-CH_2-\underset{OO^{\cdot}}{\overset{OH}{C}}-CH_2- + -CH- \longrightarrow \left[-CH_2-\underset{OH}{\overset{OH}{C}}-CH_2- + -CH- \right] + O_2$$

(1728 cm⁻¹) → $-CH_2-\underset{O}{\overset{}{C}}-CH_2-$ + H$_2$O $\xleftarrow[\text{UNSTABLE}]{\text{GEM DIOL}}$ $CH_2-\underset{OH}{\overset{OH}{C}}-CH_2- + -C-$

Scheme 2. Formation of Aldehyde C=O (Patterson-Jones[28])

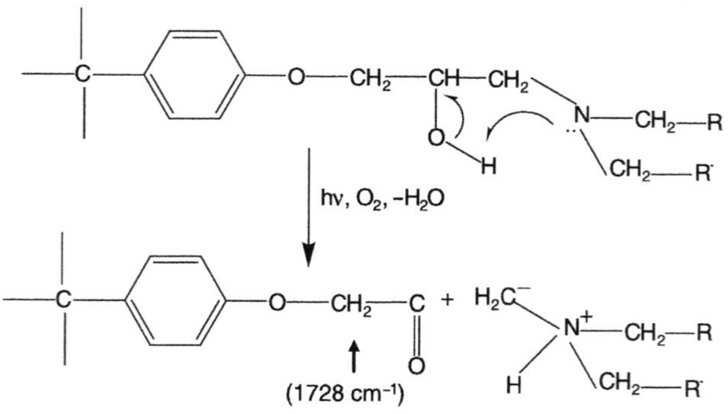

hv, O$_2$, −H$_2$O

(1728 cm⁻¹)

Figure 10—Proposed mechanisms for attribution of IR absorption band at 1728 cm⁻¹: formation of ketone C=O group (Scheme 1) and formation of aldehyde C=O group (Scheme 2).

it is uncertain why changes of other degradation modes, such as the loss of 1250 cm^{-1} band, which has been attributed to chain scission of C–O stretching of aryl ether, and the loss of 1510 cm^{-1}, which has been attributed to the chain scission of the benzene ring, also decreases substantially as the RH increases from 0% RH to 25% RH. One possibility is that when the formation of the 1728 cm^{-1} band degradation products is impeded, other degradation reactions such as chain scission could also be affected.

EFFECT OF CRITICAL EXPOSURE PARAMETERS ON MECHANISM OF CHEMICAL DEGRADATION: A simple approach to link or compare the mechanisms of chemical degradation under different exposure conditions can be achieved by plotting the characteristic IR bands associated with photodegradation against one another. Gerlock et al.[29] used this method to compare the weathering chemistry of automotive coatings exposed to the accelerated laboratory environments with different light sources. Basically, for different exposure conditions, when the peak ratios show a similar trend, the degradation mechanisms tend to be similar, even though the kinetics of the degradation are quite different. As discussed above, the exposure conditions such as intensity of UV radiation, the wavelength of the spectra UV, the temperature, and the relative humidity all play important, but different roles on the total amount of chemical degradation and/or the efficiency of the chemical damage with respect to per dosage. In this part, the effect of these main exposure parameters on the mechanism of the chemical degradation is investigated. *Figure* 11 shows the effect of the wavelength on the ratios of different chemical changes,

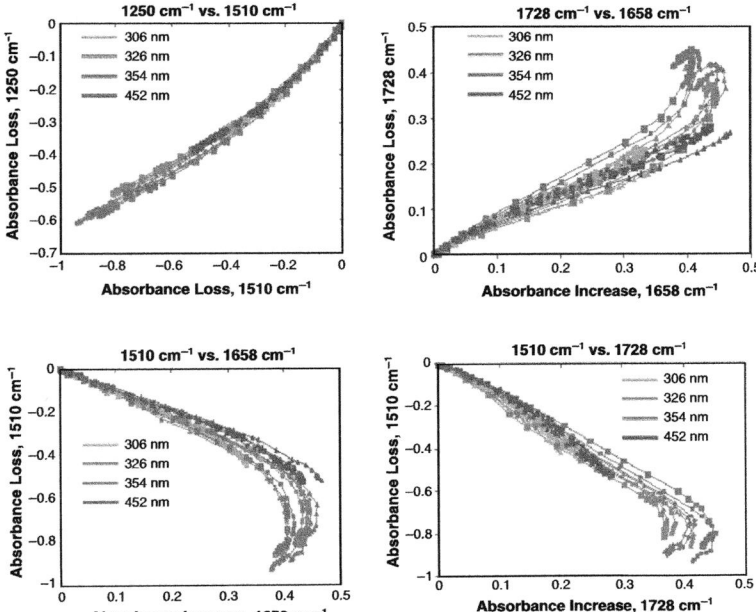

Figure 11—Effect of wavelength of BP filters on the ratios of different chemical changes. Specimens were exposed to SPHERE with four different filters (306 nm, 326 nm, 354 nm, and 452 nm) and 100% irradiance level at 45°C and 25% RH. Results from four replicates are shown.

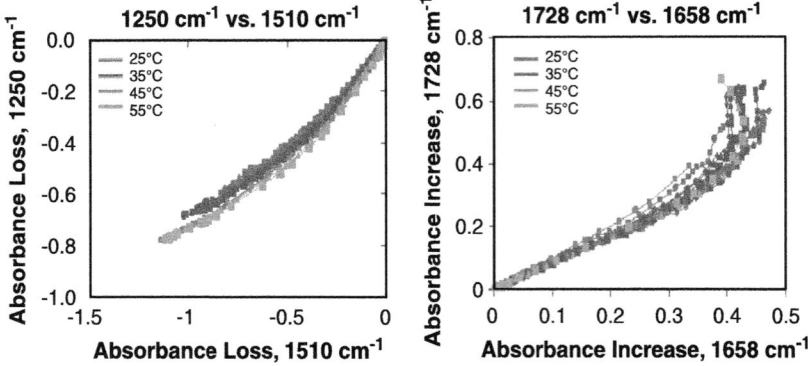

Figure 12—Effect of temperature on the ratios of different chemical changes. Specimens were exposed to SPHERE with 353 nm filter and 100% irradiance level at 25% RH and four different temperatures (25°C, 35°C, 45°C, and 55°C). Four replicates are shown for each specimen.

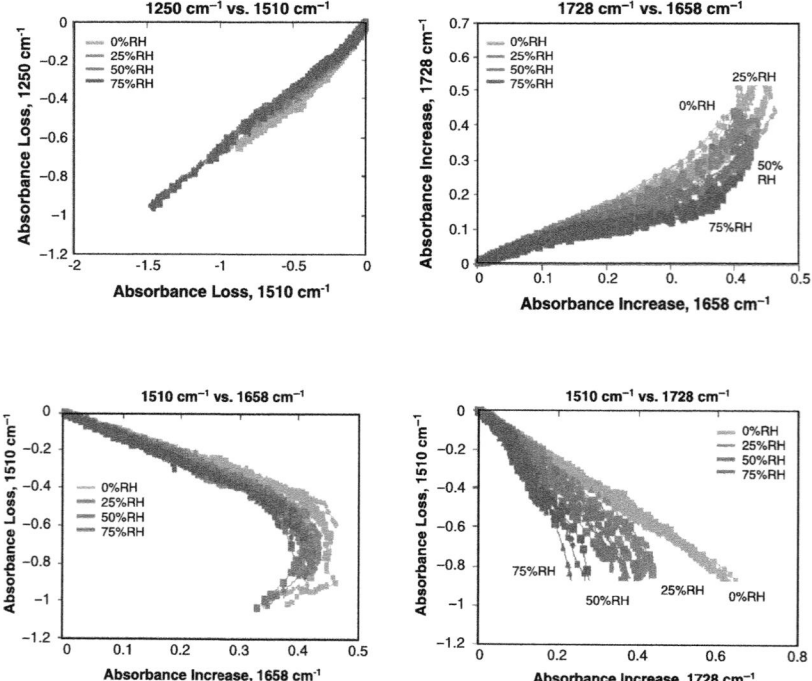

Figure 13—Effect of relative humidity on the ratios of different chemical changes. Specimens were exposed to SPHERE at 100% irradiance level at 35°C with four different BP filters (306 nm, 326 nm, 353 nm, and 450 nm) and four different RHs (0% RH, 25% RH, 50% RH, and 75% RH). Four replicates are shown for each specimen.

which are 1250 cm^{-1} versus 1510 cm^{-1}, 1728 cm^{-1} versus 1658 cm^{-1}, 1510 cm^{-1} versus 1658 cm^{-1}, and 1510 cm^{-1} versus 1728 cm^{-1}. The specimens were exposed to SPHERE with 100% irradiance level and four BP filters (306 nm, 326 nm, 354 nm, and 452 nm) at 45°C and 25% RH. Results from four replicates are shown. Apparently, the curves for all wavelengths overlay to an almost single line, except the ends of the curves corresponding to severe degradation are slightly scattered. At severe degradation levels, carbonyl products (1658 cm^{-1} and 1728 cm^{-1}) start to level off in their amounts due to secondary chemical reactions and physical depletion. In those cases, the absorbance loss at 1510 cm^{-1} continues to grow as degradation proceeds, but as the absorbance increase at 1658 cm^{-1} or 1728 cm^{-1} starts to decrease. Therefore, the curves of the ratios based on these bands deviate at the late stages of degradation except those neither associated with 1658 cm^{-1} nor with 1728 cm^{-1}. Although different wavelengths of the spectra UV have different quantum efficiencies for photodegradation, the mechanism of chemical degradation seems independent of the wavelength of the UV irradiation for this amine-cured epoxy, as shown in *Figure* 11. This data has laid the groundwork for an on-going study to more rigorously validate the additivity law.

The effect of temperature on the degradation mechanism is shown in *Figure* 12, where the ratios of different chemical changes for specimens exposed to SPHERE with 353 nm BP filters, 100% ND filters at four temperatures (25°C, 35°C, 45°C, and 55°C) and 25% RH are presented. Four replicates are displayed for each specimen. Sixteen curves are displayed in each ratio plot. Similar to the wavelength effect, the different temperatures do not substantially alter the ratios of the same pair of chemical changes from 25°C to 55°C. Similar phenomenon is observed for the intensity effect of UV irradiance and the spectra wavelength on the photodegradation. The details will be presented in a future publication.

However, an obvious effect of relative humidity on the ratios of chemical changes has been observed, particularly for the pairs associated with the 1728 cm^{-1} bands. As shown in *Figure* 13, for both 1250 cm^{-1} versus 1510 cm^{-1} plot and 1510 cm^{-1} versus 1658 cm^{-1} plot, the curves obtained from specimens exposed to SPHERE under different humidities superimpose, while for the 1728 cm^{-1} versus 1658 cm^{-1} plot and the 1510 cm^{-1} versus 1728 cm^{-1} plot, the curves do not superimpose; instead, they sequentially drop with increasing RH. This phenomenon can be explained by the fact that the band at 1728 cm^{-1} is more sensitive to the humidity than any of the other three bands. These results suggest that the relative humidity influences the 1728 cm^{-1} degradation mechanism.

Outdoor Exposure

Outdoor UV exposures were carried out in Gaithersburg, MD, from 2002 through 2006. Twenty groups of specimens were placed in a covered outdoor environmental chamber starting at different times of the year. Some groups were exposed outside the covered chamber to serve as a real outdoor exposure. The temperature and RH in the chamber and outside the chamber were continuously recorded every minute and the solar spectrum was recorded every 12 min. Typical temperature and RH data for specimens exposed inside the outdoor exposure chamber and outside the chamber are shown in *Figures* 14 and 15, respectively. Both the temperature and the humidity changed dramatically inside and outside the rooftop chamber. The temperature inside the chamber is

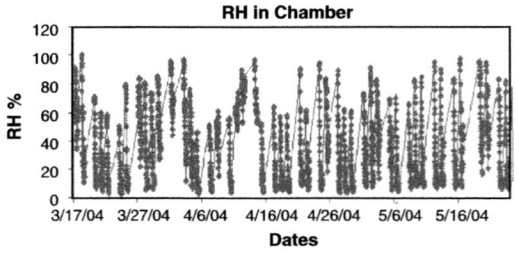

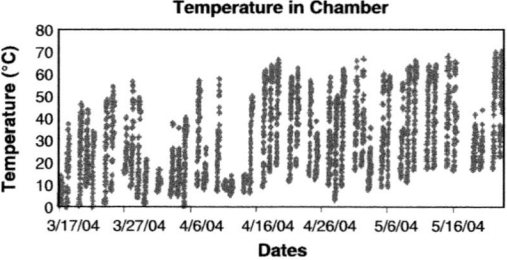

Figure 14—Relative humidity and temperature inside rooftop exposure chamber in Gaithersburg, MD, for exposure dates from March 17, 2004, to May 26, 2004.

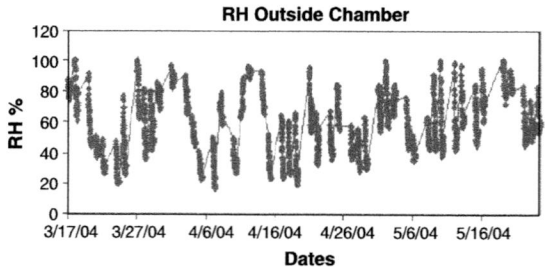

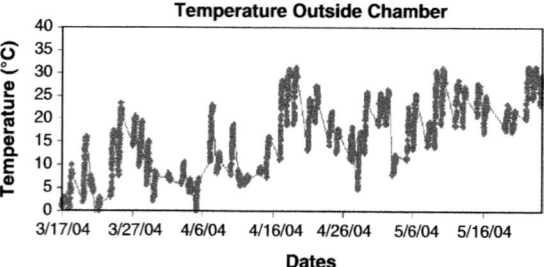

Figure 15—Relative humidity and temperature outside the rooftop exposure chamber in Gaithersburg, MD, for exposure dates from March 17, 2004, to May 26, 2004.

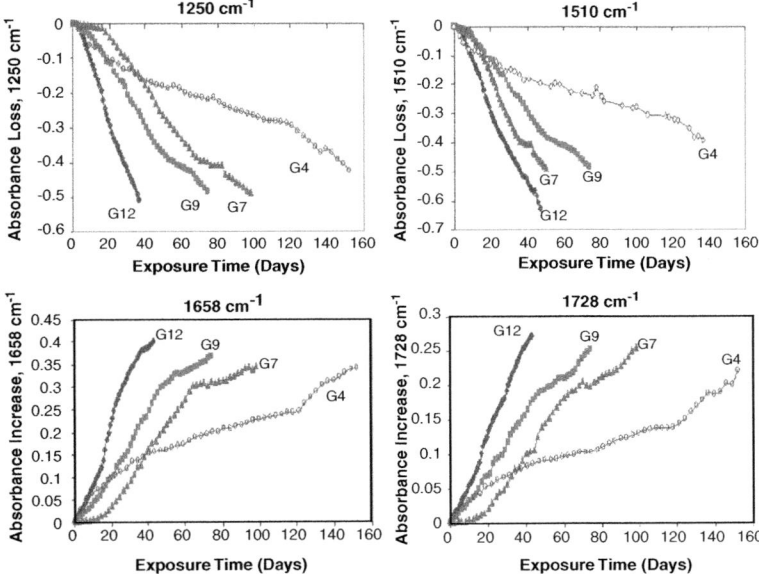

Figure 16—FTIR intensity changes of the bands at 1250 cm⁻¹, 1658 cm⁻¹, 1510 cm⁻¹, and 1728 cm⁻¹ as a function of exposure time for four groups of epoxy specimens exposed to outdoor environmental chamber in Gaithersburg, MD, started from 6/06/03 for G12, 3/04/03 for G9, 12/24/02 for G7, and 9/30/02 for G4. Results are averaged from four replicates. The uncertainty is within 5%.

sometimes as much as 25°C higher than the surrounding ambient temperature in the daytime. For the humidity, both inside and outside chambers can reach 100% RH, mostly at night. However, the environment in the chamber tends to be drier than the outside during the daytime, which is consistent with higher temperatures for the chamber in the daytime as compared to the ambient condition.

The chemical degradation of amine-cured epoxy exposed in the outdoor chamber was monitored by FTIR spectroscopy in transmission mode, in the same way as these measurements were made for SPHERE exposed specimens. Chemical changes as a function of exposure time for four groups exposed to the outdoor environmental chamber are presented in *Figure* 16. These four groups were selected as the representatives of the four seasons. The starting time for each exposure is 9/30/2002 for G4, 12/24/2002 for G7, 3/04/2003 for G9, and 6/06/2003 for G12. Due to the different exposure periods, the four groups exhibit different behaviors in the chemical changes with exposure time. The initial change rate of each group depends on the starting dates; the summer group (G12) is fastest and the winter group (G4) is the slowest in a long-time range of exposure. When the intensities of the degradation bands are plotted as a function of dosage, the effect of the solar spectra of the different seasons can be minimized, and the differences of the curves are mainly due to temperature and humidity effects on the degradation efficiency with respect to dosage. The different efficiencies of different ranges of wavelength are

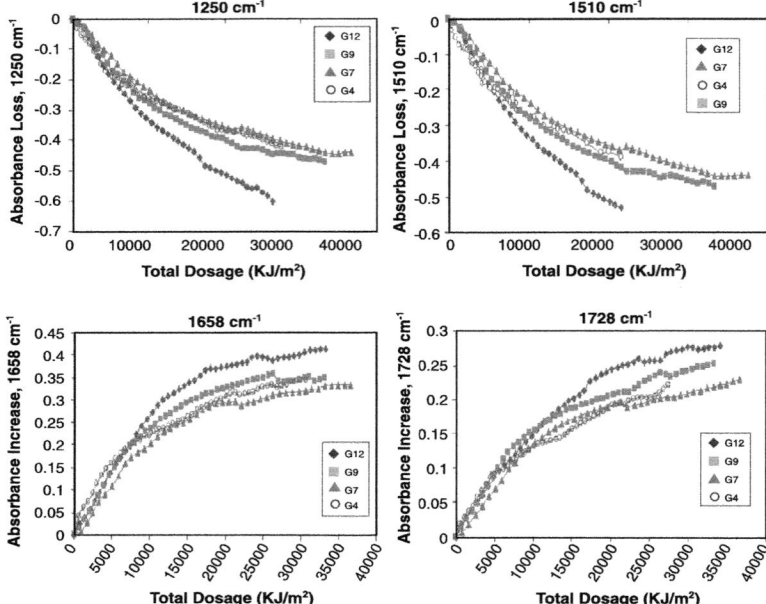

Figure 17—FTIR intensity changes of the bands at 1250 cm⁻¹, 1658 cm⁻¹, 1510 cm⁻¹, and 1728 cm⁻¹ as a function of dosage for four groups of epoxy specimens exposed to covered outdoor chamber in Gaithersburg, MD, started from 6/06/03 for G12, 3/04/03 for G9, 12/24/02 for G7, and 9/30/02 for G4. Results are averaged from four replicates. The uncertainty is within 5%.

taken into account for the total dosage calculation. As shown in *Figure* 17, for the same amount of damage, it needs the least dosage for summer group (G12), and the most dosage for winter group (G4). Moreover, the curves in *Figure* 17 are much tighter than those in *Figure* 16, suggesting that the damage is dominated by the dosage rather than the exposure time. It appears that the damage-dosage plot is more meaningful than the damage-time plot when degradation at different outdoor environments are compared.

To understand the influence of the covered environments on the chemical degradation, the results obtained from the specimens exposed to the outdoor chamber are compared with those exposed to the real outdoors, as shown in *Figure* 18. It appears that the changes of the IR bands are quite similar under the two conditions for G11, except the intensity of the band at 1658 cm⁻¹ for specimens inside the chamber is slightly enhanced compared to those exposed to the outdoors. Similar results are observed for the other group where specimens were exposed to both outdoor covered chamber and real outdoor. Because the average temperature for the covered chamber tends to be higher than the one in the real outdoor, the increase in the 1658 cm⁻¹ product is probably due to the elevated temperature, which is consistent with the laboratory exposure data.

In summary, the chemical degradation for samples exposed to outdoor environments is similar to those exposed to the SPHERE laboratory chamber. However, due to the time-dependent changing and the cycling of the outdoor environments, it is necessary to

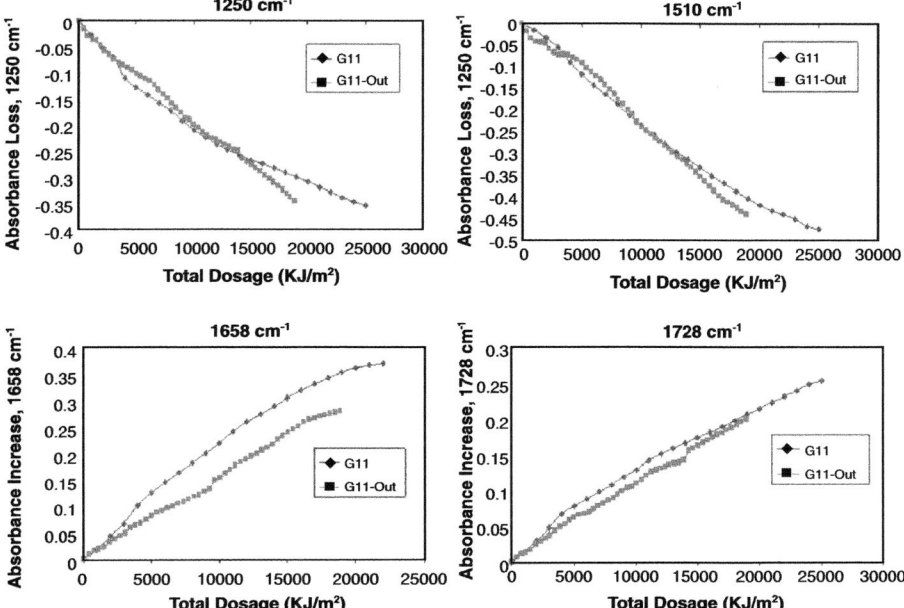

Figure 18—FTIR intensity changes of the bands at 1250 cm^{-1}, 1658 cm^{-1}, 1510 cm^{-1}, and 1728 cm^{-1} as a function of dosage for epoxy specimens exposed to outdoor chamber (G11) and real outdoor (G11-Out) from 5/06/03 in Gaithersburg, MD. Results are averaged from four replicates. The uncertainty is within 5%.

monitor the environmental factors as well as the degradation in an accurate, precise, and timely way. Further, since the solar spectra, the temperature, and the RH of the outdoors are neither repeated nor able to reproduce themselves at different times or at different places, it is impractical to try to simulate the outdoor conditions by changing the parameters of the laboratory conditions. However, if outdoor exposure is just treated as another laboratory-like experiment, and is monitored in the same manner and with the same degree of accuracy and precision as variables characterized in the laboratory, by using the total effective dosage as a metric, the linkage between the laboratory and the outdoor exposures can be made.

Linking Field and Laboratory Exposure Results

The ability to link field and laboratory exposure results is critical to predict the service life of a polymeric material in its intended service environment. In this section, three strategies will be used to link the field and laboratory exposure results. At present, the only measure of degradation that will be used to establish this linkage is chemical damage. The first strategy employed was to use the ratios of the intensities of IR bands corresponding to the different chemical changes to evaluate if the degradation mechanisms between the laboratory exposure and field exposure are the same. The second strategy is a model-free heuristic approach. A computer program is written to estimate the damage

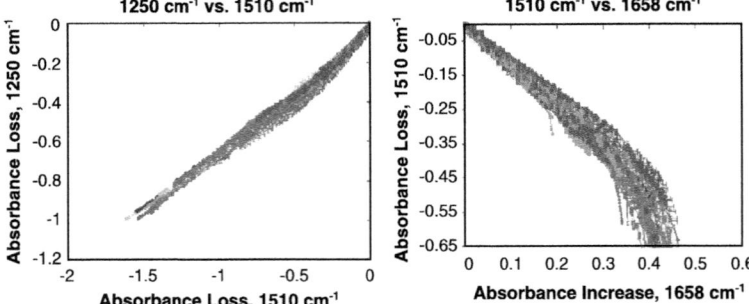

Figure 19—Plots of absorbance loss at 1250 cm⁻¹ versus absorbance loss at 1510 cm⁻¹ (left) and absorbance loss at 1510 cm⁻¹ versus absorbance increase at 1658 cm⁻¹ (right) for SPHERE exposed specimens and outdoor exposed specimens. The SPHERE exposure parameters include four temperatures (25°C, 35°C, 45°C, and 55°C), four relative humidities (0% RH, 25% RH, 50% RH, and 75% RH), four wavelengths (306 nm, 326 nm, 353 nm, and 450 nm) and 100% irradiance level, except only 0% RH for 25°C. Outdoor specimens include 14 groups (G4–G17) exposed in outdoor chamber. Four replicates are shown for each specimen.

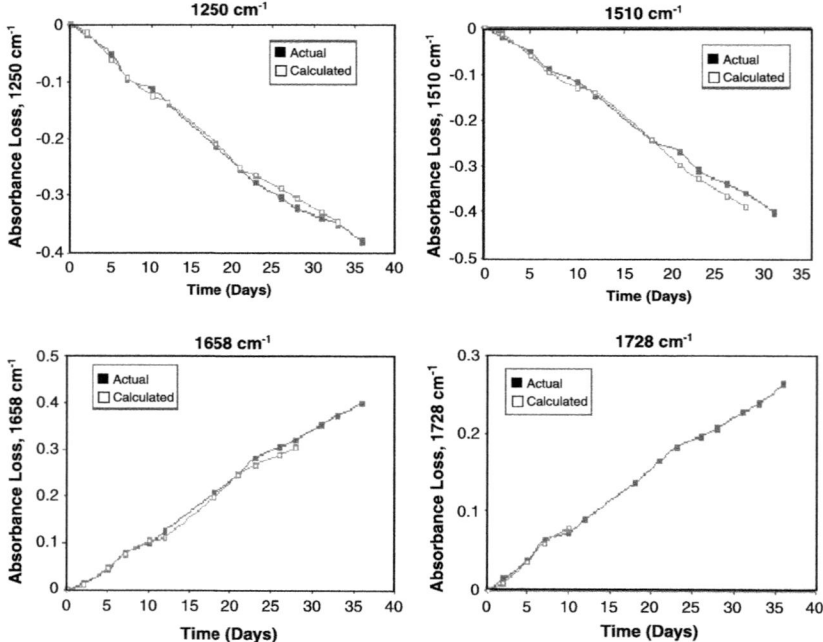

Figure 20—Comparison of the predicted damage and observed damage for IR bands at 1250 cm⁻¹, 1510 cm⁻¹, 1658 cm⁻¹, and 1728 cm⁻¹ for one specimen in G17 that was exposed to outdoor chamber from 7/19/04.

of outdoor exposed specimens based on the damage/dosage curves from the SPHERE exposure. The outdoor environmental data used were taken at 12-min intervals, including spectral dosage, temperature, and RH. The third strategy is using a prediction model[30] to calculate the damage of the specimens exposed outdoors. This model is based on total effective dosage model and the accumulated damage. Similar to the second strategy, the damage was accumulated over each 12-min interval. The calculated outdoor damage is compared to the observed outdoor damage to verify if field damage is predicted from laboratory results.

USING CHEMICAL RATIO AS A METRICS FOR DEGRADATION MECHANISM COMPARISON: *Figure* 19 presents the ratios of chemical changes (1250 cm^{-1} versus 1510 cm^{-1}, left; 1510 cm^{-1} versus 1658 cm^{-1}, right) for specimens exposed to SPHERE at 32 different conditions and specimens exposed to the outdoor chamber from 14 different times of the four years. A total of 184 curves are displayed in each plot, including 128 curves from laboratory exposure and 56 from outdoor exposure. The superimposition of these curves is observed for both plots, indicating that not only the degradation mechanisms within the 32 different laboratory conditions are the same, but they are also the same as those from the 14 outdoor exposures. As we know, the outdoor environment is time-dependent and cycling; it seems that these environmental variables do not significantly affect the mechanism of chemical degradation of the amine-cured epoxy.

MODEL-FREE HEURISTIC APPROACH: This approach uses a computer program to estimate the damage of the outdoor exposed specimen from the actual damage/dosage curves based on the SPHERE exposure. In this approach, the damage was accumulated over each 12-min interval. The solar spectra, temperature, and RH inside/outside the outdoor chamber for each time slice were input to the program. When the outdoor temperature or RH was not included in the conditions performed in the SPHERE, the closest set of conditions available was used. In the process, the solar radiation was separated into the dose passing through the four filter ranges so that the incremental dose was known for each filter range for each time slice. No filters were used for outdoor exposures. In the estimation of the evolution of the damage of an outdoor specimen, the latest estimate of the damage (with a starting value of zero) was used to find the corresponding amount of dosage from the damage/dosage curve of the closest SPHERE condition through one of the four filters. The incremental solar dosage from the outdoor time slice was used to move along the curve to a new level of dosage in the SPHERE exposure curve and the corresponding damage was found. This new damage level was then used to repeat the process for the next filter. The process was cycled over the filters and subsequent 12-min intervals of outdoor exposure until the time period between specimen spectra had been covered. The additivity law was assumed to be obeyed, thus the damage was summed up from each filter range. The estimated damage and the observed damage for the outdoor exposed specimens can be plotted against outdoor exposure time or the total dosage. When the dosage is selected, the different efficiencies of different ranges of wavelength should be taken into account for the total dosage calculation. A typical example is shown in *Figure* 20, corresponding to G17, which was exposed to the outdoor chamber from 7/12/04. Another example is shown in *Figure* 21, corresponding to a group exposed to real outdoor (G11-Out). As can be seen, for both exposures, the

24

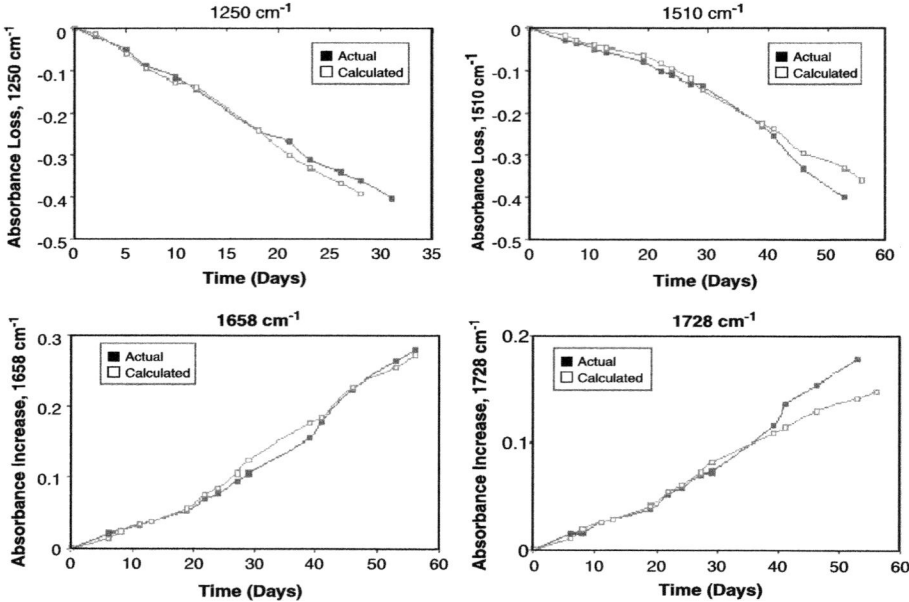

Figure 21—Comparison of the predicted damage (calculated) and observed damage (actual) for IR bands at 1250 cm⁻¹, 1510 cm⁻¹, 1658 cm⁻¹, and 1728 cm⁻¹ for one specimen in G11 that was exposed to real outdoor from 5/06/03.

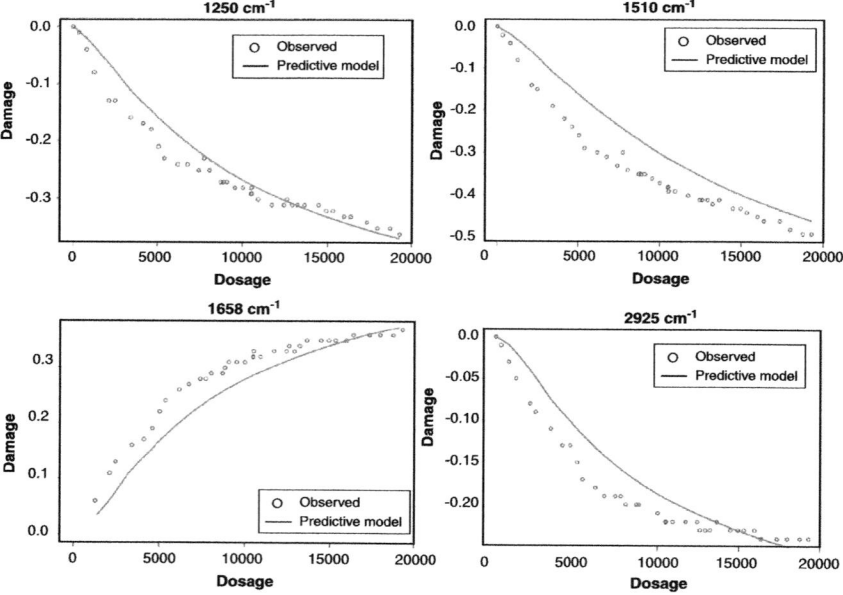

Figure 22—Comparison of the predicted damage and observed damage for IR bands at 1250 cm⁻¹, 1510 cm⁻¹, 1658 cm⁻¹, and 2925 cm⁻¹ for one specimen in G3-10 that was exposed to outdoor chamber from 8/30/02.

calculated damage from all four FTIR peaks agrees well with the actual damage as a function of outdoor exposure time, which includes the nighttime.

CUMULATIVE DAMAGE MODEL[30]: The cumulative damage model can be expressed as

$$g'(t) = \frac{d\text{Đ}[\tau, \xi(\tau)]}{d(t)} = \frac{1}{d(t) \times \sigma}[\text{Đ}(\infty) - \text{Đ}(0)] \left[\frac{\exp(z)}{(1 + \exp(z))^2} \right] \qquad (1)$$

where $\xi(\tau) = \lceil D_{\text{Inst}}(\tau), \text{temp}(\tau), \text{RH}(\tau) \rceil, \text{Đ}(0)$ is the standardized level of damage at time 0, Đ(∞) is the long-term asymptote, and z is as defined in equation (2)

$$z = \frac{\log[d(t)] - \mu}{\sigma} \qquad (2)$$

Here $d(t)$ is the effective total dosage, and μ, and o are parameters that describe the location and steepness of the damage curve, respectively. Then the prediction equation for the cumulative amount of damage at time t, based on the incremental values of dosage is

$$Dosage_{CUM}(t) = \sum_{i=0}^{t} \Delta d(i)$$

$$Damage_{CUM}(t) = \sum_{i=0}^{t} \Delta\text{Đ}(i)$$

where $\Delta d(t)) = d(t) - d(t-1)$ and $\Delta\text{Đ}(t) = g'(t) * \Delta d(t)$.

The details about the cumulative damage model can be found in reference 30. The parameters in the model for temperature, RH, wavelength, or the light intensity are estimated from the SPHERE exposure results. However, these parameters should be a function of the environment and additional unknown parameters. The outdoor environmental data are taken at 12-min intervals, including spectral dosage, temperature, and RH. To test the predictive model, the calculated outdoor damage is compared to the observed outdoor damage with respect to total dosage. A few typical examples are shown in *Figure 22*. The results reveal that the calculated damage from the predictive model generally agrees with the observed damage. Work on improving the prediction methods is being carried out, and a better prediction model will be presented in the future.

In summary, all three approaches have successfully linked the laboratory results to the outdoor data, and the chemical changes of the model epoxy coating exposed to outdoor environments are effectively predicted from the SPHERE data by both the model-free empirical approach and the accumulated damage prediction model. Further work on validation of these approaches using some physical properties such as yellowing index and gloss retention will be investigated.

CONCLUSIONS

A linkage between laboratory and field exposure results for a model epoxy coating system has been established using total effective dosage model, assuming the validity of the additivity law and reciprocity law. Precise, accurate, and time-based measurements on both exposure environments and degradation properties have been carried out. Based on the data from well-controlled SPHERE exposures, effects of critical environmental conditions on chemical degradation of the UV-exposed epoxy materials have been investigated. It has been found that exposure conditions such as intensity of UV irradiation, the wavelength of the spectral UV, the temperature, and the relative humidity all play important but different roles on the efficiency of the chemical damage with respect to per unit dosage. High humidity tends to retard the formation of photooxidation products related to the IR band at 1728 cm^{-1}, but other parameters such as different wavelengths and different light intensities do not influence the mechanism of the chemical degradation. The outdoor exposure was carried out on the roof of a NIST laboratory located in Gaithersburg, MD. Due to the cyclic changes in each outdoor weathering variable, the temperature and RH of the outdoor exposure were continuously recorded every minute while the solar spectrum was recorded every 12 min. Chemical changes in the specimens exposed outdoors were measured in the same way as for the SPHERE exposed specimens. It was found that the mechanism of chemical degradation for samples exposed to outdoor environments was exactly the same as the degradation mechanisms observed in laboratory exposed specimens. Chemical degradation depends on the total effective dosage of exposed specimens exposed in different seasons in addition to the specimen temperature and humidity. Lastly, three approaches have been used to link the chemical degradation data from both the SPHERE and the outdoor exposures. Successful linkages have been made between the laboratory results and the outdoor data via all three approaches. Both the model-free empirical approach and the accumulated damage prediction model can be used to predict the outdoor performance of the epoxy material from the SPHERE data.

ACKNOWLEDGMENTS

This research is part of a Government/Industry consortium on Service Life Prediction of Coatings at NIST. Companies involved in this consortium include AKZO Nobel, Arkema, Atlas Materials Testing Solutions, Dow Chemical, DuPont Automotives, Duron Inc., Eastman Chemicals, Millennium Inorganic Chemicals, PPG, and Sherwin-Williams Co., USDOT Federal Highway Administration, Wright Patterson AFB, USDA Forest Products Laboratory in Madison, WI, and the Smithsonian Environmental Research Center located in Edgewater, MD. We also gratefully acknowledge help from Bouchra Kidah, David Martin, Aziz Rezig and Cyril Clerici from BFRL, NIST, for their help on data collection and evaluation.

References

(1) Martin, J.W., Nguyen, T., and Wood, K.A., "Unresolved Issues Related to Predicting the Service Life of Polymeric Materials," *Service Life Prediction: Challenging the Status Quo*, Martin, J.W., Ryntz, R.A., and Dickie, R.A. (Eds.), Federation of Societies for Coatings Technology, Blue Bell, PA, p. 13, 2005.
(2) Hofmann, A.W., "Remarks on the Changes of Gutta Percha Under Tropical Influences," *J. Chem. Soc.*, 14, 87 (1860).

(3) Russell, W.J. and Abney, W. de W., "Repot to the Science and Art Department of the Committee of the Council of Education on the Action of Light on Water Colors," H.M. Stationary Office, London (1888).

(4) Blakely, R.R., "Evaluation of Paint Durability—Natural and Accelerated," *Prog. Org. Coat.*, 13, 279 (1985).

(5) Fischer, R.M., Ketola, W.D., and Murray, W.P., "Inherent Variability in Accelerated Exposure Methods," *Prog. Org. Coat.*, 19, 165 (1991).

(6) Riedl, A., "What Makes a Xenon Weathering Instrument High-End?," *Atlas Sunspots*, 33 (70), 1 (2003).

(7) Sullivan, C.J. and Cooper, C.F., "Polyester Weatherability: Coupling Frontier Molecular Orbital Calculations of Oxidative Stability with Accelerated Testing," *J. Coat. Technol.*, 67, No. 847, p. 53 (1995).

(8) Searle, N.D., Giesecke, P. Kinmonth, R., and Hirt, R.C., "Ultraviolet Spectra Distributions and Aging Characteristics of Xenon Arcs AND Filters," *Appl. Opt.*, 3, 963 (1964).

(9) Kinmonth, R.A. and Norton, J.E., "Effect of Spectral Energy Distribution on Degradation of Organic Coatings," *J. Coat. Technol.*, 49, No. 633, p. 37 (1977).

(10) Meeker, W.Q. and Escobar, L.A., *Statistical Methods for Reliability Data*, John Wiley & Sons, New York, 1998.

(11) Whittaker, I.C. and Besumer, P.M., "A Reliability Analysis Approach to Fatigue Life Variability of Aircraft Structures," Air Force Materials Laboratory Technical Report, AFML-TR-69-65, 1969.

(12) Martin, J.W. and McKnight, M.E. "Prediction of the Service Life of Coatings on Steel. 2. Quantitative Prediction of the Service Life of a Coating System," *J. Coat. Technol.*, 57, No. 724, p. 31 (1985).

(13) Martin, J.W., Saunders, S.C., Floyd, F.L., and Wineburg, J.P. "Methodologies for Predicting the Service Lives of Coating Systems," Federation of Societies for Coatings Technology, Blue Bell, PA, 1996.

(14) Schutyser, P. and Perera, D.Y., "Use of Reliability-based Methodology for Appearance Measurements," *Service Life Prediction of Organic Coatings: A Systems Approach*, Bauer, D.R. and Martin, J.W. (Eds.), ACS Symposium Series 772, American Chemical Society, Oxford Press, NY, p. 198, 1999.

(15) Tait, W.S., "Reliability Engineering: The Commonality between Airplanes, Light Bulbs, and Coated Steel," *Service Life Prediction of Organic Coatings: A Systems Approach*, Bauer, D.R. and Martin, J.W. (Eds.), ACS Symposium Series 772, American Chemical Society, Oxford Press, NY, p. 186, 1999.

(16) Guseva, O., Brunner, S., and Richner, P., "Service Life Prediction for Aircraft Coatings," *Polym. Degrad. Stab.*, 82, 1 (2003).

(17) Nelson, H.A. and Schmutz, F.C., "Accelerated Weathering: A Consideration of Some Fundamentals Governing its Application," *Ind. Eng. Chem.*, 18 (12), 1222 (1926).

(18) Duffie, J.A. and Beckman, W.A., "Solar Engineering of Thermal Processes," Second Edition. New York, Wiley-Interscience, 1991.

(19) Chin, J.W., Byrd, W.E., Embree, E., Martin, J.W., and Tate, J.D., "Ultraviolet Chambers Based on Integrating Spheres for Use in Artificial Weathering," *J. Coat. Technol.*, 74, No. 929, p. 39 (2002).

(20) Chin, J.W., Byrd, W.E., Embree, E., and Martin, J.W., "Integrating Sphere Sources for UV Exposure," *Service Life Prediction: Methodology and Metrologies*, Martin, J.W. and Bauer, D. (Eds.), Oxford Press, NY, p. 144, 2001.

(21) Bellinger, V. and Verdu, J., "Oxidative Skeleton Breaking in Epoxy-Amine Networks," *J. Appl. Polym. Sci.*, 30, 363 (1985).

(22) Bellinger, V. and Verdu, J., "Structure-Photooxidative Stability Relationship of Amine-Crosslinked Epoxies," *Polymer Photochem.*, 5, 295-311 (1984).

(23) Rabek, J.F., *Polymer Photodegradation—Mechanisms and Experimental Methods*, Chapman & Hall, New York, p. 269-278, 1995.

(24) Kelleher, P.G., and Gesner, B.D., "Photo-Oxidation of Phenoxy Resin," *J. Appl. Polym. Sci.*, 13, 9-15 (1969).

(25) Rivaton, A., Moreau, L., and Gardette, J-L., "Photo-Oxidation of Phenoxy Resins at Long and Short Wavelengths-I. Identification of the Photoproducts," *Polym. Deg. Stab.*, 58, 321-332 (1997).

(26) Nguyen, T., Martin, J.W., Byrd, E., and Embree, N., "Relating Laboratory and Outdoor Exposure of Coatings: II. Effects of Relative Humidity on Photodegradation and the Apparent Quantum Yield of Acrylic-Melamine Coatings," *J. Coat. Technol.*, 74, No. 932, p. 65 (2002).

(27) Chin, J., Nguyen, T., Byrd, W.E., and Martin J.W., "Validation of the Reciprocity Law for Coating Photodegradation," *J. Coat Technol. Res.*, 2, No. 7, p. 499 (2005).

(28) Patterson-Jones, J.C., "Mechanism of Thermal Degradation of Aromatic Amine-Cured Glycidyl Ether Type Epoxide Resins," *J. Appl. Polym. Sci.*, 19, 1539-1547 (1975).

(29) Gerlock, J.L., Peters, C.A., Kucherov, A.V., Misovski, T., Seubert, C.M., Carter, R.O. III, and Nichols, M.E., "Testing Accelerated Weathering Tests for Appropriate Weathering Chemistry: Ozone Filtered Xenon Arc," *J. Coat Technol.*, 75, No. 936, 35 (2003).

(30) Vaca-Trigo, I., and Meeker, W.Q., "A Statistical Model for Linking Field and Laboratory Exposure Results for a Model Coating," *Proc. International Symposium on Service Life Prediction: Global Perspectives*, Key Largo, FL, 2006.

Chapter 2

A Statistical Model for Linking Field and Laboratory Exposure Results for a Model Coating

Iliana Vaca-Trigo and William Q. Meeker

Iowa State University, Dept. of Statistics, Ames, IA 50011-1210

Today's manufacturers need accelerated test (AT) methods that can usefully predict service life in a timely manner. For example, automobile manufacturers would like to develop a three-month test to predict 10-year field reliability of a coating system (an acceleration factor of 40). Developing a methodology to simulate outdoor weathering is a particularly challenging task and most previous attempts to establish an adequate correlation between laboratory tests and field experience have met with failure. Difficulties arise, for example, because the intensity and the frequency spectrum of ultraviolet (UV) radiation from the Sun are highly variable, both temporally and spatially and because there is often little understanding of how environmental variables affect chemical degradation processes.

    This chapter describes the statistical aspects of a cooperative project being conducted at the U.S. National Institute of Standards and Technology (NIST) to generate necessary experimental data and the development of a model relating cumulative damage to environmental variables like UV spectrum and intensity, as well as temperature and relative humidity. The parameters of the cumulative damage are estimated from the laboratory data. The adequacy of the model predictions are assessed by comparing with specimens tested in an outdoor environment for which the environmental variables were carefully measured.

INTRODUCTION

Background

Photodegradation, caused by UV radiation, is a primary cause of failure for paints and coatings (as well as all other products made from organic materials) exposed to sunlight. Other variables that affect degradation rates include temperature and humidity. Manufacturers of such paints and coatings have had difficulty in using laboratory tests to predict field experience for their products. Historically, most of the laboratory tests attempt to accelerate time by "speeding up the clock." This is done by increasing the average levels of experimental factors like UV radiation, temperature, and humidity, and cycling these experimental factors more rapidly than what is seen in actual use, in an attempt to simulate and accelerate outdoor aging. Such experiments violate the basic rules of good experimental design. For example, varying important factors together tends to confound the effects of the factors. Also, levels of the accelerating variables that are too high may induce new failure modes. For these reasons, such accelerated tests provide little fundamental understanding of the underlying degradation mechanisms, and conclusions from them can be seriously incorrect. Because experience has shown that the results of these tests are unreliable, standard product evaluation for paints and coatings still requires outdoor testing in places like Florida (where it is hot and humid) and Arizona (where it is hot and dry). Outdoor testing, however, is costly and takes too much time.

Martin et al.[1] and Martin[2] provide a detailed description of issues relating to prediction of service life (SL) for paints and coatings. In general, the accelerated test methodology for photodegradation is much more complicated than those typically used for electronic and mechanical devices (e.g., as described in Nelson[3] and in Chapters 18–21 of Meeker and Escobar.[4]) This is because of the complicated chemical/physical failure mechanisms involved and the highly variable use environment.

Motivation

Accelerated test (AT) methods have proven to be useful for predicting the SL of materials in certain applications. These range from jet engine turbine disk materials to highly sophisticated microelectronics (these successful applications are described, for example, in Gillen and Mead,[5] Joyce et al.,[6] Starke et al.,[7] and the many examples cited in Nelson.[3]) In other areas of application, however, AT methods often yield predictions that do not correlate well with field data. This is particularly true for products exposed to outdoor weathering, such as organic paints and coatings used on automobiles, bridges, buildings, and other outdoor structures (e.g., Martin et al.[1] and Wernstål and Carlsson.[8]) For this reason, conventional laboratory AT methods are not trusted for outdoor-use products and potential users of such tests have been forced to rely on expensive, time-consuming outdoor testing.

Traditional applications in reliability and service life prediction based on accelerated test results involve chemical degradation that is accelerated by increasing variables like temperature, humidity, and current density or voltage stress, using statistical models that are motivated by knowledge from physical chemistry. The research described in this chapter is a natural extension of previous work in this area to the more complicated area of photodegradation.

EXPERIMENTAL DATA

Degradation (or damage) at time t, denoted by $Đ(t)$, usually depends on environmental variables like UV, temperature, and relative humidity that vary *over time*. Laboratory tests are conducted in well-controlled environments, usually holding these variables constant (although in other experiments such variables are purposely changed during an experiment, as in step-stress accelerated tests). Interest often centers, however, on life in a variable environment.

Table 1—Bandpass Filter Characteristics

Range		Nominal Filter Midpoint
303 nm	309 nm	306 nm
320 nm	332 nm	326 nm
334 nm	372 nm	353 nm
372 nm	532 nm	452 nm

Time Scale for Photodegradation

It is important to choose an appropriate time scale to describe the behavior of a failure mechanism (e.g., number of miles for an automobile engine bearing or number of cycles for fatigue caused by cyclic stress). The appropriate time scale for photodegradation is photon dosage. In our data sets, dosage is given in units of $KJ/m^2/nm$ and is a number that is proportional to the number of photons absorbed into the experimental specimens.

Indoor Data

In the current phase of the NIST research program, the goal has been to develop a service life prediction methodology using a crosslinked epoxy amine coating system as a simple model. The methodology described in this chapter is being developed, however, to allow easy generalization to service life prediction of other types of materials that will be exposed to outdoor weathering.

Researchers at NIST have conducted weathering experiments in both the indoor laboratory, as well as in outdoor exposure facilities. Indoor data are being taken in temperature/humidity-controlled chambers illuminated by controlled UV light from the NIST Sphere (described in Martin et al.[2] and Chin et al.[9]).

Indoor data received from NIST consist of the variables:

- Specimen Number (SA) identifying the testing chamber number and a number of a particular specimen within the chamber.
- Damage number (DA) for four peaks in the measured FTIR spectra. The heights of the peaks correspond to the amount of particular chemical products and these were measured systematically, over time, and have units cm^{-1}. One of the studied damage numbers was the peak at $1510 \ cm^{-1}$, which corresponds to benzene ring mass loss. Other peaks being used as potentially useful responses include $1250 \ cm^{-1}$ (aromatic C–O), $1658 \ cm^{-1}$ (oxidation products), and $2925 \ cm^{-1}$ (CH mass loss).
- Bandpass Filter (FI) is the center wavelength in nanometers (nm) of the bandpass filter used in exposure. *Table* 1 also gives the range of the bandpass filters.
- Neutral Density (DE) is the nominal transmittance rate of a neutral density filter ranging from 0% to 100%.
- Temperature (temp) in Celsius.
- Relative humidity (RH) which ranges from 0% to 100%.
- $DOSAGE_{Tot}$, as part of the indoor data, is a metric proportional to the total number of photons absorbed into the degrading material.
- DAMAGE values are the responses and measure the photolytic part of the chemical damage to the test specimens.

Table 2—Experimental Variables and Levels

Variable	Units	Levels
Damage number (DA)	cm^{-1}	1250, 1510, 1658, 2925
Bandpass filter (FI)	nm	306, 326, 353, 452
Neutral density (DE)	%	10, 40, 60, 100
Temperature (temp)	°C	25, 35, 45, 55
Humidity (RH)	%	0, 25, 50, 75

Table 3—Available Data

Temp	RH 0%	25%	50%	75%
25°C	x	—	—	—
35°C	x	✓	✓	✓
45°C	x	✓	✓	x
55°C	x	x	x	✓

— Data not available
✓ Data used for modeling
x Data not used for modeling

- Wall Clock is the real clock time when the data is recorded, as the number of days since January 1, 1900.

Table 2 shows the levels of the experimental variables in the Indoor data. Not all combinations of humidity and temperature levels data were available at the time of the analysis provided here. *Table* 3 shows the combinations that we used.

Outdoor Damage Data

Outdoor exposure data on specimens made of the same material were also collected at NIST. For outdoor specimens, damage is typically measured after every few days of exposure and this information is recorded in addition to spectral irradiance and weather data (temperature and humidity). Although there was no control of experimental variables for the outdoor data, temperature, humidity, and solar data were recorded, as described in the next subsection. Specimens in the outdoor were grouped by date, with 18 groups and four replicates for each group. Each group was exposed across different months, therefore temperature and humidity change from group to group. The outdoor data will allow us to check our predictive model. This will be done by generating damage predictions based on the model derived from the indoor data. To do this, the indoor model is driven by the outdoor weather data to compute predictions that can be compared with the corresponding actual outdoor damage.

Outdoor Weather Data

SOLARNET, a solar UV data network, stores spectral irradiance data with a 12-min resolution as well as climatological data (temperature, relative humidity, etc.) as 1-min averages (described in Kaetzel[10]).

ANALYSIS AND INITIAL MODELING

Initially, extensive graphical analyses of damage versus dosage path plots were conducted to get a good understanding of the data and possible relations among variables in the data set. Plots of empirically estimated acceleration factors provided insight on the effects that experimental explanatory variables have on the response.

Acceleration factors are commonly used to describe the effect that accelerating variables or other experimental variables have on lifetime or degradation rates. Acceleration factors can be expressed as the ratio of life at "fixed test conditions" to life at "higher test conditions." Acceleration factor plots were examined for temperature, humidity, and the different UV radiation band pass filters.

Data Cleaning

An important phase of modeling is looking at the raw data to identify strange patterns, outliers, or other data anomalies that could affect the modeling efforts and possibly result in unreliable estimates. Even though data were collected under a controlled environment using sophisticated analytical devices to assure the accuracy of the data, exhaustive use of graphical assessment procedures helped to identify some potential problems. The root cause for all such problems was determined and appropriate adjustments were made to the data. For example, we detected a sharp drop in the damage rate for samples at 45°C and 75% RH. The root cause for this problem was the failure of an integrated circuit chip in the environmental controllers that caused the samples in one of the chambers to be overheated for a period of time. Similar problems were identified at 55°C and 25% RH as well as at 55°C and 50% RH. Those specimens that were subjected to this overheating were not used in the modeling process. Also, data from the bandpass filter with nominal midpoint of 353 nm did not agree with the data from the other bandpass filters when fitting a model to estimate the effect of wavelength on damage rates. For this reason, these data were also ignored in the modeling.

Another potential data complication is a change of direction of the degradation path. For example, *Figure* 1a shows that the FTIR peak at 1658 cm^{-1} increases until dosage reaches approximately 4×10^3 KJ/m^2/nm, after which the degradation paths begin to decrease. This behavior is thought to be caused by physical and chemical changes in the

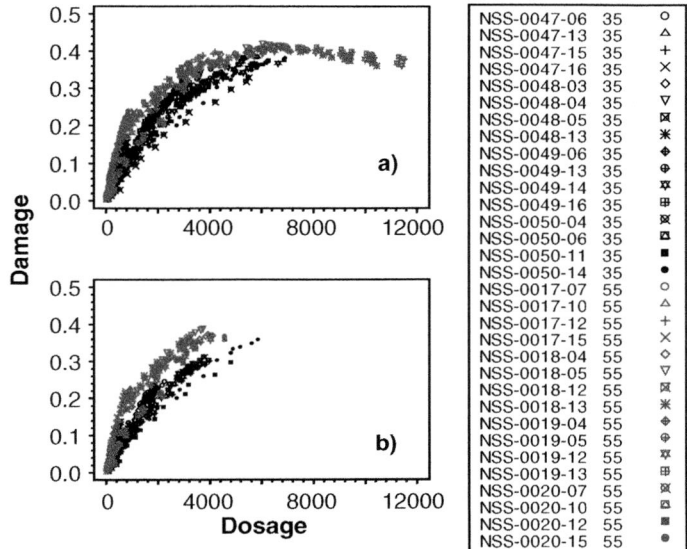

Figure 1—Illustration of data cleaning for the FTIR peak at 1658 cm^{-1} for units exposed with 326 nm nominal bandpass filter midpoint and 75% RH. (a) Original data paths. (b) Data paths after deleting outliers and increasing tails.

specimens. Because the turning point is far beyond the definition of failure, modeling beyond the turning point is not needed. Thus, we cut increasing/decreasing tails after the turning point for those cases where degradation paths changed direction. In addition, specimens at 0% RH were used only in the preliminary stages to understand data behavior. Because 0% RH is outside of the region of interest and because there was no apparent simple model to connect these "dry" results with the units run with humidity, the 0% RH data were not used in our modeling.

Initial Modeling

The data that have been analyzed to date seem to be consistent with both first-order and second-order kinetic models. Over the dosage range of interest, [that is up to the point where Ð(t) has reached a failure state] we have found, empirically, that the simple parsimonious functional form

$$\text{Ð}(t) = [\text{Ð}(\infty)\text{-Ð}(0)]\left[\frac{\exp(z)}{1+\exp(z)}\right] \tag{1}$$

$$z = \frac{\log\left[d(t)\right] - \mu}{\sigma} \tag{2}$$

fits the data well for all FTIR peaks of interest and at *all* combinations of the experimental factors for which we have received data. Here, $d(t)$ is the effective total dosage. Also, $\text{Ð}(0)$ is the standardized level of damage at time 0 and $\text{Ð}(\infty)$ is the long-term asymptote; while μ and σ are parameters that describe the location and steepness of the damage curve, respectively. In the overall model, time-scaling factor $\exp(\mu)$ will be a function of the environment and additional unknown parameters. When fitting data to a single path, if the asymptote cannot be estimated from the data (because the path has not begun to level off sufficiently), a good fit to the data can be obtained, without loss of generality, by setting $\text{Ð}(\infty)$ to a safe lower bound (upper bound) on the asymptote when the damage variable is decreasing (increasing). When we fit data to the overall model, we will be able to "borrow strength" from paths at other conditions where the asymptote can be identified. The NIST data on the epoxy material under study suggest that there is, approximately, a common asymptote for each FTIR peak, independent of the experimental conditions and we assume this in the overall model.

As an aid in model identification, a plot of an acceleration factor versus a particular experimental variable can be generated by fitting the model in equations (1) and (2) with a common value of σ and a different value of μ for each level of experimental variable. The acceleration factor at a given test level of the variable, relative to a specified reference level, is

$$AF(\text{test,reference}) = \frac{\exp\left(\mu_{\text{test}}\right)}{\exp\left(\mu_{\text{reference}}\right)}$$

The acceleration factors for the different levels of the experimental variables can be plotted in a manner such that the points should fall roughly along a straight line if the hypothesized model is adequate.

MODEL FOR THE EFFECT OF UV RADIATION ON PHOTODEGRADATION

Many of the ideas in this section are based on early research into the effects of light on photographic emulsions (e.g., James[11]) and the effect that UV exposure has on causing skin cancer (e.g., Blum[12]).

Model for Total Effective UV Dosage

As described in Martin et al.,[1] the appropriate time scale for photodegradation is D_{Tot}, the total *effective* UV dosage. Intuitively, this total effective dosage can be thought of as the number of photons absorbed into the degrading material *and that cause chemical change*. The total *effective* UV dosage at real time t can be computed from

$$D_{Tot}(t) = \int_0^t D_{Inst}(\tau)d\tau \tag{3}$$

where the instantaneous *effective* UV dosage D_{Inst} is

$$D_{Inst}(\tau) = \int_{\lambda_1}^{\lambda_2} D_{Inst}(\tau,\lambda)d\lambda = \int_{\lambda_1}^{\lambda_2} E_0(\tau,\lambda)\left\{1-\exp\left[-A(\lambda)\right]\right\}\phi(\lambda)d\lambda. \tag{4}$$

Here, E_0 is the spectral irradiance of the light source (both artificial and natural light sources have mixtures of light at different wavelengths, denoted by λ), $[1 - \exp(-A(\lambda))]$ is the spectral absorbance of the material being exposed (damage is caused only by photons that are absorbed into the material), and $\phi(\lambda)$ is a quasi quantum efficiency (QQE) of the absorbed radiation (allowing for the fact that photons at shorter wavelengths have higher energy and thus a higher probability of causing damage). The functions in the integrand of equation (4) can either be measured directly (E_0 and A) or estimated from experimental data ($\phi(\lambda)$). The definition of dosage in equation (4) differs from the dosage in our data (as described in the previous section on Indoor Data) because the QQE function is unknown and needs to be identified from the experimental data.

Intensity Effects and Reciprocity

The intuitive idea behind reciprocity in photodegradation is that the time to reach a certain level of degradation is inversely proportional to the rate at which photons reach the material being degraded. Reciprocity failure occurs when the coefficient of proportionality changes with light intensity.

Although reciprocity provides an adequate model for some degradation processes (particularly when the dynamic range of intensities used in experimentation and actual applications is not too large), numerous examples have been reported in which there is reciprocity failure (e.g., James[11] and Blum[12]). Light intensity can be affected by filters. Sunlight is filtered by the earth's atmosphere. In laboratory experiments, neutral density filters are used to reduce the amount of light passing to specimens (without having an important effect on the wavelength spectra), providing an assessment of the degree of reciprocity failure.

Reciprocity also implies that the effective time of exposure is

$$d(t) = \text{CF} \times \text{D}_{\text{Tot}}(t) = \text{CF} \times \left[\int_0^t \int_{\lambda_1}^{\lambda_2} \text{D}_{\text{Inst}}(\tau, \lambda) d\lambda d\tau \right] \qquad (5)$$

where CF is an acceleration or deceleration factor for UV intensity. For example, commercial outdoor test exposure sites use mirrors to achieve, say, "5 Suns" *acceleration* or CF = 5. A 50% neutral density filter in a laboratory experiment will provide *deceleration* corresponding to CF = 0.50.

When there is evidence of reciprocity failure, the effective time of exposure is often modeled by

$$d(t) = (\text{CF})^p \times \text{D}_{\text{Tot}}(t) = (\text{CF})^p \times \left[\int_0^t \int_{\lambda_1}^{\lambda_2} \text{D}_{\text{Inst}}(\tau, \lambda) d\lambda d\tau \right] \qquad (6)$$

where p is known as the Schwarzschild coefficient. This model has been shown to fit data well and experimental work in the photographic literature (e.g., James, 1977[11]) suggests that when there is reciprocity failure, the value of p does not depend on wavelength λ. A statistical test of $p = 1$ can be used to assess the reciprocity assumption.

For the NIST data on the epoxy material under study, there is no evidence of reciprocity failure. Thus, for this material, we expect to be able to use $p = 1$. Our model is, however, general enough to allow for reciprocity failure. Therefore, for modeling purposes, averages of damage values for specimens exposed at same conditions but different neutral density filters were used instead of individual paths.

Wavelength Effects

Following other work in the area of photodegradation (e.g., Miller at al.[14]), we will assume a simple log-linear model for QQE. That is,

$$\phi(\lambda) = \exp(\beta_0 + \beta_1\lambda).$$

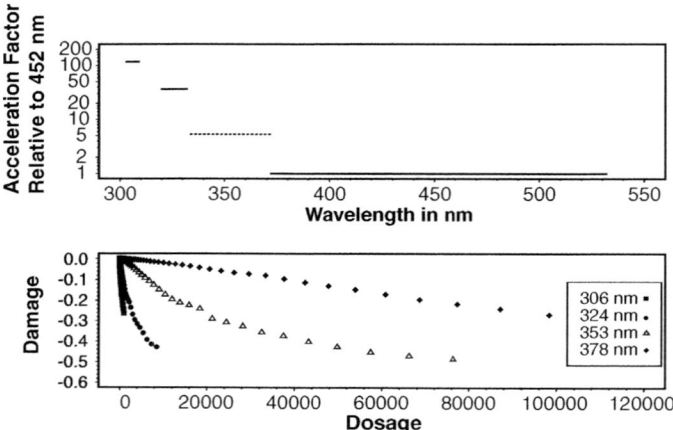

Figure 2—Quantum yield model check for the 1250 cm^{-1} FTIR peak for specimens exposed at 35°C and 25% RH.

The integral in equation (5) and subsequent integrals over wavelength are typically taken over the UV-B band (280—315 nm), as this is the range of wavelengths over which both $\phi(\lambda)$ and $E_0(\lambda, t)$ are importantly different from 0. Longer wavelengths (in the UV-A band) are not terribly harmful so that $\phi(\lambda) \approx 0$. Shorter wavelengths (in the UV-C band) have more energy, but are absorbed by ozone in the atmosphere so that $E_0(\lambda, t) \approx 0$.

An example of an acceleration factor versus wavelength plot is shown in the upper plot of *Figure* 2. The horizontal lines indicate the band pass filter width. These lines exhibit a log-linear relation for QQE except for observations corresponding to BP filter 353. Because observations from BP filter 353 were not consistent (in terms of our estimated QQY function) with the observations from the other BP filters, the 353 BP data were not used in the estimation of the parameters of the model.

The lower plot in *Figure* 2 shows degradation paths of observed damage averaged over all specimens under experimental 35°C, 25% RH, 1250 cm^{-1} FTIR peak, and a particular nominal bandpass filter midpoint. Different symbols were used to identify the bandpass filters. Filled marks and continuous lines identified data that were used in the modeling, while dashed lines and open marks were used to represent data that were available, but not used in the modeling as explained in the section on Data Cleaning. *Figure* 2 shows that, all other things being equal, wavelength has an effect on damage that tends to be stronger at shorter wavelengths.

Implicit in the model in equation (4) is the assumption of additivity. Additivity implies, in this setting, that the photoeffectiveness of a source is equal to the sum of the effectiveness of its spectral components. Experimental results obtained by NIST researchers support additivity in photodegradation of organic materials that have been studied to date.

MODEL FOR OTHER EXPERIMENTAL VARIABLES

Temperature Effects

As described, for example, in Chapter 18 of Meeker and Escobar[4] the Arrhenius equation for the reaction rate $\mathcal{R}$ can be written as

$$\mathcal{R}(\text{temp}) = \gamma_0 \exp\left(\frac{-E_a}{R \times \text{temp } \mathbf{K}}\right)$$

where temp $\mathbf{K}$ is temperature Kelvin, R is the gas constant ($R = 8.31447$ J x K^{-1} x mol^{-1}), E_a is a quasi activation energy and γ_0 is a constant specific to a product or material.

The Arrhenius rate reaction model can be used to scale time (or dosage) in the usual manner and the upper plot in *Figure* 3 shows the acceleration factor versus temperature, plotted relative to 35°C and accelerated temperatures from 35°C to 55°C. Because 25°C data were not available for all humidity levels, for sake of consistency 35°C was used as a basis level for calculating acceleration factors. Temperature was plotted on an Arrhenius scale while acceleration factor was plotted on a logarithmic scale. The acceleration factor for a temperature of 45°C is approximately 1.2. This means that the life at the use level of 35°C is approximately 1.2 times longer than the life at 45°C. The bottom plot in *Figure* 3 shows degradation paths for specimens at the 1250 cm^{-1} FTIR peak, 306 nm nominal bandpass filter midpoint, 25% RH, and at three different temperatures.

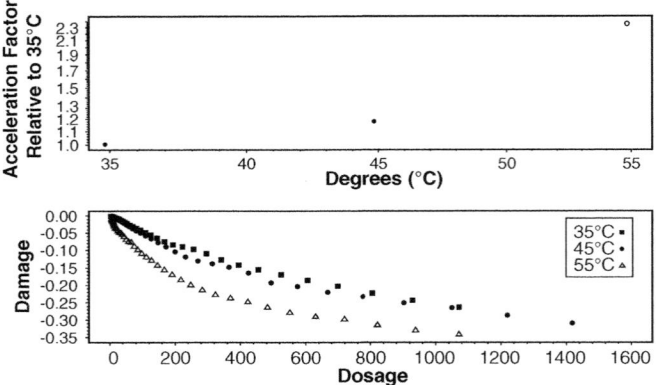

Figure 3—*Arrhenius model check for the 1250 cm⁻¹ FTIR peak for specimens exposed to 306 nm nominal bandpass filter midpoint and 25% RH.*

Figure 3 shows the effect of temperature on degradation. As expected, specimens exposed to higher temperatures tend to degrade faster than those at same conditions and lower temperatures.

Humidity Effects

Relationships between degradation rate and humidity are more complicated. Different chemical reactions respond differently to humidity and therefore damage degradation paths, for each FTIR peak will relate in an individual manner to humidity. In our initial efforts to find an appropriate model for the humidity effect presented here, our approach is more empirical than scientifically based. NIST researchers do, however, have initial hypotheses on the reasons for the observed behaviors and we expect that these will be used in subsequent modeling efforts.

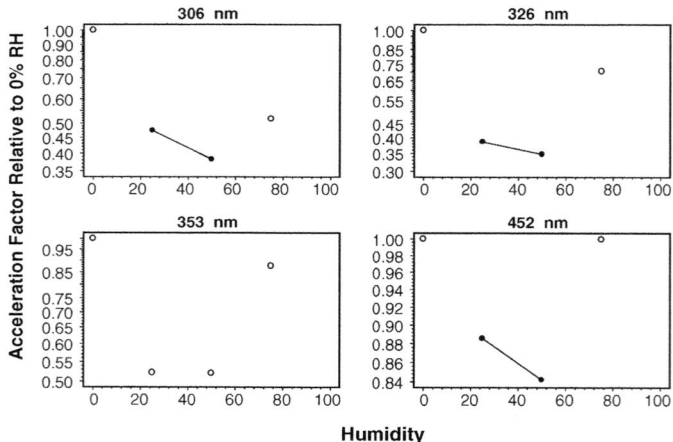

Figure 4—*Indication of linear decreasing humidity effect for the 1250 cm⁻¹ FTIR peak for specimens at 45°C.*

Figure 4 has linear axes for humidity and logarithmic axes for the acceleration factor, plotted relative to 0% RH. As seen in *Figure* 4 (for the 1250 cm⁻¹ FTIR peak), the NIST data suggest that the degradation rate decreases linearly as a function of relative humidity. Similar relationships are apparent in all of the other FTIR peaks.

Overall Model and Bandpass Filter Approximation

Combining all of the model terms in equations (2) and (6), with

$$\mu = \beta_0 + \frac{E_a}{k_B \times \text{temp K}} - \beta_2 \times RH$$

we have

$$\log\left(d(t), \text{CF}, p\right) = \log\left[D_{\text{Tot}}(t)\right] + p \times \log(\text{CF}) \tag{7}$$

where

$$D_{\text{Tot}}(t) = \sum_{\lambda \text{ in the range on the BP filter}} \text{DOSAGE}_{\text{Tot}_\lambda} \times \phi(\lambda)$$

For the indoor data we have dosage over a range of a bandpass filters. For simplicity, we assume a BP filter with rectangular shape over the given range for the filter. Therefore, $\text{DOSAGE}_{\text{tot}\lambda}$ corresponds to the value of the reported dosage divided by the range of the filter, giving the approximate dosage for the 2 nm intervals that correspond to the outdoor data.

The parameters, β_0 and β_1, for the QQY relationship, E_a and β_2, are characteristic of the material and the degradation process and in our modeling we used $p = 1$ because there was no evidence against reciprocity. As a typical example, *Figure* 5 shows fitted lines for the proposed overall model for one response and experimental condition: the 1250 cm⁻¹ FTIR peak, for specimens exposed under the 306 nm BP filter and 25% RH. The fit between the data points and the fitted model is good, considering the broadness of the response surface model. Deviations from the model are on the same order as the unit-to-unit experimental error when units were exposed at different times. We had similar results for other combinations of damage number, bandpass filter, and humidity.

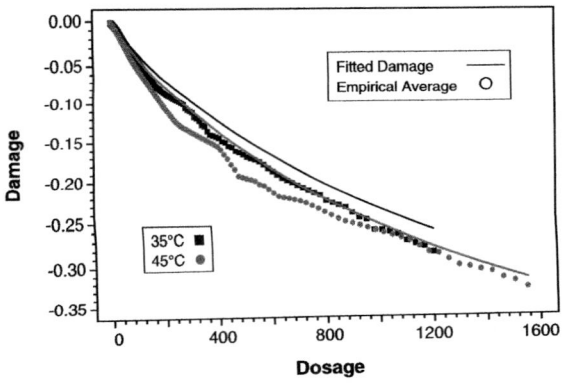

Figure 5—Indoor data versus the fitted model for the 1250 cm⁻¹ FTIR peak for specimens exposed under the 306 nm BP filter and 25% RH.

PREDICTIVE FORM OF THE CUMULATIVE DAMAGE MODEL

Cumulative Damage in a Time-Varying Environment

This section outlines the model that we used to predict total cumulative damage Đ(t) as a function of a given environmental time series realization $\xi(\tau)$. The main difference in the predictive model is that the environmental variables can be allowed to vary with time. For a given environmental profile $\xi(\tau)$, the cumulative damage at time t for a particular unit can be expressed as

$$Đ(t) = \int_0^t \frac{dĐ[\tau, \xi(\tau)]}{\tau} d\tau \tag{8}$$

where $\xi(\tau) = [D_{Inst}(\tau), \text{temp}(\tau), RH(\tau)]$.

Evaluation Total Damage in a Time-Varying Environment

The integral in equation (8) is reasonably easy to compute after appropriate discretization of the time axis. The environmental data that we will use is reported at 12-min intervals. Thus, equation (8) will be computed with a summation in which the environmental conditions will be constant over each 12-min period of time. Missing environmental data can be replaced by using a simple interpolation scheme.

For the cumulative damage model given in equation (1), the derivative of the cumulative damage with respect to dosage $d(t)$ is

$$g'(t) = \frac{dĐ[\tau, \xi(\tau)]}{d(t)} = \frac{1}{d(t) \times \sigma}[Đ(\infty) - Đ(0)] \left[\frac{\exp(z)}{(1 + \exp(z))^2} \right] \tag{9}$$

where z is as defined in equation (2) and $d(t)$ is defined in equation (7), with estimates used to replace the unknown parameters. Then the prediction equation for the cumulative amount of damage at time t, based on the incremental values of dosage, is:

$$Dosage_{CUM}(t) = \sum_{i=0}^t \Delta d(i)$$

$$Damage_{CUM}(t) = \sum_{i=0}^t \Delta Đ(i) \tag{10}$$

where $\Delta d(t) = d(t) - d(t-1)$ and $\Delta Đ(t) = g'(t) * \Delta d(t)$

To test the predictive model, first we apply it to predict cumulative damage observed in the indoor data (constant environmental conditions). As expected and as shown in *Figure* 6, the predictions from the incremental model correspond almost exactly with the fitted model and agree well with the indoor data that were obtained under a controlled environment. Although this is a useful check, it is not proof of model adequacy because we are comparing the predictions against the same data that were used to build the model.

Prediction in a Time-Varying Environment

In this section we use our predictive model in equation (10) to predict the damage observed in the outdoor exposure chambers, to check our ability to use a model estimat-

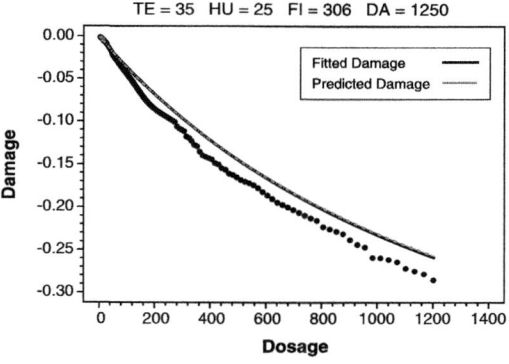

Figure 6—Comparison of the overall fitted model and the predictions for the 1250 cm⁻¹ FTIR peak for specimens exposed at 306 nm nominal bandpass filter midpoint, 35°C, and 25% RH.

ed from indoor data, to predict outdoor damage. We computed such predictions corresponding to all of the units that were tested in outdoor chambers at NIST. Here we show a few typical examples.

Our predictive model uses indoor data to estimate parameters of the model, as well as outdoor information about spectral dosage (every 2 nm), humidity, and temperature. *Figure* 7 uses lines to depict predictions for damage for different FTIR peaks for outdoor exposure group 18. The solid symbols represent the actual outdoor observations for the same group. For all four FTIR peaks, the different specimens agree well in terms of accumulated damage, as a function of dosage.

Each plot in *Figure* 8 shows damage versus dosage for four specimens from outdoor exposure groups G1, G2, G3, and G4. Each of these groups began exposure at different points in time during 2002. Variability between observations of different groups is more

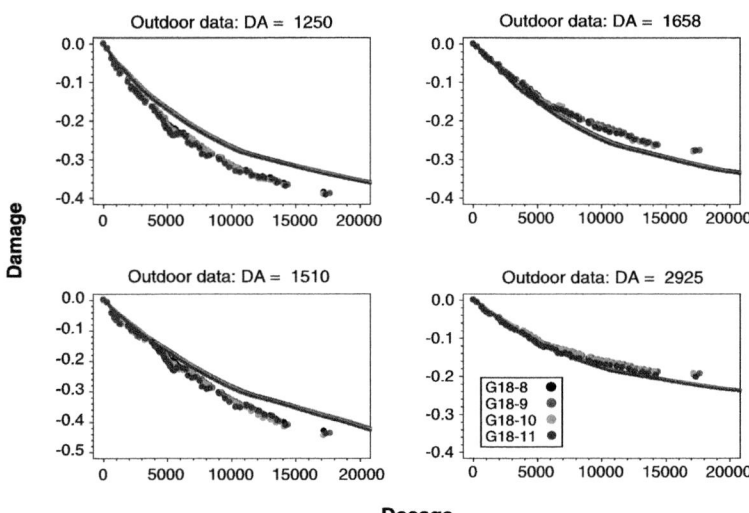

Figure 7—Comparison of the predictions for the outdoor specimens "G18-8," "G18-9," "G18-10," and "G18-11" that were exposed at same time.

42

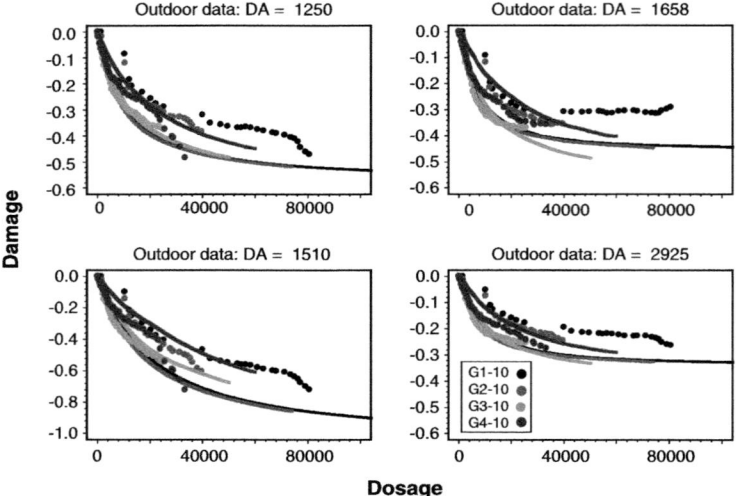

Figure 8—Comparison of predictions for the outdoor specimens "G1-10," "G2-10," "G3-10," and "G4-10" that started exposure at different times.

apparent in this plot than what we see in *Figure* 7 because specimens began outdoor exposure at different points in time. That is, variability among these specimens is larger than what we see in *Figure* 7, due to different weather conditions during the different periods of exposure.

CONCLUDING REMARKS

This chapter describes the methodology that we have developed to use indoor accelerated test data to find a model for describing the effects that environmental variables have on degradation rates. We have used this model to predict degradation rates and cumulative degradation in a time-varying environment, using outdoor weather data to drive the model. The variation between the predictions and the actual outdoor data is similar to the variability that we see in actual outdoor data. We would like to thank our many colleagues at the National Institute of Standards and Technology (NIST) who provided data, advice, and encouragement during the course of this research. These include Jonnie Chin, Brian Dickens, Xiaohong Gu, Tinh Nguyen, and Jonathan Martin. Iliana Vaca-Trigo's work on the research in this paper was partially supported by NIST Financial Assistance Award 60NANB6D6002.

ACKNOWLEDGMENTS

We would like to thank our many colleagues at the National Institute of Standards and Technology (NIST) who provided data, advice, and encouragement during the course of this research. These include Jonnie Chin, Brian Dickens, Xiaohong Gu, Tinh Nguyen, and Jonathan Martin. Iliana Vaca-Trigo's work on the research in this paper was partially supported by NIST Financial Assistance Award 60NANB6D6002.

References

(1) Martin, J.W., Saunders, S.C., Floyd, F.L., and Wineburg, J.P., "Methodologies for Predicting the Service Lives of Coating Systems," *Federation Series on Coatings Technology*, Federation of Societies for Coatings Technology, Blue Bell, PA, 1996.

(2) Martin, J.W., "A Systems Approach to the Service Life Prediction Problem for Coating Systems," in *A Systems Approach to Service Life Prediction of Organic Coatings*, Bauer, D.R. and Martin, J.W. (Eds.), American Chemical Society, Washington, D.C., 1999.

(3) Nelson, W., *Accelerated Testing: Statistical Models, Test Plans, and Data Analyses*, John Wiley & Sons, New York, 1990.

(4) Meeker, W.Q. and Escobar, L.A., *Statistical Methods for Reliability Data*, John Wiley & Sons, New York, 1998.

(5) Gillen, K.T. and Mead, K.E., "Predicting Life Expectancy and Simulating Age of Complex Equipment Using Accelerated Aging Techniques," Available from the National Technical Information Service, U.S. Department of Commerce, 5285 Port Royal Road, Springfield, VA 22151, 1980.

(6) Joyce, W.B., Liou, K-Y., Nash, F.R., Bossard, P.R., and Hartman, R.L., "Methodology of Accelerated Aging," *AT&T Technical Journal*, 64, 717–764, 1985.

(7) Starke, E.A., et al., *Accelerated Aging of Materials and Structures*, Publication NMAB-479, Washington DC: National Academy Press, 1996.

(8) Wernstål, K.M. and Carlsson, B., "Durability Assessment of Automotive Coatings—Design and Evaluation of Accelerated Tests," *J. Coat. Technol.*, 69, No. 865, 69-75, 1997.

(9) Chin, J.W., Martin, J.W., Embree, E.J., and Byrd, W.E., "Use of Integrating Spheres as Uniform Sources for Accelerated UV Weathering of Advanced Materials," *Proc. ACS Div. of Polymeric Materials: Science and Engineering*, Washington, D.C., American Chemical Society, Washington, D.C., pages 145-146, August 20-24, 2000.

(10) Kaetzel, L.J., "Data Management and Spectral Solar UV Network." In *Service Life Prediction Methodologies and Metrologies*, Martin, J.W. and Bauer, D.R. (Eds.), American Chemical Society Symposium Series 805, p. 89, 2001.

(11) James, T.H. (Ed.), *The Theory of the Photographic Process*, 4th ed., Macmillan, New York, 1977.

(12) Blum, H.F., *Carcinogenesis by Ultraviolet Light*, Princeton University Press, Princeton, NJ, 1959.

Chapter 3

Advances in Exploring Mechanistic Variations in Thermal Aging of Polymers

M. Celina* and K.T. Gillen

*Sandia National Laboratories, P.O. Box 5800, MS 1411, Dept. 1821, Albuquerque, NM 87185-1411

More confident lifetime prediction of the performance of polymeric materials requires a better understanding of how temperature may not only accelerate aging but also introduce mechanistic variation in the degradation process itself. Such effects may occur in any high stress level environments that contain a thermal reaction component, i.e., thermal aging, UV, hydrolytic, and gamma initiated degradation. The underlying reactions that govern the degradation of a material at the low stress level environment may not be represented to the same degree under accelerated conditions. Additional chemical and physical reactions can be introduced under high stress level conditions leading to anomalies and complications for lifetime prediction. Sensitive oxidation rate measurements, monitoring the consumption of an antioxidant, or chemiluminescence based wear-out experiments can be suitable avenues to probe for variations in thermal degradation processes. Under dynamic temperature conditions, knowledge of the exact thermal history and the dominant thermal reaction component, as well as its activation energy, is needed to better establish mean degradation rates and understand "real" temperature contributions.

INTRODUCTION

All polymer aging processes, natural degradation, and weathering are dependent on temperature. An increase in temperature will normally accelerate the chemical reactions or physical relaxation phenomena that govern polymer degradation reactions. It is also widely accepted that excessively accelerating degradation reactions—for example, conducting aging experiments over a few days to predict lifetimes of "years"—may introduce unknown mechanistic changes and large uncertainties in the predictive value of such experiments. Similar issues are of concern when conducting accelerated photo-degradation experiments with high short wave-length contributions or when conducting accelerated γ-irradiation experiments using high dose rates. It is not uncommon for experimentalists to be under pressure from their customers to provide fast feedback on materials performance, either via screening tests or otherwise limited studies. Lack of time, funding, and a limited understanding of the complexity of accelerated polymer

45

aging can result in aging studies with little predictive value and inadequate fundamental feedback.

There is a general trend in the literature to better address the extrapolation of accelerated aging studies.[1,2] In particular, the concept of linear Arrhenius-based extrapolations when temperature is the driving force for acceleration has attracted increased scrutiny. The awareness is growing that Arrhenius curvature in thermal aging studies is an issue that requires careful consideration. In the extreme case, a reversal in polymer stability in radiation thermal aging due to the balancing act between scission and crosslinking reactions as a function of temperature can occur. While perhaps not common, faster degradation at lower temperatures was observed, meaning extrapolations from limited high temperature aging studies would be impossible.[3,4] A secondary factor often associated with fast aging studies conducted at elevated temperatures is the occurrence of diffusion limited oxidation effects, which can lead to complications in thermal, photo- and gamma-aging conditions.[5-7] Due to fast reactions and the lack of oxygen reaching equilibrium conditions within materials, obtaining guidance from such experiments is of questionable value.[8] *Figure* 1 schematically demonstrates DLO effects in the thermal aging of a polymer for example, an elastomeric seal material, and the photo/radiation aging of a multi-layer coating. At low temperatures degradation features are more homogeneous; at high temperatures edge effects are apparent and bulk properties have changed less. Understanding oxygen diffusivity, permeation, and consumption aspects is a necessity when planning accelerated aging experiments with a goal of yielding meaningful predictive capabilities.[9,10]

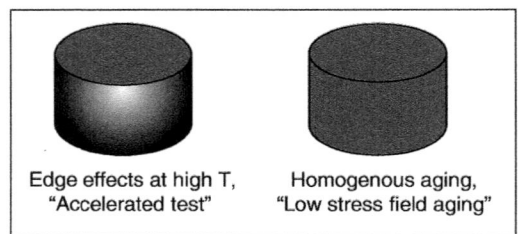

Edge effects at high T, "Accelerated test"   Homogenous aging, "Low stress field aging"

Figure 1a—Limited oxidation in the center at elevated temperatures.

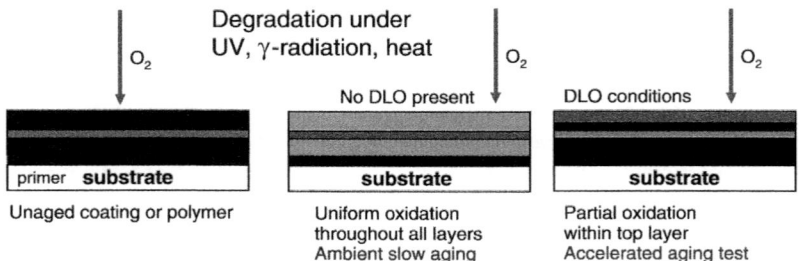

Figure 1b—Preferential degradation of top layer of coating under highly accelerated conditions.

Besides DLO conditions that can result in mechanistic variations with temperature, additional complications may occur in accelerated aging studies based on differences in the dominant chemical processes as a function of temperature. Such effects are generally difficult to measure and require comprehensive aging experiments over a large temperature range. In this conference contribution, we briefly review our latest results on developing improved analytical techniques and gaining a better understanding of temperature effects on thermal degradation. Careful multi-year aging studies were conducted to determine the subtle signatures of non-linear Arrhenius behavior and related mechanistic changes that may occur over a large temperature range.

EXPERIMENTAL

For most of the studies discussed here, a crosslinked hydroxy-terminated polybutadiene-based (HTPB) elastomer was used that was thermally aged over a large temperature range from RT to 125°C. The material is initially stabilized with 1% of a phenolic antioxidant (AO 2246). The material and processing details, as well as initial aging studies, have been reported earlier.[11] The material degradation as a function of temperature and time was monitored via oxygen consumption,[2,9,11,12] antioxidant extraction and quantification,[13] and chemiluminescence as a condition monitoring tool.[14]

DISCUSSION

Non-Linear Arrhenius Extrapolations Based on Oxidation Rate Measurements

Lifetime prediction of polymeric materials often requires extrapolation of accelerated aging data with the suitability and confidence in such approaches being the subject of ongoing discussions and extensive materials aging studies. We have been involved in long-term efforts to determine the importance of non-linear Arrhenius effects and to understand the chemical changes underlying such behaviors.

A convenient technique that allows monitoring of degradation sensitivities over a large temperature range is oxygen consumption, which measures the oxidation rate of materials.[2,9,11,12,15-17] This approach has been established as a routine analysis and has provided valuable information with predictive quality for many materials. Most importantly, various studies using this technique have shown clear evidence for faster oxidation rates than would be predicted at lower temperatures.[1,11,18,19] This led to the question as to whether other examples of Arrhenius curvature in degradation data can be found in the literature. Hence, the evidence of non-Arrhenius behavior (curvature) in various polymer degradation studies was reviewed.[2] While many of these studies certainly emphasized mechanistic variations and their importance for changes in activation energies, individual activation energies or a more detailed description of curvature were often not presented. To enable better analysis and interpretation, a simple mathematical approach describing the Arrhenius curvature was introduced. It was proposed that at minimum two competing reactions or two critical reactions with individual temperature dependence within a more complex reaction scheme should result in simple curvature. This allowed

Table 1—Two-Process Fitting Results for Some Elastomers

	E_a HT [kJ/mol]	E_a LT [kJ/mol]	T Cross-over [°C] ($k_1=k_2$)	Regression Coefficient
PU rubber, best fit[2] 119	65	62	0.9993
PU forced fit[2] 125	60	65	0.9984
PU forced fit[2] 110	70	55	0.9980
EPDM black[2] 127	78	123	0.9975
Butyl rubber[2] 100	60	55	0.9976
PP literature data:			
Richter (*Figure* 1[20]) 107	41	83	0.9994
Gijsman (*Figure* 4[21]) 156	36	85	0.9987
Gugumus (*Figure* 2[22]) 121	49	83	0.9997
Gugumus (*Figure* 9[22]) 146	41	82	0.9995

for excellent fitting of many experimental data sets as was shown for the thermal degradation rates of some elastomers (see *Table* 1). It does not require complex kinetic modeling and individual activation energies for a high and low temperature process are easily determined. *Figure* 2 shows an example of the oxidation rate shift factors for an HTPB-PU elastomer over the temperature range of 125°C to RT. The activation energy changes from 119 to 65 kJ/mol at a cross-over temperature of approximately 62°C, meaning much lower activation energies dominate the low temperature aging conditions. Reviewing independent data for the thermal degradation of polypropylene, a similar transition temperature of ~83°C was confirmed (see *Table* 1), with the high temperature process having a considerable higher activation energy (107–156 kJ/mol) than the low temperature process (35–50 kJ/mol). Such examples demonstrated the excellent fits that can be obtained by introducing, at a minimum, two active processes that will lead to curvature in Arrhenius plots.

It is important to emphasize that for accelerated aging studies where evidence of some curvature exists but limited data are available, better lifetime predictions could be made by estimating a low temperature process activation energy or allowing for a second rate dependence instead of forcing a straight line extrapolation. Since low activation energy processes can dominate at low temperatures and longer extrapolations result in larger uncertainties in lifetime predictions, experiments focused on estimating E_a values at the lowest possible temperature instead of assuming straight-line extrapolations will lead to more confident lifetime estimates.

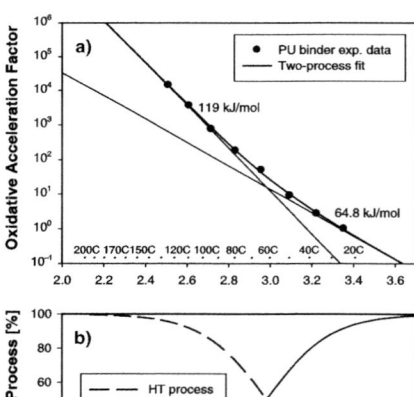

Figure 2—Example of the curvature observed in the oxidation sensitivity of a PU elastomer. The experimental data are fitted with a high and low temperature process.[2]

Condition Monitoring via Antioxidant Analysis

While oxidation rate measurements can easily cover a large temperature range, we have also explored if other techniques could be used to reveal mechanistic variations or provide data that will show temperature influences on the degradation process. An approach closely related to obtaining oxidation rates is monitoring the consumption of antioxidants that are commonly added to provide thermo-oxidative protection in polymers. A careful analysis of the consumption of stabilizers with aging exposure will provide feedback on the consumption kinetics, i.e., zero, first, second, or more complex order and how this process may depend on temperature. While such approaches appeared to be more widely used in the 1970s and 1980s—for example, for cable insulation degradation,[23,24]—few studies have recently been published despite easy availability of GC and HPLC based analytical instrumentation. Here we present a brief summary of the antioxidant consumption analysis with time and temperature of the same HTPB polymer as discussed above.[13]

Thermally aged stabilized HTPB elastomer samples were available at temperatures from 50 to 110°C. The concentrations of extractable antioxidant (AO 2246) in the polymer were quantified via AO solvent extraction and a gas chromatography based method using internal standards of stable aromatic molecules with similar retention times. Interestingly, potentially lower molecular weight AO degradation products were not observed with the GC analysis employed. The decrease in extractable AO levels as a function of time and temperature was evaluated and correlated with mechanical property changes. *Figure* 3 shows the decrease in extractable AO as a function of time and temperature and how the level of AO relates to the tensile elongation of the material. At higher aging temperatures (80 to 110°C) the levels of extractable AO decrease very rapidly in comparison with slower changes in mechanical properties (tensile elongation). This suggests excellent AO effectiveness or that some additional oxidative protection continues to be available via degraded antioxidant species, perhaps from a fraction of AO having been grafted to the polymer or chemically changed, and thereby mimicking an extractable "depletion" process. At lower aging temperatures (i.e., 50°C or 65°C) the extractable AO level decreases slowly with aging time, but despite high levels of AO the concurrent mechanical degradation process is not prevented. A superposition analysis of

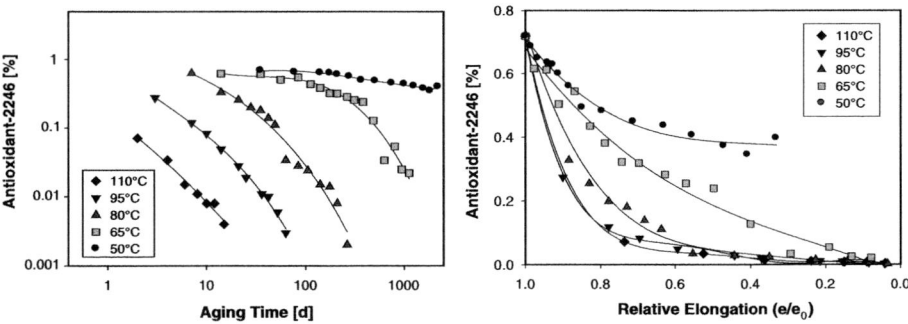

Figure 3—Decrease in extractable AO concentration at various aging temperatures and a correlation of these data with the decrease in tensile elongation.

50

the AO depletion process results in an E_a of ~135 kJ/mol. This temperature dependence is different than the thermo-oxidative aging process at lower temperatures (~65 kJ/mol), which is consistent with the AO depletion reactions becoming less important at lower temperature aging conditions. It also demonstrates that the AO reactions are related to a mechanistic variation in the thermo-oxidative degradation process and become more important at higher aging temperatures.

A corresponding time-temperature superposition of the AO depletion behavior was conducted, with the resulting shift factors yielding an activation energy of 135 kJ/mol.[13] These shift factors, referenced to 50°C, are included in a comprehensive Arrhenius plot, as shown in *Figure* 4. It is apparent that the high temperature degradation process observed for oxidation rates and reduction in mechanical properties correlates well with the AO depletion features based on similar activation energies. However, the divergence in shift factors towards high temperatures and lack of Arrhenius curvature (straight line) for the AO reactions also demonstrates that the reactions leading to loss of extractable AO are becoming more important with increasing temperature (relatively higher shift factors). Similarly, oxidation and mechanical property changes will become relatively more important at lower temperatures, independent of the reactions leading to loss of extractable AO. This is best demonstrated by comparing the curved Arrhenius plot (oxidation rates and mechanical properties) with a plot of the AO shift factors normalized to the high temperature degradation using a dashed line parallel to the shift factors originally normalized to 50°C (included in *Figure* 4). The difference between the curved Arrhenius plots and AO shift factors (dashed line) at the lower temperatures clearly shows how the ratio of AO depletion to mechanical degradation decreases at lower temperatures. For these conditions, it is apparent that the oxidative and mechanical degradation behavior is the dominant degradation process based on its reduced activation energy. This observation is similar to two competitive reactions resulting in curvature for Arrhenius plots of shift factors as discussed above. Thus, one may tentatively conclude that the cause of the changing degradation chemistry underlying the observed non-linear Arrhenius behavior for mechanical degradation and oxygen consumption is due to the drop in AO effectiveness at lower temperatures.

Importantly, these studies suggest that a condition-monitoring method using quantification of extractable AO levels in aged samples as an indicator of accumulated thermo-oxidative damage would in fact require a detailed correlation between mechanical and AO level changes at each temperature. At higher aging temperatures, a fast reduction in AO levels would sug-

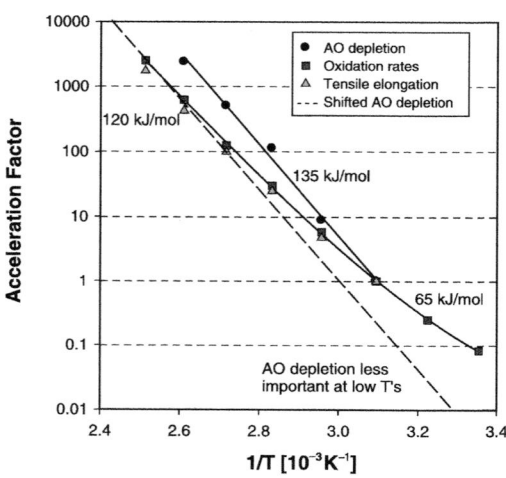

Figure 4—Arrhenius plot of shift factors.

gest rapid degradation but the material can still maintain some useful properties. At aging temperatures below 65°C, AO levels remain higher and change more slowly, but degradation of the polymer will nevertheless occur and result in mechanical failure. Furthermore, because AO levels appear to remain relatively constant during the later stages of the degradation process, measurements of AO levels would be of limited value for lower aging temperatures—precisely the conditions of most interest. These trends, as shown in *Figure* 5, would predict mechanical failure at ambient conditions at nearly initial concentrations of avail-

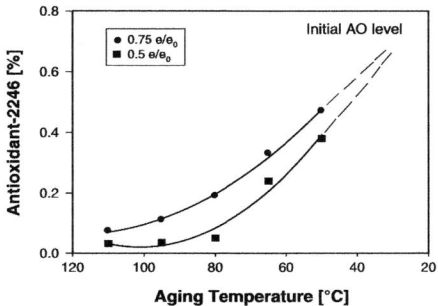

Figure 5—AO concentration as a function of aging temperature for a decrease in mechanical properties to 75% or 50% of initial elongation, suggesting a remaining AO level close to the initial concentration for aging and mechanical degradation at ambient temperatures.

able antioxidant. This is a completely unexpected result, but another example of how temperature can introduce a complexity into the aging process. It supports the following conclusions. Partial ineffectiveness of this antioxidant for the stabilization of this material results in insufficient prevention of oxidative damage at low temperatures. Alternatively, only a limited fraction of the available AO is actually involved in degradation inhibition and, furthermore, the oxidative degradation reactions may occur in centers of spatial heterogeneity despite the presence of high levels of antioxidant. Further, for condition-monitoring purposes, a universal correlation between AO levels and aging state or material condition for this material does not exist. Loss of mechanical properties and oxidative degradation is observed at lower temperatures despite significant levels of free antioxidant in the material. The antioxidant appears to be limited in its effectiveness to completely prevent degradation reactions, or only fractions of the total AO available are actually involved in the inhibition process.

Condition Monitoring via Chemiluminescence Detection

Another technique that has often been regarded as being very sensitive for measuring oxidative reactions in polymers is chemiluminescence (CL).[14,25,26] Here we briefly summarize how CL was applied as a condition-monitoring technique to assess aging-related changes in the same hydroxyl-terminated-polybutadiene based polyurethane elastomer.

Two chemiluminescence-based condition-monitoring techniques were applied to the thermally degraded HTPB samples aged between 110 and 50°C. The first CL analysis was a simple "wear-out" method relying on short-term additional isothermal aging under oxygen, which yields "wear-out" times and an initial CL rate. The second technique was a ramped temperature analysis under inert conditions used to quantify the accumulation of hydroperoxides or similar reactive degradation products in the material as a function of previous aging exposure. The feedback from these CL experiments and dependency on fractional damage in the polymer (i.e., prior aging history) were evaluated on the basis of qualifying this technique as a quick screening or condition-monitoring method for

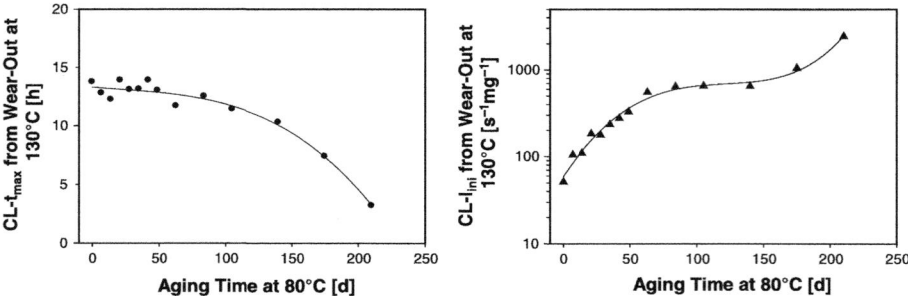

Figure 6—Examples of CL isothermal wear-out data of t_{max} and I_{ini} at 130°C of samples previously aged at 80°C.

quantification of degradation levels. These approaches also allowed for examining whether mechanistic variations with temperature may be present.

The isothermal approach yielded changes in "wear-out" time and initial CL rates that were sensitive to prior aging when correlated with mechanical property changes. An example is shown in *Figure* 6 for samples that were previously aged at 80°C and then used to conduct a quick isothermal follow-up CL oxidation experiment under oxygen at 130°C. Rather than relying on sequential and repetitive sample analyses for "wear-out" aging (using, for example, density or polymer network changes[27]), this technique allows for a convenient in-situ monitoring of the "wear-out" experiment. It delivers a time-to-maximum intensity and initial rate data from the CL experiment. While temperature ramp experiments can also identify changes or aging effects in these samples, this approach was shown to be relatively insensitive to previous aging due to a typical auto-accelerative behavior in the levels of CL active species with time.[14] Isothermal "wear-out" experiments providing measurements of the initial CL rate (I_{ini}) are the most sensitive and suitable approach for documenting material changes during the early part of thermal aging of this material. The initial rate data were obtained for all aged samples and correlated with tensile elongation as summarized in *Figure* 7. There is a trend in the data showing that this correlation depends on the original aging temperature. The I_{ini} data from the 50 and 65°C series are somewhat lower than the data from the higher aging temperatures.

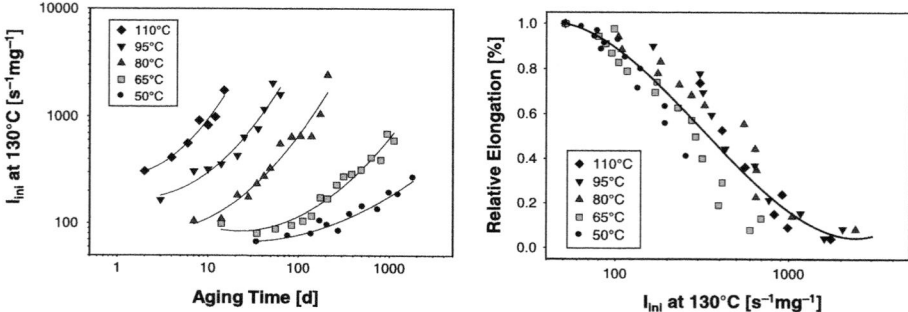

Figure 7—Initial CL intensity (I_{ini}) data from isothermal CL wear-out experiments conducted at 130°C of samples previously aged from 50 to 110°C and their correlation with tensile elongation.

This is again consistent with a mechanistic variation as a function of temperature[11,13] and a variation in the aging process that becomes more important at lower temperatures.[2] In terms of condition monitoring, it also demonstrates that fractional damage levels are easily correlated with the initial CL rate and could be used for monitoring aging effects in this material at ambient conditions.

Temperature Variations in Accelerated Aging Experiments

While accelerated aging studies using individual thermal conditions can be easily controlled to ±1–2°C, it is generally more difficult to control the temperature when UV exposure and γ-irradiation are also required. Field exposure is obviously the most variable situation in terms of temperature. This shifts our focus briefly to the complexity introduced by random temperature variations in, for example, outdoor weathering or thermal degradation of materials in actual applications. The following questions arise. Could a mechanistic framework be developed providing a better understanding of how temperature fluctuations may influence the aging process? Is an aging process unaffected if the temperature will temporarily increase by a few degrees and then drop by the same amount and for an identical length of time? Or, for example in simple terms, what is the effect on an aging process that should nominally run at 50°C, if the temperature increases to 55°C for a day and then drops to 45°C for a day to compensate for the previous increase? Is it meaningful quoting average aging temperatures if large temperature fluctuations occur, for example, in outdoor weathering? Can unexpected thermal excursions easily cancel out? What is the overall integrated effect with time? There are plenty of unanswered questions that are relevant to complex aging scenarios, and with the brief discussion below we attempt to relate acceleration factors for the thermal reactions to temperature variations in the overall aging process.

Figure 8 shows an example of possible temperature variations during an experiment at a mean temperature of 50°C. Increases and decreases in temperature will result in the acceleration or deceleration of the thermal degradation reaction that may run in parallel with UV, γ-irradiation, or hydrolytic degradation. Using a simple approximation and disregarding any curvature in the Arrhenius behavior at this point, acceleration factors are easily calculated using equation (1), and depend on the activation energy and the reaction

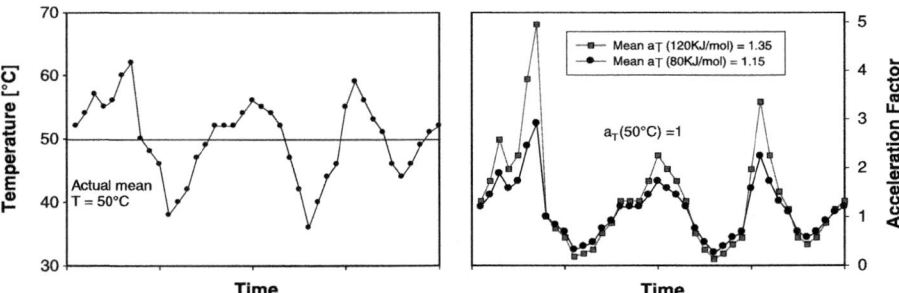

Figure 8—Example of an aging process at a mean temperature of 50°C and the corresponding acceleration factors for two activation energies (80 and 120 kJ/mol).

rate at the reference temperature T (i.e., $a_T = 1$ at 50°C). Due to the mathematical nature of the Arrhenius relationship in equation (1), absolute deviations in the acceleration factor are more pronounced towards lower temperature; for example, a 10°C increase with an E_a of 120 kJ/mol will equate to $a_T = 3.82$, but a 10°C decrease equates to $a_T = 0.24$ (i.e., 1/4.16); or, for an 80 kJ/mol process, a 20°C higher temperature equates to $a_T = 5.67$, and a 20°C lower temperature to $a_T = 0.14$ (i.e. 1/7.14). This may suggest that lower temperature deviations are more important in affecting the overall process. However, in practice a process running twice as fast for a certain amount of time cannot be compensated by then simply slowing it down to half the speed for the same amount of time. For example, if a car should be driving at an average speed of 50 km/h, and is then speeding for 1 h at 100 km/h, followed by 1 h at 25 km/h (i.e., a doubling and then half the speed), the average speed over those 2 h is 62.5 km/h. The mathematical complexity is easily apparent. One hour at three times the speed would require, at minimum, 2 h at a standstill to result in the intended average speed. Hence, in a similar sense the average rate of an aging reaction will depend on the integral of the relative acceleration factors and time intervals. For the schematic process presented in *Figure* 8, the mean acceleration factor for a 120 kJ/mol thermal degradation process is 1.35, and for the 80 kJ/mol process it is 1.15. For the 120 kJ/mol process, the aging would have been equal to being conducted at a constant "effective" temperature of 52.2°C (versus an average temperature of 50°C). This shows that the higher temperature excursions would be dominant and would accelerate the overall thermal degradation reactions. Compensating a spike in higher temperatures would require a much longer time at lower temperatures or much lower absolute deviations.

$$a_{T_2} = \frac{a_{T_1}}{e^{\frac{E_a}{R}(\frac{1}{T_2} \frac{1}{T_1})}} \tag{1}$$

If there is any recommendation to be made, temperature variations in outdoor weathering, or, for that matter, in accelerated UV experiments or other variable aging conditions, should be related to the underlying thermal degradation component and its activation energy. The effective (not the average) aging temperature that would be representative of a dynamic process should be determined. Such strategies may be useful in eliminating some of the "thermal noise" that may exist in otherwise very useful data sets. A better focus on the "real" thermal components in controlled dynamic aging environments should also allow for mechanistic variation details to be more easily established.

CONCLUSIONS

Mechanistic variations in polymer aging can occur as a function of temperature and have been observed for thermal aging in many materials.[2] The underlying reactions that govern the degradation of a material at ambient and other low stress-level environments may not be represented to the same degree under accelerated conditions. Additional chemical and physical reactions may be introduced under high stress-level conditions. While accelerated aging may provide some guidance on the expected degradation phenomena in materials, precise extrapolations and lifetime prediction will be complicated by these anomalies. More sensitive and relevant analytical techniques are required to develop better aging models and reveal subtle mechanistic variations. Further, comprehensive aging studies covering temperature spans as large as possible should be conducted to deliver suitable data sets.

In this chapter, we have briefly reviewed how sensitive oxygen uptake measurements, monitoring the consumption of an antioxidant, or chemiluminescence-based wear-out experiments could be applied to detect some variations in the thermal degradation behavior of polymers. These techniques picked up transitions between the high and low temperature aging conditions, consistent with a change in the activation energy of the acceleration shift factors. For dynamic aging experiments, it is suggested that temperature variations may be better approached with a mean rate of reaction or corrected effective temperature rather than a simplistic "average" aging temperature. Knowledge of the exact thermal history and the relevant thermal reactions coupled with predictable reaction rates based on activation energy and temperature range is required.

ACKNOWLEDGMENTS

Sandia is a multi-program laboratory operated by Sandia Corporation, a Lockheed Martin Company, for the United States Department of Energy's National Nuclear Security Administration under Contract DE-AC04-94AL85000. Gary Jones is gratefully acknowledged for assistance with the construction of the CL equipment, and Ana Trujillo for overall experimental support.

References

(1) Gillen, K.T., Celina, M., Clough, R.L., and Wise, J., "Extrapolation of Accelerated Aging Data-Arrhenius or Erroneous?," *Trends Polym. Sci.*, 5, 250 (1997).

(2) Celina, M., Gillen, K.T., and Assink, R.A., "Accelerated Aging and Lifetime Prediction: Review of Non-Arrhenius Behaviour Due to Two Competing Processes," *Polym. Degrad. Stab.*, 90, 395 (2005).

(3) Celina, M., Gillen, K.T., Wise, J., and Clough, R.L., "Anomalous Aging Phenomena in a Crosslinked Polyolefin Cable Insulation," *Radiat. Phys. Chem.*, 48, 613 (1996).

(4) Celina, M., Gillen, K.T., Clough, R.L., "Inverse Temperature and Annealing Phenomena During Degradation of Crosslinked Polyolefins," *Polym. Degrad. Stab.*, 61, 231 (1998).

(5) Cunliffe, A.V. and Davis, A., "Photo-Oxidation of Thick Polymer Samples—Part II: The Influence of Oxygen Diffusion on the Natural and Artificial Weathering of Polyolefins," *Polym. Degrad. Stab.*, 4, 17 (1982).

(6) Audouin, L., Langlois, V., Verdu, J., and de Bruijn, J.C.M., "Role of Oxygen Diffusion in Polymer Aging: Kinetic and Mechanical Aspects," *J. Mat. Sci.*, 29, 569 (1994).

(7) Gillen, K.T. and Clough, R.L., "Rigorous Experimental Confirmation of a Theoretical Model for Diffusion-Limited Oxidation," *Polymer*, 33, 4358 (1992).

(8) Gillen, K.T., Celina, M., and Keenan, M.R., "Methods for Predicting More Confident Lifetimes of Seals in Air Environments," *Rubber Chem. Technol.*, 73, 265 (2000).

(9) Gillen, K.T., Celina, M., and Bernstein, R., "Review of the Ultrasensitive Oxygen Consumption Method for Making More Reliable Extrapolated Predictions of Polymer Lifetimes," *Tech. Conf. Soc. Plast. Eng.*, 62, 2289 (2004).

(10) Celina, M. and Gillen, K.T., "Oxygen Permeability Measurements on Elastomers at Temperatures up to 225°C," *Macromolecules*, 38, 2754 (2005).

(11) Celina, M., Graham, A.C., Gillen, K.T., Assink, R.A., and Minier, L.M., "Thermal Degradation Studies of a Polyurethane Propellant Binder," *Rubber Chem. Technol.*, 73, 678 (2000).

(12) Scheirs, J., Bigger, S.W., and Billingham, N.C., "A Review of Oxygen Uptake Techniques for Measuring Polyolefin Oxidation," *Polym. Test*, 14, 211 (1995).

(13) Celina, M., Skutnik, E.J.M., Winters, S.T., Assink, R.A., and Minier, L.M., "Correlation of Antioxidant Depletion and Mechanical Performance During Thermal Degradation of an HTPB Elastomer," *Polym. Degrad. Stab.*, 91, 1870 (2006).

(14) Celina, M., Trujillo, A.B., Gillen, K.T., and Minier, L.M., "Chemiluminescence as a Condition Monitoring Method for Thermal Aging and Lifetime Prediction of an HTPB Elastomer," *Polym. Degrad. Stab.*, 91, 2365 (2006).

(15) Wise, J., Gillen, K.T., and Clough, R.L., "An Ultrasensitive Technique for Testing the Arrhenius Extrapolation Assumption for Thermally Aged Elastomers," *Polym. Degrad. Stab.*, 49, 403 (1995).

(16) Gillen, K.T., Keenan, M.R., and Wise, J., "New Method for Predicting Lifetime of Seals from Compression-Stress Relaxation Experiments," *Angew. Macromol. Chem.*, 83, 261-262 (1998).

(17) Gillen, K.T., Celina, M., and Bernstein, R., *Service Life Prediction: Challenging the Status Quo*, Martin, J.W., Ryntz, R.A., and Dickie, R.A. (Eds.), Federation of Societies for Coatings Technology, p. 67, 2005.

(18) Gillen, K.T., Celina, M., Bernstein, R., "Validation of Improved Methods for Predicting Long-Term Elastomeric Seal Lifetimes from Compression Stress-Relaxation and Oxygen Consumption Techniques," *Polym. Degrad. Stab.*, 82, 25 (2003).

(19) Gillen, K.T., Bernstein, R., and Derzon, D.K., 'Evidence of Non-Arrhenius Behavior from Laboratory Aging and 24-Year Field Aging of Polychloroprene Rubber Materials," *Polym. Degrad. Stab.*, 87, 57 (2005).

(20) Richters, P., "Initiation Process in the Oxidation of Polypropylene," *Macromolecules*, 3, 262 (1970).

(21) Gijsman, P., Hennekens, J., and Vincent, J., "The Influence of Temperature and Catalyst Residues on the Degradation of Unstabilized Polypropylene," *Polym. Degrad. Stab.*, 39, 271 (1993).

(22) Gugumus, F., "Effect of Temperature on the Lifetime of Stabilized and Unstabilized PP Films," *Polym. Degrad. Stab.*, 63, 41 (1999).

(23) Board, B.L. and Ruddell, H.J., "Investigation of Premature Depletion of Stabilizers from Solid Polyethylene Insulation," *Proc. Int. Wire Cable Symp.*, 31, 300 (1982).

(24) Ruddell, H.J., Adams, D.J., Latoszynski, P., and de Boer, B.T., "Behaviour of Four Non-Migratory Antioxidants in Solid Polyethylene Insulation," *Proc. Int. Wire Cable Symp.*, 32, 104 (1983).

(25) Celina, M., Clough, R.L., and Jones, G.D., "Initiation of Polymer Degradation via Transfer of Infectious Species," *Polym. Degrad. Stab.*, 91, 1036 (2006).

(26) Jacobson, K., Eriksson, P., Reitberger, T., and Stenberg, B., "Chemiluminescence as a Tool for Polyolefin Oxidation Studies," *Adv. Polym. Sci.*, 169, 151 (2004).

(27) Gillen, K.T. and Celina, M., "The Wear-Out Approach for Predicting the Remaining Lifetime of Materials," *Polym. Degrad. Stab.*, 71, 15 (2000).

Chapter 4

Final Report on the Subject of Accelerated Weathering: An ASTM D01.53 Ten-Year Exposure Study

D.A. Cocuzzi[1] and G.R. Pilcher[2]

[1]Akzo Nobel Coatings Inc., P.O. Box 489, Columbus, OH 43216-0489
[2]The ChemQuest Group, 111 Hoff Rd., Westerville, OH 43082

In 1992, ASTM's Task Group D01.53.03.03 Coil Coating Task Group on Accelerated Weathering was formed to answer a simple question: which accelerated test method best correlates with real-time weathering for coil coatings? A 10-year study was initiated, and coil coated panels were collected from D01.53 coater members for all current coil coatings technologies destined for exterior exposure applications. The real-time panels were weathered in southern Florida and in Phoenix, AZ. Accelerated weathering was run, including UV-condensation tests (with both "A" and "B" bulbs), Xenon Arc Weatherometer, Dew Cycle Weatherometer, and Fresnel-type weathering. The study has been completed, and results are discussed.

INTRODUCTION

In 1992, Dick Tucker, then the Technical Director of Kirby Building Systems and Chair of ASTM Subcommittee D01.53 on Coil Coated Metals, requested that his ASTM subcommittee determine which accelerated weathering technique provided the most realistic correlation for color change, gloss loss, and general appearance with real-time Florida UV-related weathering of modern coil coatings. (Corrosion prediction is not part of this study.) Mr. Tucker asked a simple question, of course, but a demanding experiment was needed to provide the answer to his question. It was decided that, to obtain the necessary data, a 10-year study would be needed. During the next two years, coil coated samples were secured, and—for the past 10 years—both real-time and accelerated tests have been running. This chapter presents the final,10-year results of this ASTM study.

It is well known that accelerated weathering devices are not able to replicate in all ways the natural weathering events that take place as a coating is exposed to heat, moisture, and sunlight. This makes durability prediction very difficult and, in fact, may lead to a number of false conclusions.[1] As a coating is exposed under natural conditions, it experiences the stresses of diurnal temperature cycling (daytime-nighttime temperature differences), moisture permeation in and out of the film (resulting from humidity, dew, and rain), as well as a host of photochemical effects as a direct result of exposure to solar

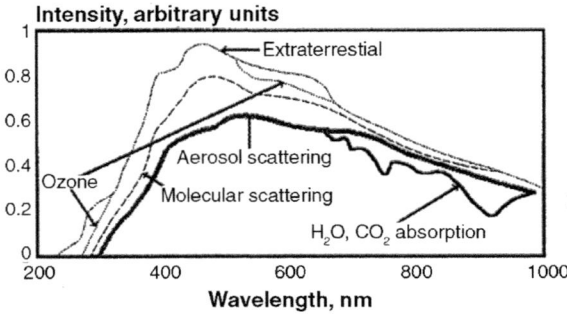

Figure 1—Solar power distribution of sunlight striking the earths surface.

radiation. Most modern accelerated weathering devices attempt to duplicate the effects of as many of these interdependent factors as possible, but none succeed completely. Not the only problem, but perhaps the most likely to lead to erroneous conclusions, is the fact that duplication of the complex solar spectrum is extremely difficult. As sunlight enters the Earth's atmosphere, ozone, water vapor, and particulate matter absorb and scatter the sunlight.

Essentially all energy <295 nm is absorbed in the atmosphere as a result of the atmospheric and extraterrestrial events shown in *Figure* 1.[2] (The UV portion of sunlight ranges from 295 nm–400 nm; visible light from 400 nm–700 nm; and infrared radiation [IR] above 700 nm.) Many light sources used in accelerated weatherometers emit radiation below the natural solar cut-off of sunlight (295 nm). This can be seen in *Figures* 2-4.

UV B-313 and FS-40 bulbs are notorious with regard to the amount of radiation emitted which has a shorter wavelength than the solar cut-off of 295 nm. UV A-340 certainly "tracks" better with normal sunshine in the UV region, as shown in *Figure* 3.

It is clear that the UV A-340 closely approximates the UV region of the solar spectrum, but it does not match the visible and IR range of sunlight. For such an irradiance match, one needs to consider Xenon arc bulbs, shown in *Figure* 4.

While the Xenon arc device more closely duplicates the solar spectrum, there are a number of "spikes" in the energy emission at certain wavelengths. Even though these occur in regions other than the UV region, they may still lead to unwanted, unexpected weathering events.

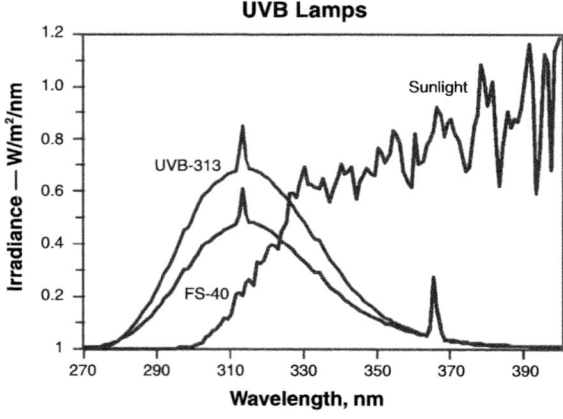

Figure 2—Power distribution of UV-B lamp.

When a coating is exposed to light which does not duplicate the solar spectrum, especially in the short wavelength UV region of the spectrum <295 nm, "unnatural" photochemistry occurs and spurious results are produced. This has been researched and reported upon extensively, especially by researchers at Ford Motor Company.[3,4] The promise of accelerated weathering, of course, is increased degradation rates (some claim between 2x and 35x times faster than real-time weathering using UV condensation devices[5]) and, therefore, reduced time needed before making a "go" or "no go" decision. This desire to know as early in the game as possible how well a product will perform falls into the realm of "service life prediction." How long will my product last in the field before it fails? How will it fail? What are the consequences of this failure?

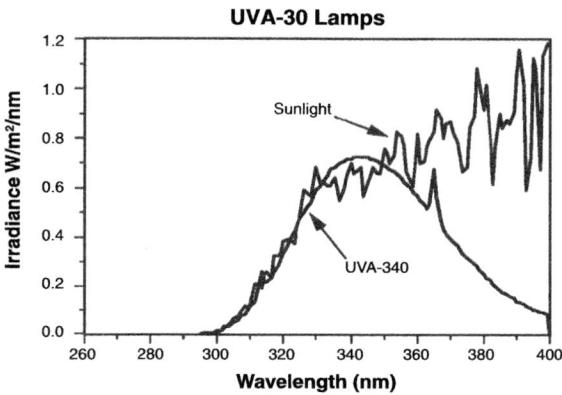

Figure 3—Power distribution of UVA-340 lamp.

Some researchers study the chemistry of degradation and base their predictions on what is happening at the molecular level,[6] while others use a relatively new approach, based upon reliability theory,[7] to predict the end of the lifetime of a material in the field. This ASTM study makes no attempt to monitor chemical changes, nor to predict service life of specific materials. It is simply an attempt to determine which accelerated weathering device exhibits the most satisfactory correlation with Florida. Such a determination, however, is not a simple matter—nor is it a task for the faint of heart.

Recognizing these difficulties, ASTM's Task Group D01.53.03.03 *Coil Coating Task Group on Accelerated Weathering* hoped to at least be able to recommend which accelerated technique would be most useful. There had been past studies involving coil coatings, but they usually contained only a few samples, or involved an insignificant number of parameters to measure. Many of these studies were curtailed after only a few years of weathering. Modern coil coatings have demonstrated a substantial durability improvement over those used years ago. Most of these coatings carry warranties that extend to over 30 years. Administering this study was clearly going to be a massive undertaking, and advice was sought in an effort to design a study that could actually be completed, with a reasonable degree of scientific rigor and statistical robustness. This was primarily accomplished by discussing the various logistical issues with authors of past studies, most notably Richard Fischer and Warren Ketola of 3M.

The basic design of the study involved gathering an abundance of samples, which represented 23 different paint systems (i.e.,

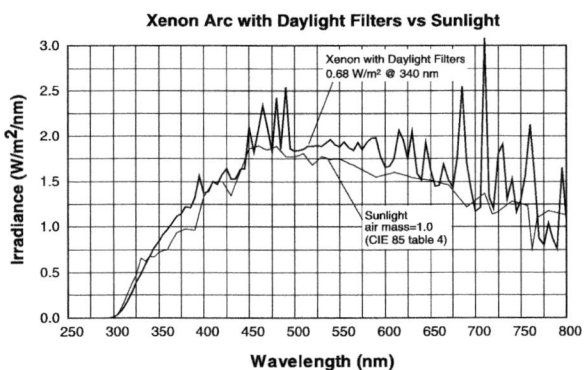

Figure 4—Power distribution of Xenon Arc lamp.

different resins and colors). Once the samples were collected, they were exposed in Florida and Arizona to begin the real-time testing. Arrangements were made with many companies to contribute accelerated testing services. To minimize measurement error, it was decided that only one laboratory would be used to take all of the readings. Duplicate sampling was used for all testing.

The question of statistical evaluation technique was tackled early on in the study. Many techniques were considered, but Spearman Rank Correlation was chosen. This is a common technique used by many doing weathering studies. Spearman Rank generates a Spearman *rho* value that designates the level of correlation achieved. This non-parametric method was developed for assessing the value of a rating system (e.g., individuals judging the performance of a group of people), and standard tables exist that define the minimum *rho* for various levels of confidence. In this study, however, we have a standard: results derived from 10-year exposure in South Florida, on open-backed racks, at 45° South. The experience of experimenters who have performed rigorous comparison testing suggested that *rho*=0.9, or better, should be the minimum level of correlation. If this value is not achieved, the likelihood of reversals (i.e., a sample looks great in an accelerated protocol, but, in reality, performs poorly in Florida) is too high. Another way of looking at the need for 0.9 *rho* is to argue that anything less means that exterior exposure is necessary, so why bother with accelerated weathering at all?

Before discussing results regarding accelerated weathering, some discussion is needed about real-time weathering of modern coil coatings and the need to carry out this ASTM study for a full 10 years. Today's coil coatings provide extremely high levels of resistance to UV, heat, and moisture. For such coatings, the first several years of weathering produce changes in the appearance, per unit of time, of the film in the form of chalking, gloss loss, and color change. This degradation may appear to be essentially linear, and it is tempting to extrapolate this data to predict chalk and fade values out 10 years, 20 years, even 30 years. As the degradation of the coating progresses, however, the change in appearance becomes less obvious, even though the changes in the chemistry of

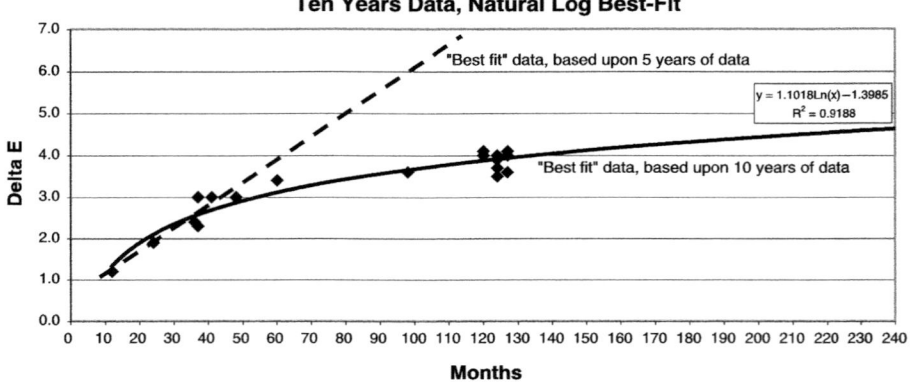

Figure 5—Example of ill-advised extrapolated data, after five years of exposure, compared to the same data series extended to 10 years of exposure.

the film may—or may not—still be proceeding at a steady state. It stands to reason, for example, that once the coating's gloss drops to <10% gloss retention, additional gloss loss is neither likely, nor likely to have a significant effect on appearance. These points can be illustrated by reviewing actual exposure data (which is not part of this ASTM study) collected over a 10-year period of time (*Figure* 5).

As can be seen in *Figure* 5, 10-years' worth of data was collected and plotted (ΔE versus length of exposure time). Using a logarithmic "best fit," this 10-year data has an R^2 value of 0.92 and may be extrapolated to 20 years with some level of confidence. This R^2 value (0.92) clearly demonstrates that the curve truly fits the data. It can be plainly seen, however, that—during the first five years—there appears to be a linear, steady state in the change in ΔE with time. With time, however, the rate of change begins to level off. Since this type of curve is clearly not linear, it is extremely dangerous to study highly durable coatings for just five years, when they are expected to provide satisfactory service in the field for 20+ years. Some coatings, of course, may continue to degrade linearly for 10 or 15 years before the chalk and fade values begin to level off, which is why the ASTM series of panels will be studied for 10 years.

The specific variables in this weathering study include 23 paint systems (various colors in polyester, silicone-modified polyester, PVDF, plastisol, and acrylic emulsion technology), tested in five accelerated devices (UV/condensation devices, with A-340 and B-313 bulb, Xenon Arc and Dew Cycle [carbon arc] weatherometers, and Fresnel weathering, with nighttime wetting), and two outdoor test sites—inland South Florida and Arizona, with a rack configuration of 45° South, open-back exposure. (See *Appendix* I for device parameters for each weathering methodology.) For the Florida and Arizona panels, and before measurements were taken, the right half of each panel was gently washed with a mild soap solution to remove debris from the surface of the panel. The term "washed" is used

Table 1—Samples Tested in This Study

Letter Code	Chemistry	Color
A	Plastisol[a]	Cream
B	Plastisol	Green
C	Plastisol	Charcoal
D	Plastisol	Charcoal
E	Plastisol	Brown
F	70% PVDF[b]	White
G	70% PVDF	Red
H	Acrylic emulsion	Low gloss brown
I	Acrylic emulsion	Brown
J	Silicone polyester	Tan
K	Silicone polyester	Stone
L	Polyester	White
M	Polyester	White
N	70% PVDF	Bronze
O	Silicone polyester	Gray
P	70% PVDF	Green
Q	70% PVDF	Terra cotta
R	Silicone polyester	Brown
S	70% PVDF	Teal
T	70% PVDF	White
U	Polyester	White
V	Silicone polyester	Light gray
W	70% PVDF	White

(a) A coil coating plastisol coating is comprised of polyvinyl chloride resin, blended with a liquid plasticizer (often di-isodecyl phthalate), pigments, and metal stabilizers used to minimize degradation due to heat and UV exposure.

(b) A coil coating PVDF coating is comprised of polyvinylidene di-fluoride resin (PVDF), blended with an acrylic resin, and is often pigmented with premium, "ceramic" pigments.

Table 2—Ranking of Exposure Series at Five Years of Exposure

Sample ID	ΔE (Real-Time Florida Weathering)	Ranking
F	0.7	1
K	0.9	2
T	1.4	3.5[a]
S	1.4	3.5[a]
L	1.6	5
U	1.9	6
A	2.4	7
M	2.7	8
P	2.8	9
Q	3.4	11.5[a]
V	3.4	11.5[a]
N	3.4	11.5[a]
W	3.4	11.5[a]
J	4.0	14
H	5.8	15
O	6.3	16
G	7.7	17
I	9.2	18
E	9.7	19
R	9.8	20
D	10.1	21
C	10.6	22
B	13.1	23

(a) Note that samples T and S were tied for third place, and that samples Q, V, N, and W were tied for tenth place.

throughout this paper to identify those measurements taken after gentle cleaning of the panel, and "unwashed" means that absolutely no cleaning took place before taking readings.

The 23 paint systems are shown in *Table* 1.

This study involved measuring five parameters for each sample. They are *color change* (using a 0°/45° color difference meter, common to the coil coating industry), *gloss change* (specular reflectance) and *chalking*. Color change and gloss change were measured on an undisturbed panel and also after a portion of the panel was washed, using a dilute solution of mild detergent, a soft sponge, and minimal pressure. This technique of washing a panel is common and is used in an effort to remove loosely-bound, friable material from the surface of the panel. Chalking is defined as the tendency of a coating to degrade and release a certain amount of its surface as a result of running a chalking test. For this study, we used the tape-chalk method. This is an extremely aggressive method of testing a coating's tendency to chalk, but experience suggests that it is the most consistent method available.

While it is not the intent of this paper to go into detail about the logic of using Spearman's *rho* (as opposed to a Pearson correlation coefficient or Kendall's *tau*), some explanation is needed to describe the general process of calculating the value *rho*. Using our five-year data as an example, the real-time Florida samples with their respective unwashed ΔE values (for this example) are simply placed in order of their rank. The sam-

Table 3—Ranking of Exposure Series after 1,235 Light Hours of Exposure in a UV/Condensation Cabinet, Outfitted with B-313 Fluorescent Bulbs

Sample ID	ΔE (QUV B-313)	Ranking
H	0.6	1
W	0.7	2
F	0.9	3
S	1.3	4
K	1.4	5
L	1.5	6
Q	1.8	7
U	2.0	8
T	2.4	9.5
M	2.4	9.5
P	2.5	11
N	2.6	12
A	2.7	13
V	2.8	14
G	3.6	15
J	7.2	16.5
I	7.2	16.5
D	7.9	18.5
C	7.9	18.5
O	8.7	20
E	9.0	21
B	10.3	22
R	12.6	23

ple with the smallest ΔE is ranked #1, the sample with the second-smallest ΔE is ranked second, etc. It does not matter how *much* difference there is between first and second place. This is a ranking tool, not a rating tool. (Note: If two panels have an identical ΔE, then a tie is declared. For example, in a Spearman Rank test, when a tie occurs between the third and fourth best panels, each gets a value of 3.5, and the next ranked panel skips "4" and becomes "5." Note also that it is important that the accelerated testing be halted when the general amount of color change reaches that same general level in Florida.) At five years, the unwashed ΔE ranking of the set of panels exposed in Florida is shown in *Table* 2.

As previously mentioned, while the Florida panels were collecting five years of exposure to the sun, heat, and rain, the various accelerated tests were being run. *Table* 3 compares the ΔE ranking values taken from a UV-condensation weathering device, using B-313 fluorescent bulbs, after 2,469 machine hours of exposure (1,235 light hours) in this cabinet.

If we inspect the data, it can be seen that the best panel in Florida is sample F, but sample F is only the third best panel in UV B-313. The best panel tested in UV B-313 is sample H, but sample H is the ninth *worst* panel in Florida testing. Many other observations may be made, but Spearman Rank correlation may now be used to quantify the comparisons among all 23 samples. To begin this process, simply combine both tables, as in *Table* 4.

Table 4—Rank Comparison between Five-Year Florida and 1,235 Light Hours B-313 Exposure

Sample ID	Ranking (Real-Time Florida)	Ranking (QUV B-313)
F	1	3
K	2	5
T	3.5	9.5
S	3.5	4
L	5	6
U	6	8
A	7	13
M	8	9.5
P	9	11
Q	11.5	7
V	11.5	14
N	11.5	12
W	11.5	2
J	14	16.5
H	15	1
O	16	20
G	17	15
I	18	16.5
E	19	21
R	20	23
D	21	18.5
C	22	18.5
B	23	22

Table 5—Spearman Rank Calculation

Sample ID	Ranking (Real-Time Florida)	Ranking (QUV B-313)	Difference in Ranking	Square of the Difference in Ranking
F	1	3	-2	4
K	2	5	-3	9
T	3.5	9.5	-6	36
S	3.5	4	-0.5	0.25
L	5	6	-1	1
U	6	8	-2	4
A	7	13	-6	36
M	8	9.5	-1.5	2.25
P	9	11	-2	4
Q	11.5	7	4.5	20.25
V	11.5	14	-2.5	6.25
N	11.5	12	-0.5	0.25
W	11.5	2	9.5	90.25
J	14	16.5	-2.5	6.25
H	15	1	14	196
O	16	20	-4	16
G	17	15	2	4
I	18	16.5	1.5	2.25
E	19	21	-2	4
R	20	23	-3	9
D	21	18.5	2.5	6.25
C	22	18.5	3.5	12.25
B	23	22	1	1
Sum of the squares				470.5

The Spearman Rank calculation treats the rank of a panel (which is an ordinal value) as a cardinal number, and equation (1) describes the calculation of *rho*:

$$rho = 1 - \frac{6 \bullet \sum (rank_{Florida} - rank_{QUV-B})^2}{N(N^2 - 1)} \tag{1}$$

where N is the number of samples (23 for this study). Staying with the Florida-UV B-313 example, the numerical difference between the rankings of each sample is calculated, then each ranking difference is squared, as shown in *Table 5*. Since there are 23 samples, $N(N^2-1) = 23(23^2-1) = 12144$. To calculate *rho* using the values already established, do the following:

$$rho = 1 - 6(470.5)/12144 = 1 - 0.23 = 0.77$$

Since we will be discussing *rho* values, it is worthwhile to first discuss some points about this value. Two sets of data that correlate perfectly have a *rho* value of 1.00. A *rho* value of less than 1.00 represents less-than-perfect correlation. In weathering studies such as this, it is commonly accepted that a *rho* value of >0.9 is necessary before one can declare an acceptable level of correlation between any two weathering techniques.[8] One must, however, be cautious about assuming too much about *rho* values. The Spearman *rho* does not have "linearity." In other words, a *rho* value of 0.8 is *not* twice as good as a *rho* value of 0.4. In our study, for example, the *rho* value for ΔE in a UV B-313 cabinet is 0.77, whereas the *rho* value in a UV A-340 cabinet is 0.84. You cannot declare that the UV A-340 cabinet is 10% better (or 10% more predictive) than the UV B-313 cabinet, only that it is better.

DISCUSSION OF RESULTS

In the building products market for coil coatings, extensive performance warranties exist. While these warranties most always are associated with "washed" results, it is the "unwashed" readings that are the most interesting. Unlike automobiles, a building is rarely going to be washed. Because of this, loosely bound dirt, mildew, and other environmental factors are quite important. Many of these factors (e.g., mildew growth) are not UV related, as are the usual considerations when one typical considers outdoor durability.

Table 6 represents all of the available data from this ASTM study. In all cases, 10-year, Florida real-time data were considered the standard, and all other data were compared to it. In addition to comparing accelerated performance to 10-year real-time data, we also compared earlier real-time exposure intervals to the 10-year real-time data.

Using a *rho* value of 0.90 as the minimum level of acceptable correlation, one can quickly see that no accelerated device produces results that correlate sufficiently. While this may seem disappointing, the ASTM committee compared early real-time test results (three-, five-, and seven-years) versus 10-year data to determine the level of correlation. What we found was that seven-year results had an acceptable level of correlation to 10-year results. The earlier results (five- and three-year results) did not correlate as well as might be expected.

Table 6—Data Table of 10-Year Real-Time Results, Compared with Various Other Testing Protocols

		Spearman Correlation Coefficient (*rho*) 10-Year Florida 45° South Data versus . . .							
Parameter	Condition	7-yr FL Data	5-yr FL Data	3-yr FL Data	2-yr FL Data	Fresnel	QUV-A	QUV-B	Dew Cycle
Gloss	Unwashed	0.87	0.78	0.65	0.53	0.67	0.63	0.72	0.49
Retention	Washed	0.94	0.89	0.78	0.66	—	—	—	—
ΔE	Unwashed	0.94	0.79	0.78	0.61	0.70	0.69	0.53	0.40
	Washed	0.89	0.85	0.74	0.69	—	—	—	—
Chalk	Unwashed	0.61	0.48	0.38	0.39	0.34	0.20	0.34	—

CONCLUSIONS

No accelerated technique was able to suitably correlate with real-time weathering. We have found that current data predicts the data that one will see in two additional years of exposure—at least with regard to color change and gloss change. For reasons not clearly understood, this correlation is not seen when it comes to the chalking parameter.

References

(1) Pilcher, G.R., "Meeting the Challenge of Radical Change: Coatings R&D as We Enter the 21st Century," *J. Coat. Technol.*, *72*, No. 921, 135-143 (2001).
(2) Wypych, G., *Handbook of Material Weathering*, 2nd Ed., p. 45, 1995.
(3) Gerlock, J.L., "Nitroxide Decay Assay Measurements of Free Radical Photo-initiation Rates in Polyester Urethane Coatings," *Polym. Degrad. Stab.*, 26, 241-254 (1989).
(4) Bauer, D.R., Paputa Peck, M., and Carter, R., "Evaluation of Accelerated Weathering Test for a Polyester–Urethane Coating Using Photoacoustic Infrared Spectroscopy," *J. Coat. Technol.*, *59*, No. 755, 123-129 (1987).
(5) Fedor, G. and Brennan, P., "Correlation between Natural Weathering and Fluorescent UV Exposures," *Durability Testing of Non-Metallic Materials*, ASTM STP 1294, Herling, R.J. (Ed.), ASTM, pp. 91-105 (1996).
(6) Bauer, D.R., "Predicting In-Service Weatherability of Automative Coatings: New Approach," *J. Coat. Technol.*, *69*, No. 864, 85-96 (1997).
(7) Martin, J.W., Saunders, S.C., Floyd, F.L., and Wineburg, J.P., "Predicting the Service Lives of Coatings Systems," *Federation Series on Coating Technology*, FSCT, Blue Bell, PA, June 1996.
(8) Fischer R.M. and Ketola, W.D., "Accelerated Weathering Test Design and Data Analysis," Chapter 17, *Handbook of Polymer Degradation and Stabilization*, 2nd Ed., Revised and Expanded, Marcel Dekker, Inc., New York, 2000.

APPENDIX I—DEVICE PARAMETERS FOR WEATHERING METHODOLOGIES

FLORIDA WEATHERING: 45° S, open-rack configuration (no backing), exposed at Q-Lab/Weathering Research Service, Homestead, FL (26° N latitude, sea level)

ARIZONA WEATHERING: 45° S, open-rack configuration (no backing), exposed at Q-Lab/Weathering Research Service, Buckeye, AZ (33° N latitude, 1,055 feet elevation)

FRESNEL-TYPE WEATHERING: Service supplied by both Q-Lab/Weathering Research in Buckeye, AZ, and Atlas/DSET Laboratories, Black Canyon Stage, AZ (34° N latitude, 2,000 feet elevation)

XENON ARC WEATHERING:
Modified SAE1960
Black panel temperature=70°C
Dry bulb temperature=47°C
Irradiance=0.85W/m^2
Lamp Wattage 12,000W
Relative Humidity=50%
Filters: CIRA inner/sodium lime outer

UV/CONDENSATION, WITH A-340 BULBS: 0.77 W/m^2; 8 h of UV light @ 60°C, followed by 4 h of condensation (dark cycle) @ 50°C

UV/CONDENSATION, WITH B-313 BULBS: 0.63 W/m^2; 4 h of UV light @ 60°C, following by 4 h of condensation (dark cycle) @ 50°C

DEW CYCLE (UNFILTERED CARBON ARC):
One hour of light @ 63°C, followed by one hour of condensation (dark cycle) @ 28°C
Wet bulb (light cycle)=35°C
Wet bulb (dark cycle)=29°C
Dry bulb (light cycle)=46°C
Dry bulb (dark cycle)=29°C
Water pressure=15–18 psi

APPENDIX II—ASTM STANDARDS ASSOCIATED WITH WEATHERING AND WEATHERING DEVICES

D 523 Standard Test Methods for Specular Gloss
D 822 Standard Practice for Conducting Tests on Paint and Related Coatings and Materials Using Filtered Open-Flame Carbon-Arc Apparatus
D 1014 Standard Practice for Conducting Exterior Exposure Tests of Paints on Steel
D 2244 Standard Test Method for Calculation of Color Differences from Instrumentally Measured Color Coordinates
D 3361 Standard Practice for Operating Light- and Water-Exposure Apparatus (Unfiltered Open-Flame Carbon-Arc Type) for Testing Paint, Varnish, Lacquer, and Related Products Using the Dew Cycle
D 4141 Standard Practice for Conducting Accelerated Outdoor Exposure Tests of Coatings
D 4214 Standard Test Methods for Evaluating Degree of Chalking of Exterior Paint Films
D 4587 Standard Practice for Conducting Tests on Paint and Related Coatings and Materials Using a Fluorescent UV-Condensation Light- and Water-Exposure Apparatus

D 5031 Standard Practice for Conducting Tests on Paints and Related Coatings and Materials Using Enclosed Carbon-Arc Light and Water Exposure Apparatus

G 7 Standard Practice for Atmospheric Environmental Exposure Testing of Nonmetallic Materials

G 151 Standard Practice for Exposing Nonmetallic Materials in Accelerated Test Devices that Use Laboratory Light Sources

G 152 Standard Practice for Operating Open Flame Carbon Arc Light Apparatus for Exposure of Nonmetallic Materials

G 153 Standard Practice for Operating Enclosed Carbon Arc Light Apparatus for Exposure of Nonmetallic Materials

G 154 Standard Practice for Operating Fluorescent Light Apparatus for UV Exposure of Nonmetallic Materials

G 155 Standard Practice for Operating Xenon Arc Light Apparatus for Exposure of Nonmetallic Materials

Advances in Field and Laboratory Exposure

Chapter 5

An Analysis of the Effect of Irradiance on the Weathering of Polymeric Materials

Kenneth M. White, Richard M. Fischer, and Warren D. Ketola

3M Company, Weathering Resource Center, 3M Center 235-BB-44
St Paul, MN 55144-1000

A physical model is proposed that describes the rate of photodegradation of a dyed polymer film as a function of the irradiance employed during exposure to radiation. Development of the model is based on photochemical behavior reported in dyes of similar chemical classes. The model accounts for the lack of reciprocity observed in the material at high levels of irradiance, where the degradation rate exhibits saturation behavior instead of a linear dependence on irradiance. The model is found to correctly predict the initial photodegradation rate for exposure at low irradiance, where linear dependence is evident. Additional phenomena that result from the radiant exposure are reported and discussed in relation to the proposed model.

INTRODUCTION

Exposure to solar radiation is a key component of weathering that causes degradation of polymer-based materials. The degradation process is dependent both on that portion of the solar spectrum that produces the chemical or physical reaction that leads to deterioration of the material and on the intensity[1] of the radiation that the material encounters. In accelerated weathering tests, this dependence on intensity may be investigated by exposing materials to various levels of irradiance and recording the corresponding time-to-failure values. Based on these results, attempts may be made to predict time-to-failure in the field by extrapolating to irradiance levels that a material will experience over the course of its lifetime. In order to develop an extrapolation that is reliable, it is important to have a model that adequately accounts for the rate of photodegradation as a function of irradiance.

It is intuitive that increasing the stress imparted by light shining on a degradable substance would shorten the time to effect a specified change in the substance (such as failure) by an equivalent factor. This principle was expressed for the photographic process via the law of reciprocity, wherein it was found that a constant degree of blackening of a photographic plate would be achieved whenever the product of the intensity of light (I) and the time of exposure (t) was constant.[2] If one determined the average rate of the photochemical reaction over the time of exposure, the rate obtained would be directly proportional to the light intensity [curve (a) in *Figure* 1]. Several polymer systems have, in

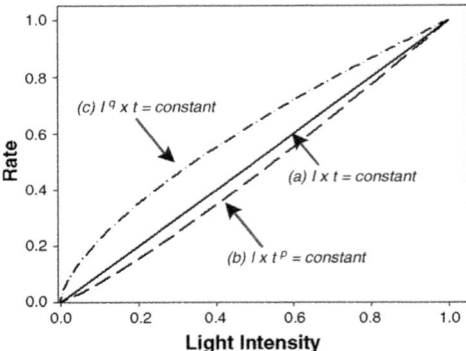

Figure 1—A comparison of models that have been used to describe rates of photoinduced changes in materials as a function of intensity and time: (a) law of reciprocity; (b) Schwarzschild's law (p = 0.86); (c) power curve (q = 0.65 in this example). The three models have been normalized for purposes of comparison by setting the constant in each expression equal to 1.

fact, been reported to deteriorate in accordance with the law of reciprocity.[3-5] Nevertheless, numerous other materials have exhibited degradation behaviors that deviate significantly from a linear response to intensity.[4-6]

An alternative model that has been discussed in consideration of a nonlinear response to light is that embodied in Schwarzschild's law, which is expressed as $I \cdot t^{\,p}$ = constant. Schwarzschild discovered this relationship for his photographic emulsions,[7] for which the value of p was observed to be 0.86. Subsequent studies confirmed, for the most part, that p was less than 1, although it has been found to vary as a function of emulsion type, wavelength, and intensity.[8] These observations are indicative of the phenomenon that the photochemical reaction in photographic emulsions is slower at low light intensities than what would be expected from a linear relationship[9] [curve (b) in *Figure* 1].

In contrast, deviations from linearity for polymeric materials have generally exhibited behavior in which photodegradation rates are *faster* than what would be expected from reciprocity determined at high irradiance. This tendency is more closely represented by curve (c) in *Figure* 1, which is based on the expression $I^{\,q} \cdot t$ = constant. Here, the exponent q is also less than one, but it is applied to the intensity factor instead of the time factor. This model has been used in several studies to analyze material degradation, for which the value of q has been reported to be material dependent.[4-6,10,11] One explanation for this intensity dependence has its basis in solid state band theory, wherein photoinduced electronic transitions that give rise to free electrons can lead to chemical degradation.[5,12] Such a theory accounts for q values that range from 1 down to 0.5. In the chemical breakdown of organic material systems containing TiO_2, an $I^{0.5}$ dependence has been reported when the exciting ultraviolet (UV) radiation is above a certain threshold intensity.[13] Below this level, the dependence on intensity is linear, which is what has been observed in the oxidation of a dye in the presence of a titania photocatalyst.[14]

The universal application of the band theory model to photodegradation of polymer materials by solar radiation is not certain, however. In films comprising organic dyes at modest concentrations, for example, the molecular interactions required for the formation of bands in the electronic structure of the system are unlikely to exist. Since attainment of durable color in transparent, dyed polymer films is important in a number of different applications, a model that appropriately explains and reliably predicts the effect of irradiance on the degradation of color in systems of this type is sought. The study of transparent films facilitates a quantitative examination by enabling the use of UV-visible absorption spectroscopy to directly measure the dye concentration as the photo-degradation progresses. Furthermore, evaluating the rate of this process, as opposed to obtaining single-point time-to-failure values, makes it possible to use greater amounts of data in constructing the model and provides the opportunity to assess any unusual behaviors that may be encountered.

EXPERIMENT

The material system selected for this study comprised a fluorescent yellow-green dye, known as solvent yellow 98 (SY98, see *Figure* 2), dispersed in a transparent, plasticized polyvinylchloride (PVC) film. The film was "double-polished" when it was calendered, such that both surfaces provided near-specular reflection losses that could easily be accounted for when analyzing the absorption spectra. The dye has a strong absorption band between 400 and 500 nm, which is primarily responsible for its tendency to fade under exposure to light—either from solar radiation or xenon arc sources. SY98 also has absorption bands in the near-UV spectral region, but they are very weak relative to the absorption band in the visible region. This, combined with the presence of a UV-absorber in the film, significantly reduces photodegradation from other than visible wavelengths and allows us to ignore UV radiation in the data analysis.

The film was cut into approximately 1.25-in. square specimens, each of which was held between two aluminum plates that provided a 1-in. square window on either side of the film. The specimens were tested in an Atlas Ci5000 Xenon Arc Weather-Ometer® that employed a quartz inner filter and a 3M proprietary outer filter[15] to approximate daylight. The weathering cycle consisted of an 8-h segment with light only, followed by a 4-h dark segment, a portion of which included water spray. The surface temperature of the film specimens during the light segment was maintained near 71°C, as determined by direct measurement with a thermocouple.

Four different levels of irradiance were employed in the experiment: 0.3, 0.5, 0.75, and 1.0 W/m^2/nm measured at 340 nm. At each level, three replicate film specimens were exposed in the first round of tests, and one or two replicate specimens were similarly exposed in a second round of tests. In both rounds, additional film specimens were placed in holders that blocked the light by means of an aluminum plate. These holders were designed so that specimens were still able to experience the effects of heat and moisture, in order to test for any degradation that might occur in the absence of radiation.

The UV-visible absorption spectrum of each film specimen (400–800 nm for round one, 300–800 nm for round two) was measured prior to exposure and at regular time intervals thereafter. Spectra were recorded using a Shimadzu UV 2401PC spectrophotometer equipped with a holder assembly that ensured repeatable positioning of a specimen each time it was measured. Film specimens that were kept in the dark at room

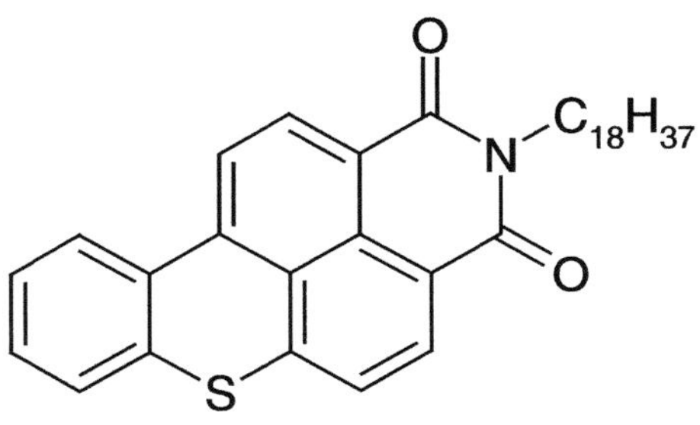

Figure 2—Molecular structure of solvent yellow 98.

temperature were also measured at each interval. Their spectra revealed that the repeatability of the absorption measurement was very near to the photometric repeatability specified for the spectrophotometer (± 0.001 absorbance units).

For each spectrum recorded as the photodegradation experiment progressed, the peak absorbance at 455 nm was determined. This value was corrected for losses in the spectrophotometer due to reflection at both film surfaces by subtracting the absorbance reading obtained at 800 nm, where no dye absorption was detected. The observed losses were in agreement with reflectivity values predicted from the refractive index of PVC. The peak absorbance was also corrected by subtracting the residual absorbance that persisted at 455 nm after the film had been completely bleached. This method of correcting for residual absorbance gave the same result as scaling the initial absorption band to fit each subsequent one and determining the fraction lost as a function of time.

RESULTS AND DISCUSSION

Films Blocked from Radiation

None of the film specimens that were prevented from receiving radiation from the xenon arc source showed signs of significant fading due solely to heat and moisture. However, each specimen did exhibit an increase in peak absorbance that was noticeable within the first 24 h of weathering exposure. The increase was on the order of 1 to 2% (*Figure* 3) and persisted more or less throughout the course of the experiment, with some specimens exhibiting a very slight decay from the initial rise in absorbance and others a very slight increase as time went on. The phenomenon of "negative fading" has been observed in systems in which dye molecules are initially aggregated, but then break up due to heat from illumination or some other source to give dispersed dyes that yield higher absorbance.[16,17] It has been suggested that this effect is reversible,[17] however, and no such occurrence was observed in the specimens studied here. A film specimen that remained at room temperature in the dark for two months after the weathering exposure retained the same increase in peak absorbance that it had exhibited immediately after the exposure had ended.

A more likely cause of the increase in absorbance was traced to shrinkage of the films. To investigate this, three film specimens that were approximately 68 mm × 50 mm × 0.010 in. were prepared. For each specimen, the length and width were measured to the nearest 0.3 mm using a ruler, the thickness was meas-

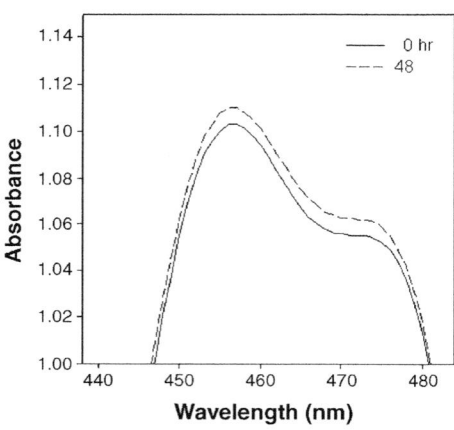

Figure 3—An example of peak absorbance measured from a film specimen prior to testing and after 48 h of weathering. This specimen was in a holder that blocked xenon arc radiation, so it only experienced the effects of heat and moisture. The increase in absorbance is ascribed to film shrinkage.

ured to the nearest 0.0001 in. using a digital thickness gauge, and the absorbance spectrum was recorded. The specimens were then heated in an oven at 70°C for 65 h, after which each of the measurements on the films was repeated. Shrinkage of the films yielded an average 2% increase in thickness and 1% decrease in volume, the latter translating to a 1% increase in dye concentration. The increases in thickness (pathlength) and concentration predict a 3% increase in absorbance by the dyed film. The average increase measured in the peak absorbance of the specimens was 2.4%, in good agreement with the prediction, when considering the experimental error of the measurements. The occurrence of shrinkage also accounts

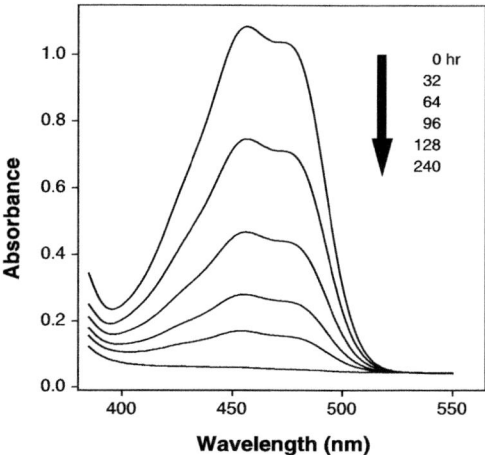

Figure 4—Absorption spectra measured from a film specimen as a function of time of exposure to the xenon arc source. The irradiance level to which this particular specimen was exposed was 1.0 W/m²/nm (at 340 nm).

for the observed irreversibility of the absorbance increase. As a result of this effect, the initial peak absorbance values for film specimens undergoing the photodegradation tests were corrected for shrinkage during analysis of the results.

Films Exposed to Radiation

The decay in the absorption spectrum of a film specimen that resulted from one of the accelerated weathering trials is displayed in *Figure* 4. In the analysis, time was recorded as the number of hours of exposure to the xenon arc source, owing to the observation that bleaching of the dye required the presence of radiation. The disappearance of the absorption band is attributed to photodegradation of the dye molecules into colorless products.

The rate of photodegradation was found to be highly dependent on the irradiance level of the light source, as seen in *Figure* 5 in which peak absorbance is plotted as a function of time of exposure. Intensity

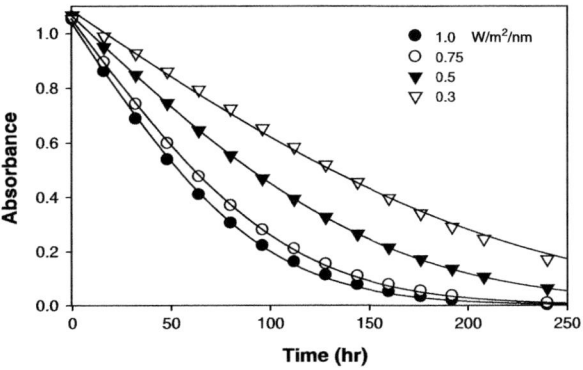

Figure 5—Corrected peak absorbance at 455 nm measured from four different film specimens as a function of time of exposure to the xenon arc source. Irradiance levels shown are at 340 nm. Solid lines were obtained by fitting the data using the solution to equation (2).

dependence is included in the rate equation proposed by Kaminow et al.[18] for a photo-bleaching reaction that is first-order in dye concentration:

$$-\frac{dD(z,t)}{dt} = \sigma \, \phi \, n(z,t)D(z,t) \tag{1}$$

where $D(z,t)$ is the dye concentration at film depth z and time t, σ is the absorption cross section of the dye, ϕ is the quantum efficiency for photodegradation, and $n(z,t)$ is the photon intensity. Other expressions that have been reported[19] can be shown to be mathematically equivalent. Here, we assume the spectral distribution of n to be the same for each level of irradiance employed in the trials and find that its wavelength dependence, as well as those of σ and ϕ, are adequately treated by invoking a monochromatic approximation.[20] This also means that hv in the relationship $n = I/hv$ is constant, so we obtain

$$-\frac{dD(z,t)}{dt} = \sigma^* \phi \, I(z,t)D(z,t) \tag{2}$$

where $\sigma^* = \sigma/hv$. The spatial dependence of D in equation (2) accounts for the limited mobility of the dye in the polymer host, which results in a non-uniform distribution of dye molecules in the film as photodegradation progresses. This is portrayed in *Figure 6*, which was generated from a solution to equation (2) that was fitted to the data obtained for one of the film specimens exposed in the study. At the commencement of the exposure, the dye concentration is uniform throughout the film depth and the light intensity falls off exponentially, in accordance with Lambert's law. As photodegradation proceeds, the dye concentration declines most rapidly at the radiation incident surface of the film, where the intensity is greatest, but more slowly at increasing depths into the film, where the radiation is attenuated. This gives rise to a concentration profile that increases with increasing film depth, which in turn causes the light intensity to depart from its exponential profile.

In order to develop a rate model based on the observed data, a closed-form solution to equation (2) was initially employed.[18] It yields a time constant for the decay of peak absorbance that can be expressed as $\beta = \sigma^* \phi \, I_0$, where I_0 is the incident intensity. Larger time

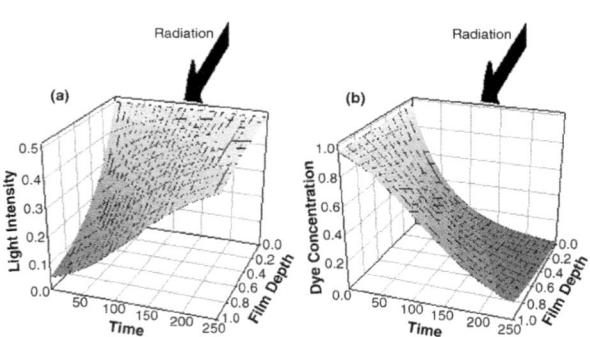

Figure 6—Modeled representation of the distribution of (a) light intensity and (b) dye concentration in the polymer film as a function of time of exposure to the irradiating source. The backside of each graph represents the film surface where radiation enters the specimen; the near side is the exiting surface. Axis scales are in arbitrary units.

constants correspond to faster decay rates. The peak absorbance data for each film specimen were fit using this solution. Examples are shown by the solid curves in *Figure* 5 that correspond to representative specimens at each of the four irradiance levels. The quality of the fits provides evidence that the photodegradation is indeed first-order in the dye concentration, as indicated by the model. The β values obtained from the fits for all film specimens are plotted in *Figure* 7. Although the linear dependence of β on light intensity in this model demands reciprocity, it is clearly seen from the figure that

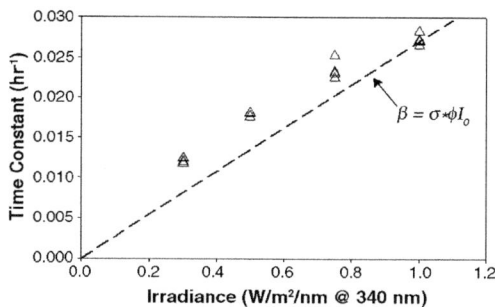

Figure 7—Time constant (β) values for photodegradation of dyed films obtained by using equation (2) to fit the decay in peak absorbance observed for each specimen. The dashed line indicates values that would be expected based on reciprocity, as determined at the highest irradiance tested.

this is not the case. Hence, it was necessary to identify another model for the photodegradation rate that retains the attribute of first-order kinetics and at the same time accounts for the nonlinear dependence on irradiance.

Modeling Irradiance Dependence of Photodegradation

An explanation for the kinetics and mechanisms involved in the photodegradation of a dye should include a description of the various electronic states that are potentially accessible to the dye molecule once it has been excited by an absorbed photon.[21] In the absence of this information for the SY98 dye examined in this study, we turned to data that have been reported for the class of dyes that have the xanthene structure, which is similar to the thioxanthene component found in the chemical structure of SY98. The photophysics of xanthene dyes can be depicted by means of a Jablonski diagram (*Figure* 8), which includes the rate of excitation (k_{abs}) from the electronic ground state (S_0) to the first excited singlet state (S_1) via absorption of light, the rate of relaxation from there back to the ground state (k_S) via both radiative and non-radiative processes, the rate of intersystem crossing (k_{ISC}) to the lowest triplet state (T_1), and the rate of relaxation from the triplet state to the ground state (k_T), also via radiative and non-radiative processes. For steady-state conditions of continuous, constant irradiation, the molecular concentration in the triplet state can be shown to be[22,23]

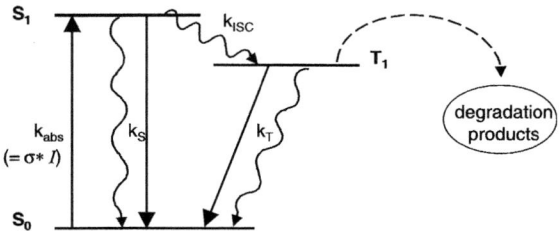

Figure 8—Jablonski diagram typical of those referenced when describing the photodegradation of dyes. In this particular scheme, the degradation pathway of the dye is shown to originate from the triplet state (T_1).

$$D_T = \frac{k_{ISC}\sigma * I}{\sigma * I \ (k_{ISC} + k_T) + k_T (k_S + k_{ISC})} D \tag{3}$$

where D is the total dye concentration and $k_{abs} = \sigma * I$. The existence of I in both the numerator and denominator of equation (3) leads to a saturation effect in the triplet state at higher intensities. This has been observed experimentally for rhodamine 6G,[22,24] a dye having the xanthene structure. Furthermore, the photodegradation pathways for several xanthene-based dyes are reported to originate from the triplet state, including pathways that are first-order in the triplet concentration.[25,26] Thus, if depletion of the triplet state by a chemical reaction that renders the dye colorless is slow compared to relaxation from that state,[27] the overall rate of dye degradation can be expressed generally as

$$-\frac{dD(z,t)}{dt} = \frac{a\sigma * I(z,t)}{b\sigma * I(z,t) + c} D(z,t) \tag{4}$$

where a, b, and c are all constants.

Since a closed-form solution to equation (4) was not available, the peak absorbance data for each of the film specimens were fit numerically by means of an iterative process.[20] An analog to the time constant β can be defined as

$$k_{eff} = \frac{a\sigma * I_0}{b\sigma * I_0 + c} \tag{5}$$

which is a pseudo-first-order rate constant that describes the rate of dye degradation near the radiation incident surface of the film. The values of k_{eff} that resulted from each of the fits are plotted in *Figure 9* as a function of the incident irradiance. Their dependence on the irradiance is very similar to that seen previously for β. The difference, however, is that now we have an expression that describes the dependence on irradiance. By fitting all of the results simultaneously, a single set of parameters was obtained: $a = 0.02$, $b = 0.44$, and $c = 1.14$. This was used in equation (5) to calculate the solid curve in *Figure 9*, which shows how the photodegradation rate varies with irradiance.

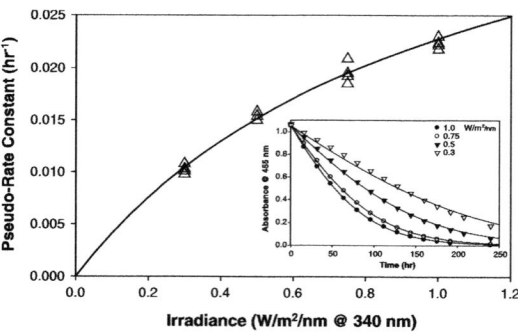

Figure 9—Pseudo-rate constants (k_{eff}) of photodegradation, obtained by using equation (4) to fit the decay in corrected peak absorbance for each of the film specimens weathered in the study. Selected fits are shown in the inset. Uncertainty in the averages calculated for the k_{eff} values is less than 6%. The solid line is the plot of equation (5) using the single set of fitting parameters that was obtained by simultaneously fitting all of the results.

Assessing the Model

In considering the utility of the model that has been derived, it is important to examine its predictability, assess whether it is realistic, and characterize its attributes. The low irradiance region is often of particular interest when making use of accelerated weathering results to estimate material lifetimes under actual service conditions. To see how well the model would predict the photodegradation rate in this region, additional film specimens were weathered at an irradiance level of 0.08 W/m²/nm at 340 nm. This level was attained by inserting two neutral density filters between the specimens and the xenon arc source that originally produced 0.3 W/m²/nm at the specimen plane. The surface temperature

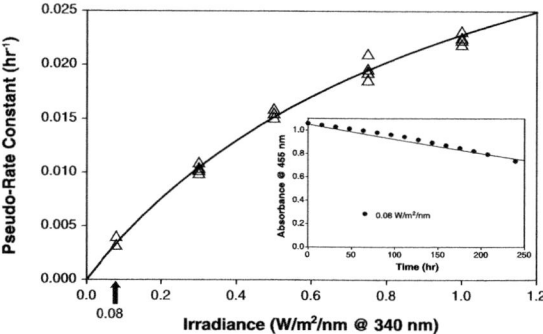

Figure 10—Plot of pseudo-rate constants (k_{eff}) of photodegradation from Figure 9, showing agreement between the values predicted by the model and those obtained for film specimens weathered at low irradiance (0.08 W/m²/nm). A plot of the decay in corrected peak absorbance of one of the film specimens and the associated fit is shown in the inset.

measured from a film was about 5°C warmer than specimens weathered without the attenuators, probably because of heating due to absorption of radiation by the filters nearby. The decay of the peak absorbance measured from one of the specimens over time, along with the fit to the data, is displayed in *Figure* 10. The figure also plots the k_{eff} values determined from the fit for the specimens tested. The values exhibit excellent agreement with the model's prediction.

The validity of the model depends, in part, on whether the tendency toward saturation in the triplet state is attainable at the irradiance levels employed in the weathering experiments. A calculation of σ^*I, based on the absorption coefficient of the dye and the monochromatic approximation used herein, yields a value on the order of 1 s⁻¹. Saturation effects would begin to be observed if the two terms in the denominator of equation (3) were roughly on the same order of magnitude. This situation could exist if the values of σ^*I and k_T were comparable and if k_S was on the same order of magnitude or less than k_{ISC}. Although these rate data are not available for SY98, those published for other aromatic hydrocarbons indicate that for several molecules, k_T can have a value around 1 s⁻¹ and that k_S and k_{ISC} are often comparable ($\sim10^6 - 10^8$ s⁻¹).[22,28]

A significant aspect of the model lies in the fact that it can

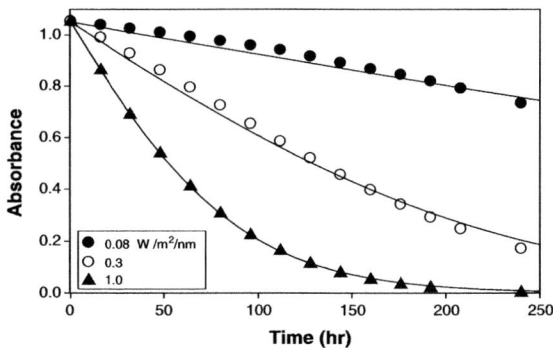

Figure 11—Decay in corrected peak absorbance data obtained for film specimens exposed at three different levels of irradiance. An acceleration in the decay rate becomes apparent at lower irradiance.

account for both reciprocity conformance and reciprocity failure, depending on the level of irradiance employed. It can be seen from equation (3) that at low irradiance, the denominator will be approximately constant, so the rate will be linear in intensity in this operating region. In this case, an analysis that makes use of equation (2) would be appropriate and the associated β value should exhibit reciprocity behavior. This situation would also exist—even at the higher irradiances attainable in a weathering device—if the intersystem crossing rate in the molecular system was relatively small or the relaxation rate from the triplet state was particularly large. This attribute of the model is key, since provision for a linear intensity response should exist when explaining photochemical reactions.

At the same time, the model predicts reciprocity failure at higher levels of irradiance. This was not only observed for the dyed polymer film in this study, but has also been seen in the weathering of other polymers.[4,6] The saturation behavior in the irradiance dependence that translates to reciprocity failure means that acceleration factors cannot be directly calculated in these cases: tripling the irradiance will not yield a tripling of the rate of photodegradation, for example. This also places a limitation on the degree to which one can expect to increase the output of the irradiating source in a weathering experiment and still obtain a significant increase in acceleration of the rate of photodegradation.

It is interesting to take note of some phenomena observed in the data that are yet to be explained. Close examination of the rates of decay of the peak absorbance (*Figure* 11) reveals that for specimens exposed to lower levels of irradiance, the rate accelerated over time during the initial period of the exposure. This effect is also readily observed in the photobleaching of the xanthene dye fluorescein, dispersed in polyvinylalcohol.[29] A similar observation in other dyes has been attributed to the breakdown of aggregated dye particles as photodegradation progresses.[16] It has been suggested that in such systems, the time required to photobleach a given fraction of dye should increase when the initial concentration of the dye in the substrate is increased.[30] Alternatively, this means the initial rate of photodegradation should increase as the initial dye concentration drops. To test this, we produced film specimens at lower initial concentration via a dye migration process by sandwiching dyed film between two PVC films that contained no dye and heating them to yield three films, each with one-third the original dye concentration. The middle film was cut into specimens and weathered at each of the four irradiance levels employed in the second round of tests described herein. The decay in peak absorbance for each specimen was fit as before, using equation (4). The resulting values of k_{eff} were found to be about

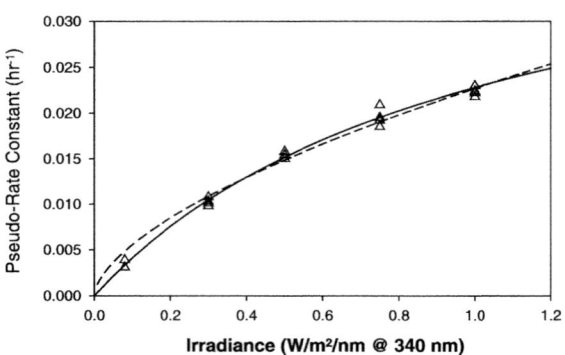

Figure 12—Pseudo-rate constants (k_{eff}) of photodegradation for the dyed films, showing a comparison of the model derived in this study (solid line) and a fit to the results based on rate ∝ I^q (dashed line). The value of q for the latter fit was 0.61.

20% lower than the corresponding values obtained at higher concentration, contrary to the anticipated result. At the same time, the acceleration seen previously in the photodegradation rate of films exposed to the lower irradiances was not apparent in these data. Given these results, it is not possible to ascertain whether or not aggregation of dyes occurs in these films. A satisfactory explanation for the acceleration in the rate of photodegradation at lower irradiance is yet to be established.

CONCLUSION

This weathering study of a dyed polymer film has yielded a model that is able to rationalize the effect of the level of irradiance on the rate of photodegradation. Based on photochemical principles derived for organic molecules, the model accounts for both the first-order kinetics observed in dye depletion during the exposure and the saturation behavior observed in the photodegradation rate as the irradiance of the xenon arc source is increased. The requirement for linearity at low irradiance and the ability to potentially predict rates of photodegradation at other levels of irradiance are also provided by the model.

The applicability of any model depends on whether it can survive further scrutiny. One of the first investigations of interest would be to determine if this model offers an advantage over previous ones in predicting service life. A first step in this effort is displayed in *Figure* 12, where the current model is compared to the model in which photodegradation rate is proportional to I^q. Fitting the pseudo-first-order rate constants from this study by use of the latter model yielded a value for q of 0.61. Although the two models generate similar curves, the results obtained for the model derived in this study appear to provide a better fit to the data. The major difference in this example lies in the low irradiance region, where the rates of photodegradation can differ by a factor of two or more. How this difference affects service life prediction depends largely on the service environment to which the material system of interest is exposed.

The application of photochemical studies to polymeric systems that have been investigated for durability when exposed to light would also be beneficial. This would offer the opportunity to verify assumptions made regarding the photochemical properties of the dye system examined in this study. With verification in hand, one could expand such studies to other polymeric systems in order to determine the extent to which the model derived herein could be used to interpret photodegradation in other materials.

ACKNOWLEDGMENT

The authors wish to thank D. Burns, R. DeVoe, and L. Pavelka of 3M Company for helpful discussions in the analysis of the experimental results.

References

(1) The term *intensity*, as used in the physical literature, is often used to describe the quantity that is technically referred to as *irradiance* in the field of radiometry. For the purposes of this study, irradiance is the quantity of interest, but intensity will be utilized occasionally as an equivalent term

since that is the term used in many of the referenced studies. See also McCluney, W.R., *Introduction to Radiometry and Photometry*, Artech House, Boston, 1994.

(2) Helmick, P.S., *Phys. Rev.,* 11, 372 (1918).

(3) Chin, J., Nguyen, T., Byrd, E., and Martin, J., *J. Coat. Technol. Res.*, 2, 499 (2005).

(4) Fischer, R.M. and Ketola, W.D., in *Service Life Prediction: Challenging the Status Quo*, Martin, J.W., Ryntz, R.A., and Dickie, R.A. (Eds.), Federation of Societies for Coatings Technology, Blue Bell, PA, p. 79, 2005.

(5) Martin, J.W., Chin, J.W., and Nguyen, T., *Prog. Org. Coat.*, 47, 292 (2003).

(6) Hardcastle, H.K., in *Service Life Prediction: Challenging the Status Quo*, Martin, J.W., Ryntz, R.A., and Dickie, R.A. (Eds.), Federation of Societies for Coatings Technology, Blue Bell, PA, p. 217, 2005.

(7) Schwarzschild, K., *Astrophys. J.*, 11, 89 (1900).

(8) Helmick, P.S., *Phys. Rev.*, 17, 135 (1921). Note the apparent error in the table on page 142, in which values for $1/p$ are listed instead of values of p, based on the expression for p given on page 142 and the parameters given in the table on page 140.

(9) John, D.H.O. and Field, G.T.J., *Photographic Chemistry*, Reinhold, New York, 1963.

(10) Jorgensen, G., Bingham, C., King, D., Lewandowski, A., Netter, J., Terwilliger, K., and Adamsons, K., in *Service Life Prediction: Methodology and Metrologies*, Martin, J.W. and Bauer, D.R. (Eds.), ACS Symposium Series 805, American Chemical Society, Washington, D.C., p. 100, 2002.

(11) Hardcastle, H.K., "A Characterization of the Relationship Between Light Intensity and Degradation Rate for Weathering Durability," Atlas Technical Conference for Accelerated Ageing and Evaluation, Ile des Embiez, France, May 11-12, 2006.

(12) Rose, A., *Concepts in Photoconductivity and Allied Problems*, Interscience, New York, 1963.

(13) Herrmann, J.-M., *Catal. Today*, 53, 115 (1999).

(14) So, C.M., Cheng, M.Y., Yu, J.C., and Wong, P.K., *Chemosphere*, 46, 905 (2002).

(15) Fischer, R.M., Guth, B.D., Ketola, W.D., Riley, J.W., U.S. Patent No. 6,859,309.

(16) Giles, C.H. and Forrester, S.D., in *Photochemistry of Dyed and Pigmented Polymers*, Allen, N.S. and McKellar, J.F. (Eds.), Applied Science, London, p. 51, 1980.

(17) Giles, C.H., Baxter, G., and Rahman, S.M.K., *Text. Res. J.*, 31, 831 (1961).

(18) Kaminow, I.P., Stulz, L.W., Chandross, E.A., and Pryde, C.A., *Appl. Opt.*, 11, 1563 (1972).

(19) Pickett, J.E., in *Service Life Prediction: Methodology and Metrologies*, Martin, J.W. and Bauer, D.R. (Eds.), ACS Symposium Series 805, American Chemical Society, Washington D.C., p. 250, 2002.

(20) Pickett, J.E. and Moore, J.E., *Polym. Degrad. Stab.*, 42, 231 (1993).

(21) Kramer, H.E.A., *Chimia*, 40, 160 (1986).

(22) Yamashita, M. and Kashiwagi, H., *J. Phys. Chem.*, 78, 2006 (1974).

(23) Deschenes, L.A. and Vanden Bout, D.A., *Chem. Phys. Lett.*, 365, 387 (2002).

(24) Yamashita, M. and Kashiwagi, H., *IEEE J. Quantum Electron.*, QE-12, 90 (1976).

(25) Leaver, I.H., in *Photochemistry of Dyed and Pigmented Polymers*, Allen, N.S. and McKellar, J.F. (Eds.), Applied Science, London, p. 161, 1980.

(26) Korobov, V.E. and Chibisov, A.K., *J. Photochem.*, 9, 411 (1978).

(27) Yeow, E.K.L., Melnikov, S.M., Bell, T.D.M., De Schryver, F.C., and Hofkens, J., *J. Phys. Chem. A*, 110, 1726 (2006).

(28) Turro, N.J., *Modern Molecular Photochemistry*, Benjamin/Cummings, Menlo Park, CA, 1978.

(29) Talhavini, M. and Atvars, T.D.Z., *J. Photochem. Photobiol., A*, 114, 65 (1998).

(30) Baxter, G., Giles, C.H., McKee, M.N., and Macaulay, N., *J. Soc. Dyers Colour.*, 71, 218 (1955).

Chapter 6

A New Approach to Characterizing Weathering Reciprocity in Xenon Arc Weathering Devices

Kurt P. Scott and Henry K. Hardcastle III

Atlas Material Testing Technology LLC, 4114 Ravenswood Ave., Chicago, IL 60613

This chapter presents the method and results of a study investigating the effect of varying light irradiance on material weathering degradation rates. An evaluation of the reciprocity law for polystyrene standard reference material and commercially available polycarbonate sheet is presented. The method utilizes the commercially available Atlas Ci5000 Weather-Ometer® xenon arc accelerated weathering device.

INTRODUCTION

The materials industry has a strong need to accelerate the effects of weather for faster evaluation of material durability. Methods of accelerating the effects of weather include intensifying weathering variables such as ultraviolet (UV) irradiance, temperature, and moisture. For many years, the materials industry has used accelerated weathering test methods with UV irradiance increased moderately over natural levels. Recently, NIST (National Institute for Standards and Technology) and NREL (National Renewable Energy Laboratory) have proposed the use of ultra-high UV irradiance exposures to accelerate the degradation of materials caused by weathering.[1,2] This approach to accelerated weathering generally assumes that a reciprocal relationship exists between UV irradiance and duration of exposure. For example, if reciprocity is obeyed, a doubling of the UV irradiance would reduce by half the exposure time needed to achieve the same level of degradation. The study in this chapter was undertaken to develop standard accelerated weathering methods to investigate the validity of reciprocity, as well as to characterize the relationship between UV irradiance and exposure times for such tests.

The so-called "reciprocity law," first expounded by Bunsen and Roscoe in 1859, describes a reciprocal relationship between light irradiance and exposure duration.[3] The reciprocity law states that, for a given level of damage, the result of a photochemical reaction depends simply on the total energy received by a sample; that is, the product of irradiance and time, and is independent of the two factors separately. Widely studied for photographic processes, the reciprocity law concepts may also be applied to material weathering studies.[4] Material scientists apply the reciprocity concept by using equal radiant exposures to relate tests conducted at various irradiance levels; for example, a highly accelerated laboratory test and a natural, real-time, outdoor Florida exposure. The pre-

84

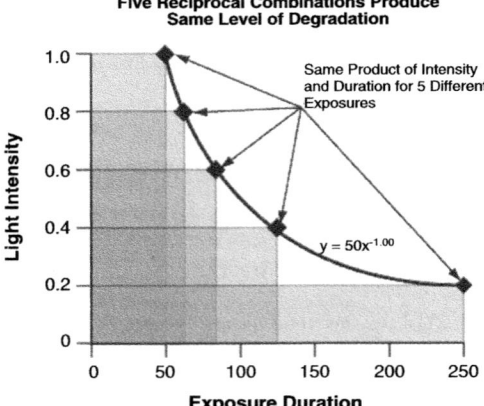

**Five Reciprocal Combinations Produce
Same Level of Degradation**

Same Product of Intensity
and Duration for 5 Different
Exposures

$y = 50x^{-1.00}$

*Figure 1—Diagrammatic representation of the
Reciprocity Law.*

vailing desire to relate highly accelerated weathering exposures back to real-time, natural, or end-use exposures warrants an empirical investigation of the reciprocity law for weathering studies.

There are a number of important considerations that must be taken into account for such an investigation. First, a laboratory xenon arc accelerated weathering device was selected for this investigation for several reasons. The use of "daylight-filtered" xenon arc light source was chosen to provide the best comparisons to results obtained using a sunlight source in previous studies.[5,6] Second, it was desired to use an artificial steady-state light source in order to compare results to natural intermittent exposure patterns obtained in the previous outdoor studies cited, and to investigate the possibility of intermittency effects. Intermittency effects are said to occur when a continuous exposure produces a different level of degradation from that produced by an exposure given in a number of discrete increments, despite each being subjected to equal total radiant dosages.[7] Natural end-use weathering exposures are inherently intermittent, for irradiance levels, temperature, and other degradation variables as well. Third, it was desired to determine if existing xenon arc accelerated weathering devices could be used in lieu of the highly specialized, custom-made, complex devices for addressing simple weathering research hypotheses. For these considerations, the Atlas Ci5000 Weather-Ometer® accelerated weathering test apparatus was selected for this investigation.

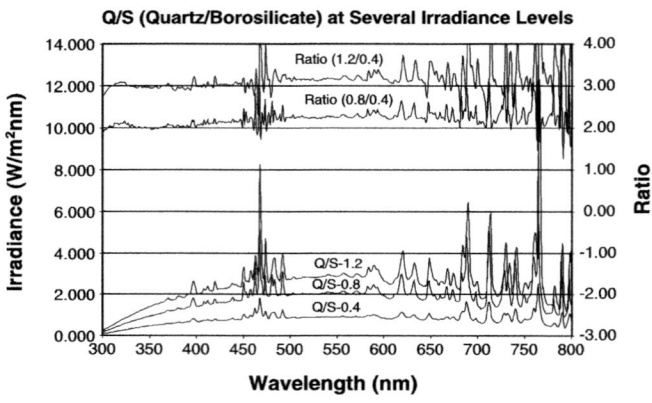

The reciprocity law itself, diagrammatically shown in *Figure* 1, represents the research question or hypothesis for this study. If accelerated weathering exposures obey reciprocity, then similar levels of degradation should be observed for the same radiant exposures under different irradiances levels. If accelerated weathering expo-

*Figure 2—Spectral integrity at different irradiance levels. Lines at top of
the graph represent a mathematical division of the spectral output at 1.2
and 0.8 by that at 0.4 W/m².*

sures do not obey strict reciprocity, then it was desired that the method provide an empirically derived estimate of the function relating irradiance and exposure duration. It was also desired that the method would experimentally block other variables that could potentially influence material degradation, including specimen temperature, variations in xenon spectral power distribution, as well as any other typical temporal exposure variables.

This chapter presents the method and results of a study to investigate the effect of changing irradiance on artificial weathering degradation rates. Two instrument enhancements were utilized to verify that the weathering variables discussed above were successfully blocked. One enhancement—the Full Spectrum Monitoring (FSM)™—measures the spectral power distribution of the instrument light source in real-time and the other— the Specific Specimen Surface Temperature (S3T)™—measures in situ specimen temperatures in real-time. Details of these enhancements may be found in references 8 and 9, respectively.

TEST INSTRUMENTS

The Ci5000 Weather-Ometer utilizes a water-cooled, precision long xenon arc lamp. A desired spectral power distribution (SPD) is produced by the optical filtering of the lamp's output to closely match solar radiation, for example. Standard quartz inner and borosilicate outer filters were used in these exposures. ASTM G-155[10] and SAE J1885[11] more thoroughly describe the design and operation of the water-cooled device as shown in *Figure* 2. The test program followed the SAE J1885, with two significant exceptions. The Ci5000, programmed for separate trials, was operated at three different irradiance levels (0.4, 0.8, and 1.2 W/(m²·nm) at 340 nm) instead of the 0.55 W/(m²·nm) prescribed, and the exposure was set for a continuous light cycle at a fixed black panel exposure temperature (the standard dark cycle was excluded). The chosen operational parameter combinations are readily controlled within the range of capabilities of the Ci5000 device.

BLOCKING TEMPERATURE VARIABLES AND SPECTRAL INTEGRITY AT DIFFERENT IRRADIANCE LEVELS

In exposures under different irradiance levels of simulated (or natural) sunlight, material surface temperatures typically vary with irradiance level (*Table* 1). However, confounding of temperature and irradiance variables is undesirable for reciprocity studies involving irradiance, and represents an important experimental design consideration.

Table 1—Calculation of Specimen Exposure Duration in Hours—Conversions Based on Irradiance and Radiant Dosage

Exposure Trial Designation	Irradiance W/(m²·nm) 340 nm	Radiant Dosage Measurement Interval kJ/m² 340 nm	Exposure Duration for Required Radiation Exposure (Hours)						
1 and 4 ....0.4		Every 37.6	26.1	52.2	78.3	104.4	130.6	156.7	182.8
2 and 5 ....0.8		Every 37.6	13.1	26.1	39.2	52.2	65.3	78.3	91.4
3......................1.2		Every 37.6	8.7	17.4	26.1	34.8	43.5	52.2	60.9

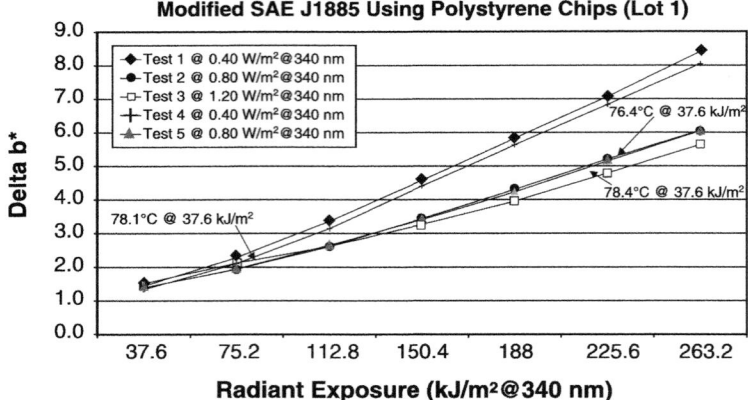

Figure 3—Degradation curves for polystyrene under each experiment condition.

Independent control of irradiance and material temperature was achieved using the ASTM black panel thermometer control system of the Ci5000 device, wherein a temperature sensor is attached directly to a black panel exposed in the chamber adjacent to test specimens. When heated by incident light, the panel sends a signal proportional to the black panel's surface temperature to the chamber's (temperature) control system. Cooling airflow is adjusted by the control system to maintain the black panel temperature sensor at a user-selected setting, and is independent of the irradiance received from the xenon arc, within a nominal operating range.

Spectral output of a source may vary with input power to a light source, depending on the sophistication of design and/or proper functionality of the source's power supply unit. The objective in this experiment was to maintain a constant relative spectral power distribution (shape of the spectral curve) as the irradiance level was changed. The onboard FSM was used to verify that spectral integrity was maintained at the different irradiance levels (*Figure 2*).

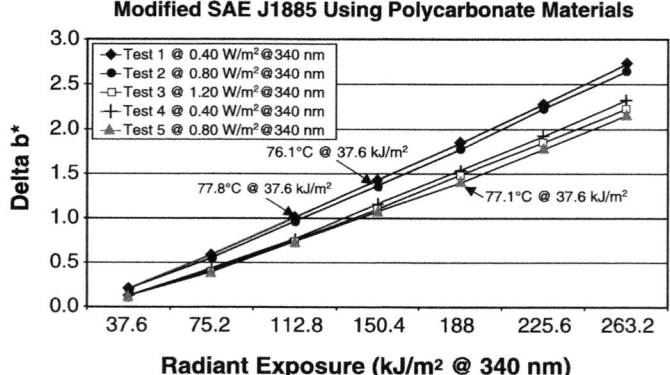

Figure 4—Degradation curves for polycarbonate under each experiment condition.

Table 2—Experimental Conditions, Degradation Functions, and Time to Failure Calculations Obtained from the Exposure Trials for Polystyrene

Exposure Trial Designation	Irradiance Setting at 340 nm $W/(m^2 \cdot nm)$	Nominal Specimen Temperature °C	Linear Regressions for Observed Degradation Data[a]	Time to Failure ($\Delta b^*=5$) From Observed Data Regressions
1 ....0.4		77	y=1.175x+0.011	111
2 ....0.8		77	y=0.793x+0.393	76
3 ....1.2		77	y=0.679x+0.701	55
4 ....0.4		77	y=1.144x-0.076	116
5 ....0.8		77	y=0.783x+0.415	76

(a) All $R^2 > 0.99$

INCREMENTALLY OFFSETTING IRRADIANCE

The Ci5000 incorporates a closed-loop control system, which allows the selection of different levels of irradiance to be received by samples from its xenon source. For this test, the irradiance level from the xenon arc is monitored at 340 nm by a system incorporating a photo-receptor located in the exposure chamber. The signal from the light receiver is passed through a 340 nm radiometer sensor and fed into the control system where it is compared to, and automatically adjusted to maintain the operator-selected irradiance level. This system allows several exposures to be performed at different irradiance levels while maintaining a specific black panel temperature.

COMBINING INCREMENTAL IRRADIANCE LEVELS AND TEMPERATURE CONTROL

In this study, incrementally offset irradiance levels were combined with controlled temperatures in five xenon arc exposure trials. In this way, appropriate, designed experimental blocking of the specimen temperature variable was achieved while allowing independent adjustment of the irradiance variable at three different controlled levels. The Specific Specimen Surface Temperature $S^3T^{®9}$ was used to verify that the temperatures

Table 3—Experimental Conditions, Degradation Functions, and Time to Failure Calculations Obtained from the Exposure Trials for Polycarbonate

Exposure Trial Designation	Irradiance Setting at 340 nm $W/(m^2 \cdot nm)$	Nominal Specimen Temperature °C	Linear Regressions for Observed Degradation Data[a]	Time to Failure ($\Delta b^*=2$) From Observed Data Regressions
1 ....0.4		77	y=0.421x−0.243	139
2 ....0.8		77	y=0.410x−0.249	72
3 ....1.2		77	y=0.354x−0.280	56
4 ....0.4		77	y=0.369x−0.293	162
5 ....0.8		77	y=0.340−0.265	87

(a) All $R^2 > 0.99$

Table 4—Reciprocal Ratios, Expected Times to Failure, and Observed Times to Failure for Polystyrene

Exposure Trial Designation	Irradiance— W/(m²·nm) @ 340 nm	Reciprocal Ratios		Expected—Time to Failure[a]	Observed— Time of Failure[a]
1...............................0.4		1.2/0.4	3	165	111
2...............................0.8		1.2/0.8	1.5	83	76
3...............................1.2		**1.2/1.2**	**1**	**55**	**55**
4...............................0.4		1.2/0.4	3	165	116
5...............................0.8		1.2/0.8	1.5	83	77

(a) Comparisons are made with the high irradiance test (trial 3), which produced failure in approximately 55 h. Therefore, we would expect the time to failure in the low irradiance, trial 1, to be 3x longer than trial 3, or 165 h.

of the specimens of interest did not change appreciable at the different irradiance levels. Specimen temperature data are shown in *Tables* 2 and 3.

This allowed successful characterization of the reciprocal relationships between exposure irradiance and exposure duration, independent of temperature.

DEGRADATION MEASUREMENTS

For the five trials, specimens were removed from exposure for measurement at the same radiant dosage of 37.6 kJ/(m²·nm) at 340 nm. Since the exposures were conducted at static irradiance levels with no dark cycles, a simple calculation allows radiant exposure to be expressed in test duration (time) using the equation: $(kJ//(m^2·nm)) = (W/m^2·nm) * 3.6 * time (h)$.

MATERIAL AND MEASUREMENTS OF DEGRADATION

Polystyrene

The polystyrene standard reference material was utilized for this study. Test Fabrics, Inc. produces the polystyrene specimens traditionally used for xenon arc weathering

Table 5—Reciprocal Ratios, Expected Times to Failure, and Observed Times to Failure for Polycarbonate

Exposure Trial Designation	Irradiance— W/(m²·nm) @ 340 nm	Reciprocal Ratios		Expected—Time to Failure[a]	Observed— Time of Failure[a]
1...............................0.4		1.2/0.4	3	168	139
2...............................0.8		1.2/0.8	1.5	84	72
3...............................1.2		**1.2/1.2**	**1**	**56**	**56**
4...............................0.4		1.2/0.4	3	168	162
5...............................0.8		1/2/0.8	1.5	84	87

(1) Comparisons are made with the high irradiance test (trial 3), which produced failure in approximately 56 h. Therefore, we would expect the time to failure in the low irradiance, trial 1, to be 3x longer than trial 3, or 168 h.

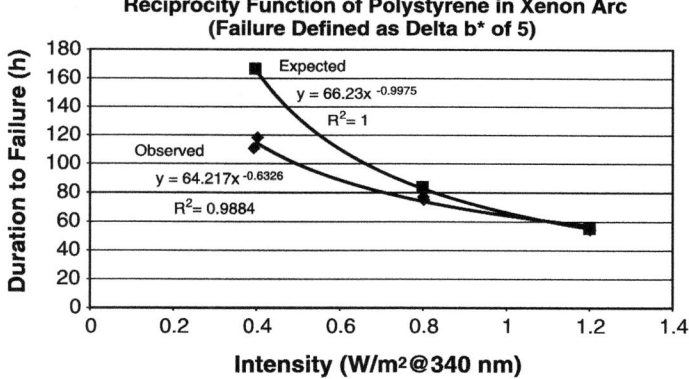

Figure 5—Expected and observed reciprocity for exposures of polystyrene.

exposure characterization. Previous experiments have shown that the polystyrene specimens are sensitive to both irradiance and exposure temperature. The specimens are readily available in single lots manufactured with an appropriate level of quality control and exhibit an approximately linear degradation function region for development of yellowness. At each increment of radiant exposure, the polystyrene specimens were removed from the exposure devices and measured for yellowness as indicated by delta b* on the CIE Lab scale. The measurements were done in reflectance mode, per SAE J1885. Failure was arbitrarily defined as 5 delta b* units for these materials.

Polycarbonate

Also for the study, a commercially available sheet polycarbonate was obtained from a retail source. Multiple specimens of appropriate size were cut from a single sheet and randomized to mitigate any inherent bias in quality. The side of the specimens not protected by UV inhibitors was oriented towards the light source during exposures. These

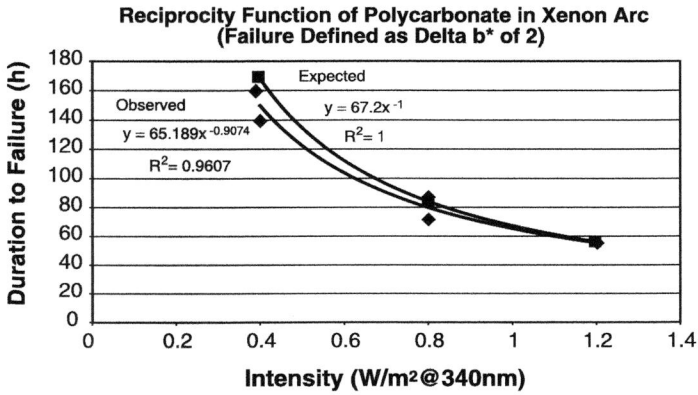

Figure 6—Expected and observed reciprocity for exposures of polycarbonate.

Table 6—Reciprocity Functions Obtained from the Exposure of Polystyrene and Polycarbonate

Material	Reciprocity Function	
	Expected	Observed
Polystyrene ....y = 66.23x^{-1}		y = 64.22x$^{-0.63}$
Polycarbonate.............y = 67.20x^{-1}		y = 65.19x$^{-0.91}$

samples were measured using the same method and schedule as the polystyrenes. However, for these materials, failure was defined as delta b* of 2.0.

EXPOSURE RESULTS

The five controlled exposures were conducted in Atlas' Engineering Laboratory in Chicago, IL, during the fall of 2005. *Figures* 3 and 4 show the degradation curves for each of the experimental conditions summarized in *Tables* 2 and 3 for polystyrene and polycarbonate, respectively. Linear regression functions were determined for each degradation curve and failure points for each trial were determined from these regressions.

RECIPROCITY ANALYSIS

If reciprocity is obeyed, to attain the same level of degradation, we would expect an exposure at 0.4 W/(m^2·nm) at 340 nm, to take (1.2/0.4 =) three times as long as an exposure at 1.2 W/(m^2·nm) at 340 nm. Conversely, it should take an exposure at 1.2 W/(m^2·nm) at 340 nm, (0.4/1.2 =) one-third the duration of an exposure at 0.4 W/(m^2·nm) at 340 nm, to reach the same failure point. *Tables* 4 and 5 summarize these reciprocal ratios, the expected relative times to failure, and observed relative times to failure for polystyrene and polycarbonate, respectively. Inspection of these tables illustrates considerable departure from reciprocity for the development of yellowness in these two materials. The data shown in *Tables* 4 and 5 are graphed in *Figures* 5 and 6, respectively. These graphs characterize the expected and observed relationship between relative irradiance levels of exposure and time to failure. The fitted power functions provide the exponent characterizing the true reciprocal relationship. The exponent should be close to 1.0 for relationships that obey reciprocity. The actual reciprocal functions obtained in this experiment are shown in *Table* 6.

SUMMARY OF OBSERVATIONS

(1) Development of yellowness in the polystyrene reference material and commercially available polycarbonate showed departures from the reciprocal relationship typically assumed to exist between irradiance level and exposure duration.

(2) Specifically, state of the art laboratory devices such as the Ci5000 Weather-Ometer® may characterize reciprocity by controlling irradiance and temperature

variables independently at static levels. This ability makes steady state, (non-intermittent) laboratory weathering exposures desirable in addition to the natural weathering characterizations for some materials.

(3) The design of this experiment provided degradation information for different static levels of simulated solar irradiance.

(4) Advanced laboratory weathering instrument enhancements that measure in situ specimen temperature and xenon light spectra were helpful to ensure critical variables were blocked in this experiment.

ACKNOWLEDGMENTS

Dr. Jacob Zhang and Lamont Elliott of the Atlas Engineering Laboratory are acknowledged for conducting the laboratory tests and assisting in some analysis.

References

(1) Chin, J.W., Byrd, E., Embree, E., and Martin, J.W., *Service Life Prediction: Methodology and Metrologies*, Martin, J.W. and Bauer, D.R. (Eds.), Chapter 8, American Chemical Society, Washington, D.C., 2002.

(2) Jorgensen, G., Bingham, C., King, D., et al, *Service Life Prediction: Methodology and Metrologies*, Martin, J.W. and Bauer, D.R. (Eds.), Chapter 6, American Chemical Society, Washington, D.C., 2002.

(3) Bunsen, R. and Roscoe, H.E., *Photochemische Untersuchungen, Poggendorff's Annalen 1855*: 96: 373-394, 1857: 100: 43-88 and 481-516, 1857: 101: 235-263, 1859: 108: 193-273.

(4) Martin, J.W., Nguyen, T., Byrd, E., et al., *Service Life Prediction Methodology and Metrologies*, Martin, J.W. and Bauer, D.R. (Eds.), Chapter 7, American Chemical Society, Washington, D.C., 2002.

(5) Hardcastle, H.K., "A New Approach to Characterizing Reciprocity," in *Service Life Prediction – Challenging the Status Quo*, Martin et al (Eds.), Chapter 15 Federation of Societies for Coatings Technology, Blue Bell, PA, 2005.

(6) Hardcastle, H.K., "A Characterization of the Relationship Between Light Intensity and Degradation Rate for Weathering Durability," Natural and Artificial Aging of Polymers – 2nd European Weathering Symposium, Reichert, T. (Ed.), *Gesellschaft fur Umweltsimulation*, Pfinztal – Germany.

(7) Mees, C.E., *The Theory of the Photographic Process*, The Macmillan Company, New York, p. 236, 1942.

(8) Scott, K.P, "Full Spectrum Monitoring: Next Generation Light Control for Laboratory Accelerated Weathering," in *Service Life Prediction—Challenging the Status Quo*, Martin, et al. (Ed.), Chapter 19, Federation of Societies for Coatings Technology, Blue Bell, PA, 2005.

(9) Scott, K.P., "Innovations in Laboratory Instrument Technology Revolutionize Weatherability Testing," 6th International Symposium on Weatherability (6th ISW), Material Life Society, Japan.

(10) ASTM G 155-05, Standard Practice for Operating Xenon Arc Light Apparatus for Exposure of Non-metallic Materials, *2005 Annual Book of ASTM Standards*, American Society for Testing and Materials, West Conshohocken, PA, 2005.

(11) SAE J1885, Operating Xenon Arc Light Apparatus for Exposure of Non-metallic Materials, *2005 SAE Handbook*, Society of Automotive Engineers, Warrendale, PA, 2005.

Chapter 7

Controlled Temperature Natural Weathering

Anthony Buxton,[1] Jon A. Graystone,[1] Richard Holman,[1] and Francesco Macchi[2]

[1] Paint Research Association, 14 Castle Mews, High St., Hampton, Middlesex TW12 2NP UK
[2] Rhopoint Instrumentation Ltd., 12 Beeching Rd., Bexhill on Sea, East Sussex TN39 3LG UK

Attempting to quantify and predict the exterior service life performance of a coating is an essential and frequently undertaken activity in the technical sections of the paint and surface coating industry. Such work, usually termed weathering or durability studies, is carried out in an attempt to underpin new product development, or to provide a measure of assurance of the in-use reliability for already commercialized products. Determining or predicting service life is often the rate-determining step of a new product launch. Any failure to eliminate premature failure under service conditions may lead to liability claims or even lawsuits. There are many approaches to service life prediction; methodologies have been variously described as descriptive, scientific, and reliability-based. Each of these will have many variants in terms of the mathematical, statistical, and modeling tools that can be deployed to aid interpretation. Practical work is typically based on field and laboratory exposures and the correlation or relationship between them. Natural weathering has an inherently high variance and therefore requires a dosage model to aid interpretation and establish relationships with laboratory testing. Establishing such a relationship can be time-consuming and expensive, particularly as it is not possible to control natural weathering. However, there are possibilities for introducing some systematic variability into natural weathering and using this to investigate dosage factors. This chapter reviews some possibilities and describes the development of a simple exposure rack enabling normal weathering at three controlled temperature regimes.

INTRODUCTION

The development of new or improved products necessarily requires meeting a wide range of economic, operational, and performance criteria. Each presents its own difficulties but issues involving durability carry the added problem of a long time scale and thus become a potential bottleneck in bringing products to the marketplace. Durability can be defined in many ways but implies the period of time that a product in its service environment will survive before requiring replacement or maintenance. Good durability is thus

93

equated with a long service life. However, it may be noted that the mode of failure is also important and in wood coatings ease of maintenance may be judged more important than a longer life if the latter carries the penalty of a more catastrophic failure mode.

A particular difficulty with many durability issues is that failure is seldom unequivocal as with, for example, an electrical fuse, but has an ill-defined and often subjective end-point. Moreover, there are several quite different failure modes, any one of which might be judged dominant by the dissatisfied end-user.

Clearly there is great commercial advantage in predicting service life from short-term measurements and indeed many products will enter the market after a shorter period of development and testing than the expected lifetime of the product itself. For this to be possible without unacceptable risks, some strategy of service life prediction (SLP) must be deployed. There are many such strategies, each of which requires the deployment of deterministic or probabilistic techniques. A central tenant to any scientific approach is the concept of causality, often expressed in terms of correlations. Mathematically, correlations can be quantified through a coefficient, though the assumptions underlying the validity of correlation coefficients should not be overlooked. Within a particular SLP strategy there is a choice of tactics. One possibility is to identify a key critical outcome property that correlates well with the anticipated failure mode. This might be measured directly, e.g., as an elastic modulus, or indirectly through analytical techniques which detect chemical degradation of the polymer backbone (such as ATR-FTIR). An alternative approach is to concentrate less on the underlying properties but focus instead on the actual performance features valued by the user, which in the case of coatings might include gloss, color, film integrity, corrosion resistance, and many others. In an ideal world causal properties stay constant and are unaffected by the service environment; in reality, of course, they change as a function of time, but in a manner dependent on the conditions. From this perspective there are alternative and not necessarily exclusive strategies that might be deployed to estimate service life on a shortened time scale, in comparison to full natural exposure. One is to increase the severity of the conditions, or challenge, in order to accelerate the rate of change; another is to extrapolate forward from measurements taken over a relatively short period so that behavior in the longer term can be predicted. Both approaches offer many choices in the protocols to carry out the investigations and the tools for analysis.

For more than a century, accounts of the successes, failures, and problems of determining durability have appeared in the scientific literature. More recently, conferences[1–3] dedicated to the topic have provided a forum for discussing many different facets of the problem including a critical reappraisal of some established practices. Extensive reviews highlight some of the unresolved issues for linkages between field exposure and laboratory testing, and the ability to predict service life. Martin et al[4] and others have categorized prediction methodologies that have been described respectively as:

- Descriptive
- Scientific measurement based
- Reliability-based

While there is overlap between these, some important differences in emphasis are summarized in *Table* 1.

Table 1—Service Life Prediction Methodologies

Defining Characteristics	Descriptive	Scientific Measurement	Reliability-Based
Information sources	Field and laboratory exposures	Field and laboratory exposures	Field and laboratory exposures
Standard of comparison	Field exposure results Standard products	Field exposure results Chemical origins	Laboratory Field as laboratory
Mathematical and scientific tools	Correlation Coefficients Ranking Time factors Acceleration factors	Chemistry of degradation Modeling Correlations Statistics	Reliability theory Dosage and cumulative Damage models Statistics Prediction Database analysis
Laboratory methods	Simulated weathering Protocols, using devices with lamps Macroscopic Physical effects	Analytical techniques Simulation Chemistry causes Relatively few variables	Statistical design around failure modes Controlled quantified dosage Many variables
Field methods	Exposure sites Environmental records Reference standards Visual standards	Exposure sites Environmental records Reference standards Instrumental standards	Quantified dosage measurement and damage characterization
Issues	Damage and dosage metrics. Light sources Poor correlations Poor repeatability	Correlations with field exposure Polymer specific protocols Isolated failure modes	Reciprocity Additivity Solar source Associates costs

All three methodologies as summarized above, albeit with some overlap, are used within industry—either by default, or because they are judged appropriate to a specific problem. Scientific advances relating in particular to photodegradation have been identified, which might benefit all three methodologies.[4] The strong claim for the reliability methodology is based on the potential to link field and laboratory results by an explicit dosage model applied independently to each, rather than by correlation. However, there remains a strong degree of scepticism that anything is better than real time exposure,[5] and published work continues to criticize correlation methods,[6,7] without fully considering other options.

A problem facing manufacturers in durability and service life studies is the high costs involved when applying any methodology rigorously. The reliability-based methodology is demanding in the amount of data that is required and may require specialist skills for analysis. Payback may be in the longer term rather than solving immediate problems. At the same time it must be recognized that many companies have launched successful products using less than rigorous methodologies, though clearly the risks will be different between, say, architectural, automotive, and coil coating markets. There are two significant factors that have enabled progress in the general coatings industry:

- The use of well characterized, known commercial products as performance standards
- Considerable past experience of specific technologies and products under field conditions

Using established products as control standards in durability investigations is an integral part of much industrial development. It also highlights the fact that in many investigations the objective is not an absolute determination of service life, but rather a relative ranking methodology. Confidence that a new product is at least as good as, or preferably better than an established one, is often a sufficient condition for sales and marketing, even though the magnitude of the difference cannot be quantified. The second point is also important and is a de facto application of Bayesian statistics (i.e., using prior knowledge[8]), even if not formally applied. Experience-based knowledge of a product's behavior over a wide range, national or international, of service life conditions, is invaluable in assessing and interpreting field or laboratory experiments, which contain as a control, or benchmark, one or more products whose behavior is well established. One might call this a "pragmatic methodology"!

Clearly the "pragmatic methodology" is less capable if a completely new technology is introduced. Within the coatings industry there is at present a current need to change products and technology in response to environmental and other legislation. Compliant products with a shorter track record are replacing better known established technologies.

One of the risks facing a manufacturer using ranking methods of assessment arises when relative rank orders of performance against a standard change. Within the coatings industry there is "core knowledge," or at least a strong belief, that the rank order of coating performance can vary between climatic zones, and some manufacturers claim to formulate differently to allow for this. This belief is very plausible if the dominant failure mode changes; for example, fungal growth can be an acute problem in hot humid conditions while in a dry sunny climate gloss loss or chalking become more problematical. Even when the final failure mode appears similar, a different underlying mechanism can operate. For example, a program to develop exterior clear coatings for wood reports light stabilizers and absorbers did show the expected durability improvement in some, but not other, resin systems.[9] This was attributed to the predominance of hydrolytic degradation over photodegradation. Where photodegradation was the dominant mechanism the light stabilizers were effective. Such systems are likely to show rank order changes in different climatic conditions.

In recent years, PRA has undertaken projects involving corrosion,[10] dirt pick-up,[11] and mold growth.[12] In every case, rank reversals for specific products were noted between sites and in ways that were impossible to simulate by laboratory testing alone.

If rank order changes are an accepted reality then the coating industry is risking what statisticians might call a "Type 1" or "Type 2" error. Type 1 errors result in a failure to recognize potential problems when conditions change. In contrast, Type 2 errors mistake poor test discrimination for real effects.

Rank order reversals, when they do occur, also emphasize that the notion of severity in climatic conditions is to some degree relative rather than absolute. It also highlights the need for "robust products" that are insensitive to variation and hence the need to understand competing mechanisms if more robust products are to be developed.

Against this background, PRA has initiated work that sets out to introduce some aspects of the reliability methodology, in particular the application of dosage factors, with a test protocol that is relatively inexpensive and forms a link between laboratory and field exposure. The tactic employed is to combine natural weathering field exposure with additional variables. It is recognized that this approach does not allow for the rigorous exploration of dosage factors offered by methods such as the NIST "SPHERE,"[13] but it does in principle allow for some exploration of the sensitivity to dosage factors, and can extend the range of conditions experienced during field exposure while retaining the character of the exposure site. This extends the options for treating the outdoor environment as a laboratory test. In principle, the approach can simulate some aspects of local climatic variation without introducing unnatural factors and also offers some decoupling of confounded factors such as sunshine exposure and temperature. The work is at a relatively early stage but has led to the development of an exposure rack with three temperature-controlled zones. Before describing the development of this apparatus, a brief overview of natural variation in three major dosage variables is given.

NATURAL WEATHERING DOSAGE FACTORS

There are several factors that will degrade a particular substrate/coating combination. Corrosion is strongly influenced by salt and acidic pollutants; general appearance will be degraded by dirt and mold spores. Hail damage is a major problem in some climatic regions. However, a major contributory factor to the degradation of coatings and other materials is the combined effects of the intensity of solar radiation, temperature, and time of wetness. Around the world it is possible to classify climate in terms of the total dosage of these factors over long periods of time. Weather, in contrast to climate, shows considerable variation on a short-time basis and cannot be considered reproducible except in a broad statistical sense. Fortunately, modern data logging equipment makes it possible to record the three major variables enabling the development of cumulative damage models. Such models will require the dose response, e.g., temperature, to be measured for the target substrate, and not the environment in general. While weathering factors can be characterized accurately, they are usually regarded as outside the control of the experimenter, save by exposing at different times or different sites. Therefore, control of variables is usually carried out in laboratories in devices that may be described as artificial or accelerated weathering. Normally, the intensity of the dosage factor is considerably increased, particularly radiation. There is, however, another way of varying the dose of all three factors and that is by altering the angle of exposure. If this is done in a systematic manner then there is a basis for exploring aspects of the dose response relationship, which might be used to predict performance, even though the intensity of the dose is not outside the normally experienced range.

In-service Irradiance Variation

Although the solar flux generated by the sun is effectively constant, the amount reaching any terrestrial object will be modified by angle of incidence and distance travelled through the atmosphere. During passage, absorption and scattering processes from air, ozone and other gases, and water vapor will modify it, leading to altitude and local differences. The angle of incidence of sunlight may be calculated,[14] taking into account factors that include latitude and the slope of the exposed surface. For a horizontal surface at a given latitude, the angle of incidence will vary continuously throughout the day and according to the time of year in a complex manner with high angles of incidence at the beginning and end of the day, and lowest around midday. As the exposed surface changes from the vertical, while maintaining a constant compass direction, the angles of incidence are significantly modified, and again if the exposure azimuth (compass direction) is changed. Graphical representations of the angle of incidence have been published,[14] which will also vary as a function of latitude. Because the influence of incident angle on irradiation will be modified by atmospheric conditions, dosage is normally measured rather than calculated at reference weathering sites.

A surface exposed to the sun will receive maximum radiation at the time when the angle of the sun's rays is normal (90°) to the surface. For a fixed surface during the day the intensity will be distributed around a maximum according to the time of day and latitude. In the northern hemisphere fixed test panels are usually faced south at a specific angle. Maximum radiant energy will be received at the "latitude angle," e.g., 25° for southern Florida, 33° for Phoenix, AZ, 51° for London. By convention, many coatings are tested at 45° which is considered a good compromise and more severe than a vertical exposure. A fixed 5° angle is more severe in summer months than 45° and will also collect more dew and rainfall—however, 45° would receive more solar energy in the winter months. Compared to vertical exposure, 45° has a significant accelerating effect and for wood coatings has been shown to range between a factor of 1.5 to 2.0 depending on the composition.[15] This will reflect time of wetness differences, as well as irradiance.

The solar radiation (and temperature) can be further increased in various ways that include backed mountings, focusing mirrors, and continual or continuous variation of the exposure angle (e.g., EMMAqua®). A wide gamut of test configurations is thus available with options of spraying with water and salt solutions. Clearly it is not possible to say that any particular one of these conditions is "right" to evaluate service life—it will depend on the objective. For automotive paints a 5° exposure, perhaps with backing, might be chosen to simulate conditions inside a car. Miami and Phoenix exposure sites are valued as representing more severe conditions (though not to local residents), particularly for photodegradation, since they shorten the time required to experience high accumulated UV dosage relative to sites which receive less sunshine. The use of mirrors and variable angle racks speeds this further. Although the objective of variable angle or enhanced exposure is often either to speed the time to failure, or test the worse case scenario, simultaneous exposure at different angles also provides a means to establish acceleration or other dosage factors that might then be used predicatively.

Hardcastle,[14] for example, has shown a strong correlation between "angle of exposure" and Yellowness Index for a polystyrene reference material. At the same time, differences in the character of the color change development are apparent between the Florida and Arizona sites, resulting from other climatic differences, such as time of wetness. In another example, coil coated metal cylinders oriented East–West were monitored at specific points corresponding to a range of exposure angles. Once again, the importance of exposure angle is shown, but with differences between sites and according to the response metric used; color and gloss behaved differently. The fortuitous octagonal cross section of lampposts in Fort Lauderdale, FL, offered a unique opportunity to study color and gloss changes of a coating in an end-use environment. Measurements on the different faces of the posts showed systematic changes in color and gloss as a function of the azimuth exposure angle. Such observations underline the possibilities for broadening the dosage information that can be gleaned from a single site and ideally this would be combined with systematic "time-of-wetness" and temperature step changes.

In-service Time of Wetness (TOW) Variation

Water is a major contributor to coating degradation, with specific effects according to the nature of the substrate. Thus, wood is prone to movement and decay, ferrous metals to corrosion, while masonry paints are highly alkaline. Apart from substrate-water interactions, coatings exposed externally are subject to hydrolysis, causing scission, while photolysis leads to free radical photooxidation, further exacerbated by the presence of water. Water is also a carrier for other materials and may contain salt or acidic materials. In consequence, exterior-weathering investigations will usually record rainfall data. It is now recognized that this is not sufficient as a dosage factor, since dew formation may account for more water on a surface than rainfall. Moreover, the nature of the specimen, including its insulation value and residual heat content, will markedly influence the TOW. Although rising temperatures, e.g., of a panel as sunshine increases, will evaporate water, it may also increase water absorption. Differences in TOW between laboratory and field exposures are thus a potential source of poor correlation between laboratory and field exposures, and may be difficult to manage even in controlled experiments. Using load cells, Boisseau et al.[7] recorded ~250 L water on a 24 cm² panel in 56 days. In contrast, 300 h of the SAE J1960/J2527 cycle, approximately equivalent to three months of outdoor weathering, used only 24 L of water. Furthermore, the water in "weatherometers" is usually deionized and not necessarily representative of rain and dew.

TOW varies greatly between climatic zones and, in a humid climate like Miami, coatings could be wet 50% of the total time, compared to an arid climate like Phoenix where 5% is more typical. Like UV dosage, TOW is greatly influenced by the design of test panels on weathering sites.[16] Backed panels undergoing radiative cooling, i.e., at nightfall, do not gain heat from ambient air as fast as unbacked panels and will fall below the dew point for longer periods, thus remaining wetter. Clearly the angle of exposure will affect water run-off; at low angles (0°–5°), water will collect in droplets. This is very noticeable when water repellents are present, as is common with exterior wood finishes. The combined effect of high UV and long TOW leads to a faster rate of photo-hydrolysis and may be sometimes seen on the flat surface of wooden balustrades. Similarly, the amount of corrosion in marine environments of a flat mild steel specimen was three to four times greater than a vertical panel. In general total corrosion increased incrementally as the angle of exposure was decreased.[17]

TOW is thus useful as a means of accelerating or maximizing a particular mode of service life degradation, and if varied systematically can be used to investigate dosage relationships. Fischer and Ketola[18] investigated TOW as a water-stress multiplier by comparing a range of materials exposed at Phoenix, Miami, and St. Paul (MN). In most cases failure was faster in the wet environment (typically 12%) with some materials showing a greater sensitivity (68% faster). TOW is likely to interact with other dosage factors but may also be confounded with UV and temperature effects.

In-service Temperature Variation

Although the UV portion of sunlight is very damaging to polymers it accounts for only 5% of the total energy; around 51% is radiated in the infrared range. Not surprisingly this has a major effect on the temperature of exposed coatings. White paints reflect a large proportion of infrared radiation, but much more is absorbed by darker colors including black; dark colored paints can consequently become up to 30° hotter than pale ones, unless the latter contain infrared reflecting pigments. Chlorophyll reflects in the infrared region, otherwise leaves could not withstand sun exposure without shrivelling. The actual temperature of a coated colored surface will clearly depend on the angle of exposure modified by other factors such as the specific heat and mass of the coated substrate as well as insulation and ventilation (wind speed). Ideally, the temperature of an exposed coating in service should be measured using data logging and appropriate sensors. Models have also been developed to predict temperature based on heat and radiative transfer mechanisms.[19-20]

Temperature is expected to have an accelerating effect on chemical reactions. Quantitatively, models such as the Arrhenius equation are used to predict the relationship between reaction rate and temperature:

$$k=A*exp(-E_a/R*T)$$

where k is the rate coefficient, A is a constant, E_a is the activation energy, R is the universal gas constant, and T is the temperature (in degrees Kelvin).

The equation predicts a doubling of reaction rate with a 10° rise in temperature, and this is often the case with chemical reactions. However, weathering degradation is a complex process with mechanisms such as light and moisture stress, which follow different kinetics. The study by Fischer and Ketola[18] showed an average acceleration factor of 1.41 (SD 0.23) for a 10°C temperature increase—the maximum recorded was 1.89. Although this is less than the doubling predicted by Arrhenius, it still represents a significant increase.

The previous discussion emphasizes that natural weathering comprises a number of factors which, acting together, bring about coating degradation. The three primary damage factors are solar irradiation, time of wetness, and temperature. These differ substantially according to geographical location, and each will vary in response to meteorological and other local conditions. Such inherent variability complicates comparison between sites, and between field exposure and laboratory testing. Quantification of in-service exposure environments provides a means of developing a dosage model for inter-site and inter-laboratory comparisons. To create such a model requires exposure and quantification at different sites, and is time consuming. Furthermore, particular combinations of exposure conditions may occur relatively rarely and interactions may not be recognized. An alternative strategy to multi-site exposure is to bring about controlled variability at a single site and this might be brought about, as noted above, by altering the angle of exposure and changing TOW in a systematic way. Such tactics are widely used as a means of maximizing degradation (e.g., the exposure options offered by commercial weathering sites), but less studied as a means of building up a dosage model. This is an area worthy of further study as it offers the prospect of effectively creating climatic variation conditions characteristic of a specific site within a single exposure period. Such an approach could be achieved by simultaneous exposures for a given site at different angles, and with controlled TOW, which are relatively straightforward to achieve. Control of temperature is

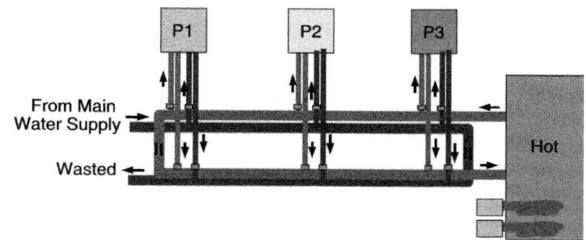

Figure 1—Schematic diagram of prototype 1 (P1 = test panel mounting 1).

more difficult and would be confounded with orientation effects. The following account describes the setting up of a fixed angle exposure rack with three controlled temperature zones, and some preliminary results. Further work is planned to cover other variables.

CONTROLLED TEMPERATURE EXPOSURE RACK

The core feature of the hardware development for a controlled natural weathering system has been the design and construction of one or more thermally regulated plates, on to which the arrays of test panels can be fixed for exposure to the prevailing weather. The system, as constructed, allows for either a fixed or a systematically controlled temperature on each plate. Facilities are included to apply a regulated water-spray to selected panels. This water is in addition to any moisture received by the test panels, from the natural rainfall, dew, or humidity.

In the current prototype, three temperature-controlled arrays are set respectively at 25°C, 35°C, and 45°C, and each array angled at 45° to the horizon and south facing. Thus, each panel on a plate would receive the same dose, at the same rate of dose, of solar energy during the period of exposure. In fact, the engineering in the system readily allows for variation in the plate's angle to the horizon, thus permitting the incident radiation dose at each plate to be varied, if required. Local climate, in particular the humidity and aspects of the solar radiation spectrum, are comprehensively recorded.

Although simple in concept, there were a number of learning points resulting in evolutionary development. The first prototype used ordinary main tap water for cooling with a separate recirculatory heater (*Figure* 1).

Problems were experienced with lime scale build up and the failure of solenoid valves, which were subsequently replaced with pneumatic-controlled valves. A closed-loop water circuit substituted the open flow cooling system and the control software improved with a PID (proportional, integral, derivative) controller (*Figure* 2).

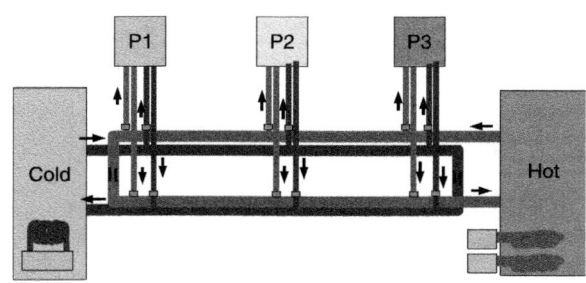

Figure 2—Schematic of prototype 2.

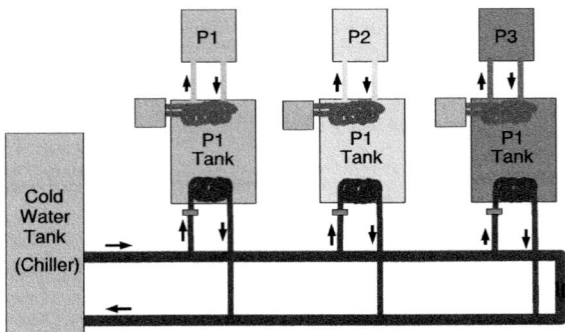

Figure 3—Schematic of prototype 3.

The second prototype gave much improved temperature control but failures in the pneumatic valves eventually occurred and there was still scope for better temperature control. A third prototype (*Figure* 3) was re-engineered by a commercial partner to incorporate the following features:

- Aluminum plates with even internal temperature distribution
- Separate water circuits to avoid mixing hot and cold water
- Isolated water circuit loops for each plate, each with its own immersion heater
- Continuous water flow
- Reduced energy losses
- Better control of energy input

Each test plate is composed of two aluminum panels with a channel machined to allow the water to flow between them. To reduce temperature gradient difference between the inlet and outlet on each plate a special dual spiral design of the channels has been introduced; this is shown in *Figure* 4. The in and out channels run parallel across the plate and their proximity helps to compensate for any temperature gradient that might occur. Thermal dissipation from the back of the plate is reduced with insulating material. A tank has been put in the water loop to add thermal inertia to the circuit with a pump circulating the water from the tank to the plate and back to the tank. A 3KW immersion heater supplies energy to the water and excess energy is removed via a heat exchanger through which cold water is circulating from an external closed circuit cold-water chiller unit common to the three plates. The temperature is measured on every plate by the control software, running on a local PC. This control software uses the acquired data and issues commands to the actuators in the cabinets—all actions and temperatures are logged. An automatic spray system

Figure 4—Diagram of the spiral channel used.

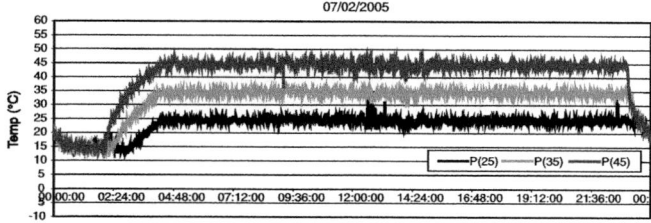

Figure 5—Example of general control of the system for 24-hour period.

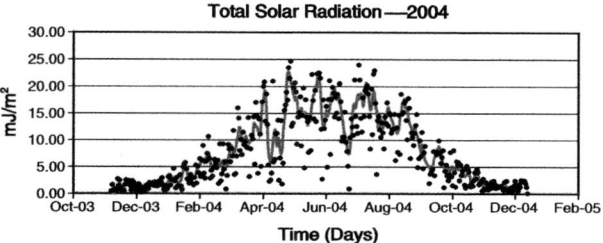

Figure 6—Total solar radiation measured using a pyranometer over 2004.

enables spraying of all or part of the test panel mountings at pre-determined intervals (*Figure 5*).

With all the modifications in place and some further tuning to the electronics, the system was sufficiently stable to require only a weekly check of water levels, and to maintain the required temperature separation over normal UK weather conditions.

External Data Logging

Various aspects of natural weather conditions were collected using a Skye Instruments DataHog2 logger; these include UVA, UVB, %RH, air temp, black and white panel temperatures, total solar radiation (via pyranometry), and wetness. Wind speed and total rainfall were considered of lesser interest and were not monitored. The collection devices used and the sampling times are summarized in *Table 2*.

Table 2—Collection Devices and Sampling Time for Monitoring Natural Weather Conditions

Sensor	Orientation	Sampling Interval/min	Logging Interval/min	Threshold Value
RH	NA	5	20	NA
Air temperature	NA	5	20	NA
Black panel sensor	45° South	5	20	NA
White panel sensor	45° South	5	20	NA
Pyranometer (350–1100 nm)	Horizontal	5	20	0000.18
UVA	Horizontal	5	20	0.0015
UVB	Horizontal	5	20	0.0015
Wetness	45° South	5	20	0.65789 V

Over a year weather conditions at the exposure site (Teddington, UK) have shown the air temperature ranging from –2.7 to 30.9°C and black/white panel temperatures from –5.6/–5.7°C to 59.2/39.5°C. *Figure* 6 summarizes the data recorded by the total light pyranometer and shows the amount of solar radiation received in a horizontal plane parallel to the roof.

EVALUATION OF THE WEATHERING RACK PROTOTYPES

Evaluation of the rack, in terms of its ability to separate temperature effects, was carried out at each stage of prototype development and is on-going. Reporting periods are thus relatively short. An objective, imposed by the funding body, was that the project's activities should be focused on the development of the system and the test methodology rather than examining weathering itself. This requirement has somewhat limited the scope for exploring the range of applications for which the controlled weathering unit might be used. The following investigations are therefore to explore the capability of the test rig, using coating formulations expected to show differences (*Table* 3).

Coating response was investigated using a pair of melamine-cured polyester resins of known relative performance. Each was made up into a dark blue coil coat formulation with a pigment:binder ratio of 0.6:1. The pigmentation comprised rutile titanium dioxide, black copper chromite, cobalt aluminate spinel, and chrome antimony titanate. Other normal formulating additives were included. The formulations were prepared without stabilizers, with HALS, and with a combination of UV absorber and HALS to give six formulations having different sensitivity to photodegradation and hydrolysis.

Degradation of the coatings was monitored using a selection of techniques including color, gloss 60°, atomic force microscopy (AFM), and Fourier transform infrared attenuated total reflectance (FTIR-ATR). Only color and gloss changes are reported here.

Using a Rhopoint® Nova glossmeter, 60° gloss was recorded. A total of seven measurements were recorded per panel.

The Lab values of each panel were measured on a Gretag Macbeth Color-eye® 7000A using a large sample port, D65 and specular component included. The color values reported are ΔE values, which were calculated from equation (1); the initial unexposed values of Lab were used for each panel when calculating ΔE.

$$\Delta E = \sqrt{\Delta L^2 + \Delta a^2 + \Delta b^2} \qquad (1)$$

During exposure the position of the coated panels was rotated on the heated plates.

Table 3—Investigation of Test Rig Capability

Stabilizer	Resin T1 T_g = >45	Resin T2 T_g = <45
None	T1N	T2N
1% HALS	T1YA	T2YA
2% UV absorber 1% HALS	T1YB	T2YB

EXPERIMENTAL RESULTS

Overall Effect of Temperature on Gloss

Using the second prototype rack, sets of the six formulations were exposed (45° S) with normal and additional wetting. Exposures were initiated during both winter and summer. A half-factorial experimental design was used. The overall effect on gloss and color for the three set temperatures is shown in *Figures* 7 and 8.

Results may also be expressed as a function of cumulative solar radiation (*Figure* 9).

A preliminary model of the relationship between gloss and dosage was derived using a relationship of the form:

$$rate = Ae^{\frac{-B}{T}} \tag{2}$$

In this instance for an iterative (Microsoft Excel™ Optimization function) best fit, the ATG versus ASR data at 25°C was obtained using data points inside the ASR range of 500 – 2500 MJ/M² and applied to determine values for A and B. Equation (3) was then applied using T=35°C and 45°C to generate predicated gloss versus dose profiles (*Figure* 10):

$$Gloss = A \times D - De^{\frac{-\sqrt{B^2}}{T}} \tag{3}$$

where: D = Accumulative total solar radiation dose
T = Temperature of the plate in °C
$A = 107.46 \times D^{-1.0129}$
$B = 221.88 \times A^{-0.0895}$

Although these initial results must be taken with caution, they do suggest that temperature effects can be detected and separated in a way that would be difficult for uncontrolled temperatures. For example, Schutyser and Perera[21] found increasing gloss retention in a QUV experiment to be in the order 57°C<66°C<36°C<73°C, and it is possible that the set temperatures did not correspond to the panel temperatures. The temperature

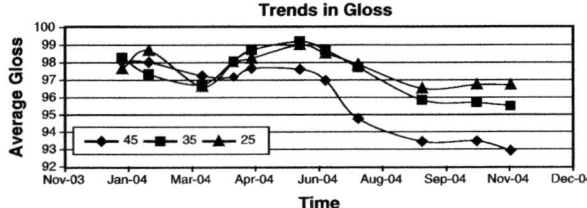

Figure 7—Trends seen in gloss 60° change (average of all panels on each plate).

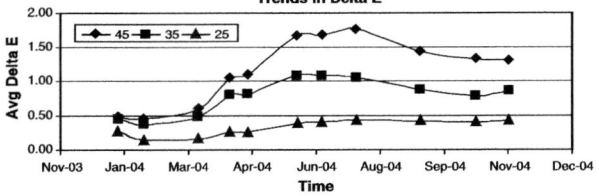

Figure 8—Trends seen in delta E color change (average of all panels on each plate).

Figure 9—Trends in average gloss for all systems with accumulative solor radiation dose.

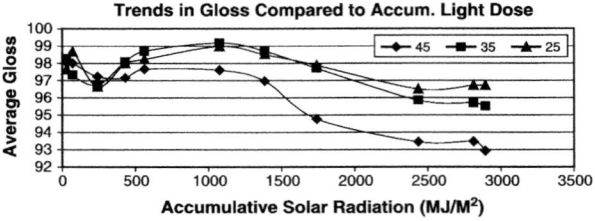

Trends in Gloss Compared to Accum. Light Dose

Table 4—ANOVA Table for 60° Gloss Results

	Sum of Squares	Degree of Freedom	Mean Squares	F Test	Probability Value
Intercept	17242.46	1	17242.46	458011011	0.000000
Time/days	0.14	12	0.01	302	0.000000
System	0.02	1	0.02	655	0.000000
Stabilizer present	0.05	1	0.05	1262	0.000000
Stabilizer type	0.00	1	0.00	47	0.000000
Plate temp	0.02	2	0.01	305	0.000000
Water	0.01	1	0.01	279	0.000000
Time/days*System	0.02	12	0.00	42	0.000000
Time/days*Stabilizer present	0.00	12	0.00	9	0.000000
System*Stabilizer present	0.00	1	0.00	3	0.069289
Time/days*Stabilizer type	0.00	12	0.00	3	0.000036
System*Stabilizer type	0.00	1	0.00	0	0.620621
Stabilizer present* Stabilizer type	0.00	1	0.00	2	0.125975
Time/days*Plate temp	0.06	24	0.00	68	0.000000
System*Plate temp	0.03	2	0.01	382	0.000000
Stabilizer present* Plate temp	0.01	2	0.00	93	0.000000
Stabilizer Type*Plate temp	0.00	2	0.00	8	0.000471
Time/days*Water	0.01	12	0.00	32	0.000000
System*Water	0.01	1	0.01	151	0.000000
Stabilizer present*Water	0.00	1	0.00	29	0.000000
Stabilizer type*Water	0.00	1	0.00	3	0.079038
Plate temp*Water	0.02	2	0.01	234	0.000000
Error	0.16	4263	0.00	—	—

Notes: To meet the conditions for ANOVA analysis, a log transformation was applied to change the data into an approximately Gaussian distribution.

If the p-value resulting from the ANOVA procedure is < 0.05 the evidence is considered statistically significant and the null hypothesis is rejected.

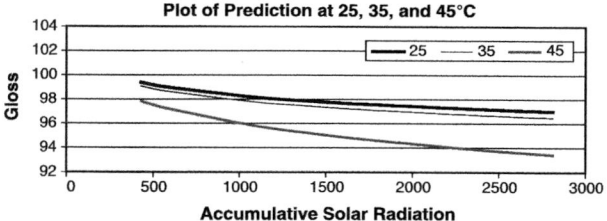

Figure 10—Plot of predicted gloss using equation (3).

rack on the other hand controls, within known limits, the panel temperature that is being studied.

The previous results are for pooled data; analysis of variance (ANOVA) was carried out to evaluate the significance of main factors and interactions. An example of the ANOVA analysis table is given in *Table* 4.

At a significance level of 99%, the main factors (temperature, ± stabilizer, ± water spray) are all significant—this is reasonably clear from the raw data. For the effect of temperature to be of practical use it is necessary to be able to detect interactions between temperature and other factors. With the present limited data this can only be achieved by statistical analysis. Results are usefully presented as graphical displays, e.g., *Figures* 11–14. The center point of any vertical bar in the interaction plot is the mean score, and the length of the bar represents the 90% Tukey confidence limit for the mean. The confidence limit denotes the outer limit of the expected value if the experiment was repeated under the same conditions. *Figure* 11 shows the main effect of "plate temperature" and supports the trend apparent in *Figures* 7 and 9.

Figure 12 is the interaction plot of "plate temperature" against "resin system." Resin T1 has a T_g greater than 45°, while the T_g of resin T2 is below 45°C. An interaction between resin and temperature is therefore to be expected, and it is encouraging that

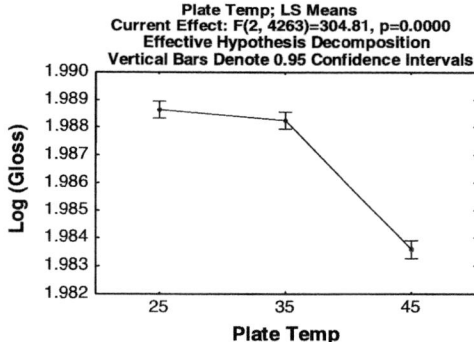

Figure 11—Graph of main factor "plate temp" (second prototype).

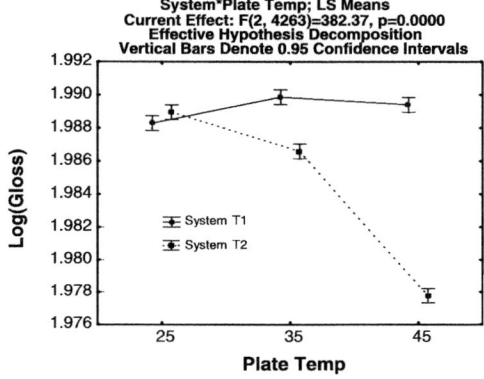

Figure 12—Interaction plot of "plate temp" vs "resin system."

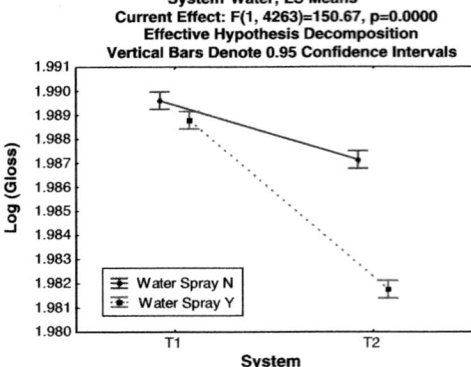

Figure 13—Interaction plot of "resin system" vs "water spray."

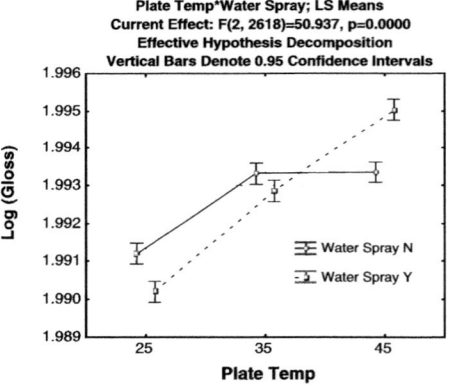

Figure 14—Plate temp*Water spray interaction.

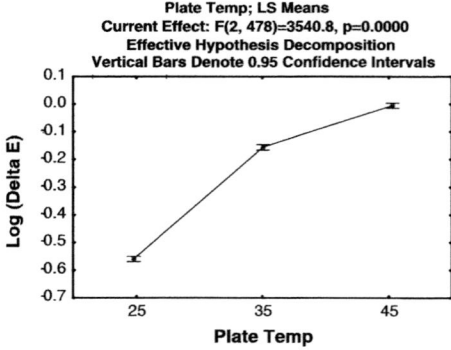

Figure 15—Main factor "plate temperature."

such an effect is detected. Several authors have stressed the significance of T_g in durability studies and accelerated weathering.[22-24]

Figure 13 shows that T2, the lower T_g resin, is more susceptible to gloss loss when subject to additional water spray. It is possible that this is indicative of hydrolysis, as the coating would also be more permeable. Temperature*water-spray interaction is only apparent at the higher temperature (see *Figure* 14).

Influence of Temperature on Color

In addition to the gloss changes, the coatings also underwent greater color changes at the higher temperature. *Table* 5 shows the ANOVA analysis for ΔE. All the main factors can be seen to be significant at the 99% level together with a number of interactions. Ranking in terms of f value shows that plate temperature is by far the most important factor. The strong temperature dependence of color change is shown in *Figure* 15. The resin system used in the paint formulation also had a marked effect, with the coatings with the softer resin giving higher ΔE values. Additional water spray, received by half of the panels, also had the effect of increasing the color change.

All the above results were obtained using the second prototype. With the commissioning of the third prototype, a further exposure series was initiated using the same formulations. A similar pattern of gloss loss was noted (*Figure* 16—compare with *Figure* 7) with somewhat smoother curves, which may reflect the better tempera-

Table 5—ANOVA Analysis for Lab Color Change ΔE

	SS	Degree of Freedom	MS	f	p
Intercept	36.10627	1	36.10627	8503.586	0.000000
Time/days	16.04739	11	1.45885	343.583	0.000000
Resin system	4.11728	1	4.11728	969.684	0.000000
Stabilizer present	0.50318	1	0.50318	118.506	0.000000
Stabilizer type	0.09304	1	0.09304	21.913	0.000004
Plate temp	30.06888	2	15.03444	3540.843	0.000000
Water spray	1.75513	1	1.75513	413.360	0.000000
Time/days*Resin system	0.20610	11	0.01874	4.413	0.000002
Time/days* Stabilizer present	0.15350	11	0.01395	3.287	0.000232
Resin system* Stabilizer present	0.01673	1	0.01673	3.941	0.047705
Time/days*Stabilizer type	0.05672	11	0.00516	1.214	0.274324
Resin system*Stabilizer type	0.04071	1	0.04071	9.588	0.002074
Stabilizer present* Stabilizer type	0.32018	1	0.32018	75.408	0.000000
Time/days*Plate temp	1.61020	22	0.07319	17.238	0.000000
Resin system*Plate temp	0.19316	2	0.09658	22.746	0.000000
Stabilizer present* Plate temp	0.26079	2	0.13040	30.710	0.000000
Stabilizer type* Plate temp	0.11553	2	0.05777	13.605	0.000002
Time/days*Water spray	0.77413	11	0.07038	16.574	0.000000
Resin system*Water spray	0.00517	1	0.00517	1.217	0.270418
Stabilizer present*Water spray	0.00644	1	0.00644	1.517	0.218641
Stabilizer type*Water spray	0.09088	1	0.09088	21.404	0.000005
Plate temp*Water spray	0.08329	2	0.04165	9.808	0.000067
Error	2.02959	478	0.00425	—	—

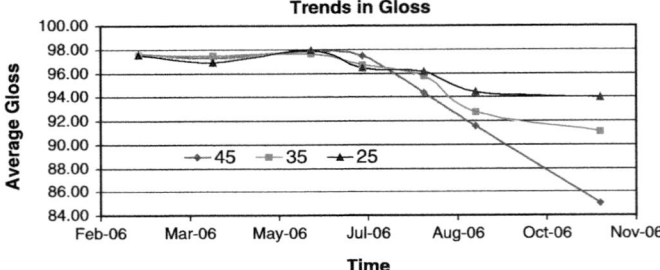

Figure 16—Trends seen in gloss 60° change (average of all panels on each plate) for third prototype.

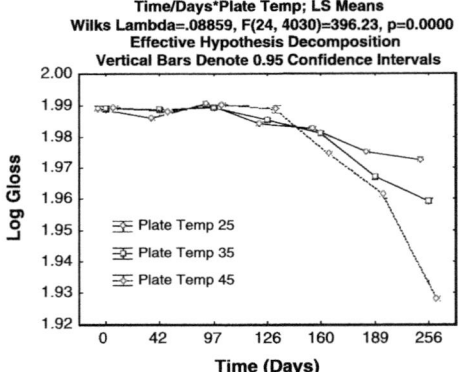

Figure 17—Main effects, plate temperature (third prototype).

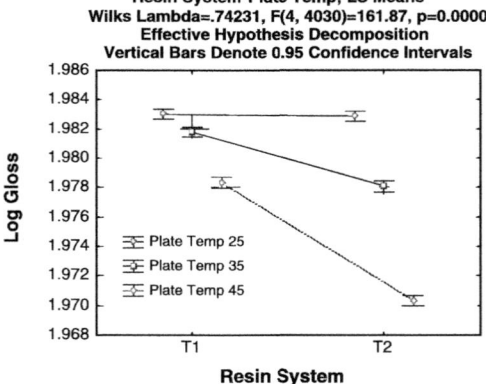

Figure 18—Plate temperature*Resin type interaction.

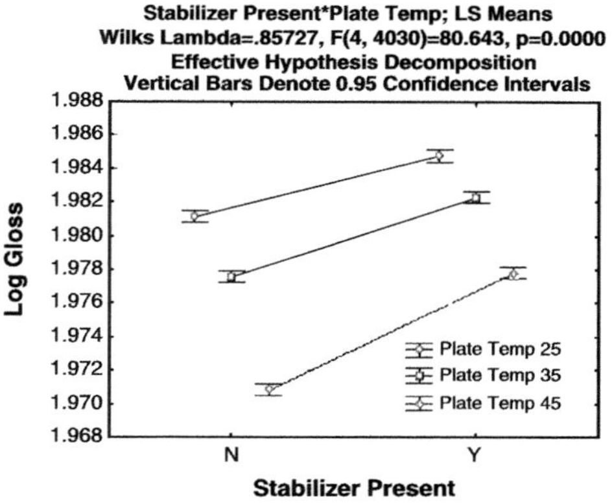

Stabilizer Present*Plate Temp; LS Means
Wilks Lambda=.85727, F(4, 4030)=80.643, p=0.0000
Effective Hypothesis Decomposition
Vertical Bars Denote 0.95 Confidence Intervals
Log Gloss
1.988
1.986
1.984
1.982
1.980
1.978
1.976
1.974
1.972
1.970
1.968
Plate Temp 25
Plate Temp 35
Plate Temp 45
N
Y
Stabilizer Present

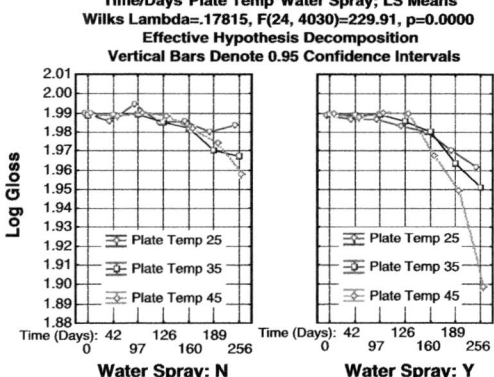

*Figure 19—Plate temperature*Stabilizer interaction.*

Time/Days*Plate Temp*Water Spray; LS Means
Wilks Lambda=.17815, F(24, 4030)=229.91, p=0.0000
Effective Hypothesis Decomposition
Vertical Bars Denote 0.95 Confidence Intervals
Log Gloss
2.01
2.00
1.99
1.98
1.97
1.96
1.95
1.94
1.93
1.92
1.91
1.90
1.89
1.88
Plate Temp 25
Plate Temp 35
Plate Temp 45
Time (Days): 42 126 189
0 97 160 256
Water Spray: N
Water Spray: Y

*Figure 20—Plate temperature*Water-spray interaction.*

Resin System*Plate Temp*Water Spray; LS Means
Wilks Lambda=.85769, F(4, 4030)=80.380, p=0.0000
Effective Hypothesis Decomposition
Vertical Bars Denote 0.95 Confidence Intervals
Log Gloss
1.990
1.985
1.980
1.975
1.970
1.965
1.960
1.955
Plate Temp 25
Plate Temp 35
Plate Temp 45
Resin System: T1 T2
Water Spray: N
Water Spray: Y

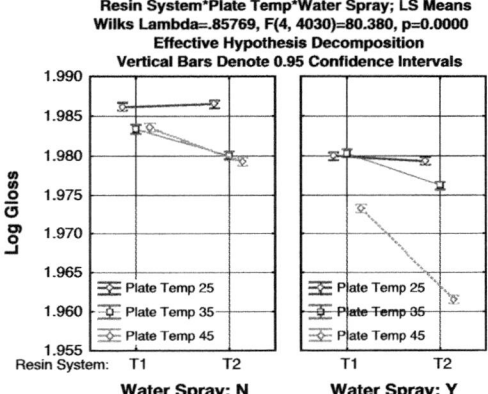

*Figure 21—Plate temperature*Water-spray*Resin type interaction.*

ture control. The initial "dormant" period during which $60°$ gloss shows no change is characteristic of many glossy coatings, although some damage may be detected using AFM[25] and other techniques.

Temperature and other interactions recorded on the third prototype followed a similar pattern to that of the second. A selection of the interaction graphs is shown in *Figures* 16–21, which provides evidence that zoning a weathering rack into temperature bands enables the separation of significant effects and interactions.

SUMMARY AND CONCLUSION

Many attempts to simulate and accelerate aspects of the natural weathering of coatings have failed to establish a good correlation. Frequent changes or reversals of rank order for performance metrics have been reported. In principle, this problem may be addressed using a proper dose-response model using a reliability-based service life prediction methodology. However, such investigations can be both time-consuming and expensive for small to medium size enterprises.

Another possibility for investigating the dosage sensitivity which, although less rigorous, can provide useful information, is to take advantage of systematic variation of natural weathering itself. Three important variables are solar irradiation dosage, time of wetness, and temperature. The first may be changed through the angle and orientation of exposure panels, and the second changed with controlled water spraying. Temperature is more difficult to control and is likely to be confounded with irradiation.

A prototype weathering rack has been developed which uses water circulation to control plates supporting panels at $25°$, $35°$, and $45°C$. Preliminary evaluation of the rack using a selected formulation set of known response has shown temperature effects and temperature interactions with other variables. It is therefore concluded that this approach is a useful way of varying conditions around natural weathering without venturing into the extremes that might cause "unnatural" results. Sensitivity to dosage factors can thus be investigated for a specific exposure site.

Further work will be required to validate the preliminary results and it is planned to broaden the range of performance metrics to include mold (fungal) growth, dirt pick-up, and corrosion. The latter are expected to show a different response to the gloss and color effects reported here.

ACKNOWLEDGMENTS

Industrial Partners: National Physical Laboratory, Beckers Industrial Coatings Ltd., and Rhopoint Instrumentation Ltd. *Funding*: UK Department of Trade & Industry (DTI) and the support of members serving on the Industrial Advisory Group (http://www.pra-world.com/research/projects/dura2.htm).

References

(1) *Service Life Prediction of Organic Coatings: A Systems Approach*, Bauer, D.R. and Martin, J.W. (Eds.), ACS Symposium Series 722, American Chemical Society, Washington, D.C., 1999.

(2) *Service Life Prediction: Methodology and Metrologies*, Bauer, D.R. and Martin, J.W. (Eds.), ACS Symposium Series 805, American Chemical Society, Washington D.C., 2002.

(3) *Service Life Prediction: Challenging the Status Quo*, Martin, J.W., Ryntz, R.A., and Dickie, R.A. (Eds.), Federation of Societies for Coatings Technology, Blue Bell, PA, 2005.

(4) Martin J.W., Nguyen T., and Wood, K.W., "Unresolved Issues Relating to Predicting the Service Life of Polymeric Materials," in *Service Life Prediction: Challenging the Status Quo*, Martin, J.W., Ryntz, R.A., and Dickie, R.A. (Eds.), Federation of Societies for Coatings Technology, Blue Bell, PA, p. 13-38, 2005.

(5) Chess J.A., Cocuzzi D.A., Pilcher G.R., and Van de Streek G.N., "Accelerated Weathering, Pseudo-Science or Superstition?" in *Service Life Prediction of Organic Coatings: A Systems Approach*, Bauer, D.R. and Martin, J.W. (Eds.), ACS Symposium Series 722, American Chemical Society, Washington D.C., p. 130-147, 1999.

(6) Hoeflaak, M., de Ruiter, B., and Maas, J.H. "Calibrating the Climate: Correlations Between Different Weathering Tests," *Eur. Coat. J.*, 3, p. 30-35 (2006).

(7) Boisseau, J., Pattison, L., Henderson, K., and Hunt, R., "The Flaws in Accelerated Weathering of Automotive OEM Coatings," *Paint Coat. Ind.*, p. 42-53 (2006).

(8) Meeker, W.Q. and Escobar, L.A., "Accelerated Life Tests: Concepts and Data Analysis," in *Service Life Prediction of Organic Coatings: A Systems Approach*, Bauer, D.R. and Martin, J.W. (Eds.), ACS Symposium Series 722, American Chemical Society, Washington D.C., p. 130-147 (1999).

(9) Friebel, S. et al., "Industrial Transparent UV-Coatings for Outdoor Wood Applications," PRA Fifth International Woodcoatings Congress, Paper 30, Prague, 2006.

(10) Holman, R., "Development of Cyclical Tests of Polymeric Coatings which Include Exposure to UV Light, Temperature Extremes and Stress," http://www.pra-world.com/research/projects/mde2.htm.

(11) Holman, R., "Extension of the Range of Relevant Environments to Validate Test Methods to Predict Microbial Attack of Water-Based Coatings," http://www.pra-publications.org.uk/research/reports/all/exmaw-summary.pdf.

(12) Kennedy, R., "Dirt Pick-Up: The Environmental Soiling of Surface Coatings," http://www.pra-world.com/research/projects/dirt.htm.

(13) Byrd, W.E., Chin, J.E, and Martin, J.W., "Integrating Sphere Sources for UV Exposure: A Novel Approach to the Artificial UV Weathering of Coatings, Plastics and Composites," in *Service Life Prediction: Methodology and Metrologies*, Bauer, D.R. and Martin, J.W. (Eds.), ACS Symposium Series 805, American Chemical Society, Washington D.C., p. 144-159 (2002).

(14) Hardcastle, H.K. "Overview of Exposure Angle Considerations for Service Life Prediction," in *Service Life Prediction or Organic Coatings: A Systems Approach*, Bauer, D.R. and Martin, J.W. (Eds.,) ACS Symposium Series 722, American Chemical Society, Washington D.C., p. 37-62 (1999).

(15) Kropf, F.W., Sell, J., and Feist, W.C., "Comparative Weathering Tests of North American and European Exterior Wood Finishes," *Forest Products J.*, Vol. 44, No. 10, p. 33-4 (1994).

(16) Crewdson, M.J. and Brennan, P. "Outdoor Weathering Basic Exposure Procedures," *J. Protect. Coat. Linings*, p. 17-25 (1995).

(17) Sawant, S.S. and Venugopal, C., "Effect of Exposure Angle on the Marine Atmospheric Corrosion of Mild steel," *Corrosion Prevention & Control*, p. 35-37 (1996).

(18) Fischer, R.M. and Ketola, W.D., "Error Analyses and Associated Risk for Accelerated Weathering Results," in *Service Life Prediction: Challenging the Status Quo*, Martin, J.W., Ryntz, R.A., and Dickie, R.A. (Eds.), Federation of Societies for Coatings Technology, Blue Bell, PA, p. 79-92, 2005.

(19) Myers, D.R., "Predicting Temperatures of Exposure Panels: Models and Empirical Data," in *Service Life Prediction or Organic Coatings: A Systems Approach*, Bauer, D.R. and Martin, J.W. (Eds.,)ACS Symposium Series 722, American Chemical Society, Washington D.C., p. 71-84. 1999.

(20) Burch, D.M. and Martin, J.W., "Predicting the Temperature and Relative Humidity of Polymer Coatings in the Field," in *Service Life Prediction or Organic Coatings: A Systems Approach*, Bauer, D.R. and Martin, J.W. (Eds.,) ACS Symposium Series 722, American Chemical Society, Washington D.C., p. 85-107, 1999.

(21) Schutyser, P. and Perera, D.Y., "Use of Reliability Methodology for Appearance Measurements," in *Service Life Prediction or Organic Coatings: A Systems Approach*, Bauer, D.R. and Martin, J.W. (Eds.,) ACS Symposium Series 722, American Chemical Society, Washington D.C., p. 198-206, 1999.

(22) Perera, D.Y. and Schutyser, P., "Effect of Physical Aging on Thermal Stress Development in Powder Coatings," *Prog. Org. Coat.*, Vol. 24, No. 1-4, p. 299-30 (1994).

(23) Aloui, F. et al., "Photostabilization of Wood Clearcoatings with UV Absorbers: Correlation with Their Effect on the Glass Transition Temperature," Journal of Physics Conference Series 40 (Statistical Physics of Ageing Phenomena and the Glass Transition Temperature), p. 118-123, 2006.

(24) Skaja, A., Fernando, D., and Croll, S., "Mechanical Property Changes and Degradation During Accelerated Weathering of Polyester-Urethane Coatings," *J. Coat. Technol. Res.*, 3 (1), 41-51 (2006).

(25) Van Landingham, M.R., Tinh, N.W., Byrd, W.E., and Martin J.W., "On the Use of the Atomic Force Microscope to Monitor Physical Degradation of Polymeric Coating Surfaces," *J. Coat. Technol.*, 73, No. 923, 43-50 (2001).

Chapter 8

Durability of Building Joint Sealants

Christopher C. White,[1] Kar Tean Tan,[1] Donald L. Hunston,[1] and R. Sam Williams[2]

[1] National Institute of Standards and Technology, 100 Bureau Dr., MS 8615, Gaithersburg, MD 20899
[2] U. S. Department of Agriculture Forest Service, Forest Products Laboratory, Madison, WI USA 53726

Predicting the service life of building joint sealants exposed to service environments in less than real time has been a need of the sealant community for many decades. Despite extensive research efforts to design laboratory accelerated tests to duplicate the failure modes occurring in field exposures, little success has been achieved using conventional durability methodologies. In response to this urgent need, we have designed a laboratory-based test methodology that used a systematic approach to study, both independently and in combination, the major environmental factors that cause aging in building joint sealants. Changes in modulus, stiffness, and stress relaxation behavior were assessed. Field exposure was conducted in Gaithersburg, MD, using a thermally-driven exposure device with capabilities for monitoring changes in the sealant load and displacement. The results of both field and laboratory exposures are presented and discussed.

INTRODUCTION

Building joint sealants are filled elastomers that are an essential component of modern construction. They serve in weatherproofing of structures by preventing unwanted moisture intrusion and subsequent water damage. Over the past two decades, rapid technological advances and tremendous market growth in the building joint sealant industry have introduced a multitude of novel sealant products into the marketplace. In service, building joint sealants are exposed to various aging factors, including temperature, humidity, solar radiation, cyclic fatigue loading, etc. The interaction of these aging factors with sealants inevitably affects their properties and determines their long-term durability.[1-4] Hence, to reduce the risk of introducing poorly performing products into the marketplace, information on long-term performance of new products is needed. However, unlike existing building joint sealant products, new sealants lack long-term performance data. Various laboratory accelerated tests have been adopted to generate durability data, and to predict the long-term performance data in less than real time. However, building joint sealants have been reported to fail prematurely in the field even though they may have performed satisfactorily using these laboratory evaluations.

115

116

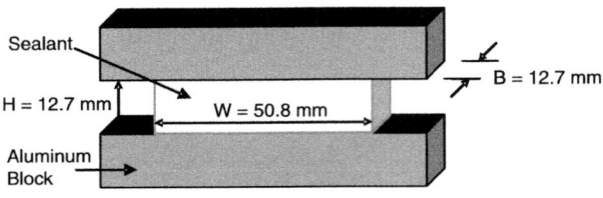

Sealant

H = 12.7 mm W = 50.8 mm

B = 12.7 mm

Aluminum Block

Figure 1—Schematic illustration of the test geometry used (not to scale).

Studies in the construction industry have shown a 50% failure rate within 10 years and a 95% failure rate within 20 years after installation.[5-7] What makes these failures particularly detrimental is that sealants are often used in areas where moisture-induced degradation is difficult to monitor and expensive to repair. Consequently, sealant failure is frequently detected only after considerable damage has occurred. In the housing market alone, premature failure of sealants and subsequent moisture intrusion damage is a significant contributor to the $65 billion–$80 billion spent annually on repair.[8] The aim of the present research, therefore, was to design a laboratory accelerated testing protocol that provides a platform for screening the relative importance of four different aging factors—temperature, humidity, radiation, and cyclic fatigue—acting independently and in combination on the degradation of building joint sealants. Since field exposure results have been designated as the standard of performance, outdoor field exposure was also carried out. This exposure was conducted in Gaithersburg, MD, using a polyvinyl chloride thermally-driven exposure device, which relies on changes in outdoor air temperature to induce cyclic fatigue deformation on sealant joints.

EXPERIMENTAL CONSIDERATIONS*

Materials and Specimen Preparation

A commercial sealant was provided by a member of a NIST/industry consortium, and formed into sealant joints conforming to the ASTM C719[9] specimen geometry shown in *Figure 1*. Specimens were prepared by extruding the sealant from a cartridge into a specimen cavity (50.8 mm x 12.7 mm x 12.7 mm) composed of aluminum supports (76.2 mm x 12.7 mm x 12.7 mm) on opposite sides with a polytetrafluoroethylene (PTFE) film on the back and PTFE spacers (12.7 mm x 12.7 mm) on each end. The specimens were cured in this fixture for 5 h at room temperature, and then removed from the fixture, keeping the PTFE spacers and the aluminum blocks intact. The samples were then allowed to cure for an additional three weeks as specified in the ASTM C719 test method.

EXPOSURE CONDITIONS AND CHARACTERIZATION

Laboratory Exposure and Characterization

Specimens were exposed to both field and laboratory accelerated conditions. The accelerated tests were conducted using a custom-made in-situ device with independent and precise control of temperature, humidity, radiation, and cyclic strain movement, and built-in characterization functionality (see *Figure* 2). A full description of the hybrid device is documented elsewhere.[10] Temperature control was achieved via a precision temperature regulator; humidity control was accomplished via the proportional mixing of dry and saturated air. Highly uniform UV radiation was attained through the use of an integrating sphere-based weathering chamber, referred to as the Simulated Photodegradation via High Energy Radiant Exposure (SPHERE).[11] Six high intensity light sources were used in the SPHERE to produce a collimated and highly uniform ultraviolet flux of approximately 480 W/m². A computer-controlled stepper motor and a precision transmission system were used to provide precise movement control that imposed mechanical deformations on the sealant specimens. There was a total of eight specimen holders, each of which was attached to a hermetically sealed load cell (Model SSM-AJ-250, Interface, Scottsdale, AZ) with a capacity of ± 113.4 kg. To precisely measure device movement, two linear variable differential transformers (LVDT) (Model HSD-750, Macro Sensors, Pennsauken, NJ) with a deflection range of ± 6.35 mm were used.

During the exposure, half of the specimens were subjected to cyclic fatigue loading while the other half remained unloaded to provide a comparison. The fatigue deformation cycle involved varying the tensile strain between 25% and 50%. A total of 1460 cycles were imposed on the specimens (38 min/cycle). Prior to and after the exposure experiment, a characterization test was run, which consisted of two loading-unloading

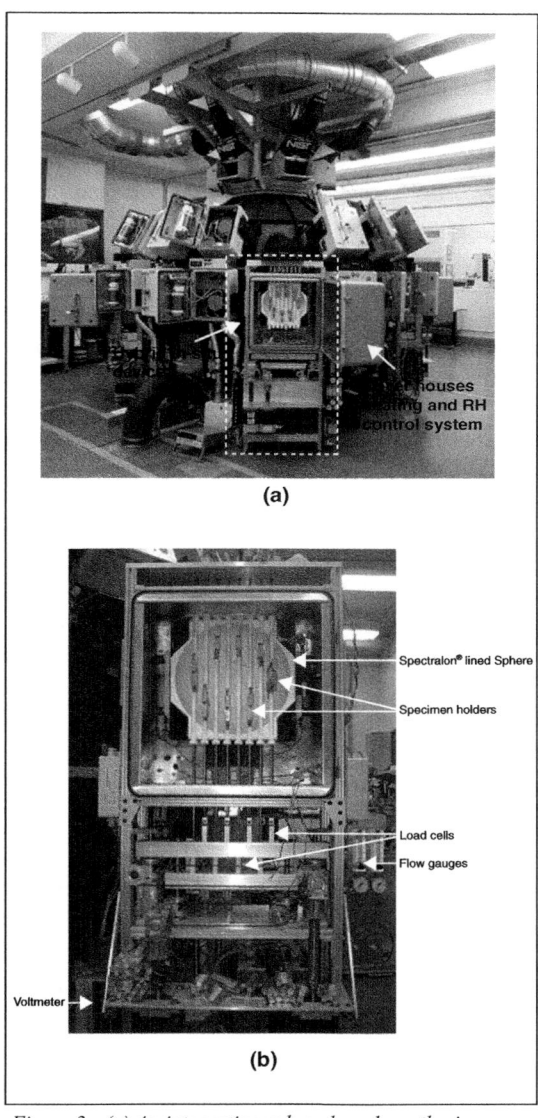

Figure 2—(a) An integrating sphere-based weathering chamber, referred to as the Simulated Photodegradation via High Energy Radiant Exposure (SPHERE), and (b) an in-situ device.

118

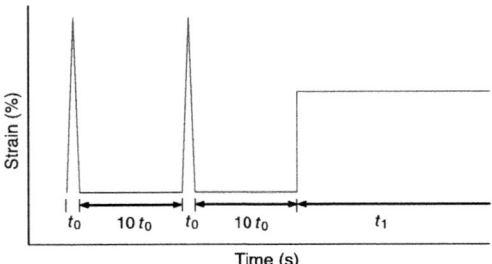

Figure 3—Strain history used for Mullins cycles and stress relaxation tests.

cycles (0% to 25% strain) and one stress relaxation measurement at 15% strain. After characterization, the specimens were allowed to recover before starting the fatigue cycling. The loading and unloading during characterization were performed at 2 cm/min which meant that the specimen was under load for a time period denoted as t_0. The time under load in the stress relaxation test was denoted as t_1. To ensure that the viscoelastic recovery from one loading was complete before the next loading was initiated, the specimen was allowed to rest for $10t_0$ or $10t_1$ between loadings (*Figure* 3). For these tests, the values for t_0 and t_1 were 20 s and 3 h, respectively. The motivation for the two loading-unloading cycles was to see if any Mullins Effect was present and to mitigate any Mullins Effect in the subsequent stress relaxation test. This Mullins Effect is a phenomenon observed in filled elastomers: the stress at a given strain is higher the first time the specimen is deformed than it is for subsequent deformations. As long as the maximum strain achieved during the first deformation is not exceeded, all subsequent loadings follow the same stress-strain curve. As a result, it is common to pre-strain a sample to high values so that the results are reproducible in subsequent testing.

For the stress relaxation measurements, specimens were loaded rapidly at 100 cm/min up to 15% strain and held at that value while load was monitored as a function of time. The time required to load a specimen was less than 1 s and the first data point was not taken until after 10 s to avoid transient effects associated with uploading. An apparent modulus, E_a, was calculated using a relationship based on the statistical theory of rubber-like elasticity[12-14]:

$$E_a(t,\lambda) = \frac{3L(t)}{WB(\lambda - \lambda^{-2})} \tag{1}$$

where W and B are the width and breadth of the sealant (see *Figure* 1), L is the load, t is the time, and λ is the extension ratio, which is given by:

$$\lambda = 1 + \frac{\Delta}{H} \tag{2}$$

where Δ is the cross-head displacement. Equation (1) assumes that the sealant is incompressible, which should be a good assumption for elastomers.

To examine the change in the specimen produced by exposure, the apparent modulus, E_a, from the characterizations before and after were compared. This was done by calculating the fractional change in apparent modulus, F. If exposure time is designated as t_e, then F is given by:

$$F = \frac{E_a(\text{at } t_e = 1 \text{ month})}{E_a(\text{at } t_e = 0)} \tag{3}$$

A second way to monitor the changes during exposure was to calculate stiffness index based on loading curve in each cycle of the fatigue deformation experiment. This stiffness index was given by the slope of the linear regression curve fit to a plot of the voltage output for a load cell versus the voltage output for a LVDT. This produced 1460 points over the one month period. Unlike the stress relaxation experiments in which changes before and after exposure were only obtained, the use of stiffness index offered the advantage of monitoring changes during exposure. Squared correlation coefficients, r^2, for fitted curves were also obtained to measure goodness-of-fit of linear regression, and hence the reliability of the indexes. A custom-written LabVIEW (National Instruments, Columbia, MD) program was developed for signal generation, analysis, and data acquisition. More information can be found elsewhere in this literature.[10]

In addition to the presence or absence of fatigue deformation cycle during exposure, specimens were subjected to one of four different environments involving combinations of temperature and RH, i.e., (a) 30°C and 0% RH, (b) 30°C and 80% RH, (c) 60°C and 0% RH, and (d) 60°C and 80% RH. In this chapter, 60°C will hereafter be denoted as "hot," and 30°C as "cold." Also, 80% RH will hereafter be denoted at "wet," and 0% RH as "dry." As a result, there were eight different exposure conditions for the sealant material, and with four replicates of each, a total of 32 specimens were examined in each test.

Field Exposure and Characterization

Field exposure was performed in Gaithersburg, using custom-made *thermally* driven outdoor exposure engines, as shown in *Figure* 4. Each engine was composed of a moving frame and a fixed support frame. The moving frame consisted of two 101.6 mm diameter polyvinyl chloride (PVC) pipes (Schedule 40), a stainless steel crosspiece, six stainless steel rods, and six specimen holders; while the fixed support frame was comprised of a wood support frame, and a fixed stainless steel crosspiece. The strain on the sealant samples was generated by thermal changes in the exposure environment combined with the relative difference in the coefficient of thermal expansion between the PVC pipe and the end grain wood frames. A full description of the devices is reported elsewhere.[15] Specimens were placed into the device at a time when the temperature was approximately 13°C (55°F), so they were under no load at that temperature. At high temperatures PVC pipes expand, causing the specimens to be loaded in tension; conversely, the specimens are loaded in compression when the pipes contract at low temperatures, as shown schematically in *Figure* 4b and c. Consequently, the engines are also known as "winter/compression" or "summer/tension" engines. The engines were mounted facing south towards the equator at an angle of 90° from the horizontal plane.

To monitor the force on each specimen, each specimen holder was attached to a hermetically sealed load cell (Model SSM-AJ-250, Interface, Scottsdale, AZ) attached with a capacity of ± 113.4 kg. The displacement was monitored with LVDTs (Model HSD-750, Macro Sensors, Pennsauken, NJ) with a deflection range of ± 6.35 mm. The LVDTs were bolted between both fixed and moving stainless steel crosspieces. These electromechanical transducers were instrumented to monitor continuously forces and deflections. Also, each engine was attached to a series of six thermocouples to record the temperatures of PVC pipes. Data from load cells, LVDTs, and thermocouples were directly fed into a Keithley 2701 Ethernet-based Data Acquisition System (Keithley Instruments,

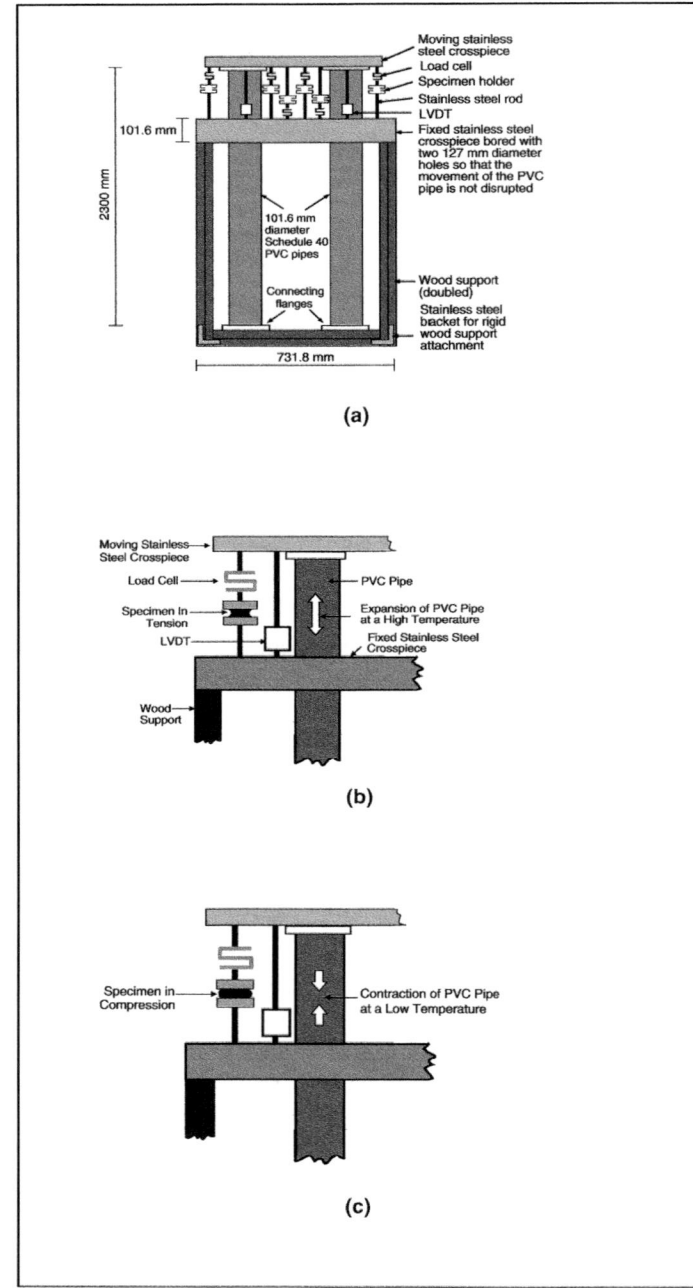

Figure 4—Schematic illustrations of (a) the thermally driven PVC engine, and the mechanism of the engine: (b) at a high temperature, the PVC pipe expands causing the specimen to be loaded in tension; and (c) at a low temperature, the pipe contracts causing the specimen to be loaded in compression.[15]

Cleveland, OH). A custom-written LabVIEW program was used to collect the voltage measurements from the Keithley system every 15 seconds, 24 hours a day. After one-minute worth of data was collected, the program averaged the values and appended the result to a tab-delimited database on a remote server. The engines allow changes in stiffness to be monitored as a function of exposure time.

RESULTS

Indoor Exposure

Figure 5 shows typical results for the two loading-unloading curves in the first characterization test on a specimen. The Mullins Effect is clearly present with higher stress levels during the first loading curve versus the second. Note that the unloading curves are the same in both cases. All subsequent loading-unloading curves will fall on top of the data in the second cycle.

Figure 6 shows a representative plot of changes in apparent moduli as a function of relaxation time for specimens under "static/hot/wet" conditions prior to exposure and after completion of fatigue deformation cycle. There were up to four replicates in each test, and the vertical bars indicate experimental uncertainty. Consequently, the difference seen between the two curves is significant. Note that there is no change in the curve shape, implying that time dependency of the apparent modulus is very similar before and after exposure. The magnitude of the apparent modulus, however, decreased significantly after exposure. Similar curves can be generated for all eight exposure conditions, but a plot with all 16 curves and their uncertainty is much cluttered. To facilitate comparison between different exposure conditions, stress relaxation data are presented as a fractional change in apparent modulus, F, as a function of relaxation time (see equation (3)). In such a graph, no change would be represented as a horizontal straight line at F=1.0. A horizontal line above or below F=1.0 indicates that exposure caused a vertical shift in the stress relaxation curve but no change in shape, i.e., the time dependence did not change. Something other than a horizontal straight line indicates a change in the

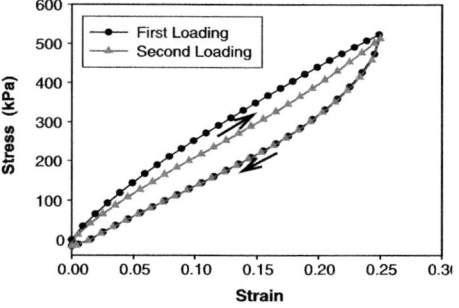

Figure 5—Loading-unloading curve for Mullins cycle.

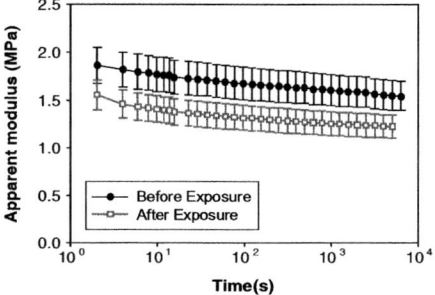

Figure 6—Variation of apparent modulus as a function of relaxation time for specimens under "static/hot/wet" conditions before and after exposures.

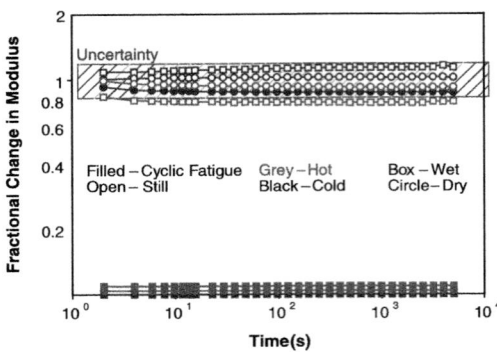

Figure 7—Stress relaxation curve for both static and cycle fatigue tests.

time dependence. The experimental uncertainty can be shown as a hashed region on each side of F=1.0 so if the point for a given curve falls within this region, there is no changes outside the experimental uncertainty. Stress relaxation data at different combinations of temperature and RH for static tests are shown in *Figure 7*. The lines at very low values of F indicate the samples have failed. The near straight lines parallel to the abscissa show that there is no change in the curve shape for all conditions.

The effect of temperature on the static performance of joints can be assessed by examining the conditions at the same RH, namely by comparing "static/cold/wet" with "static/ hot/wet," or "static/cold/dry" with "static/hot/dry." It is apparent that the temperature effect, either under a relatively dry or a moist environment, is insignificant. Also, moisture-induced deterioration in the static performance at a low temperature similarly seems unimportant, as revealed by comparing the "static/cold/dry" and "static/cold/wet" results. However, as shown in *Figure* 8, the combination of high temperature and RH produced a slight reduction in apparent modulus. All joints under static conditions remained intact and no joint failure was observed.

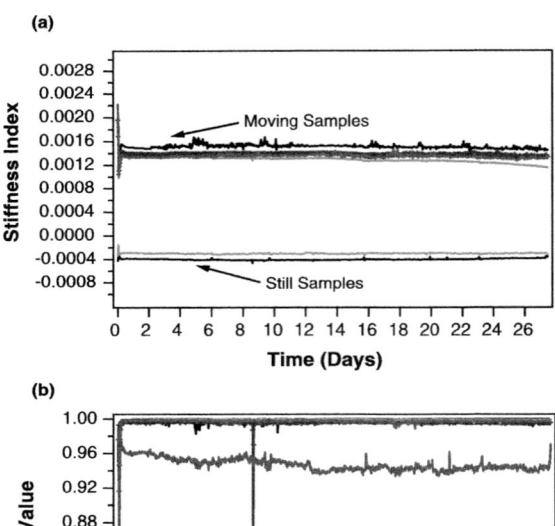

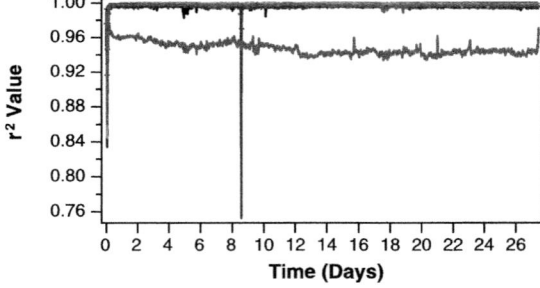

Figure 8—(a) Changes in stiffness index for samples with movement and no movement. (b) Corresponding squared correlation coefficients, r^2, as a function of exposure times in days for specimens under "fatigue/cold/dry" conditions. No visible failure is evident in the samples during the entire exposure.

Under cyclic fatigue deformation, the durability behavior differs considerably from that of static tests. The fractional changes in apparent modulus as a function of relaxation time for cyclic fatigue tests are included in *Figure* 7. In a "cold" and "dry" environment, the effect of fluctuating loads on the cyclic performance of sealant joints is insignificant. Furthermore, changes in the stiffness index of specimens, which were measured by the slopes of loading curves in the load-displacement plot for fatigue deformation cycle, were examined. The corresponding squared correlation coefficients, r^2, for the fitted slopes were also obtained to assess the goodness-of-fit of the slopes. The results for "fatigue/cold/ dry" are shown in *Figure* 8. The results for "static/cold/dry" are also included for comparison, but the slopes are always zero because there was no fatigue deformation cycle. In *Figure* 8a, the stiffness index of specimens under "fatigue/cold/dry" remained unchanged over a month of

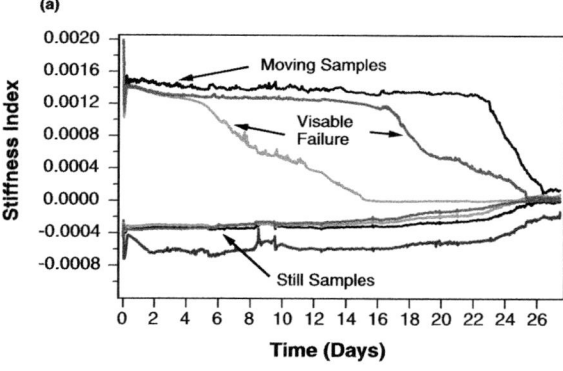

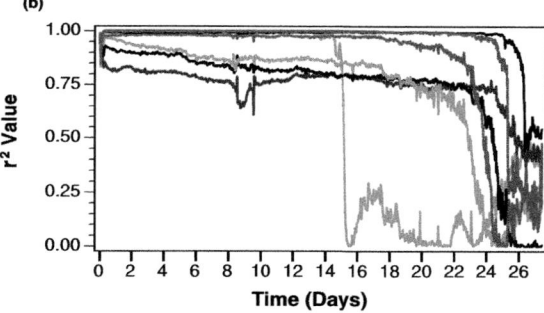

Figure 9—(a) Changes in stiffness index for both samples with movement and no movement. (b) Corresponding squared correlation coefficients, r^2, as a function of exposure times in days for specimens under "fatigue/hot/dry" conditions. Visible failure is evident in the three samples experiencing movement starting at eight days to 25 days.

exposure, which agreed with the comparable stress relaxation data. The relative high values of corresponding squared correlation coefficients confirmed the reliability of the data obtained (see *Figure* 8b).

However, as shown in *Figure* 7, the combination of cyclic fatigue with a high temperature, or with a high RH, or the combination of three aging factors, resulted in substantial changes in moduli. All joints tested under these conditions failed; thereby, the curves were plotted with ordinate magnitudes equal to zero. Changes in stiffness index and linear regression coefficients as a function of exposure time for "fatigue/ hot/dry" are shown in *Figure* 9. It can be seen that there was no change in stiffness index in early stages of exposure. With increasing exposure, the stiffness index decreased and was eventually followed by specimen failures. The stiffness index plots for "fatigue/cold/wet" and "fatigue/hot/wet" are shown in *Figures* 10 and 11, respectively. The stiffness for these specimens decreased drastically upon exposure, and then failures followed.

The loci of failure for all joints were visually observed as being cohesive within the sealant layers, indicating that the sealant itself was weak, while the interfacial adhesion between the sealant and the substrate was relatively robust. Further examination revealed

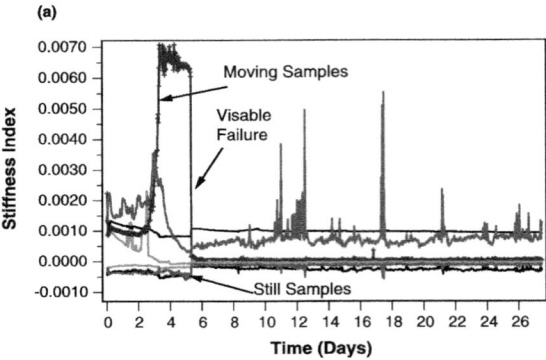

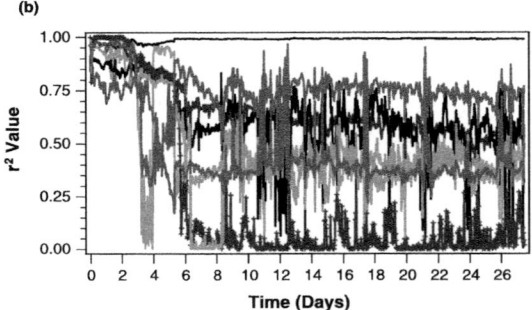

Figure 10—(a) Changes in stiffness index for both samples with movement and no movement. (b) Corresponding squared correlation coefficients, r², as a function of exposure time in days for specimens under "fatigue/cold/ wet" conditions. Visible failure is evident in the samples at times less than six days.

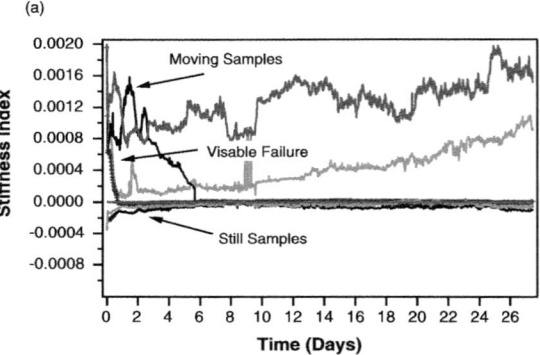

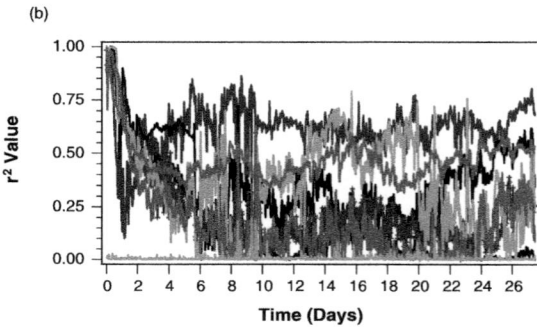

Figure 11—(a) Changes in stiffness index for both samples with movement and no movement. (b) Corresponding squared correlation coefficients, r², as a function of exposure times in days for specimens under "fatigue/hot/wet" conditions. Visible failure is evident in the samples at times less than two days.

extensive embrittlement. Therefore, it is highly likely that extensive crosslinking had taken place in the specimens, rendering them brittle and leading to premature failures. It is also clear that cyclic fatigue alone is not the critical factor leading to environmental failure of this sealant, but it is the combination of cyclic fatigue with other environmental factors (i.e., temperature or moisture) that is deleterious.

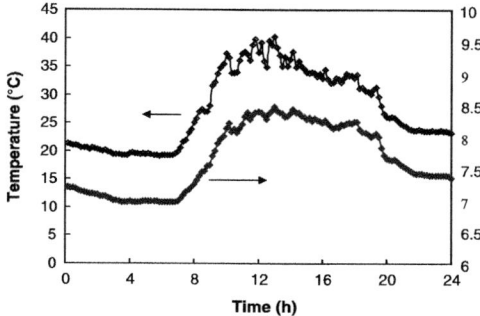

Figure 12—Variation of temperature of PVC pipe and the resulting displacement as a function of time. Measurements were made on July 1, 2004.

Field Exposure

In the case of field exposure, specimens were exposed to cyclic deformation induced by dimensional changes of PVC engines in which specimens were loaded in tension when outdoor air temperature was relatively high and in compression when the outdoor air temperature was relatively low. The evidence showing that changes in outdoor air temperature directly affect the magnitude of cyclic deformation imposed on specimens is shown in *Figure* 12. Because of temporal variations in outdoor air temperature, cyclic loading varies with the time of day. Such cyclic loading time series was what would be expected with sealants used in building structures. Load and displacement experienced by specimens were continuously monitored over one year, and the results are shown in *Figure* 13. In this plot, the data points on the right and left hand sides of the zero displacement line indicate which specimens were in tension and compression, respectively. For clarity, only data points collected over a day in each month were plotted. It should be noted, however, that data points for other days in the same month were very similar. From *Figure* 13, hysteresis in the load-displacement plot is clearly seen, demonstrating the viscoelastic nature of the sealant, which is common for all sealants although their observed degrees vary. It can

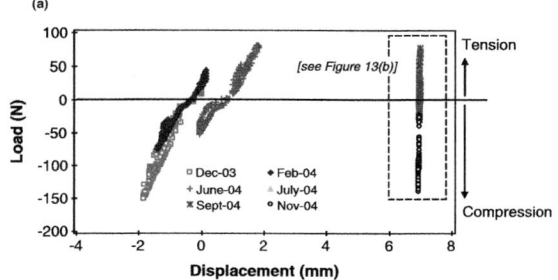

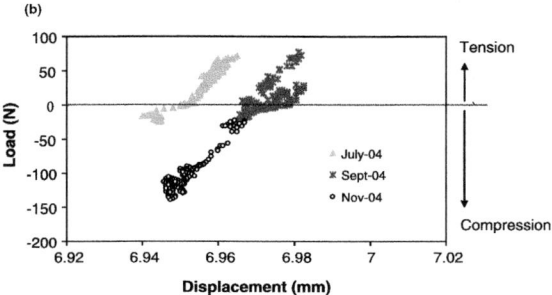

Figure 13—Load versus displacement recorded over one year of field exposure for displacement ranges of (a) −4 mm to 8 mm, and (b) 6.92 mm to 7.02 mm.

Table 1—Tabulation of Stiffness and the Corresponding Squared Correlation Coefficients, r^2, for Different Months of Exposure

Exposure Month	Stiffness (N/mm)	r^2 Coefficient
Dec. 2003	9.62	0.94
Feb. 2004	6.53	0.95
June 2004	5.98	0.97
July 2004	405.60	0.92
Sept. 2004	526.24	0.85
Nov. 2004	475.37	0.85

also be seen that specimens underwent both tensile and compressive loadings in December 2003 and February 2004, but compressive loading was found to be predominant. In contrary, specimens were mostly loaded in tension in June 2004, as shown by positive displacement values in the load-displacement plot. Interestingly, the displacement recorded in July 2004 differed significantly from that in June 2004, and the curve for July 2004 was located in the far right end of the plot, indicating that specimens had undergone a change from a mixture of compressive and tensile loading to a pure tensile loading. In later stages of exposure, specimens remained in tension.

Continuous monitoring of load and displacement allowed changes in stiffness to be examined, which was measured by the slopes of load-displacement plots (*Figure* 13). The values of stiffness are tabulated in *Table* 1, and squared correlation coefficients, r^2, for fitted curves are also included. It can be seen that r^2 value is relatively high for each month, signifying that each curve in the load-displacement plot can be fitted by a straight line with a highly reliable slope. From *Table* 1, it can be seen that the stiffness for freshly exposed specimens was 10 N/mm in December 2003. In the next seven months, the stiffness remained statistically unchanged, but in July 2004 the stiffness increased substantially to 400 N/mm. This significant increase in the stiffness shows that the sealants have undergone profound structural changes, and that summer exposure was more severe than winter exposure. Physical examination of the specimens revealed that the specimens had hardened considerably compared to unexposed specimens, indicating that extensive embrittlement had occurred. This observation indeed correlates well with the relatively high stiffness recorded. In later stages of exposure, the stiffness continued to increase to approximately 500 N/mm, and, eventually, the sealants failed cohesively.

DISCUSSION

Accurate prediction of in-service performance of sealants in less than real time has remained a modern unresolved scientific issue. At present, the generation of reliable performance data still requires long-term field exposure. Longer field exposure times are thought to reduce the risk of introducing a poorly performing product into the marketplace. However, the cost of developing new products is directly related to the product development time and the time-to-market. The more time in the pipeline, the more investment required and the smaller the eventual profit. Over the years, extensive efforts have been made to design a laboratory short-term test which provides an accurate indication of how well a building joint sealant will perform when exposed outdoors. However, these efforts have largely been unsuccessful, which arises mainly from the lack of success in relating field and laboratory results. From the results presented in this chapter, it is clearly seen that the current laboratory accelerated tests provide an excellent platform for evaluating the service life of building joint sealants. Furthermore, the present tests were successful in duplicating the same failure mode that occurred in field exposure. Specifically, extensive crosslinking was found to occur in specimens under both field and laboratory accelerated exposures, which was eventually followed by brittle fracture. The current test method therefore circumvents the problems associated with accelerating the environmental attack with the

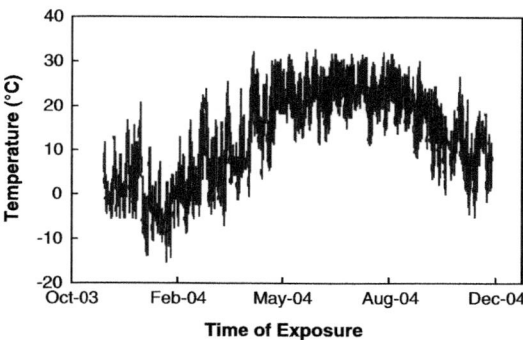

Figure 14—Changes in air temperature from December 2003 to November 2004.

use of unrealistically extreme doses of aging factors well above any likely seen in-service. Such tests often lead to unnatural failure mechanisms.

It is evident from the above discussion that the indoor laboratory accelerated test allows not only the individual effects of cyclic fatigue, temperature, and RH to be investigated independently, but also the synergistic effect of combining two or more factors. By using this test method, it has been shown that specimens were able to resist the individual influence of cyclic fatigue, high temperature, and RH, but degraded substantially when exposed to the combination of cyclic fatigue with a high temperature or RH, or the combination of these three factors. Fatigue, high temperature, and RH collectively provide strong synergism, thus accelerating the degradation mechanism and rapidly deteriorating sealant properties. Such observations correlated well with observations made under field exposure in that summer exposure was found to more aggressive in terms of environmental attack than winter exposure. This is because air temperature was generally higher in summer, as shown in *Figure* 14.

It is noteworthy that threshold type tests such as ASTM C719 have been widely adopted by the industry for selecting appropriate sealant formulations for specific applications. For example, ASTM C719 establishes the performance of sealants through the following protocol: a one-month period of static cure followed by a sequential stress regime including immersion in water (7 d), baking in an oven (7 d), exposure to UV, and, finally, mechanical cycling.[9] The samples are then visually evaluated for defects. Obviously, such a protocol assumes that no strong synergistic effect exists between the different aging factors. However, as shown by the present study, the effect of an individual factor acting alone may be different from the combined effect of two or more factors. The sealant material studied here will therefore pass ASTM C719, and, as such, will be mistakenly approved for installation on buildings, where it may fail prematurely. The existence of such synergistic effects raises serious concerns as to whether viewing the environmental effects of these factors independently is meaningful, highlighting the prime importance of accounting for such synergism in the development of scientifically meaningful accelerated durability tests.

CONCLUSIONS

A test methodology has been designed to duplicate the same failure modes occurring in in-service exposures. Such methodology employs a systematic approach in which both independent and synergistic effects of various aging factors on the durability of building joint sealants were evaluated in terms of changes in modulus, stiffness, and stress relaxation behaviors. Indoor accelerated exposures were carried out using an integrating sphere-based weathering chamber; while one-year field exposures were carried out in Gaithersburg, using a polyvinyl chloride (PVC) device, which relied on thermal response of PVC to outdoor air temperature to induce cyclic fatigue deformation on sealants. Indoor test results revealed that cyclic fatigue, high temperature, or moisture, on sealant mechanical properties acting alone did not degrade this sealant, in combination, however (e.g., cyclic fatigue deformation with temperature and/or moisture) was detrimental, resulting in extensive embrittlement and leading to premature failure. Sealants exposed to field conditions exhibited the same behavior, indicating that the accelerated test methodology provided an accurate indication of the durability of sealants exposed outdoors. The present study has clearly shown the importance of designing experiments that enable effects of various aging factors to be systematically evaluated, with test results correlating to field performance if accelerated conditions more accurately reflect the balance of field exposure conditions.

References

(1) Tan, K.T., White, C.C., Benatti, D.J., Stanley, D., and Hunston, D.L., in *Handbook of Sealant Technology*, Mittal, K.L. and Pizzi, A. (Eds.), in press.

(2) Wolf, A.T., in *Durability of Building Sealants*, Beech, J.C. and Wolf , A.T. (Eds.), p. 63-89, E & FN Spon, London, UK, 1996.

(3) Lacher, S., Williams, R.S., Halpin, C., and White, C.C., in *Service Life Prediction: Challenging the Status Quo*, Martin, J.W., Ryntz, R.A., and Dickie, R.A. (Eds.), Federation of Societies for Coatings Technology, Blue Bell, PA, p. 207, 2004.

(4) Lowe, G.B., in *Durability of Building Sealants*, Wolf, A.T. (Ed.), p. 225-234, RILEM Publications S.A.R.L., Cachan Cedex, France, 1999.

(5) Woolman, R. and Hutchinson, A., *Resealing of Buildings: A Guide to Good Practice*, Butterworth-Heinemann, Oxford, UK, 1994.

(6) Grunau, E., *Service Life of Sealants in Building Construction* (in German), Research Report, Federal Ministry for Regional Planning, Building and Urban Planning, Bonn, Germany, 1976.

(7) Chiba, R., Wakimoto, H., Kadono, F., Koji, H., Karimori, M., Hirano, E., Amaya, T., Sasatani, D., and Hosokawa, K., *Improvement System of Waterproofing by Sealants*, p. 175-199, Japan Sealant Industry Association, Tokyo, Japan, 1992.

(8) Expenditures for Residential Improvements and Repairs—1st Quarter, Current Construction Reports, U.S. Census Bureau, Department of Commerce, Washington, D.C., 2002.

(9) ASTM C719-93 (2005), "Standard Test Method for Adhesion and Cohesion of Elastomeric Joint Sealants Under Cyclic Movement (Hockman Cycle)," American Society for Testing and Materials, Philadelphia, 1993.

(10) White, C.C., Tan, K.T., Hunston, D.L., Hettenhouser, J., and Garver, J.D., in *Service Life Prediction for Sealants Phase II Report: Precision Exposure Chambers and Screening Experiments*, White, C.C. (Ed.), p. 10-37, NIST, Gaithersburg, MD, 2007.

(11) Chin, J.W., Byrd, E., Embree, N., Garver, J., Dickens, B., Finn, T., and Martin, J., *Rev. Sci. Instrum.*, 75 (11), 4951 (2004).

(12) Taylor, C.R., Greco, R., Kramer, O., and Ferry, J.D., *Trans. Soc. Rheol.*, 20, 141 (1976).

(13) Ferry, J.D., *Viscoelastic Properties of Polymers*, John Wiley & Sons, Inc., 3rd Ed., New York, 1980.

(14) Ketcham, S.A., Niemiec, J.M., and McKenna, G.B., *J. Eng. Mech.*, 669 (1996).

(15) White, C.C., Tan, K.T., O'Brien, E.P., Hunston, D.L., and Williams, R.S., in *Service Life Prediction for Sealants Phase II Report: Precision Exposure Chambers and Screening Experiments*, White, C.C. (Ed.), p. 38-72, NIST, Gaithersburg, MD, 2007.

Chapter 9

Evaluating Cyclic Fatigue of Sealants During Outdoor Testing

R. Sam Williams,[1] Steve Lacher,[1] Corey Halpin,[1] and Christopher White[2]

[1] U.S. Department of Agriculture, Forest Service, Forest Products Laboratory, Madison, WI
[2] National Institute of Standards and Technology, Gaithersburg, MD

A computer-controlled test apparatus (CCTA) and other instrumentation for subjecting sealant specimens to cyclic fatigue during outdoor exposure was developed. The CCTA enables us to use weather-induced conditions to cyclic fatigue specimens and to conduct controlled tests in-situ during the outdoor exposure. Thermally induced dimensional changes of an aluminum bar were fed to a computer that enhanced the movement, set limits on the movement, and supplied movement information to the test apparatus. As specimens moved, load/deformation and weather conditions (temperature, relative humidity, UV radiation, precipitation, and wind velocity) were measured every few minutes to give an extensive database containing these variables. In addition, controlled tests were done every four hours during the exposure. At these four-hour intervals, the computer was programmed to interrupt weather-induced cyclic fatigue and conduct a standard-strain test. The data enabled us to calculate the elastic modulus of each specimen at any time during the exposure and to construct a model that fit the observed temperature and relative humidity effects on modulus.

INTRODUCTION

Background

Sealants (caulking compounds) are used extensively in the outdoor envelope of residential construction to seal against water intrusion. They often fail shortly after installation because of improper joint design, incompatibility with the substrate, poor application practices, and excessive movement of wood and wood-based materials.

As part of ongoing work on materials used in residential construction, studies on sealants began in 2002 at the USDA Forest Service, Forest Products Laboratory, Madison, WI, in cooperation with the National Institute of Standards and Technology (NIST) in Gaithersburg, MD, and several sealant manufacturers. These studies resulted in several publications that described the construction of test apparatuses, instrumentation of indoor and outdoor test facilities, data collection systems, and data analysis methods.[1-8] This work involved detailed laboratory and field studies designed to develop service life prediction methods for sealants. Studies are continuing, and a critical part of the work is the measurement of load and deflection of specimens as they are fatigue-cycled during outdoor exposure. Measurements can be combined with measurements of the environmental "dose" (various weather factors that may cause a load/deflection or degradation) to form a dose-response relationship that expresses degradation of specimens as a function of exposure conditions rather than exposure time. The dose-response of the sealant obtained from outdoor testing accomplishes several advances over traditional time-response experiments. The critical factors causing degradation can be established, interactions among factors can be determined, and dose-response from outdoor experiments can be compared with dose-response from accelerated laboratory experiments to obtain a meaningful acceleration factor.

The overall objective of our studies was to develop a protocol for predicting the service life of sealant formulations and to identify the fundamental mechanism of early failure. This protocol included both accelerated laboratory and outdoor exposures. All outdoor weathering factors—temperature, ultraviolet (UV) radiation, relative humidity (RH), wind velocity, and rainfall—and their amount (dose) were monitored, and the sealant degradation (material response) was linked to the dose. In short, our objectives were to (1) gather real-time data of stress and strain of sealants during outdoor exposure, (2) gather real-time data on weather, including UV radiation, and (3) match sealant response to accumulated environmental stresses. Ultimately, our goal is to use the information from outdoor materials response and weather data to develop appropriate accelerated tests and the capability to determine service life in less than real time with statistical confidence and reasonable accuracy.

Weather and Weathering Factors

Weathering factors are solar radiation, abrasion, pollutants, changes in temperature and relative humidity, and wet/dry and freeze/thaw cycles. The weather at any location can be quite variable from day to day and year to year. For example, the record high in Madison, WI, on June 21, was 100°F (38°C) in 1988; the record low was 36°F (2°C) in 1992. The average high/low temperature on June 21 is 80°/57°F (27°/14°C). The average temperature is rather meaningless in terms of the material degradation if the actual temperature can vary over such a wide range. Many materials, particularly polymeric, may be at a different state (above or below the glass transition), which could greatly influence the rate and mechanism of degradation. Average

Figure 1—Severe weather and damage resulting from freezing rain (courtesy of John Walker, Boone REMC, Lebanon, IN).

weather is comprised of large variation and often includes severe weather conditions (*Figure* 1).

In reviewing the work in our laboratory and similar work reported by others who have developed accelerated test methods, two problems became obvious. First, weather is not consistent from day to day, season to season, year to year, and century to century.[9] Second, weather data have not been used in any meaningful way during outdoor exposure. That is, there has been no link between specific weather events and the response of materials to these events. Thus, a variety of weathering events that may have had a dramatic influence on degradation have only been averaged. There has been no way of knowing what the conditions were when the specimens degraded. The specimens may have degraded more or less at the same rate since the previous evaluation, or all of the degradation could have occurred during a single weather event, such as freezing rain (*Figure* 1). Periodic evaluations—the standard method used in accelerated tests—show only the change since the previous evaluation, not the conditions that caused the change. Results have often been misleading because critical factors causing the degradation were either not known or not measured.

Reliability-Based Service Life Prediction

Reliability-based service life prediction depends on accurate and precise measurement of specimen response as it degrades, measurement of degrading factors, and integration of the factors to obtain the dose. Field data can then be compared with accelerated laboratory tests using dose and response rather than some response versus time (*Figure* 2).

Traditional methods for linking outdoor weathering tests with accelerated laboratory tests usually involve a weathering device programmed with a wet/dry and UV radiation cycle to approximate the outdoor weather. UV radiation from a carbon arc[10] was used in early weathering devices; to a great extent, this source was later replaced by a xenon-arc source.[11,12] Other devices use UV fluorescence light sources.[13,14] Considerable effort has been made to develop a light source that duplicates natural sunlight. Researchers continue to search for the correct cycle that replicates outdoor conditions. Researchers want accelerated tests that predict the degradation that occurs in actual use, but it is difficult to ensure that the degradation mechanism is the same in the test and in actual use. If the mechanism of degradation is not the same, there is little chance that the accelerated test will give reliable results. The key to ensuring that the degradation mechanism is the same in various exposure conditions is to monitor the chemical changes. Other than recent work by Martin and others at the National Institute of Standards and Technology (NIST), few

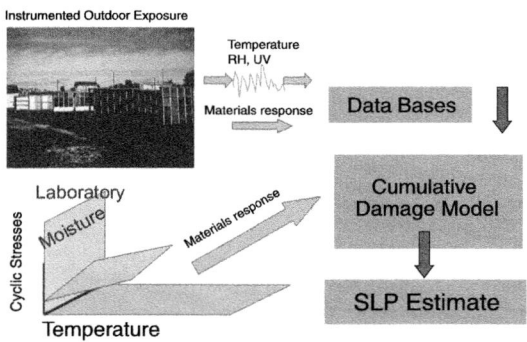

Figure 2—Reliability-based service life prediction (SLP) methodology.

attempts have been made to measure chemical changes occurring in materials exposed outdoors and link them to chemical changes in materials exposed in accelerated laboratory tests.[15-17]

Previous Sealant Tests

Sealants (caulking compounds) have become widely used in commercial and residential structures to help seal their exterior envelope against water intrusion. Sealants are formulated to accommodate cyclic stress and strain, but they may undergo cyclic fatigue in combination with other degrading factors.

General guidelines for accelerated testing of sealants are described by ASTM C 1442, Standard Practice for Conducting Tests on Sealants using Artificial Weathering Apparatus.[18] Other ASTM methods for accelerated weathering of sealants are C 718–93 (UV–Cold Box Exposure of One-Part, Elastomeric, Solvent-Release Type Sealants), C 732–95 (Aging Effects of Artificial Weathering on Latex Sealants), C 734–93 (Low-Temperature Flexibility of Latex Sealants After Artificial Weathering), C 793–97 (Effects of Accelerated Weathering on Elastomeric Joint Sealants), C 1257 (Accelerated Weathering of Solvent-Release-Type Sealants), and D 2249 (Predicting Effect of Weathering on Face Glazing and Bedding Compounds on Metal Sash).[19-24] These tests use a weathering device with UV radiation and intermittent water spray to accelerate degradation. The tests are not meant to predict the exact service life nor the mechanism of failures, but are more qualitative evaluations of sealant performance.

The Hockman Cycle, a standard that combines cyclic stress (fatigue), water immersion, and temperature change, was initially developed by A. Hockman at NIST.[25] Several devices have been used to induce cyclic movement (fatigue) of sealants during exposure. These devices generally use thermally induced dimensional change to cause fatigue in a sealant. For example, Onuoha[26] used unplasticized polyvinyl chloride (PVC) pipe to produce fatigue in one-part polyurethane and polyurethane-hybrid sealants. Racks have also been built using dissimilar materials such as wood and aluminum,[27] concrete and aluminum,[28] and steel and aluminum[29,30] to develop fatigue stresses. Manually operated devices have been used to create cycling effects.[31-33] Lacasse and Margeson[34] cycled sealants during cure. Bolte et al.[35] reported on cyclic movement of sealants during cure in natural and artificial weathering. Liu and Wang[36] used cyclic freeze-thaw cycles to test sealants used for horizontal joints in concrete. Al-Qadi et al.[37,38] also evaluated sealants used for concrete slabs using cyclic shear and static horizontal deflection. Using the RILEM TC139-DBS test method, Miyauchi et al.[39] conducted durability

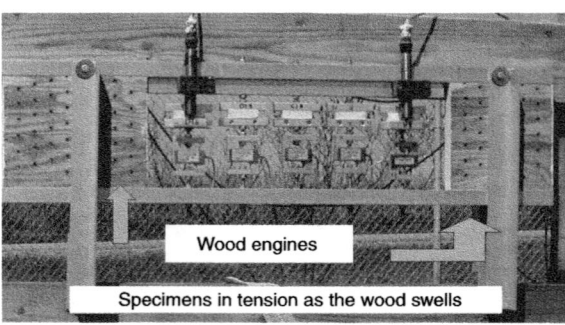

Figure 3—Sealant specimens attached to wood-driven TS cyclic test apparatus: the apparatus places the specimens in tension (T) as the wood swells (S).

tests that included cyclic fatigue in combination with exposure to UV radiation in a xenon-arc weathering device. Sharman et al.[40] evaluated sealants in outdoor exposure using a cyclic movement test apparatus. Koike et al.[41] cycled sealants in laboratory tests at temperatures of 0–80°C and calculated modulus and loss tangent. Marechal and Kalifa[42] used cyclic movement during artificial weathering to

Figure 4—Sealant specimens attached to wood-driven CS cyclic test apparatus: the apparatus places the specimens in compression (C) as the wood swells (S).

evaluate sealant performance under three different simulated climate conditions. Beech[43] reviewed test methods for sealants. Wolf [44,45] reviewed the correlation of long-term artificial aging and outdoor exposure of sealants. Wolf and Seneffe[46] reported on work within ISO TC 59/SC8 and RILEM TC139-DBS committees since 1990 that dealt with developing meaningful tests for sealants. In previous reports describing sealant tests, environmental factors were not recorded in an ongoing basis. Therefore, it was not possible to correlate the dose of degrading factors with quantifiable changes in chemical or physical properties. Evaluations were usually done on a predetermined schedule (weekly, monthly, or biannually) and involved only a visual examination. This examination might indicate an adhesive or cohesive failure, but it was impossible to determine when the specimens failed or the conditions that caused the failure.

Wood-Based Cyclic Fatigue Apparatus

In our previous cyclic fatigue studies,[5–7] the weather-induced movement of sealant specimens was obtained from the dimensional change in flat-sawn (tangential) blocks of red oak (*Quercus rubra*). Precipitation and/or changes in relative humidity caused the wood to change moisture content, thus giving the required dimensional change. Two designs (TS and CS) were developed for conducting these exposure tests. In TS (*Figure* 3), the test specimens are placed in tension (T) by an increase in moisture content of the wood (as the wood swells [S]). This subjects the

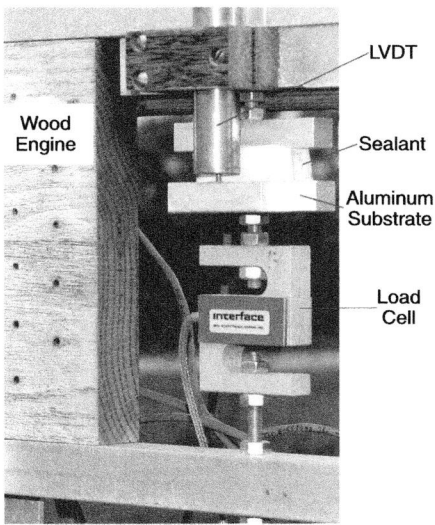

Figure 5—Close-up of specimens on the wood driven test apparatus showing the specimen, load cell, and the linear variable differential transformer (LVDT).

134

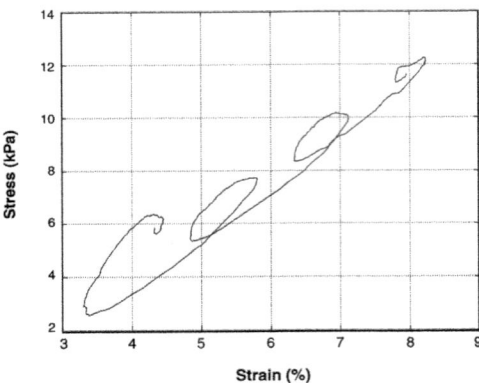

Figure 6—Stress-strain plot of a sealant specimen showing diurnal cycles and superimposed on an overall decrease in tension over four days. The beginning of the four-day period is at the top.

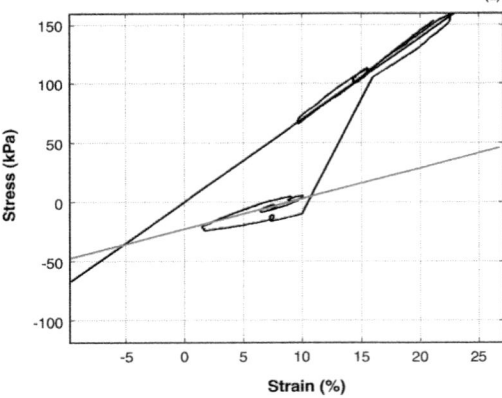

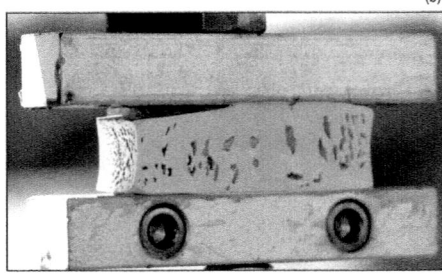

Figure 7—Failure of a sealant specimen: (a) stress-strain plot showing the failure (failure occurred on May 16, 2003, at about 1:15 pm), (b) photograph showing an adhesive failure. Note the failure shows a slope because the two data points were taken 15 min apart.

specimen to stresses similar to that obtained via thermal expansion or contraction of metals, glass, ceramics, or polymers. This apparatus simulates the thermally driven designs previously used, but it is driven primarily by changes in relative humidity and water, not temperature. In CS (*Figure 4*), the test specimens are placed in compression (C) as the wood swells, which subjects the specimens to stresses similar to that obtained when sealants are used on wood. The wood moved against an aluminum frame to produce a cyclic movement. The aluminum was attached to the bar at the top of the apparatus; the load cell and specimen are uni-axial (*Figure 5*). All fasteners in the apparatus were stainless steel. Each specimen could be attached to a linear variable differential transformer (LVDT). However, we found that deflection could be determined using only four LVDTs.

When we began measuring dimensional changes and loads on specimens, we thought we could determine changes in "apparent modulus," adhesive and cohesive failures, and the effect of temperature, strain, and strain rate by analyzing the vast amount of data we obtained. *Figure 6* shows the stress-strain plot over four days as a specimen was cycled following a period of rain. This particular test apparatus placed the specimen in decreasing tension as the wood dried (*Figure 3*). The diurnal cycles are superimposed on the overall decrease in tensile strain. The best-fit linear slope of stress-strain linear portions or this type of plot

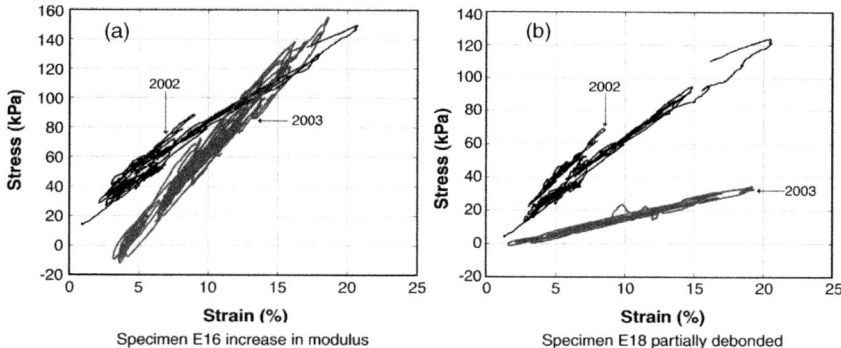

Figure 8—Stress-strain plots for two sealant specimens. Data were collected for three months a year apart. (a) sealant shows an increase in elastic modulus, (b) sealant failed in adhesion.

gives an "apparent elastic modulus" (henceforth referred to as elastic modulus). It is an apparent modulus because we did not take into account the change in cross-sectional area or the effect of cohesive or adhesive failure of specimens as they degraded. Cohesive and adhesive failures were easy to observe, and this was reported earlier (*Figures* 7a and b).[5-7] *Figure* 7a shows several cycles with an abrupt change in the slope of the stress-strain plot. This abrupt change occurred at 1:15 pm on May 16, 2003, following a day of rain that placed the specimen close to 25% strain. *Figure* 7b shows the failed specimen. General trends could be determined by comparing plots over time (*Figure* 8). The data could also be plotted to show the modulus versus net cumulative strain energy (*Figure* 9) or cumulative UV radiation (*Figure* 10). Determining modulus in real time proved to be much more difficult because it required selecting data on an equivalent basis of temperature, RH, and strain and calculating modulus from the relatively small number of data points that were taken under similar environmental conditions. Analysis of more than three years of data from previous work[5-7] measuring load-deflection response showed no two times when the temperature, strain, and strain rates on the specimens were the same. During three years, the weather never repeated itself, so the specimens were never at the same condition more than one time in three years. The data were too variable to detect

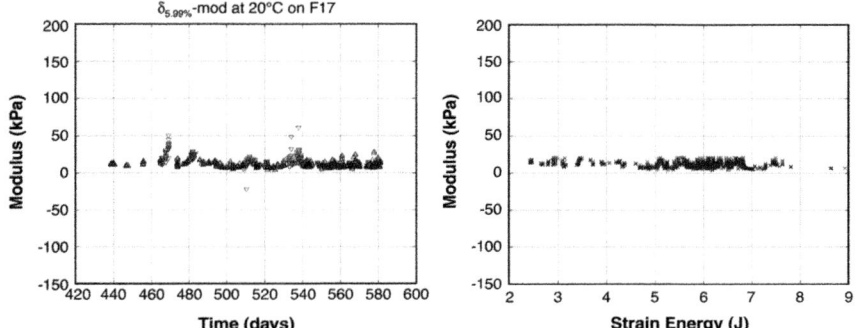

Figure 9—Elastic modulus versus cumulative strain energy and versus time.

136

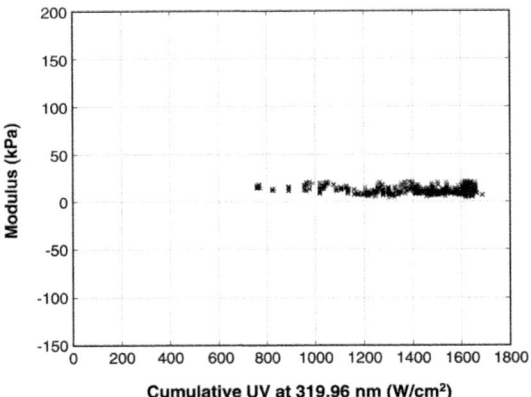

Cumulative UV at 319.96 nm (W/cm²)

Figure 10—Elastic modulus versus cumulative UV radiation dose at 320 nm.

modulus changes with adequate certainty. We were aware that data scatter could result from the sealant's sensitivity to strain history. This shortcoming of the original uncontrolled cyclic fatigue experiments led to the development of a computer-controlled test apparatus (CCTA).[47] The CCTA made it possible to conduct controlled measurements at the beginning of the exposure to determine specimen quality and properties at various temperatures, strains, and strain rates as the specimens aged. Greater control of the experiment enabled us to discover material properties that were obscured in previous work. The CCTA is capable of temporarily suspending routine exposure testing and performing a precisely controlled stress-strain test (henceforth referred to as "standard strain test") to measure elastic modulus.

EXPERIMENTAL

Weather Station, Test Apparatus, and Data Acquisition

The Forest Products Laboratory field site, which is 5 km west of Madison, WI, has been fully instrumented to collect weather data during outdoor exposure of materials. The weather station, test apparatus, and data acquisition system have been described in detail in previous publications.[5-7,47] We measured temperature wind velocity, relative humidity, rainfall, and UV irradiance at 18 different wavelengths in the range of 290 to 324 nm.

Computer-Controlled Cyclic Test Apparatus

Figure 11—Computer-controlled test apparatus (CCTA) with 18 sealant specimens and load cells mounted. Inset: Close-up of aluminum bar and linear variable differential transformer that provides input to the CCTA.

The intensity at each wavelength can be integrated to get dose. The dose for other weather factors can also be calculated from the data (total rainfall, degree days, time at dew point). This work involves detailed laboratory and field studies designed to develop SLP methods for sealants. The data from our studies show a clear link between the materials response during weathering and the weather conditions causing this response. The cyclic fatigue is caused by the changes in weather, and the load and deflection of each specimen is recorded along with all weather data. Data is collected every five minutes.

Sealant specimens were mounted on a CCTA capable of holding up to 20 specimens (*Figure* 11). Cyclic motion was obtained from thermally induced changes in dimension of an aluminum bar painted black (*Figure* 12). As the bar changed length, the measurement was fed to a computer. The computer multiplied the measured dimensional changes to

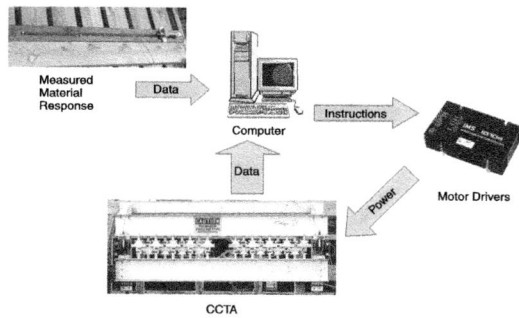

Figure 12—Schematic showing (clockwise) the aluminum bar to obtain temperature-driven dimensional change, computer to modify the dimensional change to meet test objectives, motor driver to control the stepper-motors in test apparatus, the test apparatus, and the feed-back loop to ensure that the movement is correct.

meet the target deflections on the specimens. Limits were placed on the maximum movement to ensure the strains did not exceed ±25%. The modified signal drove linear actuators (stepper motors with jack screws) on the apparatus to move one of the two parallel beams, thus applying a deflection to the specimens (*Figure* 13). A feedback loop to the computer ensured that the deflection on the specimen was correct. That is, if the linear actuator failed to apply the proper deflection, the feedback loop corrected this discrepancy. The load on each specimen was measured and the deflection was either measured directly or calculated from the measured movement of the apparatus (*Figure* 14).

All specimens were subjected to a standard strain test every 4 h during the exposure to determine changes in modulus caused by temperature, RH, strain history, and degradation. The standard strain test was used to determine the elastic modulus of specimens by imposing a specific strain profile and measuring the resulting stress (*Figure* 15). Each standard strain test consisted of first bringing all specimens to 0 strain for 60 s, then imposing the strain profile consisting of a constant rate ramp up to 10% tensile strain, a 60-s hold, a constant rate ramp down to 10% compressive strain, a 60-s hold, and a constant rate ramp back to zero strain. This cycle was repeated to eliminate the Mullins effect (a property of elastomers causing a stiffer modulus measured on the first cycle than on subsequent cycles).

Figure 13—Nine sealant specimens and load cells mounted on the computer-controlled test apparatus.

138

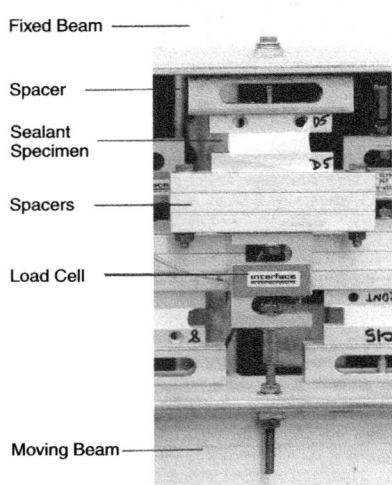

Fixed Beam ————

Spacer ————

Sealant
Specimen ————

Spacers ————

Load Cell ————

Moving Beam————

Figure 14—Close-up of a specimen mounted in the test apparatus showing the stationary beam, spacers, sealant specimen, load cell, and the moving beam.

Sealants

Two sealants, one low modulus and one high modulus, were selected from the seven different experimental sealants formulated for these studies. The formulations were designed to have a range of properties because the intent of this research is to establish a protocol for future work. The sealants were selected to give different load deflection characteristics and have short service lives (generally less than three years).[5-7] For the study reported here, six specimens from the A-formulation and seven specimens from the B-formulation were evaluated periodically during two years of outdoor exposure using our standard strain test as described above. The specimens for the two formulations are labelled as A1, A2, A3, A4, A5, and A6, and B1, B2, B3, B4, B5, B6, and B7. Our industry partners prepared the specimens. We did not control their manufacturing nor did we know the specimen formulations. Therefore, we are treating them as distinct experimental units, rather than experimental replicates.

RESULTS AND DISCUSSION

Cyclic Fatigue

To determine chemical changes (additional crosslinking or bond cleavage) in sealants during exposure, we attempted to measure changes of physical properties as they age. We

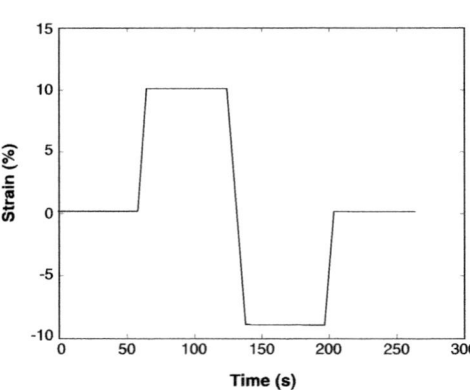

Figure 15—An example of the strain versus time profile imposed on specimens during the standard strain test.

chose this approach to avoid the difficulty of directly measuring chemical changes in situ during exposure. We measured elastic modulus. Specimen degradation, whether crosslinking or bond cleavage, could be expressed as a change in modulus of the specimens. As specimens continue to crosslink, we expected the modulus to increase; conversely, as bonds cleaved the specimen modulus should decrease. If both crosslinking and bond cleavage occur, it would be difficult to determine these changes by measuring the changes in modulus. In addition, other factors such as stress relaxation, strain hardening, and Mullins effect could affect our measurements. As we did not know the chemistry, the measurement of modulus was the first attempt to quantify possible chemical changes in situ during the test. Scatter in the measurement data could also obscure changes in modulus; therefore, all other sources of experimental variability must be identified and accounted for. Possible sources of variability include, but are not limited to, temperature, water content, and strain history of the specimens as well as imposed strain rate, measurement errors, and electronic noise.

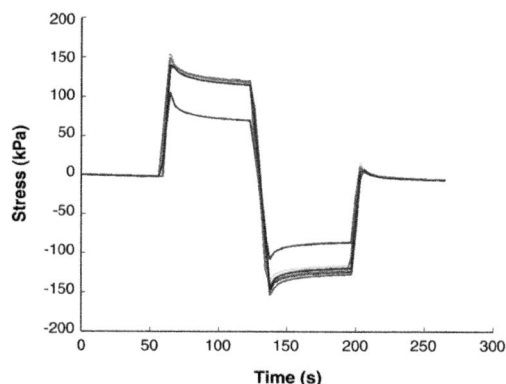

Figure 16—Stress response versus time of B-formulation specimens to the standard test strain profile. Test was performed at the beginning of outdoor exposure.

Temperature, water content, relative humidity, and strain history of the specimens were controlled by environmental conditions and varied with the weather. Temperature, relative humidity, and strain history were measured and recorded. Currently, there are no means of measuring water content, though it is believed to depend on relative humidity and strain history. We minimized measurement and signal errors by selecting force transducers sensitive to ±1.25 N (0.3 lb) and LVDTs sensitive to ±0.2 mm (0.0007 in.), shielding all signal cables, and calibrating all instrumentation. We decreased experimental error by holding controllable factors such as short-term strain history and strain rate constant. The standard strain test always used the same strain rate. We measured initial modulus on all specimens prior to exposure to determine specimen-to-specimen variation.

The strain versus time profile of the standard strain test is shown in *Figure* 15 and stress response versus time for seven B-formulation specimens measured at the beginning of the exposure is shown in *Figure* 16. The plots for most of the specimens overlap considerably, but one is much different. This was our first indication of considerable variability among the industry-prepared specimens within each formulation group. Data from these two time series (stress versus time and strain versus time) were combined to create a stress versus strain plot for each specimen for every test (e.g., spec-

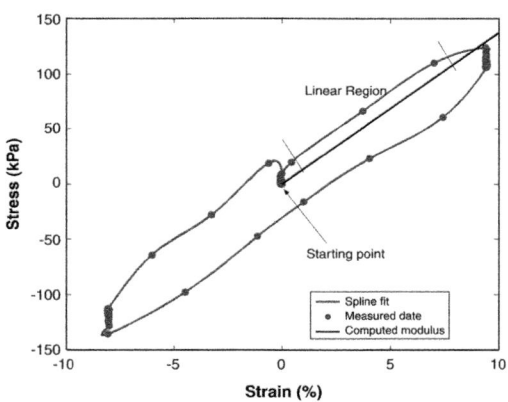

Figure 17—Stress versus strain for sealant specimen B7 for a standard test showing linear portion of the curve used in calculating the apparent elastic modulus. The line is a computed best-fit estimate of the slope.

140

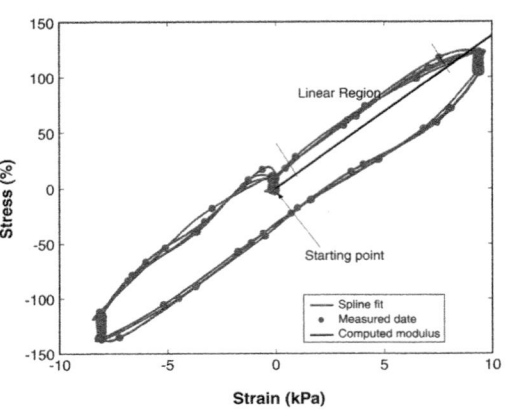

Figure 18—Stress versus strain for sealant specimen B7 showing four consecutive standard tests. The first curve shows a slightly higher maximum stress, whereas the next three curves are similar to each other. The line is a computed best-fit estimate of the slope.

Figure 19—Apparent elastic modulus calculated from (a) second standard test versus first standard test, (b) third standard test versus second standard test. Data points indicating perfect experimental replication lie along a line having slope equal 1. Plot b shows a tighter grouping than plot a (especially in the higher modulus B specimens), indicating better replication.

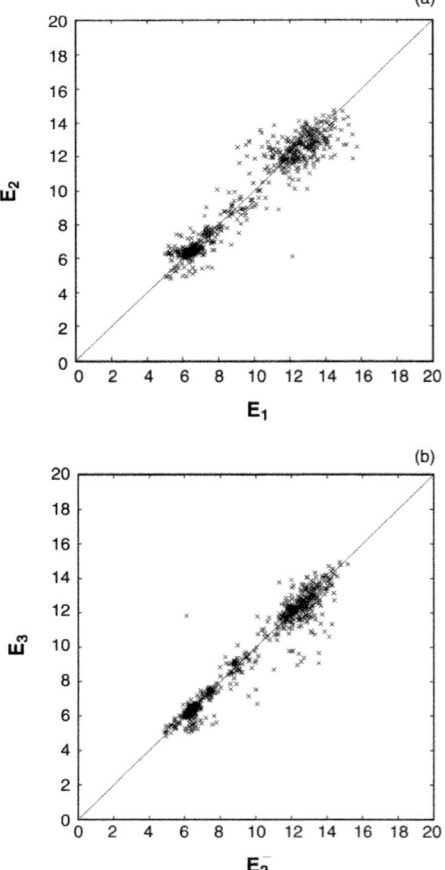

imen B7, *Figure* 17). This plot shows a fairly constant slope during the initial tension phase of the standard test. A linear regression of this portion of the stress versus strain plot was used to compute the apparent elastic modulus. Similar analyses were completed for all specimens, and elastic moduli calculated from these standard tests were less variable than those calculated from linear portions of weather-induced stress-strain data obtained in previous studies.[5-7] Three problems were encountered in analyzing the data from these uncontrolled tests. We could not control the initial strain, the strain rate, or account for the Mullins Effect. We had no control over the previous stress-stress-cycles, therefore dissimilar strain histories gave variable and often misleading measurements.

Mullins Effect is well known to occur with viscoelastic materials such as sealants and was previously shown to occur with these particular sealants.[48,49] We characterized the Mullins Effect on the standard strain test by conducting four successive standard tests (*Figure* 18). The moduli extracted from the linear portions of the plots were different for successive tests. Plots of the modulus of the second test versus the modulus of the first test (E2 versus E1) and third test versus second test (E3 versus E2) were used to show these differences (*Figures* 19a and 19b). Points that fall on the diagonal line indicate exact replication. If there were no measurement errors in these experiments, points above the line would indicate stiffening and those below the line would indicate softening of the specimens in the subsequent test.

Comparison of the plots in *Figures* 19a and 19b show a clear shift of data to below the line, indicating softening of specimens between tests, which is what would be expected from the Mullins Effect.[48,49] The points above the line are clustered close to the diagonal line and we attribute the points above the line to experimental error. There may be other factors that could be responsible for the points above the line, such as compressive set in the specimens prior to the test. Additional studies are under way to investigate this and other possible causes for this apparent variation. Histograms of this comparison show slightly decreased scatter in the data (*Figure* 20). The observed change between the second through the fourth test was small; therefore, we are confident that two cycles were sufficient to minimize the Mullins effect in our data.

As we developed test methods that took into account the Mullins Effect, considerable differences became appar-

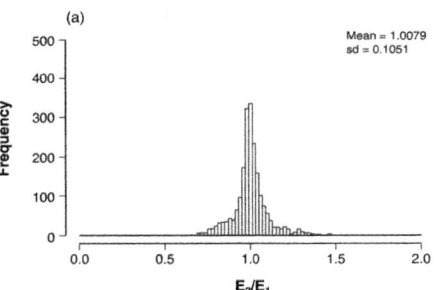

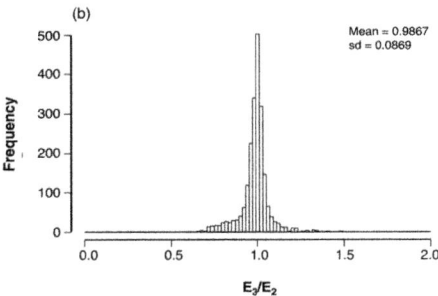

Figure 20—Histograms of (a) E2/E1 and (b) E3/E2. A value of 1.0 indicates perfect experimental replication. As in Figure 10, the data indicate that tests 2 and 3 are better replicates than tests 1 and 2.

142

ent among the specimens within the same formulations. The standard strain test performed shortly after the specimens were placed in test showed distinct differences among the specimens. As mentioned previously, the plot of the B specimens (*Figure* 16) clearly shows that one of the specimens is different. We have no explanation for this, but it indicates that we may have considerable specimen-to-specimen variability. This high variability in the specimens may make it difficult to average the data for each formulation to calculate a material property such as modulus. To determine if specimens were statistically distinct, we performed an analysis of variance (ANOVA), and plotted confidence intervals for all pair-wise differences between specimens. The A specimen contrasts are seen in *Figure* 21a, and the B specimen contrasts are seen in *Figure* 21b. The horizontal lines show a 95% confidence interval for the difference between each pair of specimens. If that line crosses the dotted vertical line toward the left of the plot (difference = 0), then those specimens are not significantly different. The 95% confidence limits were computed taking into account the number of comparisons performed, so the 95% refers to a family-wise error rate, not a test-wise error rate. Among the A-formulation specimens, A1 and A3 are not distinct. All other A specimens are distinct. Among B-formulation specimens, B2 and B5 are not distinct. All other B specimens are distinct. Box and whisker plots show the variability in the data for the A and B specimens (*Figures* 22a and 22b). Outliers are marked with a circle (determination of outliers was done automatically by the R statistical software). The whiskers extend the full range of the non-outlier data. The box extends from the 25th to the 75th percentile of the non-outlier data. The thick line indicates the mean.

The ANOVA suggests that all our specimens are statistically distinct. We did not have control over the preparation of the specimens; our industry partners provided them. For this reason, we did not feel it was appropriate to pool data from multiple specimens in our analysis. If they are not experimental replicates, and we cannot say with certainty that they are, then conclusions from pooled

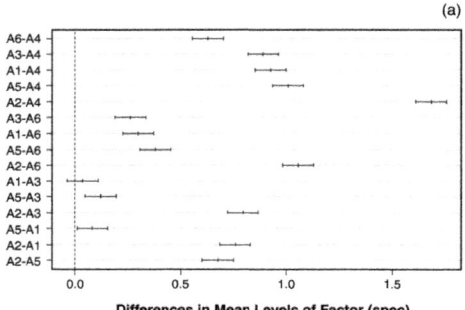

(a)

Differences in Mean Levels of Factor (spec)

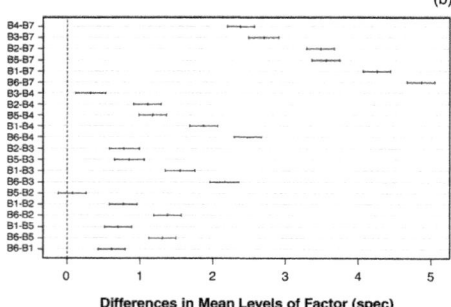

(b)

Differences in Mean Levels of Factor (spec)

Figure 21—Results of multiple comparison test used to determine whether individual specimens are statistically different from each other. Horizontal bars indicate 95% confidence intervals of the difference between pairs of specimens. A difference of zero (dashed line) indicates 95% confidence that two specimens have the same apparent elastic modulus. (a) depicts A-formulation specimens, (b) depicts B-formulation specimens.

analysis could be misleading. From this point on in the discussion, all analyses are for individual specimen.

Temperature and relative humidity (RH) are important environmental factors for these specimens. Therefore, we sorted the data on the basis of these factors and created "bins" containing data that fall within a specified range of conditions (measurements taken at similar temperature and RH). *Table* 1 shows the number of data points within each bin for specimen B7. Taking the data from the fullest bin and plotting it versus time (*Figure* 23), there is little we can conclude. Two periods, roughly six months apart, have the same conditions. Within each period, the data has considerable scatter. To draw any meaningful conclusions from this data, we will use statistical techniques.

Statistical Analysis

Statistical analysis was used to show the variability in the specimens and to determine the relationship between weathering stressors and material degradation. Specifically, we attempted to detect changes in elastic modulus as the specimens were exposed outdoors. The difficulty lies in separating changes in modulus caused by material degradation from variation in modulus caused by environmental conditions. The weather did not repeat itself, but the CCTA provided the capability to repeat standard tests—thus eliminating initial strain, strain rate, and Mullins Effect as variables. The binning of data according to specific temperature and relative humidity ranges gave sufficient data to construct a model to explain these effects on modulus over the range of weather conditions observed during the exposure. The model also includes factors for energy inputs into each specimen from UV radiation and cumulative strain. The simplest model to characterize these sealants uses a linear combination of these independent environmental factors. We recognize that there are interactions among the factors, particularly temperature and RH; however, this model gives a starting point. Future work will need to

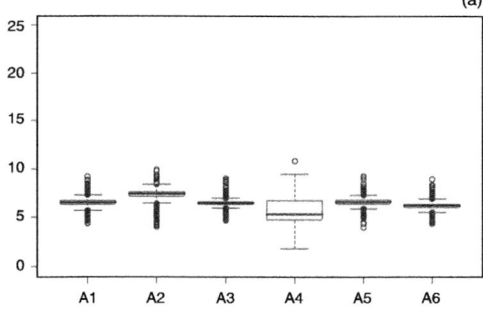

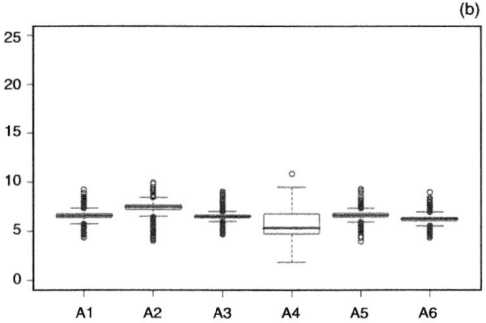

Figure 22—Box and whisker plots of apparent elastic modulus for all data. (a) A–formulation specimens and (b) B–formulation specimens. The heavy horizontal lines indicate average modulus, the rectangular boxes enclose values falling between the 25th and 75th percentiles and the whiskers indicate the range of the data excluding outliers. Outliers are depicted as circles.

Table 1—Number of Standard Tests Binned by Environmental Factors (Specimen B7). Entries with 10 or More Data Points Are Displayed in Bold

Temperature (°C)	RH (%) 35	40	45	50	55	60	65	70	75	80	85	90	95	100
-10														
-7.5							1		5	5	1	5	1	5
-5							2	2	5	7	4	9	5	**11**
-2.5						2	3	4	3	7	9	**10**	**17**	**10**
0					3	7	6	1	**13**	**18**	**13**	**15**	**21**	**34**
2.5				2	3	6	3	5	**14**	6	9	8	9	**21**
5		1	1	4	7	4	5	**12**	6	6	**14**	9	**19**	**51**
7.5				1	3	3	1	1	2	4	4	3	6	**19**
10	3	4	3	**11**	9	6	5	7	4	3	**17**	**14**	4	**17**
12.5		2		2			3	5	4	3	2	5	2	**16**
15		1	4	3	4	7	6	3	4	5	3	4	7	**39**
17.5	1		3	4	1	5	3	2	4	8	6	9	9	**29**
20		1	3	3	3	5	1	1	7	7	5	**13**	**17**	**26**
22.5	1	3	1		2	2	5	5	9	**14**	**12**	4	6	**38**
25	1		3	4	6	**10**	9	**10**	9	**11**	5	**11**	8	**15**
27.5			4	2	3	8	**10**	3	7	2	7	4	2	2
30			3		2	4	6	4	7	4	2	2		
32.5			1			2	3	3	2					
35					1	1	1							

be directed toward developing an understanding of other environmental factors causing degradation and the non-linear response of these materials to those factors.

Other investigators are working on developing constitutive models of elastomers.[50-52] These models have been successful for particular materials tested under laboratory conditions using viscoelastic materials having known chemistries with a limited number of additives. It is difficult to extend these models to describe the behavior of the sealants used in our work because the chemistry of the sealants was not known and the deflection of the specimens varied because of the weather.

Our initial analysis was a multiple linear regression using ordinary least squares to fit the data[53,54]:

$$sqrt(mod) = b0 + b1*Cold + temp\ K*(b2+b3*Cold) +$$
$$RH*(b4+b5*Cold) + b6*CSE + b7*CUVE +$$
$$b8*Time + N(0,MSE)$$

where
- sqrt(mod) is the square root of the modulus
- b0 through b8 represent the regression coefficients
- temp K denotes the ambient temperature in Kelvins
- Cold is a {0, 1} variable, which will be 1 when the ambient air temperature is less than 273 K
- RH denotes the relative humidity

- CSE denotes cumulative strain energy in joules applied to the specimen; this is a measure of the net physical work done on the specimen through cyclic loading/unloading
- CUVE denotes the cumulative UV energy in joules applied to the specimen
- Time denotes the exposure time in seconds
- N (0, MSE) denotes normally distributed error with a constant variance over the range of the data

The terms for above and below 0°C are included because we have observed a change in modulus at 0°C for many of the sealants being evaluated in this work. This is discussed below.

This model stipulates that the shape of the response surface for these materials is a pair of planes, one above freezing and one below freezing. Stepwise selection was used to reduce these models by removing statistically non-significant terms. Results of this process are summarized for each specimen (*Table* 2). The R^2 values indicate the percentage of the variation in the data that the model explains. The *p*-values indicate how well the model explains that variation (0.05 or lower indicates statistically significant explanatory power). The b0 and b1 terms are intercepts at 0 K. They are required for building the model, but have no physical meaning independent of the model. The b2 term indicates the change in modulus for a 1 K change in temperature when the temperature is above freezing. The sum of b2 and b3 indicate the change in modulus for a 1 K change in temperature when the ambient temperature is below freezing. The b3 term can then be interpreted as the additional effect of temperature below freezing. The b4 term indicates the change in modulus for a 1% change in RH when the ambient temperature is above freezing. The sum of b4 and b5 indicate the change in modulus for a 1% RH change when the ambient temperature is below freezing. The b5 term is the additional effect of RH below freezing. The b6 term indicates the change in modulus caused by cyclic loading. The b7 term indicates the change in modulus caused by UV degradation. The b8 term indicates time-dependent degradation of the specimen.

Calculation of the coefficients (*Table* 2) was straightforward using all the data in all the bins (*Table* 1; approximately 1250 tests) for a particular specimen. Using these coef-

Table 2—Estimated Model Parameters Produced by Multiple Regression. Coefficients That Were Not Significantly Different from Zero Have Been Omitted

	b0	b1 (cold)	b2 (temp)	b3 (cold*temp)	b4 (RH)	b5 (cold*RH)	b6 (CSE)	b7 (CUVE)	b8 (time)	R^2	p
A1	3.35E+00		2.86E-03				-4.49E-03	-2.32E-10	5.94E-09	0.11	<0.0001
A2	3.88E+00	-1.39E-02	4.12E-03				-4.29E-03		1.22E-09	0.36	<0.0001
A3	3.35E+00	8.16E-01	2.75E-03	-3.04E-03			-6.06E-03	-3.13E-10	1.12E-08	0.31	<0.0001
A4	1.42E+00	-2.81E-02			-9.62E-04		-4.56E-03	4.11E-10	1.57E-08	0.67	<0.0001
A5	3.79E+00	-1.48E-02	1.89E-03		-1.70E-04		-6.88E-03	-2.68E-10	1.16E-08	0.28	<0.0001
A6	3.44E+00	4.84E-01	2.31E-03	-1.80E-03			4.01E-03	-1.96E-10	3.75E-09	0.38	<0.0001
B1	5.83E+00	4.05E+00	-1.10E-02	-1.48E-02	-9.52E-04			-7.33E-10	1.15E-08	0.73	<0.0001
B2	7.28E+00	4.06E+00	-1.20E-02	-1.53E-02	-1.35E-03	1.30E-03		-7.02E-11		0.72	<0.0001
B3	6.82E+00	4.40E+00	-8.50E-03	-1.61E-02	-9.36E-04		-7.50E-03	-1.34E-09	4.56E-08	0.75	<0.0001
B4	7.51E+00	3.74E+00	-9.58E-03	-1.37E-02	-8.91E-04		-4.65E-03	-1.37E-09	2.79E-08	0.77	<0.0001
B5	6.12E+00	3.90E+00	-1.14E-02	-1.42E-02	-9.74E-04			-6.46E-10	8.57E-09	0.82	<0.0001
B6	6.31E+00	5.02E+00	-1.23E-02	-1.83E-02	-8.72E-04			-3.96E-10	9.04E-09	0.73	<0.0001
B7	5.41E+00	3.92E+00	-1.19E-02	-1.42E-02	-1.12E-03			-4.89E-10	1.10E-08	0.67	<0.0001

Table 3—Example Calculation of Initial Modulus Using the Model from *Table* 4 on Specimen B7

Coefficients		Terms		Results	
b0	5.41E+00				5.41
b1	3.92E+00	cold	0	b1* cold	0.00
b2	-1.19E-02	temp	278	b2* temp.	-3.31
b3	-1.42E-02	cold-temp	0	b3* (cold*temp)	0.00
b4	-1.12E-03	RH	100	b4*RH	-0.11
b5		cold*RH	0	b5* (cold*RH)	0.00
b6		CSE	0	b6*CSE	0.00
b7	-4.89E-10	CUVE	0	b7*CUVE	0.00
b8	1.10E-08	Time	104032200	b8*Time	1.15
				sqrt(mod)	3.14

Predicted Modulus 9.84 kPa
Measured Modulus was 10.50 kPa

Note: The value for "time" (104032200) for the b8 corresponds to time in seconds from the beginning of the experiment. The predicted modulus from this calculation is 9.84 kPa and the measured modulus for the first three months of exposure is 10.50 kPa.

ficients, a modulus can be calculated for any specimen. Simply measure the required factors, multiply them by the appropriate b-coefficient, sum the result, then take the square to predict the modulus. An example of this calculation for specimen B7 is presented in *Table* 3. Some of these coefficients can be independently interpreted as rates of change. For example, b8 gives the rate of time-dependent degradation in kPa/s. The two intercept terms (b0, b1) lack physical meaning outside the model. The temperature and RH related terms can be combined to compute above/below-freezing rates of change. For example, b2 gives the rate of change with temperature above freezing in kPa/K and (b2 + b3) gives the same for the below freezing case.

To verify that these models were functioning correctly, we checked the model predictions against the mean modulus computed within a number of bins. The results are seen in *Table* 4. There is a good correspondence between predicted modulus and measured values. Similar correspondence results were obtained for other specimens at various temperatures and RHs. All of the models had significant p-values, indicating that the correlation between our data and the model did not arise simply by chance (*Table* 2). The A specimens are not well characterized (R^2 values range between

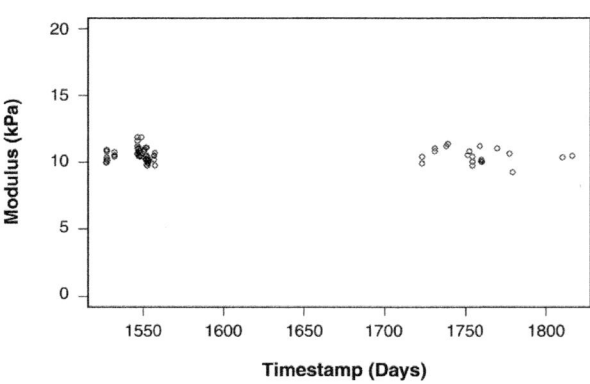

Figure 23—Modulus versus time for specimen B7 selected for 5 ± 1.25°C and 97.5%–100% RH. The 0 on the time axis is the date we brought our weather station on line about five years ago.

Table 4—Comparison of Model Output and Date Mean in Selected Bins for Specimen B7 and Specimen A6

Specimen	Temp (°C)	RH (%)	N	Predicted Modulus (kPa)	Predicted Modulus (kPa)	% Error
A6	−2.5	95	17	6.3	6.24	0.96
A6	0	95	21	6.25	6.22	0.55
A6	0	100	35	6.22	6.26	−0.54
A6	5	95	19	6.32	6.35	−0.54
A6	10	55	10	6.38	6.36	0.3
B7	−2.5	95	17	11.29	11.15	1.26
B7	0	95	21	11.07	11.02	0.48
B7	0	100	35	11.09	11.49	−3.48
B7	5	95	19	10.53	10.58	−0.43
B7	10	55	9	10.31	10.13	1.8

Both the "measured" and the "predicted" refer to whole-exposure averages.

0.11 and 0.38). Among the B specimens, R^2 values range from 0.67 to 0.82. These R^2 values indicate that the B specimens are much better characterized by these models. The Bs appear to be more sensitive to environmental factors than the As.

This modeling effort is meant only as a first pass. We are certain that the true behavior of these materials is non-linear. We are also certain that there are very important non-linear interactions between environmental factors. Unfortunately, we do not yet understand the nature of those relationships well enough to construct a better model than the one we present here. Future effort will focus on improving the accuracy and increasing the complexity of our models. We will work to generalize short-exposure laboratory condition models developed by other investigators such that they can handle environmental variation and incorporate a time-series component. However, note that the simple model could explain as much as 80% of the behavior of the B specimens. Most of the specimens responded to RH, but not all. Many specimens showed a distinct difference in modulus above and below 0°C. *Figure* 24 shows this change in modulus for the B7 specimen.

The ability to conduct precise standard tests allowed us to identify experimental anomalies and take steps to correct them or to eliminate the data from the data set. Changes in elastic modulus

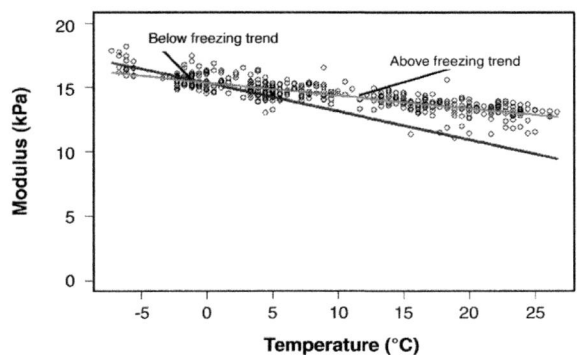

Figure 24—Modulus versus temperature for specimen B7 selected for 97.5%–100% RH, with lines showing predicted values above and below freezing. Note that the best-fit slope is not identical above/below 0°C.

caused by weathering have been detected and can be partially explained in terms of energy flux into the specimen from cyclic UV exposure and loading/unloading (*Table* 2, columns b6 and b7). There is some residual correlation with exposure time (*Table* 2, column b8) even after strain and UV energy inputs are considered. It is our belief that once a sufficient dose/damage relationship is found, we will be able to eliminate time and account for the degradation in terms of the energy input into the specimens and the environmental factors.

Though improvements have been made in our data analysis, further improvement may be possible by increasing the speed of measurement. The current system has a maximum measurement rate of 5.3 measurements/ specimen/second for 18 specimens in the apparatus. We plan to install a new faster data acquisition and control system. This system has an accurate electronic pulse generator that will improve the timing of step commands to the motors. This will increase the linearity of the standard test strain ramp-up and improve reproducibility of the tests. The new system will be capable of measuring load and displacement hundreds of times per second. Faster measurement means more data points can be taken during the standard strain test and used for calculating modulus. Together, the fast accurate pulse generator and the high data acquisition rate will allow modification of the standard test to incorporate a higher rate of strain ramp-up. Faster strain application decreases the amount of relaxation during a test, resulting in less viscous and more elastic response of the elastomeric specimens. These improvements are expected to decrease variability of the calculated modulus, enabling detection of smaller material changes caused by environmental conditions and weathering.

CONCLUSIONS

We have shown that data collected from the CCTA can be used to develop models to help us to understand the effects of temperature, RH, cyclic fatigue, and UV exposure. Based on these models, we can make statements about the rate of degradation for given doses of UV or cyclic loading. This is a promising first step on the path toward understanding the factors that cause these specimens to age.

We have demonstrated that the response of sealants and the weather-causing cyclic movement can be monitored continuously. From our earlier studies, measurement of load and deflection during this cyclic movement could be used to determine the elastic modulus; however, a controlled standard test using the computer-controlled CCTA gave much better results. The standard strain test made it possible to evaluate specimen quality at the beginning of the test and to determine the variability in a group of replicate specimens. The data from the standard test were used to construct a model to evaluate the effect of temperature and relative humidity on the elastic modulus. As one would expect from these viscoelastic sealant materials, the elastic modulus was dependent on temperature and RH, and we were able to detect changes in modulus as the specimens aged. A plot of elastic modulus versus temperature showed a change in modulus at 0°C, indicating a change in the plasticizing effect of moisture above and below freezing.

ACKNOWLEDGMENTS

We thank Russell Goodrich and Patrick Neale (Smithsonian Environmental Research Center, Edgewater, MD) for their help in setting up, calibrating, and monitoring the UV radiometer; Dave Eustice and Larry Zehner (Forest Products Laboratory) for building the test apparatus; and all the members of the FPL shops for their help with the initial stages of setting up the experiment.

References

(1) White, C.C., Martin, J., Weber, S., Shultz, L., and Williams, R.S., "Reliability-Based Method for Service Life Prediction of Materials," *Proc. SPE/ANTEC National Meeting*, 2002.

(2) White, C.C., Hunston, D.L., and Williams, R.S., "Challenges in Characterizing Sealant," *Proc. ANTEC Conference from Society of Plastics Engineering*, Nashville, TN, 2003.

(3) White, C., Embree, N., Buch, C., and Williams, R.S., "Design, Development, and Testing of a Hybrid in situ Testing Device for Building Joint Sealant," *Rev. Sci. Instrum.*, 76, 045111 (2005).

(4) White, C.C. and Hunston, D.L, "Issues Related to the Mechanical Property Characterization of Sealants," *Durability of Building and Construction Sealants and Adhesives*, ASTM Special Technical Publication 1453, p. 325-334, 2004.

(5) Williams, R.S., Sanandi, A., Halpin, C., and White, C., "Merging Weather Data with Materials Response Data during Outdoor Exposure," *Proc. Third International Woodcoatings Congress*, The Hague, The Netherlands, 2002.

(6) Williams R.S., Lacher, S., Halpin, C., and White, C., "Development of Test Apparatus for Service Life Prediction of Sealant Formulations and Evaluation of Data from 18 Months of Outdoor Weathering," *Proc. Fourth International Woodcoatings Congress*, The Hague, The Netherlands, 2004.

(7) Williams, R.S., Lacher, L., Halpin, C., and White, C., "Evaluating Weather Factors and Material Response during Outdoor Exposure to Determine Accelerated Test Protocols for Predicting Service Life," in *Service Life Prediction: Challenging the Status Quo*, Martin, J.W., Ryntz, R.A., and Dickie, R.A. (Eds.), Federation of Societies for Coatings Technology, Blue Bell, PA, p. 171-185, 2005.

(8) Williams, R.S., Lacher, S., Halpin, C., and White, C., "Evaluation of Materials During Outdoor Testing Using a Computer-Controlled Test Apparatus," *Proc. Fifth International Woodcoatings Congress*, Prague, Czech Republic, 2006.

(9) Martin, J.W. and Bauer, D.R. (Eds.), *Service Life Prediction: Methodology and Metrologies*, American Chemical Society Series 805, Oxford Press, New York, 2001.

(10) D 822 96, Standard Practice of Conducting Tests on Paint and Related Coatings and Materials using Filtered Open-Flame Carbon-Arc Light and Water Exposure Apparatus, *Annual Book of ASTM Standards*, Vol. 04.07, ASTM International, West Conshohocken, PA, 2000.

(11) D 5031 96, Standard Practice of Conducting Tests on Paint and Related Coatings and Materials using Enclosed Carbon-Arc Light and Water Exposure Apparatus, *Annual Book of ASTM Standards*, Vol. 06.01, ASTM International, West Conshohocken, PA, 2001.

(12) G 26 96, Standard Practice for Operating Light-Exposure Apparatus (Xenon-Arc) With and Without Water for Exposure of Nonmetallic Materials, *Annual Book of ASTM Standards*, Vol. 06.01, ASTM International, West Conshohocken, PA, 2001.

(13) G 53 96, Standard Practice for Operating Light- and Water-Exposure Apparatus (Fluorescent UV-Condensing Type) for Exposure of Nonmetallic Materials, *Annual Book of ASTM Standards*, Vol. 06.01, ASTM International, West Conshohocken, PA, 2001.

(14) D 4587 91, Standard Practice of Conducting Tests on Paint and Related Coatings and Materials using a Fluorescent UV-Condensation Light- and Water-Exposure Apparatus, *Annual Book of ASTM Standards*, Vol. 06.01, ASTM International, West Conshohocken, PA, 2001.

(15) Martin, J.W., " A Critical Review of the Role of Field-Exposure Experiments in Predicting the Service Life of Coatings," *Plastics and Coatings Durability, Stabilization, Testing*, Ryntz, R.A. (Ed.), Ch. 2. Hanser Gardner Publications, Inc., Cincinnati, OH, p. 33, 2001.

(16) Martin, J.W., "Repeatability and Repeatability of Field Exposure Results," *Service Life Prediction Methodologies and Metrologies*, Martin, J.W. and Bauer, D.R. (Eds.), ACS Symposium Series 805, American Chemical Society, Oxford Press, NY, p. 1, 2001.

150

(17) Nguyen, T., Rezig, A., Gu, X., Kidah, B., and Martin, J.W., "Degradation Kinetics and Mechanisms of Aliphatic Amine-Cured Epoxy Coatings Exposed to Xenon Arc UV and Outdoors," *Proc. Western Coatings Conference*, Las Vegas, NV, 2005.

(18) C 1442 99, Standard Practice for Conducting Tests on Sealants using Artificial Weathering Apparatus, *Annual Book of ASTM Standards*, Vol. 04.07, ASTM International, West Conshohocken, PA, 2000.

(19) C 718 93, Standard Test Method for Ultraviolet (UV)-Cold Box Exposure of One-Part, Elastomeric, Solvent-Release Type Sealants, *Annual Book of ASTM Standards*, Vol. 04.07, ASTM International, West Conshohocken, PA, 2000.

(20) C 732 95, Standard Test Method for Aging Effects of Artificial Weathering on Latex Sealants, *Annual Book of ASTM Standards*, Vol. 04.07, ASTM International, West Conshohocken, PA, 2000.

(21) C 734 93, Standard Test Method for Low-Temperature Flexibility of Latex Sealants After Artificial Weathering, *Annual Book of ASTM Standards*, Vol. 04.07, ASTM International, West Conshohocken, PA, 2000.

(22) C 793 97, Standard Test Method for Effects of Accelerated Weathering on Elastomeric Joint Sealants, *Annual Book of ASTM Standards*, Vol. 04.07, ASTM International, West Conshohocken, PA, 2000.

(23) C 1257 94, Standard Test Method for Accelerated Weathering of Solvent-Release-Type Sealants, *Annual Book of ASTM Standards*, Vol. 04.07, ASTM International, West Conshohocken, PA, 2000.

(24) D 2249 94, Standard Test Method for Predicting Effect of Weathering on Face Glazing and Bedding Compounds on Metal Sash, *Annual Book of ASTM Standards*, Vol. 04.07, ASTM International, West Conshohocken, PA, 2000.

(25) ASTM C 719 93, Standard Test Method for Adhesion and Cohesion of Elastomeric Joint Sealants Under Cyclic Movement (Hockman Cycle), 2005.

(26) Onuoha, U.O., "Durability of One-Part Polyurethane and Polyurethane Hybrid Sealants," *Durability of Building Sealants*, Wolf, A.T. (Ed.), RILEM Pub., France, p. 235–251, 1999.

(27) Brown, N.G., "Assessment of Joint Sealant by Outdoor Exposure in Cyclic Movement Testers," Report No. 01.1-2, CSIRO Division of Building Research, Highett, Victoria, Australia, p. 16-22, 1965.

(28) Burstrom, P.G., "Durability and Ageing of Sealants," In *Durability of Building Materials and Components*, ASTM STP 691, Sereda, P.J. and Litvan, G.G. (Eds.), ASTM Pub., Philadelphia, PA, p. 643–657, 1980.

(29) Karpati, K.K., Solvason, M.R., and Sereda,, P.J., "Weathering Rack for Sealants," *J. Coat. Technol.*, 49, No. 626, p. 44–47 (1977).

(30) Karpati, K.K., "Device for Weathering Sealants Undergoing Cyclic Movements," *J. Coat. Technol.*, 50, No. 641, p. 27–30 (1978).

(31) Lacasse, M.A., "Advances in Test Methods to Assess the Long Term Performance of Sealants," *Science and Technology of Building Seals, Sealants, Glazing and Waterproofing*, Myers, J.C. (Ed.), Vol. III, ASTM STP 1254, ASTM International, Philadelphia, PA, 1994.

(32) Lacasse, M.A., Giffin, G.B., and Margeson, J.C., "Evaluation of Cyclic Fatigue as a Means of Assessing the Performance of Construction Joint Sealants," ASTM Special Technical Publication, n, 1271, *Science and Technology of Building Seals, Sealants, Glazing, and Waterproofing*, ASTM International, Philadelphia, PA, 1996; *Proc. 5th Symposium on the Science and Technology of Building Seals, Sealants, Glazing, and Waterproofing*, Phoenix, AZ, Vol. 5, p, 266-281, 1995.

(33) Lacasse, M.A., Giffin, G.B., and Margeson, J.C., "Laboratory Cyclic Fatigue Test of Silicone Sealant Mini-Specimens,"ASTM Special Technical Publication, n. 1334, *Proc. 7th Symposium on the Science and Technology of Building Seals, Sealants, Glazing, and Waterproofing*, San Diego, CA, p. 51-65, 1998.

(34) Lacasse, M.A. and Margeson, J.C., "Movement during Cure of Latex Building Joint Sealants," *Proc. 6th Annual Symposium on the Science and Technology of Building Seals, Sealants, Glazing, and Waterproofing*, Fort Lauderdale, FL, Vol. 6, p. 129–145, 1999.

(35) Bolte, H., Boettger, T., and Wolf, A.T., "Preliminary Report on the Ageing Behavior of Sealants Exposed to Dynamic Natural and Artificial Weathering," *RILEM Proceedings PRO 10 Durability of Building and Construction Sealants*, p. 151–172, 2000.

(36) Liu, X-X, and Wang, S.T., "Joint Sealant Fatigue Property of Airport Concrete Pavement," *J. Traffic Transport. Engineering*, 6 (1), 44–47 (2006).

(37) Al-Qadi, I.L., Abo-Qudais, S., and Khuri, R.E., "Method to Evaluate Rigid-Pavement Joint Sealant under Cyclic Shear and Constant Horizontal Deflection," *Transport. Res. Rec.*, 1680, p. 30–35 (1999).

(38) Al-Qadi, I.L., and Abo-Qudais, S., ASTM Special Technical Publication, n. 1254, pt 3; *Proc. 3rd Symposium on the Science and Technology of Building Seals and Sealants*, Fort Myers, FL, p. 85–94, 1993.

(39) Miyauchi, H., Enomoto, N., Sugiyama, S., and Tanaka, K., "Artificial Weathering and Cyclic Movement Test Results Based on the RILEM TC 139-DBS Durability Test Method for Construction Sealants," ASTM Special Technical Publication, n 1453, *Durability of Building and Construction Sealants and Adhesives*, American Society for Testing and Materials, West Conshohocken, PA, p. 206–212, 2004.

(40) Sharman, W.R., Fry, J.I., and Whitney, R.S., "Six Years Natural Weathering of Sealants," *Durability of Building Materials*, 2 (1), 79-90 (1983).

(41) Koike, M., Tanaka, K., and Oda, S-I., "Dynamic Properties of Polysulfide and Silicone Sealants in Cyclic Tests of Longer Periods," Report of the Research Laboratory of Engineering Materials, Tokyo Institute of Technology, 7, 159-165 (1982).

(42) Marechal, J.C., and Kalifa, P., "The Prediction of Long-Term Sealant Performance from Dynamic Accelerated Weathering," *Durability of Building Materials and Components 7, Proc. International Conference on Durability of Building Materials and Components*, 7th, Stockholm, p. 54–64, 1996.

(43) Beech, J.C., "Test Methods for the Movement Capability of Building Sealants: The State of the Art, *Materials and Structures*, 18 (108), 473–482 (1985).

(44) Wolf, A.T., "Aging Resistance of Building and Construction Sealants, Part 1," *RILEM Proc. Durability of Building Sealants*, p. 63–89 (1996).

(45) Wolf, A.T., "Attempts at Correlating Accelerated Laboratory and Natural Outdoor Ageing Results," *Durability of Building Sealants*, Wolf, A.T. (Ed.), RILEM Pub., France. p. 181–201, 1999.

(46) Wolf, A.T. and Seneffe, B., "Development of RILEM Durability Test Methods for Curtainwall Sealants," *Proc. International Conference on Durability of Building Materials and Components*, 9, Brisbane, Australia, p. 105/1-105/9, 2002.

(47) Lacher, S., Williams, R.S., Halpin C., and White C., "Development of a Powered Outdoor Sealant Fatigue Test Apparatus," In: *Service Life Prediction, Challenging the Status Quo*, Martin, J.W., Ryntz, R.A., and Dickie, R.A. (Eds.), Federation of Societies for Coatings Technology, Blue Bell, PA, p. 207–216, 2005.

(48) Hunston, D.L., White, C.C., and Williams, R.S., "Viscoelastic Characterization of Sealant Materials," *Proc. 26th Annual Meeting Adhesion Society*, Myrtle Beach, SC, 2003.

(49) Hunston, D.L., White, C.C., and Williams R.S., "Mechanical Behavior of Caulks and Sealants," *Proc. 225th ACS National Meeting*, New Orleans, LA, 2003.

(50) Abraham, F., Alshuth, T., Jerrams, S., "The Dependence on Mean Stress and Stress Amplitude of the Fatigue Life of Elastomers," Deutsches Institut für Kautschuktechnologie e. V. Publikation, 134, 2001.

(51) Bergström, J.S. and Boyce, M.C., "Time-Dependence of Elastomeric Materials: Experiments and Modeling," *10th International Conference on Deformation, Yield and Fracture of Polymers*, Cambridge, UK, April 1997.

(52) Bergström, J.S., "Constitutive Modeling of Elastomers—Accuracy of Predictions and Numerical Efficiency," PolymerFEM.com, 2005.

(53) Wilkinson, G.N. and Rogers, C.E., "Symbolic Descriptions of Factorial Models for Analysis of Variance," *Applied Statistics,* 22:392-9 (1973).

(54) Chambers, J.M., "Linear Models," Chapter 4 of *Statistical Models in S*, Chambers, J.M. and Hastie, T.J. (Eds.), Wadsworth & Brooks/Cole, 1992.

Chapter 10

Lifetime Predictions for Hardcoated Polycarbonate

James E. Pickett[1] and Jonathan R. Sargent[2]

[1] GE Global Research, 1 Research Cir., Niskayuna, NY 12309
[2] Exatec, LLC, Wixom, MI 48393

Weatherable, abrasion-resistant coatings can be applied to polycarbonate to make them suitable for many glazing applications. An additional highly abrasion-resistant layer can be applied using an expanding argon plasma and silicone precursors to make polycarbonate suitable for many automotive glazing applications that also demand 10-year weatherability. Predictions for Miami weathering performance were made from xenon arc exposures. Analysis of the spectral power distributions of Florida sunlight and xenon arc lamps suggested that approximately 3700 kJ/m²/nm at 340 nm of borosilicate-filtered xenon arc would give a UV dose equal to a year in Miami in the range of 290 to 350 nm, the critical range for polycarbonate. The higher temperature of xenon arc conditions reduces this correlation to approximately 3000 kJ/m²/nm at 340 nm for clear PC and 2500 kJ/m²/nm at 340 nm for darker, "privacy" color PC. We predict about eight years 45° south or horizontal Florida lifetime for a clear system with a 7 µm UV-blocking layer, and about 10 years for an 8 µm layer. Privacy grades absorb more light and get warmer, reducing the delamination lifetimes by about 20%. We can accelerate weathering by an additional factor of 1.9 by moving clear samples 4 in. (10 cm) closer to the xenon arc lamp in our weathering device. Microcracking can be accelerated by exposure under applied strain, but the more advanced silicone hardcoat + plasma system still does not microcrack before delamination at even 0.6% strain.

INTRODUCTION

   Polycarbonate (PC) is an attractive material for many glazing applications because of its low weight compared with glass, and high impact resistance compared with other transparent polymers. Polycarbonate glazing usually is coated to impart weatherability and scratch resistance. Early abrasion-resistant coatings lacked UV absorbers and failed by delamination after about 12 months of 45° south exposure in Florida. The highest performing wet coatings consist of UV absorbers co-polymerized with a silicone/colloidal silica nanocomposite. Such first generation silicone hardcoats (SHC) typically would weather in Florida for three to five years before the underlying PC began to turn visibly yellow and the coating would delaminate. Some coating formulations also were subject to microcracking after 18 months to two years of Florida exposure. The stability of the UV absorber was found to be the limiting factor, and a second generation SHC with an improved UV absorber lasts at least six years in Florida without delamination at a 5 μm thickness. These coatings typically have sufficient scratch resistance for architectural glazing with Taber haze values of approximately 10–15% after 500 cycles under a 100 g load.

Automotive glazing applications for side and rear windows have more stringent abrasion resistance requirements: < 2% haze after 1000 cycles of the Taber abrasion test.[1] Such abrasion resistance is difficult to achieve in a wet coating while maintaining other properties. One solution is to apply a glass-like layer on the SHC by a plasma-enhanced chemical vapor deposition (PECVD) process. Exatec, LLC was created as a joint venture by the General Electric Company and Bayer to develop and license technology for applying such highly abrasion-resistant coatings onto PC for automotive glazing applications.* Such coatings meet all the physical property requirements of the ANSI standard[1] for side and rear windows. However, the weathering lifetime of these coatings must be established. Since 7–10 years of outdoor weathering cannot be obtained in less than 7–10 years, we must rely on accelerated tests.

The coating matrices themselves are quite resistant toward weathering since they are composed of silicones and silica. The UV absorber is subject to slow photodegradation,[2] and this can lead to cracking of the coating. In addition, the UV transmitted by the coating causes photooxidation of the PC surface, resulting in yellowing and eventually causing enough loss of molecular weight to allow delamination near the coating/PC interface. The delamination lifetime of the coating is determined by the initial UV transmission of the coating, the rate of UV absorber loss, and the amount of transmitted UV dose required to sufficiently damage the PC surface.[3] The surface degradation can affect physical properties as well. Nichols and Peters have reported the changes in mechanical properties of a hardcoated polycarbonate after xenon arc weathering.[4]

Accelerated weathering has a checkered reputation for reliability, and lifetime predictions must be made very cautiously. Bauer has made a rational analysis of acceleration factors for coating testing,[5,6] taking into account UV, temperature, and moisture effects on typical automotive coatings. His analysis suggested that the SAE J1960 cycle using borosilicate-filtered xenon arc exposure (0.55 W/m^2/nm at 340 nm, 2 h light/1 h dark) should be a 4.5× acceleration over Florida exposure in terms of UV dose. One year of operation of such a cycle applies 11,564 kJ/m^2/nm at 340 nm, implying that 2570 kJ/m^2/nm at 340 nm is the equivalent of one year Florida exposure. He points out that the unnatural spectral distribution of the borosilicate-filtered xenon arc can accelerate different materials to different degrees. Seasonal average daily high temperatures were used to adjust for temperature effects. His predicted acceleration factors for UV absorber loss, acrylic coatings, and aromatic ester coatings closely matched experimental data. Recently, we have had some success in developing accelerated weathering protocols that predict color shifts and gloss retention for engineering thermoplastics.[7,8] Our experience with SHC-coated PC led us to believe that we could make rational lifetime predictions for plasma-coated materials.

---

* Since September 2007, Exatec is wholly owned by SABIC Innovative Plastics. Exatec® 900 employs SHX™ silicone hardcoat, a second-generation SHC, as the UV protective layer with PECVD-deposited abrasion layers.

There are several principles we can apply to strive for predictive testing. First, we recognize that weathering is the result of a complex interaction among any number of environmental variables, but primarily sunlight, temperature, and moisture. Since the interactions are difficult to predict, successful accelerated testing will simulate as many of these variables as realistically as possible. Second, all variables cannot be applied at natural levels, or no acceleration will result. Something, usually the light dose rate, must be increased. It should be known that increasing the environmental stress does not result in gross changes in the degradation or failure mechanisms. Third, the stresses must be applied rationally; that is, the degree of acceleration should be predictable. It does no good to devise a test if one does not have a good estimate of the correlation before all the outdoor data come in. This can come from both the rationality of the test design and experience with exposing similar materials under similar conditions. Of course, "similar" is a vague term, and this adds risk to the lifetime prediction of new products. Finally, a good test must be acceptable to all parties and preferably be run on generally available equipment.

The approach of environmental simulation cannot be successful without considerable understanding of the effects of individual stresses and their interactions on a material. Test design, interpretation of results, and confidence in the results are made possible by controlled experiments showing that a material obeys reciprocity (that the degradation rate varies linearly with light intensity), finding the activation energy (effect of temperature on degradation rate), investigating the effects of moisture and humidity, etc. There is always risk in applying knowledge of simple materials to a final product. Real products usually are articles composed of several materials to give the desired properties, and weatherability is really a system property, not a material property. While knowledge of individual material performance greatly aids the design of weatherable systems and their testing, final products must always be exposed to conditions as close to the end use environment as possible to see how the components work together.

One goal of this work was to determine whether our accelerated weathering exposure conditions were likely to be predictive for plasma-deposited abrasion layers on weatherable silicone hardcoats. Secondly, since we were expecting that at least one year of accelerated exposure would be required to cause failure, we sought methods for further accelerating the weathering process to determine more quickly if future compositional or process changes might affect the weathering performance of the product.

EXPERIMENTAL SECTION

Samples

Unless otherwise specified, the samples in this study were prepared at Exatec, LLC in Wixom, MI. Injection molded panels of either clear (92% transmission) or privacy color (17% transmission) PC measuring 73 cm x 73 cm x 4 mm were flow coated with either the first or second generation silicone hardcoat (SHC) after application of the appropriate acrylic primer. These coatings were prepared at different pH and have different UV absorbers. The coatings had thicknesses of approximately 5 μm near the top, 7 μm in the middle, and 8.3 μm near the bottom of the panels. The tops deliberately were made thinner than specification for optimal weathering to test a wide range of thicknesses. Some

Table 1—Settings for the Accelerated Weathering Exposure Conditions

	"CRD"	"Gmod"
Instrument	Ci35a	Ci35a[a]
Irradiance (W/m²/nm @ 340 nm)	0.77	0.75
Light (min)	160	102
Dark/dry (min)	5	—
Dark/spray (min)	15	—
Light/spray (min)	—	18
Black panel temp. (°C)	70	65
Dry bulb temp. (°C)	45	40
RH (%)	50	50
Inner filter	Type S boro	Type S boro
Outer filter	Type S boro	Type S boro
kJ/m²/h at 340 nm	2.46	2.70

(a) Originally an Hi35a (3-Sun) retrofitted to be identical to a Ci35a.

panels of each color and SHC type were passed through an expanding argon plasma reactor and received approximately 3 μm of silica-like overcoat from a silicone precursor using an Exatec proprietary process. Sample coupons were cut from top, middle, and bottom sections for weathering studies. The complete sample set therefore contained 24 unique combinations: two colors, two wetcoats, with and without plasma overcoat, and three wetcoat thicknesses (2 x 2 x 2 x 3).

Weathering Conditions

The samples were exposed at GE Global Research in Atlas Ci35a Xenon Arc Weather-Ometers according to the conditions in *Table* 1. The "CRD"* conditions were originally devised to provide approximately one Florida day's worth of light followed by a brief dark cool-down and a dark water spray. This gives one wet period and temperature cycle per simulated "day." The irradiance was increased to as high as possible while maintaining temperature control and reasonable lamp lifetime. The "Gmod" conditions are a modification of ASTM G155 Cycle 1[9] (prev. ASTM G26 Method A) with the irradiance increased to achieve maximum acceleration while maintaining temperature control. It is also similar to ISO 4892-2A. (Workers at Exatec have found that the larger Atlas Ci5000 Weather-Ometers must be run at lower humidity and higher temperature at this irradiance setting.) All exposures in this work are expressed in units of kJ/m²/nm measured at 340 nm. For brevity, this will be expressed simply as kJ/m² in the text.

The higher irradiance experiments were run under "Gmod" conditions with the spray nozzles moved several inches closer to the lamp. Sample holders were designed to hang on the inside center ring of the rack either 6.35 cm (2.5 in.) or 10.16 cm (4 in.) closer to the lamp than the normal sample position. Samples on these racks were calculated to receive 1.5x and 2x higher irradiance than those in the normal sample position. Radiant energy measurement at these positions is reported for the radiometer calibrated at the nor-

---

*"CRD" stands for "Corporate Research and Development," the former name of GE Global Research.

Figure 1—Schematic side view of four-point strain jig.

mal sample position—that is, not reflecting the higher irradiance of the nearer position.

Humidity experiments were performed by placing samples in sealed aluminum chambers with quartz front windows. The chambers contained either calcium sulfate (Drierite) desiccant or water to maintain ~0% and 100% relative humidity, respectively. Chambers were placed on the bottom rack of a Weather-Ometer under "CRD" conditions. The sample surface temperature was found to be 75°C. Samples were removed for inspection at approximately two week intervals.

The applied strain experiments were performed on 4.3 cm x 13 cm samples cut from near the bottom of the coated panels. The edges of the samples were smoothed with fine-grit sandpaper to minimize crack initiation. The samples were mounted three across in four-point strain jigs designed by Michael Takemori of GE Global Research as a modification of an earlier three-point design by Don LeGrand.[10] A schematic side view is shown in *Figure* 1. The jigs were mounted on the bottom tier of the sample rack with the convex side facing the lamp. Samples were tested at 0%, 0.3%, 0.4%, and 0.6% strain, which was adjusted by screwing down the end rods to different degrees. Strain for cracking at zero exposure was determined by applying a strain to samples, exposing them backwards in the Weather-Ometer for 24 h, and checking for cracking. If no cracking was observed, the strain was increased by 0.1% and the procedure repeated until cracking occurred. In this way, the samples saw the same additional strain that might occur during temperature cycling as the other strained samples.

Sample Analysis

Color measurements were made on a GretagMacbeth ColorEye 7000A spectrometer in transmission mode using the CIE color scale and yellowness index as defined by ASTM D 1925. Delamination was defined as when the first visible bits of coating spontaneously detached from the surface. Microcracking failure was defined as when defects progressed from points to visible cracks when illuminated from the back and viewed against a dark background.

RESULTS AND DISCUSSION

Expectations for Correlations

The xenon arc light source provides a broad continuum of light that can be filtered to make an approximation to sunlight. Water-cooled xenon arc lamps have concentric inner and outer filters of various types. *Figure* 2 shows the spectral power distributions for Miami sunlight and three common combinations of filters used in xenon arc weathering equipment, all normalized to equal energy at 340 nm. It is apparent that the quartz/borosilicate combination has excessive energy at wavelengths < 300 nm compared to natural sunlight. This is the combination specified in the original SAE J1960 automotive exterior weathering protocol. The combination of borosilicate inner and outer filters has an overall good approximation to sunlight with a slight excess of energy at wavelengths < 300 nm. The CIRA (quartz with an IR-reflecting coating)/soda lime combination has a good cutoff at 300 nm, but is slightly deficient in the wavelengths between 300 and 340 nm. All of the xenon sources have excess energy in the range of 340 to 400 nm compared with sunlight. (Energy of sunlight in this range is attenuated by absorption due to metals in the sun's atmosphere.)

In order to derive an acceleration factor for xenon arc exposure, one must know how long it takes to apply the same number of photons in some critical wavelength range compared to outdoors. Unfortunately, because the spectral power distribution of no available xenon arc source exactly matches sunlight, there is no single answer to the question. Long-term measurements in Miami of broad-band UV show that the average annual dose of radiation received by a sample facing south at a 45° angle is 280 MJ/m² in the wavelength range of 295–385 nm.[11] We can normalize the Florida spectral power distribution shown in *Figure 2* and integrate it to give the annual UV doses through various wavelength ranges shown in *Table 2*. Xenon arc weathering commonly is monitored using a radiometer with a narrow band pass filter centered at 340 nm. *Table 2* shows the xenon arc dose measured at 340 nm that gives an equivalent total exposure over each wavelength range calculated from the spectral power distributions shown in *Figure 2*. For example, to apply 280 MJ/m² at wavelengths ≤ 385 nm requires 3194 kJ/m² of Q/B xenon exposure, 3269 kJ/m² of B/B, or 3495 kJ/m² of CIRA/SL exposure measured at 340 nm. (These values differ somewhat from those of Bauer,[5] who may have used a wavelength range ≤ 385 nm for outdoor UV and ≤ 400 nm for xenon arc for his calculations.) These would be the correlation factors for xenon arc to Florida if one were concerned about this wavelength range, there was uniform acceleration at all wavelengths, and there were no temperature effects. However, *Table 2* shows that the correlation is different if one chooses a different wavelength range. Quite simply, there is no single answer for what the correlation between xenon arc and Florida should be based on the spectral power distribution of the light source. Since most polymers have sensitivities that extend through at least 360 nm,[12] somewhere between 3000 and 4000 kJ/m² at 340 nm of xenon arc exposure should give about the same UV dose as one year in Florida. (Rigorously, the analysis should be number of

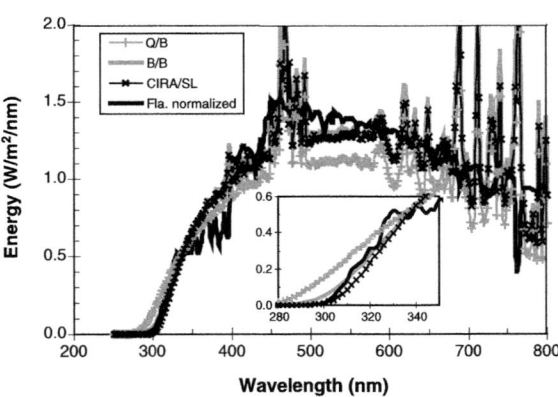

Figure 2—Spectral power distributions of Miami sunlight and a xenon arc lamp equipped with quartz/borosilicate (Q/B), borosilicate/borosilicate (B/B), and CIRA/soda lime glass (CIRA/SL) filters. Data from Atlas Material Testing Technology, LLC.

Table 2—Calculations Showing the Integrated Annual Radiant Energy Through Various Wavelength Ranges in Florida and the Xenon Arc Dose (Measured at 340 nm) Required to Deliver the Same Amount of Energy Through that Wavelength Range

Wavelength Range	Florida Total	Equivalent Xenon Dose Measured at 340 nm (kJ/m²/nm)		
(nm)	MJ/m²	Q/B	B/B	CIRA/SL
≤ 300	0.35	117	562	3293
≤ 310	5.84	857	2375	5969
≤ 320	22.9	1812	3556	5663
≤ 330	52.3	2569	4030	5201
≤ 340	87.3	2938	3967	4593
≤ 350	124	3058	3730	4104
≤ 360	163	3107	3528	3783
≤ 370	212	3205	3453	3653
≤ 380	260	3236	3360	3530
≤ 385	**280**	**3194**	**3269**	**3495**
≤ 400	359	3179	3138	3285
≤ 450	779	3843	3578	3727
≤ 500	1286	3974	3591	3760
≤ 600	2228	4236	3735	3878
≤ 700	3049	4230	3681	3822
≤ 800	3691	4128	3542	3708

photons, but the results are almost identical, and measurements usually are made in terms of energy.)

Polycarbonate is known to be extremely sensitive to UV with wavelengths < 300 nm.[13] For this reason, we found the CIRA/soda lime filter combination to be most predictive for uncoated PC.[7,8] However, the UV absorber in the silicone hardcoats, especially the second generation absorber, very effectively blocks UV < 300 nm. Since the borosilicate/borosilicate filter combination has the overall best match to sunlight, we settled on that combination for the testing of UV-blocking coatings. The sensitivity of PC extends to about 350 nm, so approximately 3700 kJ/m² of B/B xenon arc should be the equivalent to one Florida year of UV. This is consistent with the results we obtained for CIRA/soda lime exposure of uncoated samples.[7,8] We do not recommend using the quartz/borosilicate filter combination because of its very high output < 300 nm. We have found that the light intensity can be increased for PC without having the degradation rate deviate from linearity.[14]

Temperature Effects

A second important factor is temperature. We have found that the activation energy for PC photodegradation is 4±1 kcal/mol.[14] From the Arrhenius law, this results in a rate increase of approximately 25% for each 10°C temperature increase. Temperatures in xenon arc weathering chambers cannot be set at will for all irradiance levels because of inherent limitations of the devices. The average daily high air temperature at a Miami test site is approximately 30°C while the average daily high black panel temperature is 46°C.[15] Typical accelerated weathering conditions have air temperatures ≥ 40°C and black panel temperatures ≥ 65°C. We would therefore expect PC photodegradation to be

Table 3—Measured Temperatures of Samples on the Middle Tier Under the Exposure Conditions as Shown in *Table* 1[a]

Color	Estimated FL/AZ (°C)	CRD (°C)	CRD Acceleration	Gmod (°C)	Gmod Acceleration
Privacy...........40		66	1.8×	60	1.6×
Clear...............32		48	1.5×	42	1.3×

(a) The estimated Florida/Arizona effective temperatures were determined as described in the text. The acceleration over Florida/Arizona was calculated using an activation energy of 5 kcal/mol.

accelerated at least 25% over Florida ambient, thereby reducing the expected correlation from approximately 3700 kJ/m^2 to around 3000 kJ/m^2 or less.

Better estimates can be made from actual measured temperatures. Thermocouples were attached to the exposed surface of dummy samples using a small dab of silicone adhesive. The samples had surface temperatures as shown in *Table* 3. These values are surprisingly difficult to measure reproducibly and vary by several degrees in different locations within the instrument. They should therefore be considered as rough guides.

To get a better estimate of outdoor temperatures, we compiled hourly-parsed temperature and irradiance data for all of 2005 at an Arizona exposure site (kindly provided by Joe Robbins of Arizona Desert Testing, LLC). Using a 5 kcal/mol activation energy, irradiance-weighted "effective temperatures" for ambient and a black panel could be determined using a cumulative damage model, and are 30°C and 42°C, respectively. Details will be published elsewhere. Since there is more damage when the samples are exposed hot and less when they are exposed cool, these are the constant temperatures at which the same amount of damage is accumulated as over a year of natural temperature variations from the same light dose. Note that these values are not much different from the average daily high ambient and black panel temperatures as used by Bauer.[5] Compiled data show that the average daily high ambient and black panel temperatures are nearly the same for Florida and Arizona,[15] so one expects the effective temperatures to be about the same as well. To a first approximation, clear PC should be a few degrees warmer than ambient while the privacy color will be a few degrees cooler than the black panel. This gives the estimated effective temperatures shown in *Table* 3. We are now in the process of determining experimentally effective temperatures in Arizona for clear and colored PC panels.

The Weather-Ometer temperatures are considerably higher than the effective temperatures outdoors. Considering the activation energy for PC, the thermal "over-acceleration" for the CRD and Gmod test conditions should be about as shown in *Table* 3. This reduces the correlation of xenon arc under Gmod conditions to Florida for PC from the estimated 3700 kJ/m^2 per year for no thermal effect to roughly 2300 kJ/m^2 for privacy and 2850 kJ/m^2 for clear. If the activation energy is 4 kcal/mol, these correlations would be 2500 and 3100 kJ/m^2, respectively.

Experience with Similar Coatings

We have been testing coatings under CRD xenon arc conditions since 1990 and have accumulated considerable laboratory and outdoor data on coated clear PC. These are compiled in *Table* 4, assuming a correlation of 2800–3000 kJ/m^2 to one year in Florida. Samples were exposed at commercial test sites near Miami, facing south at angles of either 5° or 45°. There should be only a 10% difference between the two angles, which is not significant in this experiment. One sees that the predicted failure times agree quite well with what was found for all coating types. In particular, the premature adhesion failure of an otherwise promising prototype SHC was predicted as well as the failure modes of a UV-cured acrylic, which generated haze rather than delaminating. In fact, *we have never seen a failure mode for hardcoated PC in Florida that was not anticipated by borosilicate/borosilicate-filtered xenon arc exposure.* Since the plasma-on-second generation SHC is chemistry very similar to examples in *Table* 4, we have confidence that xenon arc will be similarly predictive in this case as well.

Most of our fundamental work on PC photodegradation has been on cosmetic properties such as yellowing and gloss loss. A legitimate question is whether this information is relevant to coated PC. Unpublished work in our laboratories has shown that changes in the molecular weight distribution of thin PC films are linearly correlated to changes in yellowness. The activation energies for yellowing and gloss loss in PC are both 3–5 kcal/mol. This makes it likely that the chemistry leading to yellow products is closely related to the chemistry that causes chain scission and crosslinking. The latter processes are responsible for the surface degradation that results in coating delamination, and preliminary work in our lab shows a similar activation energy for delamination of silicone hardcoats. Physical failure in the coating, by cracking for example, is an independent phenomenon, and we have much less information about how environmental variables affect it. Results below show that relatively subtle differences in test conditions do cause differences in cracking times for some of these coatings. However, the empirical results in *Table* 4 show that, in general, cracking in Florida exposures is predicted by cracking observed in xenon arc exposures.

Table 4—Years To Delamination and Microcracking Found Upon Exposure Near Miami, FL, for Various Coatings on Clear Polycarbonate[a]

Coating	Delamination		Microcracking	
	Predicted	Found	Predicted	Found
UV-cured acrylic	>4	>3[b]	2–3	>3[b]
Prototype SHC[c]	1	1	—	—
1st generation silicone hardcoat	5	4–5	1.5–2	1.5–2
Thin 1st generation silicone hardcoat[d]	2–3	2–3		
Exatec plasma on 1st generation SHC	6	4.5+	4–5	4.5[e]
Exatec plasma on thin 1st generation SHC[d] .	2.5	2		
Margard MR-5 silicone hardcoat	4–5	3–5	3–4	3–5
Thin MR-5[d]	2–3	2–3		
Modified 1st generation silicone hardcoat ...	3.5–6	4.5–5[e]	2–3	1.5–2.5
2nd generation silicone hardcoat	6–8	5.5[e]	5	5–5.5

(a) Prediction based on CRD xenon arc conditions and assuming 2800–3000 kJ/m^2 equals one year of Miami exposure.
(b) Terminated at 3. Predict increase haze ~3, find increased haze at 3.
(c) Xenon arc predicted premature loss of tape adhesion; seen in Florida.
(d) Coating purposely made thinner than optimal will result in these lower lifetimes.
(e) Indicates ongoing test.

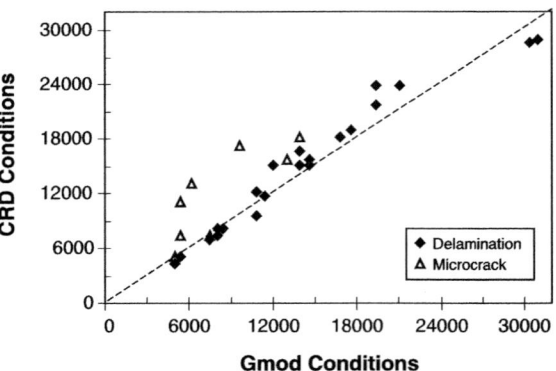

Figure 3— Failure times (kJ/m²) for microcracking and delamination of 2nd generation SHC with and without plasma overcoat under the two exposure conditions. (The dashed line is at slope = 1 as a guide for the eye.)

Effect of Exposure Conditions on Weathering Lifetimes

The two exposure conditions shown in *Table* 1 can be used to determine whether or not the coating is very sensitive to minor changes in the environmental variables. *Figure* 3 shows a plot of delamination and microcrack times for the set of 24 samples exposed under CRD and Gmod conditions. Delamination times are nearly identical under the two test conditions. By contrast, microcracking usually is faster under the Gmod conditions. Gmod conditions are actually set a little cooler than the CRD conditions, but the water spray is more frequent. In addition, since the water spray occurs during the light cycle, the air temperature rises to nearly the black panel temperature during the spray period. The combination of frequent very hot and wet periods may cause the earlier microcracking. Such simultaneous very hot and 100% relative humidity exposure is unlikely to occur very often in real service conditions. In any case, as we will see below, none of the samples of the plasma overcoat on 2nd generation SHC microcracked under either condition, making this issue moot for the coating system in which we have the most interest.

Reproducibility of Results

A set of samples prepared by Michael Mercedes in a developmental reactor at GE Global Research was exposed under CRD conditions. These were coated with the 2nd generation SHC on both sides but with the plasma overcoat applied to only one side. The

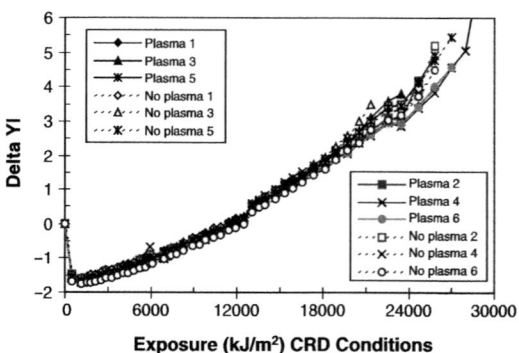

Figure 4—Exposure of clear PC with the 2nd generation silicone hardcoat with and without the plasma overcoat.

Table 5—Results of Reproducibility Study on 2nd Generation SHC With and Without a Plasma Overcoat[a]

	Delamination		Microcracking	
	mean	std. dev.	mean (n=6)	std. dev.
SHC only (n=6)...................24,500		1900	18,100	0
SHC + plasma (n=6) 25,600		1300	not observed	—

(a) Failure times are in $kJ/m^2/nm$ at 340 nm under CRD xenon arc conditions.

overcoat was applied in various thicknesses and at several angles. Six of those samples were cut in half and exposed on either the SHC-only or the SHC + plasma sides. The weathering results as evidenced by yellowing are shown in *Figure* 4. One sees that the yellowing is identical for all samples. Delamination and microcracking results are shown in *Table* 5. The reproducibility of the results is excellent.

Effect of Plasma Overcoat on Weathering

The delamination and microcracking failure times from both exposure conditions for the 1st generation SHC with and without the plasma overcoat are shown in *Figure* 5. One sees that the plasma overcoat has a small positive effect on the delamination lifetime. The effect is dramatic on microcracking. Most of the non-plasma samples microcracked around 6,000 kJ/m^2 of exposure while the plasma-treated samples generally microcracked near the delamination times, if they cracked at all. By contrast, *Figure* 6 shows that delamination times of the 2nd generation SHC seem little affected by the plasma overcoat. While many of the non-plasma samples microcracked, none of the plasma-treated samples has exhibited any cracking prior to delamination.

Humidity Effects

Samples cut from the top portions of clear panels coated with 2nd generation SHC with and without plasma overcoats were placed in closed chambers having quartz windows. One chamber was kept dry using calcium sulfate (Drierite) while the other was maintained at 100% relative humidity. Because the chambers acted as greenhouses, the sample temperatures were all 75°C. The samples in the dry and wet chambers delaminat-

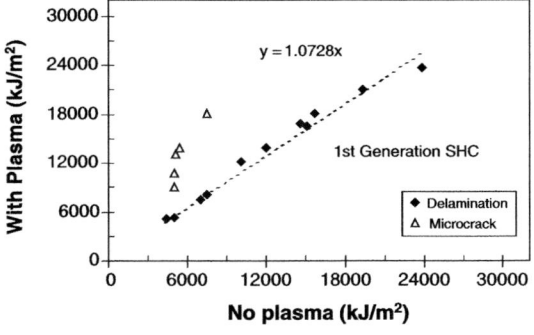

Figure 5— Delamination and micro-crack times of the 1st generation SHC with and without the plasma overcoat. Both PC colors, three SHC thicknesses, both test conditions. The dashed line is the least squares best fit for the delamination data.

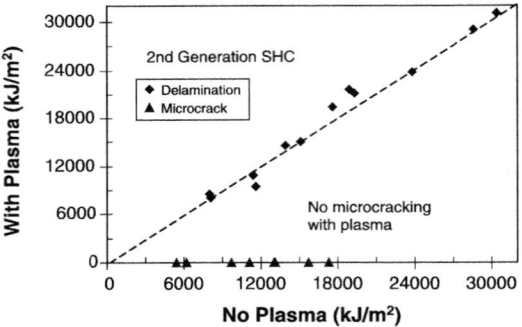

Figure 6—Delamination and microcrack times of the 2nd generation SHC with and without the plasma overcoat. Both PC colors, three SHC thicknesses, both test conditions. The dashed line is at slope = 1 as a guide for the eye.

ed within one check point of each other. No unusual cracking was noted. We have observed unusual blistering and delamination failures when samples saturated with water at 65°C or hotter are very rapidly cooled after a moderate amount of weathering, but this is extremely unlikely to occur under natural conditions. We conclude that humidity has little or no effect on coating lifetime.

Higher Acceleration: Increased Irradiance

The best coatings have lasted over one year of xenon arc exposure under high irradiance conditions. The only way to accelerate delamination times is by increased light intensity, but the Ci35a Weather-Ometers have a practical limit to light intensity for reasons of lamp wattage, lamp lifetime, and temperature control. One solution is to move samples closer to the lamps. Transparent PC samples can be moved closer without much temperature increase since they absorb very little visible or infrared radiation. The measured temperature on the surface of an unbacked clear sample at 4 in. offset was 48°C compared with 42°C at its normal position.

It was not immediately obvious to us that a line source such as a xenon arc lamp would obey the inverse square rule for distance and light intensity. A model was constructed assuming uniform output over the length of the lamp and that each point of the lamp obeyed the inverse square law. The results are shown in *Figure 7* for the sample full back (12.25 in. from the lamp center) and offset 2.5 in. and 4 in. closer. The results show we can expect fairly uniform 1.5x light intensity for samples 2.5 in. closer to the lamp and approximately 2x intensity for samples 4 in. closer with about ± 5% center to top/bottom variation.

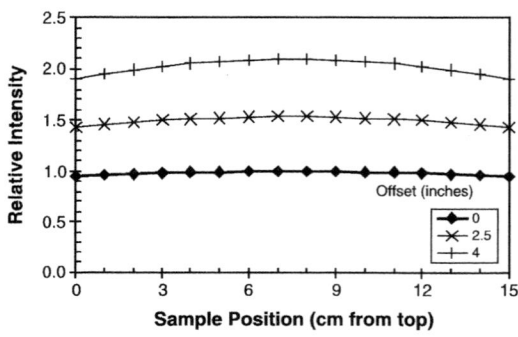

Figure 7—Model showing uniformity of light intensity on a sample holder at normal position in the center rack and offset 2.5 in. and 4 in. closer to the lamp.

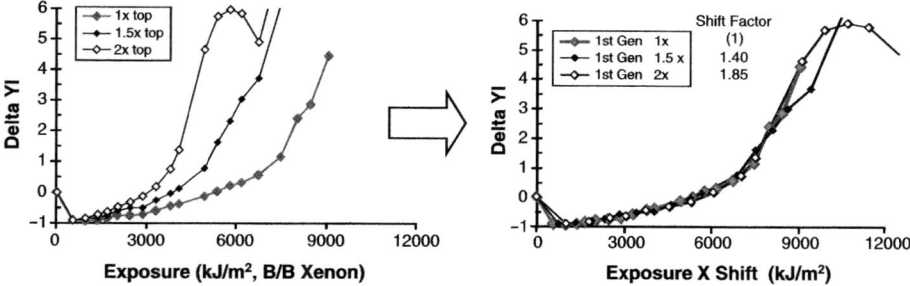

Figure 8— Yellowing results for the 1st generation SHC without plasma overcoat. The abscissas of the 1.5× and 2× samples were multiplied by the shift factors shown in the right graph legend to cause the data to superpose on the 1× data set. These shift factors are the relative rates of yellowing.

Clear samples (two SHCs, with/without plasma, three thicknesses) were exposed in the normal position and at the two offset positions. Typical yellowing results and data treatment are shown in *Figure* 8. The rates are clearly faster for the offsets (1.5×, 2×), but the non-linear data make it difficult to define a rate. The abscissas (X axis) of these data sets can be multiplied by shift factors to cause the data to superpose on the 1× set. These shift factors are the relative rates. If the data could not be superposed, no single number could define the relative rate. This is a very robust method of determining relative rates.[16,17] We see in this case that the relative rates are 1.40 and 1.85, slightly less than the expected 1.5 and 2.0. Relative yellowing rates, delamination, and microcrack times were determined for all of the clear samples, and the results are shown in *Table* 6.

Table 6 shows that the samples moved closer to the lamp underwent degradation measured by yellowing, delamination, and microcracking at approximately 90% of the

Table 6—Rates of Yellowing, Delamination, and Microcracking Times for Samples Exposed 2.5 in. and 4 in. Closer to the Lamp Relative to Samples Exposed at Normal Distance. The Expected Rate Increases were 1.5x and 2.0x, Respectively

		2.5 in. Offset			4 in. Offset		
		Yellowing	Delam	Microcrack	Yellowing	Delam	Microcrack
1st Generation SHC	top	1.40	1.39		1.85	1.83	
	mid.	1.40		1.51	1.90	1.95	1.83
	bot.	1.40	1.39	1.09	1.80	2.13	1.63
1st Generation SHC + plasma	top	1.30	1.31		1.80	1.50	
	mid.	1.35	1.40		1.70		1.99
	bot.	1.40	1.39	1.64	1.80	2.05	1.86
2nd Generation SHC	top	1.25	1.18		1.60	1.69	
	mid.	1.37	1.39	1.35	1.65	1.88	2.11
	bot.	1.50	1.50	1.20	2.00	2.43	1.80
2nd Generation SHC + plasma	top	1.35	1.28		1.75	1.87	
	mid.	1.35	1.33		1.70	1.86	
	bot.	1.50	1.47		2.00	2.23	
Mean		**1.38**	**1.37**	**1.36**	**1.80**	**1.95**	**1.87**
Std. dev.		**0.07**	**0.09**	**0.23**	**0.13**	**0.25**	**0.16**

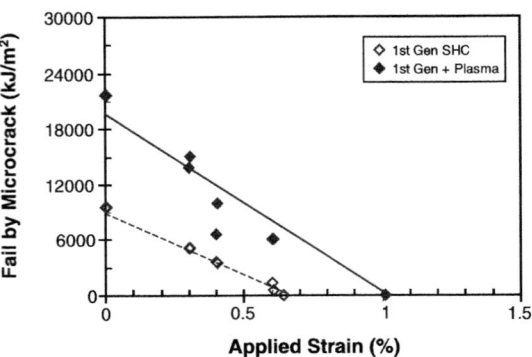

Figure 9— Effect of applied strain on microcracking for the 1st generation hardcoat with and without plasma overcoat.

expected rate increase. The ratio of the two offset rates comes out to be very close to the expected 1.33 (2/1.5). No unusual failure mechanism was observed for the samples closer to the lamp. Therefore, it appears that the samples can be moved 4 in. closer and achieve a 1.8- to 1.9-fold increase in weathering rate. This reduces typical test time to delamination failure to about 5000 h (seven months)—better than the 12–13 months otherwise required.

Higher Acceleration: Applied Strain

Many silicone hardcoats fail by cracking when the coating has degraded such that the stresses caused by temperature cycling exceed the tensile strength of the coating. This can be accelerated by applying a strain to the samples during weathering.[10] The results for this coating series are shown in *Figures* 9 and 10. The plasma overcoat increases the initial strain to microcrack for both the 1st and 2nd generation hardcoats. Indeed, the 2nd generation coating exhibited no microcracking before delamination, even at 0.6% strain. Strain has a minor effect on delamination times as shown in *Figure* 11.

The value of this technique is the ability to test how minor changes made to the substrate, coating formulation, or process might affect weathering in a timely manner. We are investigating microcrack behavior of the 2nd generation coating at higher strains. At 1%

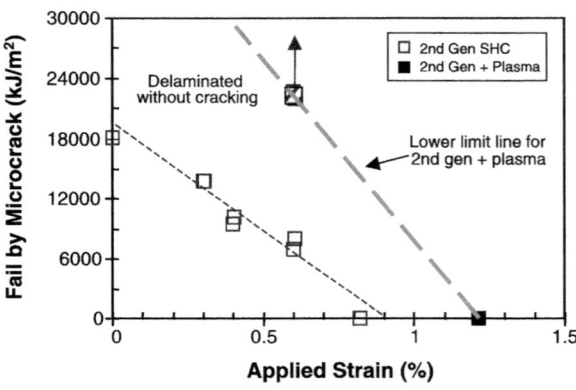

Figure 10— Effect of applied strain on microcracking for the 2nd generation hardcoat with and without plasma overcoat. The samples with the plasma overcoat showed no microcracking before delamination failure at 22,000 kJ exposure.

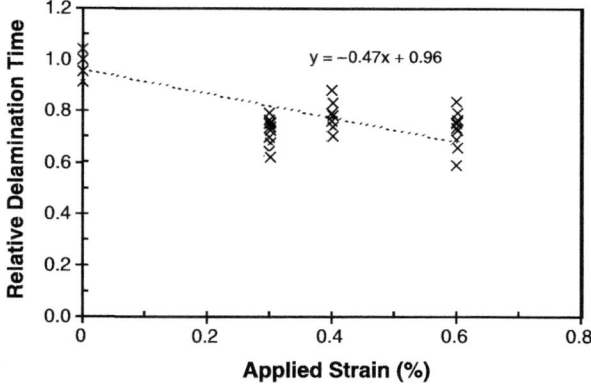

Figure 11—Effect of strain on relative delamination times for all coatings.

applied strain, we expect microcracking after 6,000 to 12,000 kJ/m² of exposure and should be able to detect poorer performing coatings that might result from changes.

Predicted Florida Lifetime for Plasma-Overcoated Samples

We are most interested in the system with the 2nd generation SHC and the plasma overcoat. Data show that its failure mode is delamination. Clear samples have begun to yellow at the delamination time, but not yet severely (Delta YI < 4). Microcracking is not expected since it has not been observed in any of these samples, even under applied strain. We have not observed the formation of more than 2–3% haze. The lifetime is therefore determined by the delamination time.

The delamination times for the coatings are determined by the thickness of the SHC, since it carries the UV absorber that protects the coating/PC interface. *Table 7* shows delamination data gathered at GE Global Research for the 2nd generation SHC on clear PC under both xenon arc exposure conditions. Given the analysis that roughly 3000 kJ/m² of borosilicate/borosilicate exposure should be the approximate equivalent of one year in Florida, and the successful predictions shown in *Table 3*, we expect this coating to last at least eight years when exposed at a 45° angle facing south for coatings with at least 7 μm of SHC, and about 10 years with 8 μm of SHC. The privacy color gets hotter

Table 7—Delamination Times (kJ/m²) for 2nd Generation Coatings on Clear PC Under Xenon Arc Exposure Conditions

	Position	SHC Thickness (μm)	CRD	Gmod
No plasma	Top[a]	4.9	11,600	11,400
	Middle[a]	7.1	23,800	19,300
	Bottom	8.3	28,600	30,400
Plasma	Top[a]	4.9	9,500	10,800
	Middle[a]	7.1	23,800	23,000
	Bottom	8.3	29,000	31,000

(a) These thicknesses are less than specified by Exatec for optimal weathering.

due to its higher light absorption (*Table* 3), and its lifetime will be reduced by approximately 20% to approximately 6.7 years and eight years for 7 and 8 µm SHC, respectively. We have undertaken further analysis to predict the service lifetimes of coatings in real automotive glazing application environments based on Florida weathering data, and this will be reported separately.[18]

CONCLUSIONS

Accelerated weathering data must be treated cautiously when making lifetime predictions. Analysis of the spectral power distributions of Florida sunlight and xenon arc sources suggests that approximately 3,700 kJ/m^2 of borosilicate/borosilicate filtered xenon arc exposure should be the equivalent of one year of Miami, FL sunshine. However, the higher temperature of xenon arc protocols reduces this by about 25% for each 10°C that the test temperature differs from the effective exposure temperature for polycarbonate materials. Empirical data on a wide range of hardcoated polycarbonate show good agreement of xenon arc and Florida results, assuming a correlation of approximately 3000 kJ/m^2 per year in Florida on clear polycarbonate for the test conditions described in this work.

The amount of exposure required to cause failure is very high, and even at the high irradiance level of 0.75 W/m^2, test times are >1 year for samples within the specification range. Samples can also be exposed using offsets to move them closer to the lamp, and an acceleration of 1.9× can be achieved using a 4 in. offset, thereby reducing test times to seven to eight months. Exposure times to microcrack failure can be reduced by applying strain to the samples.

Results from offset and normal position exposure indicate that the delamination of the 2nd generation SHC with or without plasma will occur after approximately 24,000 to 30,000 kJ/m^2 of exposure at a SHC thickness of 7 to 8 µm. The conservative correlation factor of approximately 3000 kJ/m^2 of xenon arc exposure equating to a Florida year gives a predicted lifetime of at least 10 years of direct Florida exposure for coatings with sufficient SHC thickness. Further lifetime prediction models not described in this paper[18] indicate that this translates to >10 years of lifetime for >99% of vehicles in a side window or backlight application.

Extension of these correlation factors to other materials requires some caution. The high absorption of the coating protects the underlying PC from the unnaturally short wavelength UV present in borosilicate-filtered xenon arc that is known to highly accelerate the degradation rate of PC and other aromatic polymers. Bauer's analysis of acceleration factors showed that aromatic ester coatings were accelerated about 2× more than aliphatic acrylic coatings.[5] Temperature corrections will vary if materials activation energies are much more or less than 5 kcal/mol as assumed here. These coatings and the polycarbonate substrate did not show much moisture sensitivity, so that did not have to be taken into account while other coatings and materials are known to be much more sensitive to moisture. Predictive weathering requires knowledge of how environmental variables affect particular materials, reasonable simulation of the critical variables, and corrections for known variance from natural conditions.

ACKNOWLEDGMENTS

We thank Karen Webb of GE Global Research for taking most of the data used in this paper, Mike Mercedes for preparing plasma-coated control samples, Mike Takemori for his strain jigs, and Chuck Iacovangelo and Tom Miebach for helpful discussions. We also thank Warella Browall of GE Silicones and Holly Blaydes of Exatec for weathering data on other coatings.

References

(1) ANSI/SAE Z26.1-1996, "American National Standard for Safety Glazing Materials for Glazing Motor Vehicles and Motor Vehicle Equipment on Land Highways—Safety Standard."

(2) Pickett, J.E., "Permanence of UV Absorbers in Coatings and Plastics," *Service Life Prediction: Methodology and Metrologies,* ACS Symposium Series 805, Martin, J.W. and Bauer, D.R. (Eds.), American Chemical Society, Washington, D.C., p. 250-265, 2002.

(3) Pickett, J.E., "UV Absorber Permanence and Coating Lifetimes," *J. Test. Eval.,* 32 (3), 240-245 (2004).

(4) Nichols, M.E. and Peters, C.A.,"The Effect of Weathering on the Fracture Energy of Hardcoats over Polycarbonate," *Polym. Degrad. Stab.,* 75, 439-446 (2002).

(5) Bauer, D.R., "Interpreting Weathering Acceleration Factors for Automotive Coatings Using Exposure Models," *Polym. Degrad. Stab.,* 69, 307-316 (2000).

(6) Bauer, D.R., "Global Exposure Models for Automotive Coating Photooxidation," *Polym. Degrad. Stab.,* 69, 297-306 (2000).

(7) Pickett, J.E., "Highly Predictive Accelerated Weathering of Engineering Thermoplastics," in *Service Life Prediction: Challenging the Status Quo,* Martin, J.W., Ryntz, R.A., and Dickie, R.A. (Eds.), Federation of Societies for Coatings Technology, p. 93-106, 2005.

(8) Pickett, J.E., "Highly Predictive Accelerated Weathering of Engineering Thermoplastics," *Atlas SunSpots,* 35 (73), 1-11 (Spring, 2005).

(9) ASTM G155-98, "Standard Practice for Operating Xenon Arc Light Apparatus for Exposure of Non-Metallic Materials."

(10) LeGrand, D.G. and Olszewski, W.V., U.S. Patent 4,583,407, April 22, 1986.

(11) *Weathering Testing Guidebook,* Atlas Material Testing Solutions, LLC, p. 20.

(12) Searle, N.D., "Activation Spectra of Polymers and Their Application to Stabilization and Stability Testing," in *Handbook of Polymer Degradation,* 2nd ed., Hamid, S.H. (Ed.), Marcel Dekker, p. 605-643, 2000.

(13) Andrady, A.L., Searle, N.D., and Crewdson, L.F.E., "Wavelength Sensitivity of Unstabilized and UV Stabilized Polycarbonate to Solar Simulated Radiation," *Polym. Degrad. Stab.,* 35, 235-247 (1992).

(14) Pickett, J.E. and Gardner, M.M., "Effect of Environmental Variables on the Weathering of Some Engineering Thermoplastics," *Polymer Preprints,* 42(1), 424-425 (2001).

(15) Pickett, J.E. and Gardner, M.M., "Reproducibility of Florida Weathering Data," *Polym. Degrad. Stab.,* 90 (3), 418-430 (2005).

(16) Simms, J.A., "Acceleration Shift Factor and Its Use in Evaluating Weathering Data," *J. Coat. Technol.,* 59, No. 748, 45-53 (1987).

(17) Gillen, K.T. and Clough, R.L., "Time-Temperature-Dose Rate Superposition: A Methodology for Extrapolating Accelerated Radiation Aging Data to Low Dose Rate Conditions," *Polym. Degrad. Stab.,* 24, 137-168 (1989).

(18) Sargent, J.R., Exatec, LLC, Wixom, MI. Presented as a poster at the 3rd International Symposium on Service Life Prediction, Sedona, AZ, Feb. 1-6, 2004.

Chapter 11

Degradation of Polypropylene Filament Under Elevated Temperature and High Oxygen Pressure

Y.G. Hsuan,[1] M. Li,[2] and Wai-Kuen Wong[1]

[1] Dept. of Civil, Architectural, and Environmental Engineering, Drexel University, Philadelphia, PA 19104
[2] GSE Lining Technology, Inc., Houston, TX 77073

The oxidation degradation of a stabilized polypropylene filament was evaluated in incubation conditions that combined elevated temperature and high oxygen pressure. The incubation temperatures ranged from 45 to 105°C; oxygen pressures were 1.3, 2.8, 4.9, and 6.3 MPa. The antioxidant content in the filament was measured using the high pressure oxidative induction time (HP-OIT) test, while the oxidation of the polymer was evaluated by measurement of the tensile strength. Onset of oxidation corresponded to 10% HP-OIT retained. The temperature effect on the oxidation rate obeys the Arrhenius equation; the oxidation rate is exponentially related to the pressure. Temperature and pressure were found to act independently on the oxidation of the PP filament. The service life at 25°C and one atmosphere was predicted based on a master curve that was generated by applying time-temperature superposition and time-pressure superposition principles.

INTRODUCTION

Over the past 30 years, plastics have become increasingly integrated into many civil engineering projects as engineering materials. Widely known applications include plastic pipes for pressurized and non-pressurized applications, composite bridges, geosynthetics in waste containment and reinforcement, etc. Despite the many advantages that plastics have relative to conventional construction materials, engineers are still reluctant to adopt these new materials. One of the issues is the longevity—i.e., the service life—of polymeric materials. Typically, the service lives of engineering systems range from 50 to 100 years; in some environmentally sensitive applications, service lives of hundreds of years are desired. While the best way to demonstrate the durability of a material is through case histories and/or field data, the history of the application of polymeric materials in civil engineering is relatively short. Therefore, it is necessary to use laboratory accelerated aging as an alternative method to assess the durability of these materials.

Polypropylene (PP) is widely used in plastic pipes, geotextiles, geomembranes, and geogrids. However, PP is well known for its susceptibility to oxidative degradation, leading to polymer chain scission[1,2] and eventually to reduction in engineering properties including tensile strength. Antioxidants (AO) are added to PP to delay the onset of oxi-

171

172

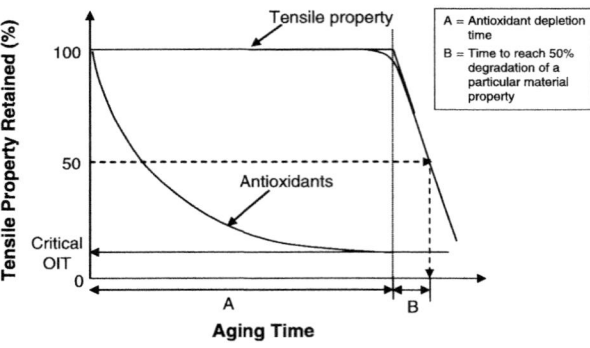

Figure 1—Changes in antioxidants and tensile properties with time.

dation. For a stabilized PP material, the oxidation process is expressed in its engineering properties (here, exemplified by tensile strength) in two stages, as illustrated in *Figure* 1. Stage A is the antioxidant depletion period in which no engineering property changes take place. Stage B refers to the oxidative degradation period during which there is a significant reduction in engineering properties.

In the majority of the aging studies, the lifetime of a material is defined as the time for a selected property to drop to 50% of its initial value. Here, PP filaments have been thermally aged at multiple temperatures and the Arrhenius equation has been used to predict the service life at a specific temperature. For polymers that do not require long service life, the thermal aging is an adequate test method. However, for polymeric construction materials from which a long service life is demanded, thermal acceleration can be very slow. In such cases, it would take an extremely long incubation time to generate meaningful data for service life prediction. Instead of assessing polymer degradation directly, some researchers have studied the depletion of AOs in the polymer at elevated temperatures.[3,4] As shown in *Figure* 1, the depletion of AOs occurs during the initial stage of the oxidation degradation and this depletion is the precursor for the polymer degradation. It has been demonstrated that the mechanical properties begin to decrease when the majority of the AO in the polymer is depleted.[5,6] For a well-stabilized polyethylene geomembrane, more than two years of incubation at temperatures from 55 to 85°C were required in order to predict the lifetime of the AOs.[3] Such a test period is too long for manufacturers and users to evaluate the performance of different antioxidant packages.

To further accelerate the oxidation processes, oxygen pressure has been applied as the second acceleration agent in addition to temperature.[7-10] A standard method is also being developed by the European Committee for Standardization to evaluate geosynthetics made from polyolefins in reinforcement applications. Test samples are immersed in water under high oxygen pressure and elevated temperature. The properties of the incubated samples are monitored with time.

In this study, the oxidative degradation of PP filaments was evaluated under four oxygen pressure levels from 1.3 to 6.3 MPa at seven temperatures from 45 to 105°C. The depletion of antioxidants was monitored by oxidative induction time (OIT), while oxidative degradation of the polymer was assessed by the tensile strength. The changes of tensile strength with time were analyzed using time-temperature superposition (TTS) and time-pressure superposition (TPS) to generate the master curve at service condition of 25°C and one atmosphere so that the service life of the PP filament can be predicted.

Table 1—Properties of As-Received PP Filament

Test Method	Mean Value	Standard Deviation
Density (kg/m^3)—ASTM D 792 901		—
Thickness (mm) 0.95		0.008
Single Filament Tensile Test—ASTM 3822		
Break strength (N)..48		5.0
Break elongation (%)...16.5		5.5
HP-OIT (min)—ASTM D 5885 ....118.3		10.1

Note: The physical, mechanical, and thermal properties listed in the table are based on the results of 5 to 10 replicates of the as-received material in each test.

TEST MATERIALS

Polypropylene filaments were taken from the weft direction of a woven slit film geotextile. *Table* 1 shows the material properties of the PP filament. The antioxidant package used to stabilize the filament was not revealed; however, the manufacturer indicated that it included hindered amine light stabilizer (HALS). Independent chemical analysis using ASTM D 5524 was performed to identify the types of antioxidants. It was found that the AO package consists of Irganox® 3114 (~ 1000 ppm), Irganox® 1076 (~ 700 ppm), and Irgafos® 168 (~ 40 ppm). Although HALS was not detected, its presence could not be completely ruled out since the detection of HALS is known to be challenging.

INCUBATION CONDITIONS

As shown in *Table* 2, the incubation series involved seven temperatures and four oxygen pressures. The incubation was carried out in specially designed high pressure cells. Each cell was equipped with a pressure gauge with accuracy of ± 70 kPa and a thermal well to measure the temperature inside the cell. The pressure cell and forced air oven are shown in *Figure* 2. Pressurization involved replacing the air inside the cell by flushing the cell with oxygen at 1 MPa at least twice. Oxygen was then introduced gradually to

Table 2—Incubation Conditions for PP Filaments

PO$_2$ (MPa)	Temperature (°C)						
	45	55	65	75	85	95	105
1.3	–	–	TS	–	—	–	TS
2.8	–	–	TS	–	TS	–	TS
4.8	–	–	TS	–	TS	–	TS
6.3	OIT/TS[a]	OIT/TS	OIT/TS	TS	OIT/TS	TS	OIT/TS

(a) TS = tensile strength; OIT = oxidative induction time

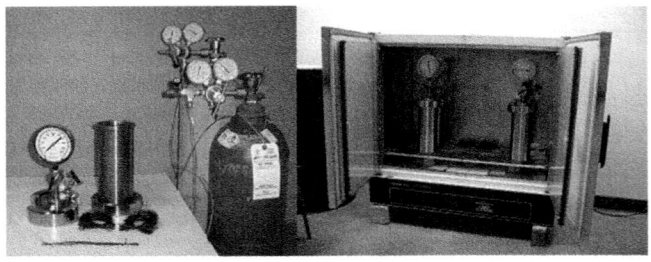

Figure 2—Pressure cell and forced air oven used for the high pressure incubation.

the desired pressure after which the inlet valve was closed to maintain a constant volume. The pressurized cells were then placed in different ovens that were set at the incubation temperatures. The pressure in the cell was monitored weekly. If the pressure decreased more than 0.1 MPa, the cell was depressurized and then repressurized to the desired value.

TEST METHODS

The samples in the pressure cells were retrieved at different time intervals to monitor changes in material properties. Two properties were tested as described below.

Oxidative Induction Time Test

The high pressure oxidative induction time (HP-OIT) test was used to measure the amount of antioxidants remaining in the tape yarns. The reason for using the HP-OIT test to evaluate the antioxidants is the possibility of HALS presence in the antioxidant package. HALS have melting points ranging from 120 to 150°C and cannot be evaluated using the standard OIT test (ASTM D 3895), which is carried out at an isothermal temperature of 200°C. The high test temperature induces an artificial depletion of HALS.[5]

The HP-OIT test was performed according to ASTM D 5885 using a TA 2920 differential scanning calorimeter (DSC) with a high pressure cell. The test uses an isothermal temperature of 150°C and 3.5 MPa oxygen pressure. The specimen is heated from room temperature to 150°C at a constant heating rate of 20°C/min under 0.035 MPa of nitrogen. The specimen is held under nitrogen at 0.035 MPa for 4 min to reach the isothermal temperature of 150°C. Oxygen is then introduced and gradually increased to 3.5 MPa, at which point the time measurement begins. The test is terminated when an exothermal peak is observed. The exothermal peak corresponds to the oxidation of the polymer.

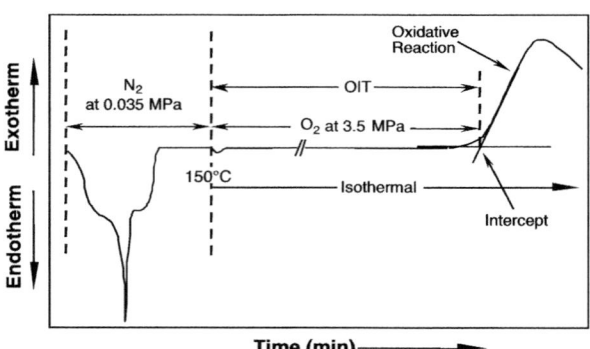

Figure 3—Typical thermal graph illustrating the obtaining of the HP-OIT value.

The onset of the exothermal peak is determined using a tangent method based on the steepest slope, as illustrated in *Figure* 3.

At least two, and sometimes as many as four, replicates are used to establish a particular value of HP-OIT. The coefficient of variation (COV) was found to be approximately 10% for the unaged filaments and approximately 20% for aged filaments.

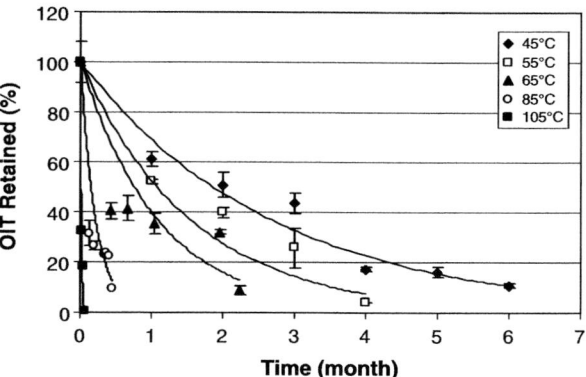

Figure 4—HP-OIT% retained versus time at 6.3 MPa at different incubation temperatures.

Tensile Properties

The tensile strength of the PP filaments was evaluated according to ASTM D 3822. The test equipment was an Instron Model 42060 equipped with a 45 kg load cell. The test was performed at a strain rate of 200 mm/min with a gauge length of 200 mm. Both tensile strength and elongation were automatically measured. Eight specimens were tested for each incubation period. The average value and standard deviation were recorded and used for the analysis.

RESULTS AND DISCUSSION

The changes in the two material properties are presented in percent retained, which is calculated according to equation (1).

$$\% \text{ retained } = \frac{\text{test value of aged sample at a specific incubation time}}{\text{test value of unaged}} *100 \qquad (1)$$

Antioxidant Depletion with Incubation Time

Antioxidant depletion was measured by the HP-OIT test at different incubation times. From previous studies,[3,10] the OIT retained was found to decrease exponentially with incubation time according to equation (2). The equation indicates a first order reaction for the depletion of antioxidants.

$$\ln(OIT\%) = -kt \qquad (2)$$

where OIT% is HP-OIT percent retained after incubation time t, t is incubation time (months), and k is a reaction rate constant which is a function of the incubation temperature and the oxygen partial pressure.

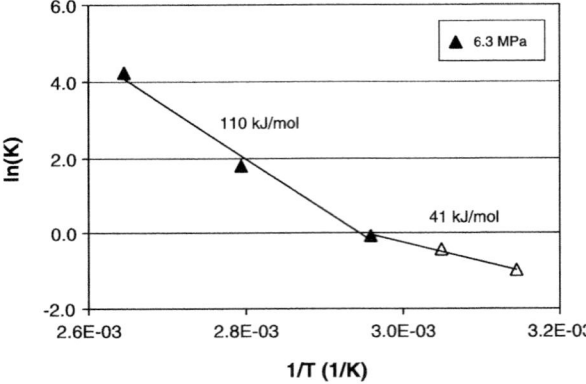

Figure 5—Plotting ln(k) against inverse incubation temperature.

Figure 4 shows the OIT depletion curve under 6.3 MPa at different incubation temperatures. An exponential curve was fitted to each set of the data from which the reaction rate (k) was determined. Figure 5 shows the Arrhenius plot, correlating the reaction rate with temperature. It can be seen that the activation energy does not remain constant over the test temperature range, from 45 to 105°C (a transition is observed at 65°C). The activation energy is much higher at temperatures above the transition.

Changes of Tensile Strength with Incubation Time

In this study, macrostructure changes caused by oxidation in the polymer were monitored through changes in tensile strength. Although the tensile elongation is known to have a good correlation to the degradation of the polymer, crimps caused by the weaving of the geotextile lead to a large variation in the break elongation of the filaments. *Figures* 6 and 7 show the decrease in retained tensile strength with time under pressure and temperature, respectively. The curves were fitted against equation (3), where the *n* values were in the range between 4 and 7 in the majority of curves.

$$\sigma = 100 * \left[1 - \left(\frac{t}{t_f} \right)^n \right]$$ (3)

where σ is percent strength retained (%), t is time (day), t_f = time to reach a strength retained value of less than 20%, n is constant.

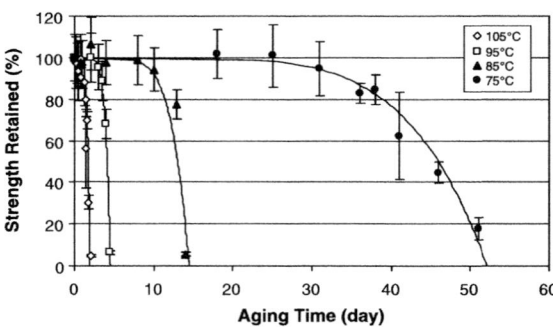

Figure 6—Effects of temperature on the tensile strength with time under pressure of 6.3 MPa.

The onset of the polymer degradation is defined when the tensile strength dropped below 100%. The onset times are correlated to the HP-OIT depletion curves in *Figure* 4 to determine the critical value. *Table* 3 shows the onset times of tensile strength reduction and the corresponding critical HP-OIT retained values under the incubation condition of 6.3 MPa. The average value for

the critical HP-OIT is approximately 10%, or ~11 minutes,
which is close to the sensitivity limit of the HP-OIT test.
This implies that the AOs in
the filament have probably
completely depleted. The AO
content in some of the incubated filaments with tensile
strength retained around 80 to
90% was further verified by
ASTM D 5524. Except for a
small amount of degraded

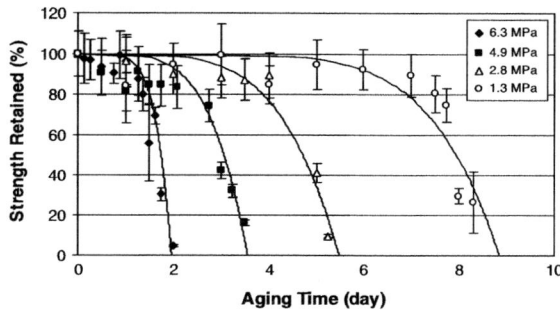

Figure 7—Effects of pressure on the tensile strength with time at 105°C.

Irgafos® 168, no other antioxidants were detected in the filaments. It seems that the HP-OIT test can effectively monitor the amount of AO package in the filament. Furthermore, the onset of tensile strength reduction can be deduced once the critical HP-OIT is estab-lished.

The service life of the PP filament, i.e., Stages A and B of *Figure* 1, is defined as the time to reach 50% tensile strength retained ($t_{50\%}$), which was deduced from equation (3) for each of the incubation conditions. The reciprocal value of $t_{50\%}$ was used as the oxi-dation rate to analyze the oxidation mechanism. *Table* 4 shows the $t_{50\%}$ values for differ-ent incubation temperatures under 6.3 MPa. The time corresponding to the degradation of the polymer, Stage B, was calculated and these values have been listed in *Table* 4. The duration of Stage B varies from 15 to 28% depending on the incubation temperature. This indicates that the service life of the PP filament is largely governed by the AO concen-tration in the polymer. Similar behavior was observed in a stabilized HDPE pressure pipe; the time that took to degrade the polymer was 8% of the total service life.[11] Clearly the efficiency of the AO package is critical to the longevity of the commercial plastics, and yet it is often the least known component in the product.

The relationships between reaction rate and incubation temperature under different pressure conditions are summarized by the Arrhenius plots in *Figure* 8. As with the antioxidant depletion curve shown in *Figure* 4, the activation energy is not constant over the whole test temperature range; there is a transition around 65°C at 6.3 MPa. From 65 to 105°C, the Arrhenius lines are roughly parallel to each other, indicating that pressure

Table 3—Onset Time of Strength Reduction and Corresponding HP-OIT Value

Incubation Condition	105°C 6.3 MPa	85°C 6.3 MPa	65°C 6.3 MPa	55°C 6.3 MPa	45°C 6.3 MPa
Onset time of strength reduction month	0.042	0.33	2.23	4.0	8.0
Critical HP-OIT retained (%)	6% (7 min)	14% (16 min)	12% (14 min)	7% (8 min)	8% (9 min)

Table 4—Time to Reach 50% Strength Retained and %

Incubation Condition	105°C 6.3 MPa	85°C 6.3 MPa	65°C 6.3 MPa	55°C 6.3 MPa	45°C 6.3 MPa
Onset time of strength reduction month	0.042	0.33	2.23	4.0	8.0
$t_{50\%}$ (Stages A & B) (month)	0.058	0.43	2.7	4.7	9.5
% total time (Stage B) (month)	28%	23%	17%	15%	16%

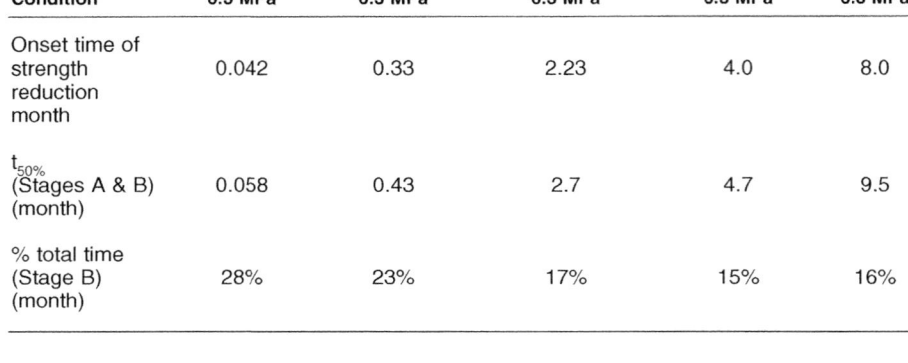

Figure 8—Arrhenius plots for different pressures.

has little effect on the temperature-controlled mechanism. For the temperature range above 65°C, the average activation energy is calculated to be 112 kJ/mol, while at temperatures below 65°C, it is only 56 kJ/mol. These two activation energy values are similar to those of the antioxidant depletion reaction. The change in activation energy with temperature has also been observed by other researchers. Gugumus[12] detected a decrease in activation energy at temperatures around 80°C for stabilized and unstabilized PP. Ding[13] found a non-linear line in the Arrhenius plot for the oxidation of PP cable in oven aging at temperatures between 30 and 90°C. The change of activation energy implies that the Arrhenius extrapolation cannot be applied to predict low temperature performance using data obtained from high temperature incubation. With 65°C as the transition point, different activation energies should be applied. Although such transition at lower oxygen pressures should be veri-

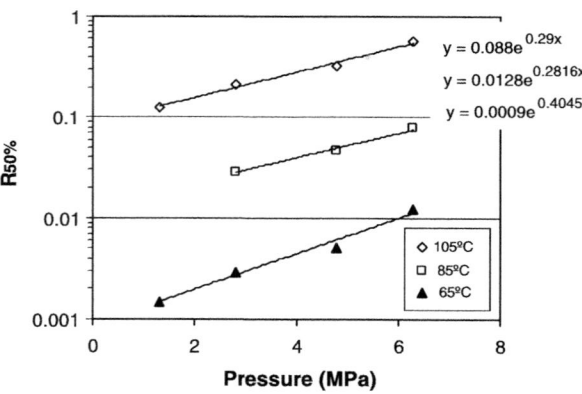

Figure 9—Relationship between reaction rates and pressure.

fied, it raises the question of the reliability of service life prediction estimates made from elevated temperature acceleration tests, particularly using temperatures above 100°C for PP. It is particularly worrisome that a high activation energy value would predict a long service life based on the Arrhenius model. For the PP filament in this study, the service life is 17 times longer for a prediction based on 112 kJ/mol than on 56 kJ/mol.

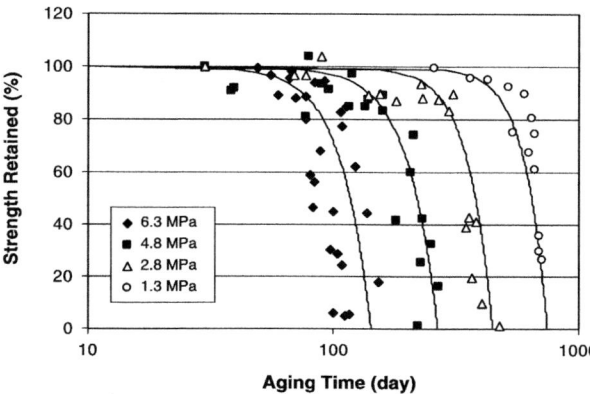

Figure 10—Strength versus time master curves at 65°C under different pressures.

The relationship between the reaction rate and pressure is shown in *Figure* 9. The reaction rate is exponentially related to the pressure, as expressed in equation (4). At 85 and 105°C, the slopes of the lines (β) are very similar while the slope at 65°C is slightly higher.

$$R = \alpha * e^{\beta P} \tag{4}$$

where R = reaction rate, P = incubation pressure, and α and β are constants

Oxidation Prediction

Commonly, the prediction of oxidative degradation is performed on a single data point $t_{50\%}$. As pointed out by Gillen et al.,[14] such a method significantly depreciates the test data since it confines the prediction to a single value. Furthermore, it introduces a greater uncertainty in the prediction due to the limited data points on which it is based. Gillen, et al.[14,15] applied the time-temperature-dose rate superposition principle to predict the tensile degradation of an elastomer. A master curve was generated by shifting test data from elevated test temperatures to a lower service temperature. Scattering of the data at the service temperature would provide the analyst with an indication of confidence value one could assign to this prediction.

In this study, the degradation of tensile strength with time is further evaluated by applying the TTS and TPS methods. The relatively parallel lines in both temperature and pressure plots of *Figures* 8 and 9 indicate that the temperature and pressure effects on the acceleration of polymer oxidation can be evaluated separately. The tensile strength versus time curves at different incubation conditions are shifted to a single condition to generate a master curve at the service temperature and pressure.

The following procedures were carried out to generate a master curve at a reference temperature and pressure of 25°C and 0.02 MPa, respectively. This master curve is pred-

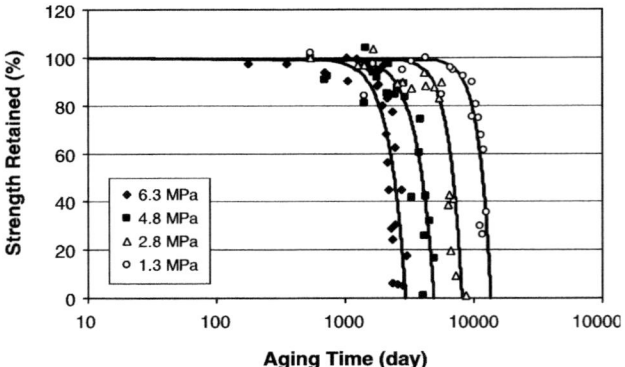

Figure 11—Strength versus time master curves at 25°C under different pressures.

icated on the assumption that the same oxidation mechanism holds from 1.3 MPa to 0.02 MPa at temperatures above and below 65°C.

(1) Apply TTS to obtain the master curve at 65°C for each pressure condition. The shift factor is expressed in equation (5) with an activation energy (E_a) of 112 kJ/mol. *Figure* 10 shows the four master curves at different pressures.

$$a_T = \exp\left[\frac{E_a}{R}\left(\frac{1}{T_{ref}} - \frac{1}{T}\right)\right]$$

(5)

(2) The four master curves in *Figure* 10 are shifted again from 65 to 25°C using equation (5) with an activation energy (E_a) of 56 kJ/mol. The four strength degradation curves at temperature of 25°C under different pressures are displayed in *Figure* 11.

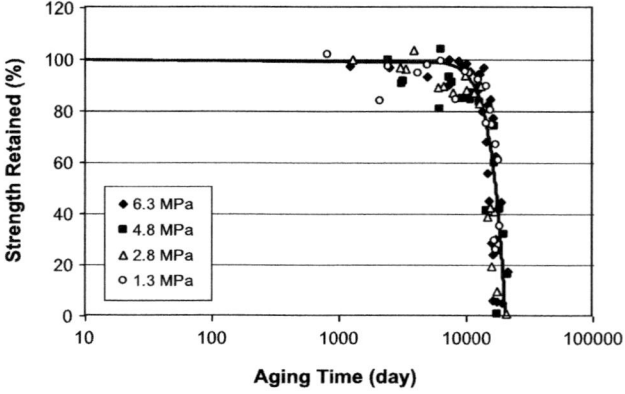

Figure 12—Master curve at 25°C and 0.02 MPa.

(3) Apply TPS, as expressed in equation (6), to shift the four different pressure curves in *Figure* 10 to a single pressure at 0.02 MPa, as shown in *Figure* 12.

$$a_P = \exp\left[\beta\left(P - P_{ref}\right)\right] \tag{6}$$

where $\beta = 0.325$ MPa^{-1} (average slope value in *Figure* 8)

The master curve provides failure times for different percentage strength retained. The predicted service life, $t_{50\%}$, is 47 years at 25°C and one atmosphere.

CONCLUSIONS

The oxidation of PP filaments with a specific antioxidant package was evaluated under high pressures and elevated temperatures. The depletion of the AOs in the filaments can be effectively monitored by the HP-OIT test. The oxidation degradation of the polymer, which was measured by the tensile strength of the filament, occurred when AOs in the polymer were essentially depleted.

The Arrhenius plots revealed that the activation energy was non-linear between 45 to 105°C under pressure of 6.3 MPa. A transition around 65°C was observed. The oxidation rate was found to increase exponentially with pressure. In addition, there was no synergistic effect between temperature and pressure, and therefore their acceleration of the oxidation process can be assessed separately. TTS and TPS were found applicable to predict the service life of the PP filament.

ACKNOWLEDGMENT

This research is supported by the National Science Foundation (Grant No. CMS 9872285) and the U.S. Federal Highway Association (Contract No. DTFH 61-98-R-00043).

References

(1) Grassie, N. and Scott, G., *Polymer Degradation and Stabilisation*, Cambridge University Press, England, p. 222, 1985.

(2) Billingham, N.C. and Calvert, P.D., Chapter 5: "The Physical Chemistry of Oxidation and Stabilization of Polyolefins," in *Developments in Polymer Stabilization*, 3rd Ed., Scott, G. (Ed.), Applied Science Publishers, London, p.139, 1980.

(3) Hsuan, Y.G. and Koerner, R.M., "Antioxidant Depletion Lifetime in High Density Polyethylene Geomembranes," *J. Geotechn. Geoenvironmental Eng.*, ASCE, Vol.124, No. 6, p. 532–541 (1998).

(4) Viebke, J. and Gedde, U.W., "Assessment of Lifetime of Hot-Water Polyethylene Pipes Based on Oxidation Induction Time Data," *Polym. Eng. Sci.*, Vol. 38, No. 8, p. 1244–1250 (1998).

(5) Hsuan, Y.G. and Guan, Z., "Antioxidant Depletion During Thermal Oxidation of High Density Polyethylene Geomembranes," *Proc. Sixth International Conference on Geosynthetics*, Atlanta, GA, p. 375-380, 1998.

(6) Karlsson, K., Eriksson, P-A, Hedenqvist, M., Ifwarson, M., Smith, S.D., and Gedde, U.W., "Molecular Structure, Morphology and Antioxidant Consumption in Polybutene-1 Pipes in Hot-Water Applications," *Polym. Eng. Sci.*, Vol. 33, No. 5, p. 303-310 (1993).

(7) Salman, A., Elias, V., and DiMillio, A., "The Effect of Oxygen Pressure, Temperature and Manufacturing Process on Laboratory Degradation of Polypropylene Geosynthetics," *Proc. Sixth International Conference on Geosynthetics*, Atlanta, GA, p. 683- 690, 1998.

(8) Schroeder, H.F., Bahr, H., Herrmann, P., Kneip, G., Lorenz, E., and Schmuecking, I., "Durability of Polyolefin Geosynthetics under Elevated Oxygen Pressure in Aqueous Liquids." *Proc. 2nd European Geosynthetics Conference* (EuroGeo 2), Italy, p. 459-464, 2000.

(9) Vink, P. and Fontijn, H.F.N., "Testing the Resistance to Oxidation of Polypropylene Geotextiles at Enhanced Oxygen Pressures," *Geotextiles and Geomembranes*, 18, p. 333-343 (2000).

(10) Li, M. and Hsuan, Y.G., "Temperature and Pressure Effects on the Degradation of Polypropylene Tape Yarns–Depletion of Antioxidants," *Geotextiles and Geomembranes*, Vol. 22, No. 6, p. 511-530 (2004).

(11) Gedde, U.W., Viebke, J., Leijstrom, L., and Ifwarson, M., "Long-Term Properties of Hot-Water Polyolefin Pipes—A Review," *Polym. Eng. Sci.*, Vol. 34, No. 24, p.773-1787 (1994).

(12) Gugumus, F., "Effect of Temperature on the Lifetime of Stabilized and Unstabilized PP Films," *Polym. Degrad. Stab.*, 63, 1, pp. 41-52 (1999).

(13) Ding, S., Ling, M.T.K., Khare, A., and Woo, L., "Activation Energies of Polymer Degradation," in *Plastics Failure—Analysis and Prevention*, Moalli, J. (Ed.), William Andrew Publishing/Plastics Design Library, p. 219-225, 2001.

(14) Gillen, K.T., Clough, R.L., and Wise, J., "Prediction of Elastomer Lifetimes from Accelerated Thermal-Aging Experiments," *Polymer Durability*, ACS, p. 557-575 (1999).

(15) Gillen, K.T. and Clough, R.L, "Time-Temperature-Dose Rate Superposition: A Methodology for Extrapolating Accelerated Radiation Aging Data to Low Dose Rate Conditions" *Polym. Degrad. Stab.*, 47, p. 137-168 (1989).

Chapter 12

Risk Analysis for Pipeline Assets—The Use of Models for Failure Prediction in Plastics Pipelines

Stewart Burn, Paul Davis, and Scott Gould

CSIRO Land and Water, Highett, Australia

A key component in any risk-based asset management strategy is the ability to predict when assets will fail. For pipe materials such as cast iron (CI) and asbestos cement (AC), sufficient failure data are available to allow statistical failure models to be developed. However, for newer materials such as polyvinylchloride (PVC) and polyethylene (PE), only limited failure data are available. Lifetime prediction for brittle polymers can be based on Linear Elastic Fracture Mechanics (LEFM) theory. LEFM uses the concept of a single parameter known as the applied Stress Intensity Factor (SIF) that character-izes the stress field at the tip of a crack in a plastics material. The dependence of crack growth rate on the applied SIF can be determined using small labora-tory scale tests on coupon samples, which can then be applied to the geometry of a flawed asset in service via a geometrical correction factor. While this approach can be applied to comparatively brittle materials such as unplasti-cized polyvinylchloride (PVC-U) and older PE materials, newer PE materials can form a large craze zone at the crack tip, which renders LEFM invalid. An extension to LEFM theory, Elastic Plastic Fracture Mechanics (EPFM), can be used to model post-yield fracture that occurs after plastic deformation in duc-tile materials. However, care must be taken to distinguish plastic fracture from crack tip blunting. In the case of crack tip blunting failure mechanism, EPFM failure predictions cannot be transferred from small test geometries to larger pipes in service and other approaches such as craze mechanics theory must be applied. When developing a failure model for new polymers with high, slow crack growth resistance, it is essential that the underlying theory (e.g., LEFM) is validated. Recent advances in failure mechanism research has led to a novel test method which isolates the craze formation and separation process in tougher pipeline materials such as the newer bimodal PE materials. By com-paring measured craze strength from this new method with material yield strength, criteria can be applied which allow selection of the appropriate life-time prediction methodology.

INTRODUCTION

Asset management of water reticulation assets entails obtaining the optimal value from the asset. This needs to occur throughout its entire lifecycle, from design through construction, operation, maintenance, renovation, and finally disposal. The critical factor that controls the latter four factors is asset failure, either through physical pipe failure or

Table 1—Average Failure Rates (per 100 km) from UKWIR National Failure Database[1]

Material Group/Year	Average Failure Rate per 100 km					
	1998	1999	2000	2001	2002	Average
Asbestos cement	16.4	17.1	15.1	15.8	15.6	16
Ductile iron	5	5.3	4.8	4.8	6.5	5.28
Cast iron	23.7	23.7	19.1	21.7	12.3	20.1
PE	3.5	2.9	3.3	3.1	3.0	3.16
PVC	9.6	9.1	7.2	7.4	3.3	7.32
Steel	5	6.1	5.8	5.7	33.1	11.14
Unknown	0.1	0	0	0.1	15.9	3.22

Note: The UK experienced a peculiarly high PVC pipe failure rate due to poorly fused products from some manufacturers.

water quality issues. In asset management methodologies, assets are managed to deliver value to the community, and, in this respect, the costs of managing the asset should be less than the benefits derived by the community. This strategy can be achieved using a risk-based methodology to determine the correct strategy to take throughout the asset's lifetime. For most water reticulation assets such as pipe networks, many water authorities are reactive and let risk manifest itself as pipe failure. However, for more critical assets such as sewer rising mains or large diameter water reticulation pipes, the consequence of failure is so large that risk cannot be allowed to be expressed as asset failure and proactive action must be taken to minimize the risk before failure occurs. Risk is a multi-dimensional concept that is influenced by many factors such as the environmental conditions, operational practices, as well as the social, political, and economic aspects. Generally, such a complex analysis of risk is not carried out and risk is often considered to be the product of the probability of failure, $Pr(f)$ and the consequence of failure $C(f)$:

$$Risk = Pr(f).C(f) \qquad (1)$$

The consequence of failure depends on a number of environmental, social, and economic factors that can normally be determined using so-called Triple Bottom Line or sustainability analysis. Generally, a multi-criteria approach is used to carry out a full sustainability analysis, which allows ranking of different scenarios. However, for most water assets this approach is currently too complex and a simple financial approach is utilized to determine the consequence of failure.

The probability of failure depends on a significant number of factors associated with pipe manufacture, pipeline design, pipeline installation, operating environment, as well as maintenance frequency and quality; and for most asset types where water authorities let assets operate to failure, the probability of failure is expressed as a failure rate. The failure rate depends on the pipeline material as well as the operating and installation characteristics. As shown in *Table 1*,[1] older materials such as cast iron (CI) and asbestos cement (AC) generally have higher failure rates compared to newer materials such as Ductile Iron (DI), Polyvinylchloride (PVC), and Polyethylene (PE).

For older materials, sufficient failure data exists to allow the development of statistical models that predict future failures relatively well. However, for newer materials the failure data is limited and alternative approaches, such as physical/probabilistic models, need to be used. This chapter details a comprehensive methodology that can be used to predict pipeline failure in plastics pipes operating under static pressures if the operating and installation conditions are known.

SERVICE LIFE PREDICTION METHODS FOR POLYMER PIPES

Polymeric materials subjected to hydrostatic pressure generally fail by one of three mechanisms as shown in *Figure* 1.

For Stage I of the hydrostatic pressure test data, the pipe is mainly subjected to mechanical overload and ductile failure occurs by plastic instability. Ductile yielding is the end result of the gradual creep extension of the pipe, which results in a thinning wall and a growing diameter, both of which contribute to the development of a steadily larger true hoop stress. The driving force in this case is the bulk, or average value of the developed hoop stress.

In the transition to Stage II, the time duration of the creep extension and ductile failure processes observed in Stage I is extended (due to the lower applied stress) and failures begin to occur by a quasi-brittle mode of slow crack growth.[2] Stage II is often referred to as beyond the "mechanical knee" in hoop stress versus time data for PE materials and corresponds to the brittle-like failures which occur from inherent defects in the pipe wall at these lower applied stress levels.[3]

In the later stages of this mechanical knee, a second transition is thought to exist to Stage III, when the heat stabilizer package of the raw material is consumed and brittle fracture failures are induced by thermal degradation. Stage III is referred to as beyond the "chemical knee" and failure times are effectively independent of applied stress levels.[3] Although *Figure* 1 represents the transitions in failure mechanism that can occur at extended times, it should be noted that it does not relate directly to possible field failures. For example, Stage I is seldom, if ever, observed in field failures and whilst Stage III is also not generally seen, there is one known exception. This occurred for PE pipe made by a single manufacturer, circa 1974, which was found about 10 years later to become brittle, causing the development of premature failures in water service applications. Laboratory tests indicated that the embrittlement was the result of oxidation due to a deficiency in stabilizer levels. This observation led to the establishment of a test requirement for PE pipes requiring that the material contained in extruded PE pipes must meet certain minimum thermal stability criteria. Since then, there has been no report of a field failure caused by the Stage III process. However, further research is required to demonstrate that Stage III failure will not occur over the extended lifetimes required by PE pipes. In practice, Stage II is the dominant failure mode that has been observed to

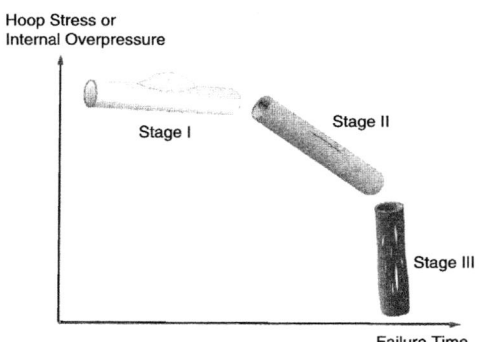

Hoop Stress or
Internal Overpressure

Stage I

Stage II

Stage III

Failure Time

Figure 1—Schematic illustration of three stages during the service lifetime of a pressurized PE pipe.[3]

occur in the field. Depending upon the polymer studied, Stage II failures can be predicted using Linear Elastic Fracture Mechanics (LEFM), Elastic Plastic Fracture Mechanics (EPFM), or Craze Mechanics (CM) service life modeling approaches.

Linear Elastic Fracture Mechanics (LEFM)

In the case of brittle fracture in materials such as PVC-U, Linear Elastic Fracture Mechanics (LEFM) theory idealizes the fracture process as a balance between an applied stress intensity factor (SIF) and the fracture toughness of the pipe material.[4] The applied SIF is determined by in-service loading conditions and the "current condition" of the pipe. Service loading can be split into contributions from operating pressure, external soil loading, residual stress, and external impact. The current condition of a pipe is characterized by the level of damage (i.e., external surface scratches and inclusions) in the pipe wall. A possible set of conditions that could lead to fracture failure in service is illustrated schematically in *Figure* 2.

In this case, the inner surface of a buried plastic pipe is subjected to tensile stresses from internal pressure, and bending stresses from deflection loads associated with soil and traffic loads. If we consider a flaw at the crown or invert of the pipe, the magnitude of the applied stress will increase at the tip of a sufficiently severe flaw and crack initiation will eventually occur. This crack will slowly grow through the pipe wall thickness until at some point the applied SIF will exceed the fracture toughness of the pipe material and brittle fracture will occur. Although it is a less severe case, an external surface flaw at the spring-line of the pipe would grow radially inward, eventually producing brittle fracture. This is a less severe case because the residual stress at the outer surface of an extruded polymer pipe is compressive in the circumferential direction, compared to tensile stresses at the inner wall. Consequently, crack initiation from a defect located at the outer surface of the pipe is less likely to occur.

In addition to those stresses illustrated in *Figure* 2, a buried pipe can be subjected to other localized stress increasing situations (i.e., rock impingement, gouges, differential settlement, poorly made fusion joints, sudden changes in geometry), which are major factors in the development and propagation of slowly growing cracks. A polymeric pipe must exhibit high resistance to the potentially adverse effect of all these "normally" occurring localized stress increasing situations, if it is to perform satisfactorily in service.

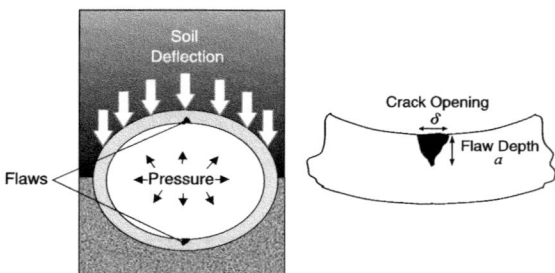

Figure 2—Expected operating conditions of a buried plastic pipe.

According to LEFM theory, the fracture process can be split into three stages, which combine to determine the total failure time under a static load[4]:

- An incubation period between load application and crack initiation
- A period of slow crack growth
- Fast brittle fracture when the applied SIF exceeds the material fracture toughness

To calculate service lifetimes associated with each of these stages, the SIF associated with a flaw in the pipe wall must be quantified. The applied SIF is determined by the in-service loading conditions and the "condition" of the pipe. Service loading can be split into contributions from operating pressure, external loading, and residual stress, whilst the condition of a pipe is characterized by the size of flaws or defects in the pipe wall. Such defects could originate from external surface scratches and/or melt inclusions. In many cases, polymer pipes are extruded and then cooled by water spray on their outside surface only. As stated above, this produces a through thickness residual stress profile comprised of a tensile stress at the pipe inner surface and a compressive stress at the pipe outer surface. Since the outer surface is likely to be subjected to a compressive residual stress, the effect of external surface scratches is less severe than defects located at the pipe inner surface. In fact, a frequently observed defect type that can cause crack initiation and failure in polymer pipes is an elliptical or "thumb-nail" shaped defect located at the inner surface of the pipe.

For a defect of radial depth, a, in the pipe wall, an applied SIF (K), can be defined which relates the remotely applied stresses to the stresses at the defect tip as detailed in equation (2):

$$K = \sigma Y \left(\pi a \right)^{1/2} \qquad (2)$$

where Y is a geometric correction factor, which accounts for the influence of pipe geometry and loading type on K. Geometric correction factors have been previously defined for plastic pipes under combined internal pressure and diametrical deflection, as well as for residual stresses.[5] For a buried pipe, a total applied SIF can be defined as

$$K_{ITOT} = K_{IP} + K_{ID} + K_{IR} \qquad (3)$$

where K_{IP} is the applied SIF from internal pressure, K_{ID} is the SIF from deflection, and K_{IR} is the SIF for residual stress. It should be noted that for simplicity, equation (3) only includes the effect of those stresses depicted by *Figure* 2. As stated above, other conditions can be present in actual buried pipe situations, such as point loads, that produce a possibly greater total stress intensity factor than that which is predicted from equation (3). Previous researchers have proposed that the incubation period between crack initiation ending and slow crack propagation beginning, occurs when the defect or crack opening displacement (COD) attains a critical value.[4,6] This critical COD, $\delta*$, is written in terms of the applied SIF as

$$\delta* = \frac{K_I^2}{E \sigma_Y} \qquad (4)$$

where E is the Young's modulus, σ_Y is yield stress, and K_I equates to K_{ITOT} from equation (3). If visco-elastic behavior is represented by the time-dependent Young's modulus $E = E_0 t^{-p}$ and $\sigma_y = \sigma_0 t^{-q}$, then the incubation time, t_i, can be written as

$$t_i = \left[\frac{E_0 \sigma_0 \delta *}{K_I^2} \right]^{\frac{1}{p+q}} \tag{5}$$

However, previous research has shown that equation (5) can significantly over-predict initiation times at low applied stress intensity factors.[7] After crack incubation ends, a period of slow crack growth occurs. Under a static load, fracture mechanics theory proposes that K can be related empirically to crack growth rate da/dt in the form

$$\frac{da}{dt} = DK^m \tag{6}$$

where D and m are empirical constants, which can be determined from laboratory fracture tests.[8] Assuming that the crack grows in a stable manner through the pipe wall, equation (6) can be integrated between the limits of defect size and pipe wall thickness to obtain the time to rupture. It is also assumed that the geometry factor Y in equation (3) remains constant during crack growth. Under internal pressure only, this is reasonable since the majority of time to rupture is spent with the crack length small (less than half the pipe wall thickness). Under these conditions, Y is effectively constant.[9] Integrating equation (6) between the limits of initial defect size and pipe wall thickness gives the time to stress rupture τ_{SR} as

$$\tau_{SR} = \left[\frac{2\left(\pi^{1/2} Y \sigma_H\right)^{-m}}{D(2-m)} \right]\left[b^{1-(m/2)} - a_o^{1-(m/2)} \right] \tag{7}$$

where σ_H is the hoop stress in the pipe wall, b is the pipe wall thickness, and a_o is the initial defect size. It can be shown that if b is much greater than a_o, and m is sufficiently large, then equation (7) reduces to

$$\tau_{SR} \cong \left[\frac{2\left(\pi^{1/2} Y \sigma_H\right)^{-m}}{D(m-2)} \right] a_o^{1-(m/2)} \tag{8}$$

The crack growth exponent m can be obtained experimentally by subjecting pipes with different inherent defect sizes to internal pressure and recording times to failure. It should be noted that the integration of equation (6) assumes that crack growth is controlled by a single applied stress intensity factor. In many cases, this generally means that secondary influences are ignored and the total SIF K_{ITOT} is assumed to be equal to the contribution from internal pressure only, K_{IP}. A more rigorous treatment that includes contributions to K_{ITOT} from deflection and residual stress is reported in references 5 and 7. In this approach, equation (6) is discretized into a number of small time steps. A repeated calculation is then performed in which, at the end of each time step, the crack length is incremented by the amount of crack growth produced by K_{ITOT}. This approach allows all contributions to K_{ITOT} to be included in the analysis. However, it should also be noted that the decrease in residual stress over time due to ageing and stress relaxation remains unaccounted for.

Sometime during this growth period and generally before the crack grows completely through the pipe wall thickness, the crack will attain a critical size and fast brittle fracture will occur. The criterion for the stress intensity at which this fast failure occurs can be expressed as

$$K_{ITOT} = K_{IC} \qquad (9)$$

where K_{IC} is the material fracture toughness that can be measured using a range of different test methods.[10]

A rigorous analysis of each of these mechanisms can determine the time associated with each stage of the failure process (i.e., incubation followed by slow crack growth and then fast brittle fracture), and combines them to give the total time to failure.[5] However, a simpler alternative approach is to simply use results from fracture toughness testing under a range of test loads.[8,11,12] This can provide an empirical relationship between fracture toughness and time in the form[5]

$$K_{IC} = At^B \qquad (10)$$

where t is time in hours, and A and B are constants. Having established such a relationship, the actual total SIF applied for a pipe in service can be obtained from equation (3), and used in equation (10) to determine a time to failure.

Elastic Plastic Fracture Mechanics (EPFM)

Fracture in more ductile materials such as polyethylene (PE) involves extensive plastic deformation and in this case an Elastic Plastic Fracture Mechanics (EPFM) approach may be applicable.[13] The elastic-plastic regime involves a number of additional complications that are not present in the elastic regime. First, there is inherent non-linearity in material deformation and the large geometry changes that accompany fracture. A further complication is the significant crack blunting that occurs prior to crack initiation and also in the stable crack growth phase that can occur prior to unstable final fracture. The various stages of ductile fracture are shown in the schematic diagram in *Figure 3*.

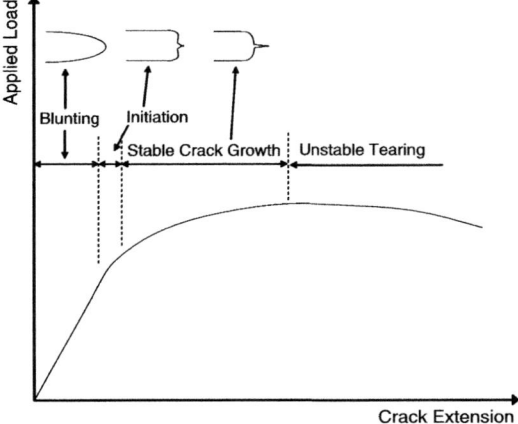

Figure 3—Schematic diagram illustrating the stages of ductile fracture.[14]

In EPFM, an engineering approach to predicting failure can be based on the crack tip characterizing parameter known as the J-integral. As outlined by Williams,[13] J is an energy release per unit area of crack face for a non-linear elastic solid. The J-integral provides an energy release rate (expressed in kJ/m^2), and corresponds to a crack driving force per unit width of crack front (expressed in Newtons/metre). The General Electric (GE) Handbook[14] contains tabulated solutions for J obtained using finite-element stress analysis for ductile materials. For example, for an internally pressurized cylinder with a radial flaw at the inner surface, the elastic-plastic crack driving force (J) is expressed as detailed in equation (11)[14]:

$$J = \frac{K_{ITOT}^2}{E} + \alpha \varepsilon_0 \sigma_Y c h_1 \left\{ p / p_0 \right\}^{n+1}$$

(11)

In equation (11), K_{ITOT} is the total applied SIF from LEFM theory; E is the material Young's modulus; c is the un-cracked ligament length; p is the applied internal pressure; p_0 is the limit pressure assuming perfect plasticity; σ_Y is the material yield strength; and the parameters α, ε_0, and n are obtained from the stress-strain relationship for the material. The dimensionless EPFM function h_1 is tabulated for a range of crack lengths and pipe sizes in the *GE Handbook*.[14]

Although the equations are omitted for simplicity, the applied SIF K_{ITOT} and the dimensionless factor h_1 in equation (11) both depend on crack length. To account for the crack tip craze zone, EPFM theory defines the "effective" crack length a_e as the original crack length a plus an additional length ϕr_y, which is some fraction of the crack tip craze zone.

$$a_e = a + \phi r_y$$

(12)

As outlined by Bernal et al.,[15] the additional crack length is given by

$$\phi = \frac{1}{1 + \left(p / p_0 \right)^2}, \quad r_y = \frac{1}{6\pi} \left[\frac{n-1}{n+1} \right] \left(\frac{K_{ITOT}}{\sigma_Y} \right)^2$$

(13)

where p is the applied internal pressure. Although there is no EPFM-based expression to predict time-dependent crack growth directly, the effect of time can be incorporated indirectly through the change in craze zone length. As outlined by Atkins and Mai,[16] the time dependence of material yield strength σ_Y follows a power law in the form

$$\sigma_Y = \sigma_0 t^{-q}$$

(14)

As time progresses, σ_Y decreases as per equation (14), and, consequently, the length of the craze zone will increase following equation (13). According to equation (12), this will also produce an increase in the effective crack length a_e. This increase in effective crack length will produce corresponding increases in the applied SIF K_{ITOT} and the dimensionless EPFM function h_1. Finally, referring back to equation (11), the net effect of the time-dependent decrease in yield strength is to increase the crack driving force J.

Having defined the crack driving force using the J parameter, a failure criterion for fracture under EPFM conditions is required. J-crack growth resistance values (often denoted as "J_R") can be determined for a particular material by following an experimental protocol described by the European Structural Integrity Society.[17] The recommended test method uses either compact tension or single-edge notch bend (SENB) specimens, which are subject to a constant displacement rate until failure. For each specimen, J_R is determined by numerically integrating under the load versus displacement plot resulting from each test. To fully characterize J_R for a particular material, a series of specimens are loaded to different final displacements and the amount of crack extension that occurs during the test is measured. By plotting crack growth Δa against the corresponding J_R value, a material J-resistance (J-R) curve can be generated. Previous tests on ductile polymers indicate that material J-R curves follow a power law form given by

$$J_R = A\Delta a^N \tag{15}$$

where A and N are empirical constants determined by fitting experimental J_R versus Δa data. A critical J_R value required for crack initiation is arbitrarily defined as $J_{0.2}$, or the value of J_R that corresponds to 0.2 mm of crack growth.[17] As prescribed in the ESIS technical committee protocol,[17] this corresponds to the material elastic-plastic crack initiation toughness J_C. Typical values for PE pipe materials are $A = 38.04$ and $N = 0.56$.[15] Substituting these values into equation (15) for a crack extension, $\Delta a = 0.2$ mm, gives an elastic-plastic crack initiation toughness of $J_C = 15.45$ N/mm.

In the elastic-plastic regime, crack initiation is assumed to occur when the crack driving force per unit width J exceeds J_C. After crack initiation, failure can occur by two possible mechanisms. First, stable crack growth through the pipe wall can occur when the effective crack length calculated from equation (12) exceeds the pipe wall thickness b. Second, unstable crack growth can occur after some amount of stable crack growth in ductile materials. After crack initiation, unstable crack growth is predicted to occur at the point of tangency between the applied J versus a curve (calculated from equation [11]) and the material J_R versus Δa curve. At this point, the gradient of the two curves is equal and the amount of stable crack growth before instability can be determined. The criterion for crack initiation and subsequent failure in the elastic-plastic regime are written as

$$\text{Crack initiation: } J = J_{0.2} = J_C \tag{16}$$

$$\text{Stable crack growth through pipe wall: } a_e = b \tag{17}$$

$$\text{Crack instability:} \begin{cases} J = J_R \\ \dfrac{\partial J}{\partial a} = \dfrac{dJ_R}{da} \end{cases} \tag{18}$$

The criteria for initiation and unstable ductile fracture are shown graphically in the schematic diagram in *Figure* 4.

Craze Mechanics (CM)

While single parameter fracture mechanics (LEFM or EPFM) may apply in some cases, recently developed polymers (such as bimodal PE materials) can form a fibrillated craze zone, ahead of a crack tip, which may compromise the validity of these theories.

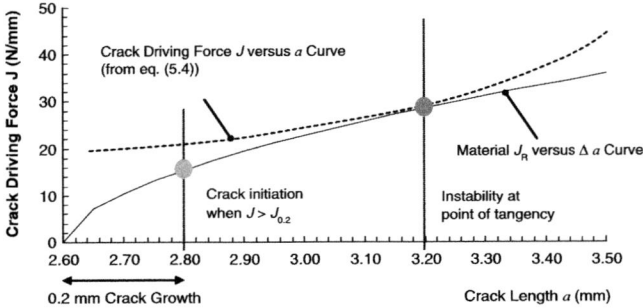

Figure 4—J-based failure criteria for elastic plastic fracture.

Figure 5 shows a schematic diagram of a typical craze developed in PE.

As shown in *Figure 5*, the craze zone is comprised of fibrils of material drawn from the bulk as the crack tip opens. At the root of the crack tip is a plastically deformed "membrane" that separates the crack tip from the fibrillated craze zone. Extending out approximately 45° from the crack tip membrane are secondary zones of shear deformation. The time-dependent deformation (and ultimate failure) of the membrane and craze zones governs the slow crack growth and brittle fracture resistance of tougher materials such as polyethylene. Further evidence of the interaction between the membrane and craze zone is shown by periodic striations on the fracture surface of PE pipe failures.[18] This is thought to indicate a "stepwise" crack propagation mechanism in which the craze and membrane sequentially grow and fail, resulting in intermittent crack propagation.

Since the extent of craze deformation is dependent on the specimen geometry, values of fracture resistance (i.e., K_{IC} or J) may not be transferable from small laboratory test specimens to larger geometries such as buried pipelines. As such, the formation of large crazes may invalidate the use of both LEFM and EPFM theories. In this respect, Kanninen et al.[19] proposed that for those tough materials that contravene LEFM and EPFM theory, detailed models of the crack tip craze should be developed to assess long-term performance.

One such modeling approach has been developed that uses a novel experimental method to isolate and measure craze behavior. The so-called Circumferentially Deep Notched Tensile (CDNT) test uses a rectangular test specimen which is machined to give a circular ligament at the mid-point that is constrained by a sharp razor notch around its circumference (see *Figure 6*). The deep circumferential notch develops a highly constrained region in the small remaining ligament. As a consequence of this geometrical discontinuity, a tri-axial stress state and high magnitude of local stress is achieved.[20] Therefore, when loaded in tension, the high constraint promotes craze growth in the ligament, which is the most severe damage mechanism likely to occur for a flawed pipe in service. By subjecting CDNT specimens to dif-

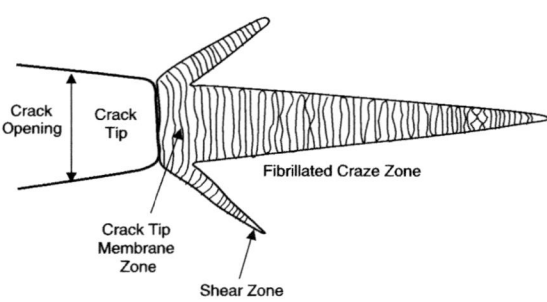

Figure 5—Schematic illustration of crack tip craze formed during slow crack growth in PE materials.

ferent static loads and recording failure time, an empirical relationship that describes that time-dependent craze strength of a material can be developed.

Duan and Williams[21] attempted to predict the time-to-crack initiation from experimentally determined long-term craze-strength data. While craze-strength data can be obtained using circumferentially deep-notched tensile specimens, the problem is to relate the uniform tensile stress state that exists in CDNT specimens to the actual stress state in the craze zone that is formed in

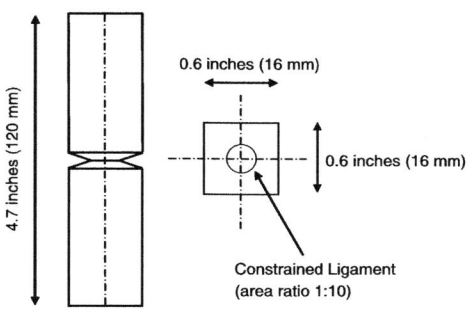

Figure 6—Circumferentially Deep Notched Tensile (CDNT) specimen.

other component geometries. From observations of plane strain single-edge notch bend (SENB) tests on tough PE materials, Duan and Williams[21] proposed that upon initiation, a crack jump was observed through the crack tip craze zone. For this sudden failure of the entire craze zone to occur, it was reasoned that two conditions must be met. First, the stress must be constant along the entire craze zone and, second, the load bearing time for different points along the craze must be the same at failure. It was proposed that these conditions can only be met if the stress redistribution time due to craze fibril creep is much shorter than the time-to-crack initiation. This was observed in long-term SENB tests, where the time for the craze to reach 85% of its final length at failure was less than one tenth of the total time-to-crack initiation.[21] It was therefore proposed that the stress redistribution time in tough materials occupies only a small portion of the total fracture initiation time. As outlined by Duan and Williams,[21] this implies that after this short stress redistribution period, a uniform tensile stress exists in the craze zone of any specimen geometry, which is the same as the tensile stress in the CDNT test. Therefore, provided the tensile stress in the craze zone is known for the component under investigation (in this case, an SENB specimen) the time-to-crack initiation could be predicted using measured craze-strength data.

For a given component geometry, the actual stress in a craze zone after creep-induced stress redistribution can be estimated using the concept of a "reference stress." According to Duan and Williams,[21] the reference stress concept originated from the creep analysis of metal rectangular beams under uniform bending moments. An analysis conducted by Schulte[22] revealed the existence of a particular location in the beam where the stress remained invariant during stress redistribution by creep. It was postulated that this value of the invariant stress could serve as a reference stress for the whole beam and any uniaxial creep and rupture tests conducted at the reference stress could reproduce the behavior of the beam.[21] According to Duan and Williams,[21] an estimate of the upper bound of the reference stress (σ_{ref}) can be derived as

$$\sigma_{ref} = \frac{P\sigma_Y}{P_U} \tag{19}$$

where P is the current load on the component, P_U is the rigid-plastic collapse load for the component, and σ_Y is the yield stress. Following the reasoning outlined above, the time dependence of reference stress is the same as the time dependence of craze stress determined using circumferentially deep notched tensile specimens under constant tensile load. For one particular pipe-grade PE material, this time-dependency was found to be

$$t_i = \left(\sigma_{ref}/21.45\right)^{-20} \tag{20}$$

where t_i is the initiation time (in hours).[21] Therefore, by using equation (19) to calculate reference stresses for SENB specimens, crack initiation times could be predicted using equation (20). Using the reference stress concept, reasonable agreement has been obtained between predicted and measured initiation times in SENB specimens under constant applied load.[20]

For a buried pipe in service, Davis et al.[23] propose that equivalent reference stress solutions can be derived for internal pressure and diametrical deflection loading. It can be shown that for a pipe under internal pressure containing a radial crack at its inner surface, the reference stress is

$$\sigma_{ref(pressure)} = \frac{-p}{\ln\left(\left[R_i + a\right]/R_o\right)} \tag{21}$$

where p is the actual applied internal pressure, R_i and R_o are the inner and outer pipe radii, respectively, and a is the radial crack length from the pipe inner surface. For the same pipe under diametrical deflection, it can be shown that the reference stress is

$$\sigma_{ref(deflect)} = \frac{\Delta E\left(b - a\right)}{0.584 D_m^{\,2}} \tag{22}$$

where Δ is the applied defection, E is tensile modulus, b is pipe wall thickness, and D_m is the pipe mean diameter. Therefore, if the reference stress is calculated for a given loading condition in service, the empirical relationship for time-dependent craze strength of a particular material can be used to predict failure.

However, it should be noted that this Craze Mechanics approach to service life prediction assumes that the service lifetime is terminated upon crack *initiation* and subsequent crack growth is ignored. Therefore, the existing Craze Mechanics approach provides a lower than expected estimate of service lifetime and further development is required to model stepwise crack growth based on craze behavior.

VALIDITY OF SINGLE PARAMETER FRACTURE MECHANICS MODELS

While LEFM theory has been used to successfully predict failure times in PVC-U and some older pipe-grade PE materials, its validity has been questioned for newer tougher polymers such as Medium Density PE (MDPE) and grades of High Density PE (HDPE). The metric for deciding the applicability of LEFM theory is the extent of crack tip plastic deformation that occurs during fracture property testing. If the stress perpendicular to the crack is σ_{yy} and the distance from the crack is r, LEFM theory predicts that

$$\sigma_{yy} = \frac{K}{\left(2\pi r\right)^{1/2}} \tag{23}$$

where K is the applied stress intensity factor. As shown in the schematic diagram in *Figure 7*, a stress singularity is observed as r approaches zero, which is characteristic of LEFM theory.

According to the $(r^{-1/2})$ dependency in equation (23), LEFM theory predicts that the stress right at the crack tip is infinite.[16] However, if this were true, bodies containing cracks would be unable to sustain any loads and have no strength at all. In reality, high crack tip stresses in ductile materials such as PE materials are relieved by craze formation (over some length as shown in *Figure 5*) and the stress singularity is removed.

As discussed above, LEFM theory ignores the details of events that take place in the craze zone and simply uses a scaling parameter (the Stress Intensity Factor K) of the stress field outside the craze zone to characterize the fracture behavior.[21] However, craze formation at the tip of a crack removes the stress singularity required by LEFM theory and can compromise its validity. LEFM is valid when the craze zone is sufficiently small to be engulfed in a stress field dominated zone, and all microscopic events happening in the craze zone can be regarded as being controlled by the macroscopic stress field. However, difficulties occur for tougher polymers (such as new PE materials), which produce larger craze zones. The separation of these craze zones absorbs a significant amount of the work available for crack propagation and confers high toughness to these materials such as the newer PE 100 materials. As discussed by Kanninen et al.,[19] the criterion for LEFM applicability is that craze length should be less than 6% of the ligament length and where this level is exceeded, alternative methodologies should be applied.

In the event that LEFM theory cannot be applied, it may be possible to model service life using an EPFM approach. However, as discussed by Davis et al.,[23] the applicability of EPFM to very tough polymers can be questioned. While EPFM theory accounts for plastic deformation that accompanies ductile fracture, actual crack growth is required to occur for the material *J-R* curves developed in EPFM to be physically meaningful. As demonstrated by Bernal et al.,[15] some tough PE materials form large crazes and exhibit extensive crack blunting, which appears as apparent crack extension in laboratory tests but merely represents plastic deformation and collapse at the crack tip. In this fully plastic regime, the use of a single parameter

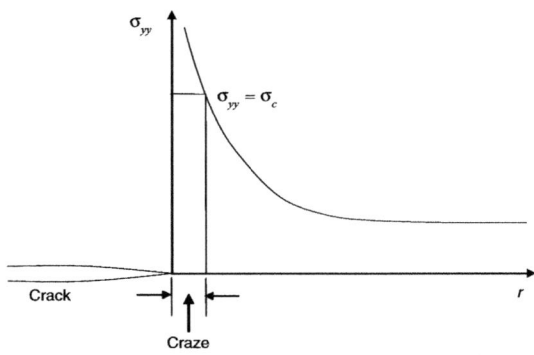

Figure 7—K dominant stress field and crack tip singularity according to LEFM theory.

fracture criterion (such as the J-integral) would be valid only if the cracked specimen exhibited a relatively high level of stress triaxiality and crack tip constraint.[15] In the absence of such constraint, critical fracture resistance values would exhibit a size and geometry dependence, and material J-R curves (determined from coupon-sized specimens) would not be transferable to other geometries such as buried pipes. In these cases, if crack blunting occurs instead of actual crack growth, the material J-R curve should be linear rather than follow a power law.

For those materials where EPFM is deemed inapplicable (i.e., crack blunting is observed in J-Integral testing instead of crack growth), a Craze Mechanics-based model should be adopted to predict pipe service lifetimes.

CRITERIA FOR SELECTING A MODELING APPROACH FOR SERVICE LIFE PREDICTION

As discussed by Broberg,[24] the material yield strength, its cohesive or craze strength, and a quantity defined as the T-stress can be used to determine the mode of failure that occurs at a crack tip in a polymer. According to Broberg[24] the cohesive strength is the stress required to separate a "unit cell" of material, where the unit cell is the smallest material unit, which contains sufficient information about the crack growth properties of the material. For the purposes of this study, it is assumed that a unit cell embodies the crack tip craze and the craze strength is the stress required for craze separation and failure. For a given crack size and structure geometry, the T-stress represents the stress that acts parallel to the crack plane (in contrast to K, which characterizes the stress perpendicular to the crack plane). As described by Wang,[25] the T-stress can substantially alter the level of crack tip stress triaxiality (the presence of tensile stress on three orthogonal axes) and influence crack tip constraint. For example, a positive T-stress strengthens the level of crack tip triaxiality and leads to high crack tip constraint. Under these conditions, brittle fracture rather than ductile yielding is promoted. In contrast, a negative T-stress reduces the level of crack tip triaxiality and leads to the loss of crack tip constraint, and under these conditions ductile deformation rather than brittle fracture is promoted.

Figure 8 illustrates Broberg's[24] failure criterion diagram, which defines how T-stress, material yield strength (σ_Y), and cohesive strength (σ_C) combine to promote different failure modes at a crack tip.

While the T-stress is calculated for particular combinations of defect size loading condition and structure geometry,[26] values of yield strength and cohesive strength must be measured. Although measuring yield strength is straightforward, it is assumed (as a first approximation) that the cohesive strength corresponds to the material craze strength, which can be measured using the CDNT test.[23]

Broberg[24] interprets the different regions in *Figure* 8 in terms of a crack tip "process zone," which is either encompassed by a region in an elastic stress state or a plastic stress state. It is assumed in this study that the process zone referred to by Broberg[24] is analogous to the crack tip craze, which controls the process of crack initiation and growth. While Broberg's[24] diagram provides an interesting insight into how relevant properties combine to dictate the mode of crack tip failure, it also provides a practical method for selecting a particular service life modeling approach for polymer pipes. Provided that values of *T*-stress, σ_Y, and σ_C can be obtained, *Figure* 8 can be used to decide if single parameter fracture mechanics theory can be applied. For example, for a given *T*-stress/σ_Y ratio, if the ratio σ_C/σ_Y is sufficiently low then a process zone develops in an elastic environment. For polymeric materials, this process zone will take the form of a crack tip craze and provided

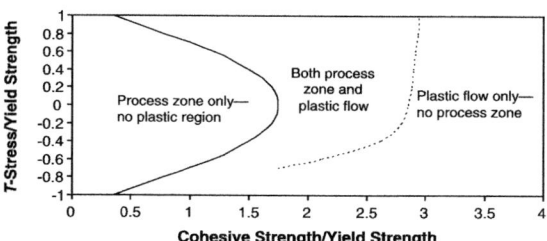

Figure 8—T-stress and cohesive strength/yield stress ratios with corresponding failure modes. Adapted from Broberg.[24]

that the craze length is less than 6% of the remaining un-cracked ligament thickness, LEFM theory can be applied.[19] In the intermediate case, a process zone develops within a plastic environment and EPFM theory may apply, provided it can be demonstrated that crack tip blunting does not occur. If crack tip blunting does occur, EPFM (and, by definition, LEFM) is invalid and a Craze Mechanics modeling approach should be adopted. Finally, if σ_C/σ_Y is sufficiently high, plastic flow develops with no process zone and failure occurs by plastic collapse rather than fracture. In this last case, the problem then reduces to one in which the main effect of the crack is to reduce the cross-section of the structure. As described by Bernal et al.,[15] ultimate failure is predicted to occur by plastic collapse of the remaining un-cracked ligament when the applied stress exceeds the material yield strength.

SERVICE LIFE MODELING FOR PVC-U, PVC-M, AND PE PIPES

To illustrate how the above criterion can be applied in practice, this section demonstrates how service life modeling approaches can be selected for PVC-U, PVC-M, and four different PE pipe materials. *Table* 2 describes the materials investigated.

Measuring Craze Strength σ_C

The craze properties of the range of polymeric materials detailed in *Table* 2 were assessed using the CDNT test as described previously. CDNT specimens were produced for the PE materials by compression moulding reground pipe material into rectangular plaques, (195 mm × 135 mm × 20 mm thick). A rectangular mould was designed to hold the required volume of raw material and allow for excess molten material to be forced out through channels under compression. The mould was compressed between two heating platens up to a temperature of 200°C and an applied platen pressure of 1 MPa. After compression for a period of 3 h to ensure a uniform temperature distribution throughout the

Table 2—Pipe Material Grades Assessed Using the CDNT Test for Craze Behavior

Sample	Origin
PVC-U	Australia
PVC-M	Australia
PE 80C (AUS)	Australia
PE 80B (AUS)	Australia
PE 100 (AUS)	Australia
PE 3408 (US)	USA

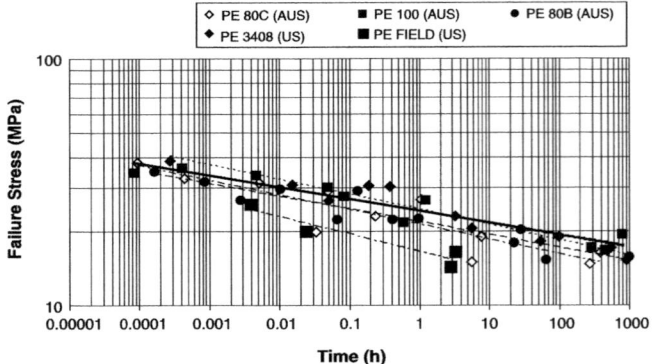

Figure 9—Craze stress σ_C *versus failure time under static load for different PE materials in the CDNT test.*[23]

molten polymer, the mold was cooled rapidly to room temperature by passing coolant through channels in the heating platens. In the case of PVC, pipes were first cut into sections approximately 200 mm by 200 mm square. These were then placed between steel plates in an air-circulating oven at 115 to 120°C for approximately 40 min. The heated sections were removed from the oven and cooled slowly under a light load in a hydraulic press. The cooling rate was controlled in such a manner to ensure that little or no distortion of the specimens was caused by residual stress. Rectangular bar specimens (16 mm × 16 mm × 120 mm long) were then cut from the PE and PVC plaques. It is acknowledged that the different thermal history imposed by compression moulding compared to pipe extrusion may result in different morphology and microstructure in CDNT specimens. However, it is also noted that erasing the previous thermal history from pipe extrusion and imposing consistent processing conditions across all specimens ensures a more accurate comparison. In the case of PE materials, it has previously been reported that while thermal history can influence crazing and slow crack growth resistance, performance is primarily governed by the molecular weight and comonomer distributions of raw materials. Since compression moulding will not change these attributes, its effect on the ranking of materials will be insignificant.

Table 3—Craze Assessment Material Properties for the Assessed Materials

Sample	Origin	Relevant Standard	1000-h Cohesive Strength (MPa)	1000-h Uniaxial Yield Strengh (MPa)	Calculated T-Stress (MPa)
PVC-U	Australia	AS 1477	32.0	35.0	6.2
PVC-M.....................	Australia	AS 4765	38.1	31.0	8.8
PE 80C (AUS)	Australia	AS 4131	14.0	10.2	4.0
PE 80B (AUS)	Australia	AS 4131	15.1	7.0	4.0
PE 100 (AUS).........	Australia	AS 4131	17.2	10.2	5.0
PE 3408 (US)	USA	AWWA C-906	16.0	8.4	5.5

Note: Yield strength data from Burn et al.[7] and Davis et al.[23]

Following the procedure outlined by Pandya and Williams,[20] a circumferential notch was introduced into CDNT specimens with a single point cutting tool of tip radius ≤ 20 μm by rotating the specimen in a four-jaw self-centering chuck on a lathe at low rotation speed. The notch tip was sharpened, also on a lathe, such that the final ligament/bulk area ratio was at least 1:10.[19] *Figure* 6 shows a schematic of a CDNT test.

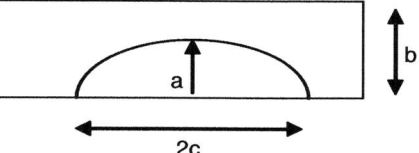

Figure 10—Elliptical surface crack. Adapted from Wang and Bell.[26]

Craze behavior was assessed under constant load conditions at 20°C, with specimens mounted in grips and connected to a pin-jointed lever arm. A mass was then applied at the end of the lever arm to apply the required static tensile stress to the ligament. Local craze extension was measured using a clip gauge extensometer.

Typical results of this test are shown in *Figure* 9 and empirical power law in the form

$$\sigma_C = A t_f^{-m} \tag{24}$$

was fitted, where σ_C is the applied stress or cohesive stress, t_f is the failure time, and A and m are material constants. Values of the cohesive stress for each material assessed are given in *Table* 3. To ensure a consistent comparison, cohesive stress values at an arbitrary failure time of 100 h are compared.

Calculating T-Stress Values

One frequently observed defect type that can cause crack initiation and failure in polymer pipes is an elliptical or "thumb-nail" shaped defect located at the inner surface of the pipe (*Figure* 10).

Values of *T*-stress for different semi-elliptical cracks have been calculated previously by Wang and Bell.[26] *Figure* 11 shows the corresponding normalized *T*-stress solutions for elliptical defects of different crack lengths. *T*-stress is normalized with respect to the remote applied stress. In the case of a buried pipe, this will be the hoop stress at the pipe inner surface.

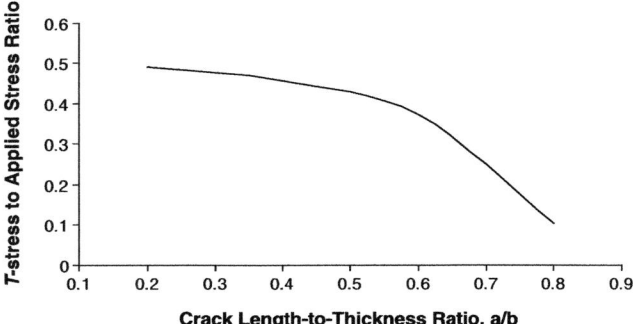

Figure 11—T-stress solutions for elliptical defects of different lengths. Aspect ratio a/2c = 0.4.[26]

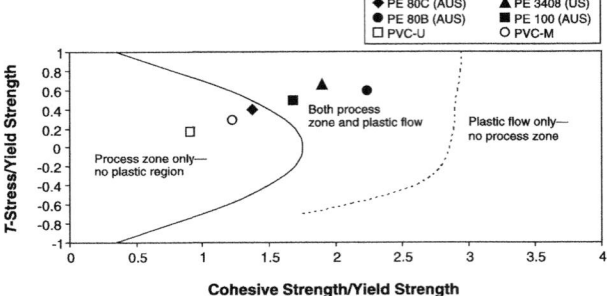

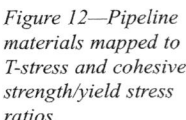

Figure 12—Pipeline materials mapped to T-stress and cohesive strength/yield stress ratios.

Figure 11 can be used to calculate the normalized *T*-stress for different crack length to wall thickness ratios.

T-stress values for each case in *Table* 3 were calculated assuming an elliptical crack with length to thickness ratio *a/b* < 0.2 at the inner surface of the pipe. For each case in this hypothetical example, *Figure* 11 indicates that the ratio of *T*-stress to remote applied tensile stress (perpendicular to the crack plane) is approximately 0.5.[26] Therefore, assuming a remote applied stress equal to the maximum allowable stress specified by the relevant performance standard, the actual *T*-stress can be calculated (*Table* 3).

Applying Broberg's Criterion for Model Selection

As discussed, the ratio of measured craze strength (determined using the CDNT test) and the material uniaxial yield strength can be combined with calculated *T*-stress values used to determine the appropriate modeling approach to adopt for each polymeric material.

Using the values of σ_Y, σ_C, and *T*-stress from *Table* 3, the deformation state at the tip of the defect in each material can be determined. *Figure* 12 shows the different crack tip deformation states that would exist based on yield strength and cohesive strength values at 1000 h. Ideally, for long term predictions, analysis at times exceeding 1000 h should be carried out to ensure that the determined failure mechanisms are still valid.

As shown in *Figure* 12, Broberg's[24] criteria suggests that PVC-U and PE 80C would exhibit a crack tip craze zone in an elastic environment, indicating that LEFM theory may be applicable. However, as discussed by Davis et al.,[23] a series of SENB tests on PE80C indicated craze lengths exceeding the 6% threshold stipulated by Kanninen et al.[19] This is also reflected by the proximity of PE 80C to the boundary between elastic deformation and plastic flow regimes in *Figure* 12. This indicates that LEFM cannot be applied to the PE 80C material.

For PVC-M, Broberg's[24] criteria suggest that the crack tip would be surrounded by an elastic environment and again indicates that LEFM theory may be applicable. However, this is in contrast to the observed ductile failure mode that is observed during fracture testing of this material, which indicates that LEFM cannot be applied.[7] However, it should be noted that, as shown in *Figure* 12, the stress state described for PVC-M is closer to that for PE 80C than PVC-U, which may indicate that some ductility can be expected to accompany fracture. The inability of Broberg's[24] criteria to clearly identify ductile failure in PVC-M may be attributed to the absence of the fibrillated craze mechanism in CDNT tests on PVC-M. While inspection of PVC-M specimens indicates stress whitening and ductility, the craze formation mechanism which is observed in PE materials does not appear to occur. Therefore, it is not clear if CDNT tests on PVC-M samples actually measure cohesive strength for use in Broberg's[24] criteria. Further work is required to validate the applicability of the CDNT test for PVC-M materials.

For PE80B, PE100, and PE 3408, the resulting crack tip craze will be embedded in a region of plastic flow indicating that although LEFM theory cannot be applied, EPFM may be valid. However, as described above, a check should also be made that crack tip blunting does not occur before EPFM theory can be adopted. As described by Williams,[13] if crack tip blunting is occurring rather than crack growth in J-integral testing, a theoretical blunting line can be defined as

$$J = 2\sigma y \Delta a \tag{25}$$

where J is the measured material J-crack growth resistance, σ_y is the material yield strength, and Δa is the recorded crack growth. *Figure* 13 compares experimental J versus Δa material curves with theoretical blunting lines for each PE material.

As shown in *Figure* 13, the experimental points of the J-R curves for each of these PE materials lie very close to their corresponding theoretical blunting lines. This indicates that observed crack growth in these tests is apparent and is actually due to crack tip blunting and associated plastic deformation. Consequently, the use of EPFM is also invalid for these materials and a Craze Mechanics modelling approach for service life prediction should be adopted.

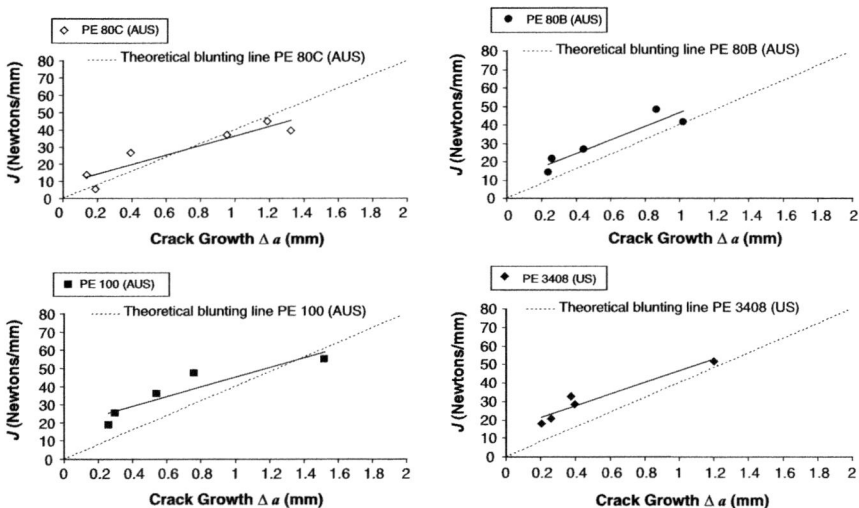

Figure 13—Comparison of J-R data with theoretical blunting line for PE materials. Data from Davis et al.[23]

Selecting Defect Size for Model Applications

The main problem with applying service life prediction methods is the selection of an appropriate defect size. Predictions from both traditional fracture mechanics and Craze Mechanics approaches are sensitive to defect size, yet there are no non-destructive testing technologies that can measure the size of defects for pipes in-service. One possible approach to obtain defect size information is to examine fracture surfaces from previous failures that have occurred in service. Careful extraction and examination of fracture surfaces can often reveal the location of crack initiation, and, in some cases, the size of the defect that was responsible. While this provides a biased estimate (i.e., cracks initiate at the largest defects), it also provides a "worst case" estimate of defect size. An example of this kind of approach is reported by Davis et al.,[27] who examined fracture surfaces from a large number of PVC-U pipes failures in Australia. While inherent variation in defect size was clearly observed, a maximum defect size of 1 mm was reported. It is proposed that this can be used as a current "worst case indicator."

Since failure rates in PE pipes are lower than those reported for PVC-U pipes (*Table* 1), it is more difficult to estimate defect size from PE pipe fracture surfaces. However, Chudnovsky et al.[28] report defect sizes found in fracture surfaces from notched tensile tests on PE. According to Chudnovsky et al.,[28] an expected maximum defect size of 0.62 mm was reported. However, it is noted that these test samples are not necessarily representative of extruded pipes and the study conducted by Barker et al.[29] is perhaps more useful. In a laboratory study of fatigue fracture resistance of PE pipes, Barker et al.[29] examined fracture surfaces from failed pipes and reported defect sizes that caused failure. Examination of the data from this previous study indicates a maximum defect size of 0.95 mm.

It is acknowledged that current pipe extrusion processes may include filtration that reduces the size of inherent defects that are included in the pipe wall. However, in the absence of available data, a maximum defect size of 1 mm is assumed for the purposes of illustrating how service life models can be applied.

Example Service Life Predictions

Broberg's[24] criterion indicates that while service life prediction for PVC-U pipes can be modelled based on LEFM theory, single parameter fracture mechanics is invalid for all PE materials. In these cases, a Craze Mechanics-based model should be adopted. Whilst precise details can be found elsewhere in the literature,[7,27] equations (2) to (6) and (19) to (22) can be used to predict service lifetimes of the PVC-U and PE pipes assessed in this study under different operating and installation conditions. For example, if we

Table 4—Lifetimes for Different Pipe Materials

Pipe Type	Lifetime Methodology	Estimated Pipe Life (years)
PVC-U	LEFM	170
PE 80C (AUS)	CM	195
PE 80B (AUS)	CM	479
PE 100 (AUS)	CM	577
PE 3408 (US)	CM	574

assume damage at the internal pipe surface of 1 mm and an operating pressure of 0.8 MPa, the time to first failure of a pipe can be determined. *Table* 4 shows examples for each pipe material considered. For PVC-U, a 100 mm diameter class PN16 pipe was analyzed,[30] whilst for PE, 110 mm SDR 11 pipes were analyzed.[30]

As shown, extended failure times are predicted for PE pipes in these hypothetical examples. As suggested previously, the realization of these extended failure times will depend on the level of heat stabilizer that is included in the raw material. If this stabilizer package is consumed within these predicted time frames, then a transition to Stage III failure may occur (see *Figure* 1). Further work is required to identify if such a transition exists in these materials at ambient temperatures.

CONCLUSIONS

A methodology has been proposed that allows the lifetimes of plastic materials to be determined, based on LEFM, EPFM, or Craze Mechanics theory. To check the validity of the relevant theory for a range of plastics materials, static craze strength and uniaxial yield strength data were combined in Broberg's[24] assessment criteria for the stress state ahead of a defect or a crack. Results indicate that the PVC-U materials can be characterized using LEFM. However, for the range of PE materials assessed, LEFM theory was found to be invalid even though Broberg's[24] methodology predicted one material (PE80B) would be suitable by assessment using LEFM procedures. In this case, application of LEFM was invalid because length of the crack tip craze exceeded the maximum threshold level of 6% of the remaining ligament.

An alternative to LEFM may be to use EPFM theory to analyze ductile fracture. In this engineering approach, ductile fracture is predicted based on the crack tip characterizing parameter known as the *J*-integral. However, comparisons with theoretical blunting lines were also made to determine the validity of the EPFM to PE. The theoretical blunting line represents the apparent increase in crack length from blunting at the crack tip due to local yielding and collapse, rather than actual crack growth. The experimental data points obtained for all PE materials were close to their corresponding theoretical blunting lines, indicating that a fully plastic regime exists at the crack tip in these materials. This indicates that EPFM may not be valid and cannot be transferred to other structures, such as pipes containing defects. For these materials, an alternative service life modeling approach, based on behavior in the crack tip craze zone using Craze Mechanics to predict failure times, should be adopted.

ACKNOWLEDGMENTS

The authors would like to thank the American Water Works Association Research Foundation (AwwaRF) for the partial funding of this work.

References

(1) MacKellar, S. and Pearson, D. "Nationally Agreed Failure Data and Analysis Methodology for Water Mains Volume 1: Overview and Findings," Report Ref. No. 03/RG/05/7, UK Water Industry Research, London, UK, 2003.

(2) Marshall, G.P., Pearson, D., and MacKellar, S., "Specification of Performance for Modified Poly(Vinyl Chloride) and Polyethylene Pipes," *Plast. Rubber Compos. Process. Appl.*, 25 (6), 276-286 (1996).

(3) Andersson, U., *Proc. of the 11th International Conference on Plastics Pipes*, Munich, Germany, Plastics and Rubber Institute, 2001.

(4) Williams, J.G., "Applications of Linear Fracture Mechanics," *Adv. Polym. Sci.*, 27, 67-120 (1978).

(5) Lu, J.P., Davis, P., and Burn, L.S., "Lifetime Predictions for ABS Pipes Subjected to Combined Pressure and Deflection Loading," *Polym. Eng. Sci.*, 43, 444 (2003).

(6) Marshall, G.P., "Design for Toughness in Polymers: 1—Fracture Mechanics," *Plast. Rubber Compos. Process. Appl.*, 2 (2), 169-182 (1982).

(7) Burn, S., Davis, P., Schiller, T., Tiganis, B., Tjandraatmadja, G., Cardy, M., Gould, S., Sadler, P., and Whittle A.J., *Long Term Performance Prediction of PVC Pipes*, AwwaRF Project No. 2879, 2005.

(8) Davis, P., Burn, L.S., and Whittle, A.J., "Investigating Crack Growth and Plasticity in the C-Ring Fracture Toughness Test," *Proc. Plastics Pipes XI*, Munich, Germany, IOM Communications, September 2001.

(9) Rooke, D.P. and Cartwright, D.J., *Compendium of Stress Intensity Factors*, London, UK, 1976.

(10) ISO/DIS 11673.2, *Determination of the Fracture Toughness Properties of Un-plasticized Poly(Vinyl Chloride) (PVC-U) Pipes*, International Organization for Standardization, Geneva, Switzerland, 1999.

(11) Burn, L.S., "Lifetime Prediction of uPVC Pipes—Experimental and Theoretical Comparisons," *Plast. Rubber Compos. Process. Appl.*, 21 (2), 99-108 (1994).

(12) Marshall, G.P., Brogden, S., and Shepherd, M.A., "Evaluation of the Surge and Fatigue Resistance of PVC and PE Pipeline Materials for Use in the U.K. Water Industry," *Proc. Plastics Pipes X Conference*, Gothenburg, Sweden, IOM Communications, September 1998.

(13) Williams, J.G., *Introduction to Elastic Plastic Fracture Mechanics*, ESIS-TC4 Polymers and Composites, Cambridge, UK, 2000.

(14) Kumar, V., German, M.D., and Shih, C.F., "An Engineering Approach for Elastic Plastic Fracture Analysis," EPRI Report NP-1931, 1981.

(15) Bernal, C., Lopez Montenegro, H., and Frontini, P., "Failure Prediction Analysis for Polyethylene Flawed Pipes," *Eng. Fract. Mech.*, 70, 2149 (2003).

(16) Atkins, A.G. and Mai, Y.W., *Elastic and Plastic Fracture*, Ellis Horwood, Chichester, UK, 1988.

(17) Hale, G.E. and Ramsteiner, F., *J-Fracture Toughness of Polymers at Slow Speed, ESIS-TC4 Polymers and Composites*, Cambridge, UK, 2000.

(18) Sehanobish, K., Moet, A., Chudnovsky, A., and Petro, P.P., "Fractographic Analysis of Field Failure in Polyethylene Pipe," *J. Mater. Sci. Lett.*, 4, 890-894 (1985).

(19) Kanninen, M.F., O'Donoghue, P.E., Popelar, C.H., and Kenner, V.H., "A Visco-Elastic Fracture Mechanics Assessment of Slow Crack Growth in Polyethylene Gas Distribution Pipe Materials," *Eng. Fract. Mech.*, 36 (6), 903-918 (1990).

(20) Pandya, K.C. and Williams, J.G., "Measurement of Cohesive Zone Parameters in Tough Polyethylene," *Polym. Eng. Sci.*, 40, 1765 (2000).

(21) Duan, D. and Williams, J.G., "Craze Testing for Tough Polyethylene," *J. Mat. Sci.*, 3, 635 (1998).

(22) Schulte, C.A., in ASTM STP060, American Society for Teting & Materials, Philadelphia, PA, 1960.

(23) Davis, P., Burn, S., and Gould S., *Long Term Performance Prediction of PE Pipes*, AwwaRF Project No. 2879, 2007.

(24) Broberg, K.B., "Influence Of T-Stress, Cohesive Strength, and Yield Strength on the Competition between Decohesion and Plastic Flow in a Crack Edge Vicinity," *Int. J. Fract.*, 4, 11 (1999).

(25) Wang, X., "Elastic T-Stress for Cracks in Test Specimens Subjected to Non-Uniform Stress Distributions," *Eng. Fract. Mech.*, 69, 1339 (2002).

(26) Wang, X. and Bell, R., "Elastic T-Stress Solutions for Semi-Elliptical Surface Cracks in Finite Thickness Plates Subject to Non-Uniform Stress Distributions," *Eng. Fract. Mech.*, 71, 1477 (2004).

(27) Davis, P., Burn, S., Moglia, M., and Gould, S., "A Physical Probabilistic Model to Predict Failure Rates in Buried PVC Pipelines," *Reliability Eng. System Safety*, 92, 1258-1266 (2007).

(28) Chudnovsky, A., Baron, D., and Schulkin, Y., "Lifetime Prediction and Reliability Evaluation for Gas Piping Grade Polyethylene," *Proc. 14th Intern. Plastic Fuel Gas Pipe Symposium*, San Diego, CA, American Gas Association, p. 190-196, 1995.

(29) Barker, M.B., Bowman, J., and Bevis, M., "The Performance and Causes of Failure of Polyethylene Pipes Subjected to Constant and Fluctuating Internal Pressures," *J. Mater. Sci.*, 18, 1095-1118 (1983).

(30) *PVC Pipes and Fittings for Pressure Applications*, AS/NZS 1477, Standards Australia/Standards New Zealand, 1999.

(31) *Polyethylene (PE) Pipes for Pressure Pipes and Fittings*, AS/NZS 4130, Standards Australia/Standards New Zealand, 2003.

Chapter 13

Predicting the Creep Behavior of High Density Polyethylene Geogrid Using Stepped Isothermal Method

S.-S. Yeo and Y. G. Hsuan

Dept. of Civil, Architectural, and Environmental Engineering,
Drexel University, Philadelphia, PA 19104

The creep behavior of a high density polyethylene (HDPE) geogrid used for soil reinforcement applications was evaluated using two accelerated creep tests: Time-Temperature Superposition (TTS) and the Stepped Isothermal Method (SIM). TTS has become a well-accepted method to evaluate viscoelastic behavior of polymeric materials, while SIM is a relatively new accelerated creep test that was initially developed to evaluate polyethylene-terephthalate (PET) geogrids. However, the applicability of SIM has not been extensively studied for HDPE geogrids.

In this chapter, variations in the three test parameters of SIM (i.e., temperature increment, dwell time, and applied load) were investigated. As recommended in ASTM D 6992, a temperature increment of $7°C$ and a dwell time of 10^4 seconds were found to be suitable for the testing of HDPE geogrids. On the other hand, a new analytical procedure was implemented to generate the creep master curve. The creep properties obtained from SIM and TTS were similar at 20 and 30% ultimate tensile strength (UTS). The geogrid exhibited primary creep at 10 and 20% of UTS, while secondary creep was detected at 30 and 40% UTS. Furthermore, the activation energies at different applied loads from both accelerated creep methods were determined and their values ranged from 140 to 200 kJ/mol.

INTRODUCTION

Geogrid is a two-dimensional grid-like polymeric material which consists of connected parallel sets of intersecting ribs with apertures to allow strike-through of surrounding soil, as depicted in *Figure* 1. The primary function of the geogrid is reinforcement for walls, slopes, foundations, and roads; they are subjected to constant stresses and undergo creep deformation during their service life.[1-3] To account for such deformation, a creep reduction factor is incorporated into the design to ensure the integrity of the structure and to minimize the deformation. Since the reduction factor is calculated from the creep properties of the geogrids, the long-term creep behavior must be appropriately evaluated, particularly for design lives of 50 to 100 years.

Figure 1—Various types of geogrids used in reinforcement applications.

Creep refers to a time-dependent deformation process occurring at tensile loads less than the initial tensile strength of a material.[4] The creep property varies with the type of polymer, exposure temperature, glass transition temperature, T_g, and melting temperature, T_m, of the polymer. Geogrids are commonly made from four types of polymers: high density polyethylene (HDPE), polypropylene (PP), polyester-terephthalate (PET), and polyvinyl alcohol (PVA). Depending on the manufacturing processing, polymer is drawn to different degrees, leading to various creep responses among the geogrids. Ideally, the creep behavior of geogrids should be evaluated according to ASTM D 5262, which requires a minimum of 10,000 hours (~ 1.1 years) testing time to obtain data at a laboratory ambient condition. The creep data is then extended to one log cycle (e.g., from 10,000 to 100,000 h) with limited confidence. However, this extrapolation falls far short of the expected design life for these materials, which is typically between 50 and 100 years[5] or equivalently 400,000 to 800,000 h. Therefore, accelerated creep tests are commonly utilized to evaluate the creep behavior. The time-temperature superposition (TTS) principle, in which the testing time is reduced by elevating the test temperature, has been widely applied in laboratory accelerated creep tests. The result of the TTS creep test yields a master curve with duration much longer than the individual creep test at elevated temperatures.[4,6,7] According to the Boltzmann superposition principle (BSP), in a creep test the total strain of the test specimen is equal to the sum of the strain of each independent event.[8]

In the last ten years, the Stepped Isothermal Method (SIM) has been introduced to evaluate the creep property of geogrids. SIM combines both TTS and BSP by subjecting a single specimen to a series of temperature steps under a constant load to generate a sequence of creep responses.[9-11] A similar test procedure has also been used to assess the creep behavior of polymethyl-methacrylate (PMMA) by Sherby and Dorn.[12] The major difference of SIM and TTS is the accumulated strain in the test specimen. In SIM, the induced creep strain at each temperature step is not removed, but accumulated in the test specimen. A new specimen is used for each test temperature in TTS; thus, test specimens do not have a strain history.

The validity of SIM has been investigated by comparing its creep data with those obtained from the TTS tests on different types of geogrids. Thornton et al.[10] compared SIM and TTS with PET yarns and found good agreement between the results for these two test protocols. They also recommended temperature increments and dwell times (i.e., isothermal duration) for testing PET geogrids and these parameters have been incorporated into ASTM D 6992. Baker and Thornton[13] evaluated the creep properties of the PP woven geotextile. They found good agreement in the creep data between SIM and TTS. Lothspeich and Thornton[14] used SIM to evaluate the creep behavior of HDPE, PP, PET, PVA, and polyamide (PA) geogrids. They found that HDPE and PP were more sensitive to creep deformation than the others; however, they did not verify creep data with TTS.

The applicability of SIM on the HDPE geogrid has not been thoroughly evaluated and a comprehensive assessment is needed. The HDPE geogrid is in a rubbery state at the test temperature range, while the PET geogrid is in a glassy state. Furthermore, the HDPE geogrid has a relatively low melting point compared to geogrids made from other types of polymers. Therefore, plastic deformation would dominate the creep behavior of HDPE, particularly at high testing temperatures. The accumulated strain that is induced in SIM may have a significant effect on the overall creep behavior of the HDPE geogrid, leading to a different creep deformation than from the TTS test.

In this study, the creep behavior of the HDPE geogrid was evaluated using both SIM and TTS. The effects of temperature steps, dwell times, and applied loads on the data of SIM were investigated. Furthermore, the analytical procedure employed to generate the SIM creep master curve was different than that described in ASTM D 6992. The resulting creep master curves from both SIM and TTS are compared and the differences are identified and discussed in this chapter.

TEST MATERIAL AND APPARATUS

A unidirectional HDPE geogrid was used in this study. The geogrid is made by drawing a HDPE sheet with pre-punched holes in one direction to create the highly oriented ribs. The draw ratio varies from product to product with a minimum of 8 to 1. *Table 1* shows the physical and mechanical properties of the geogrid.

The tensile and creep tests were performed using Instron® 5583 with Merlin® software for load control and strain measurement, as depicted in *Figure 2*. The deformation of the specimen was determined by the cross-head movement, which was then divided by the initial gauge length to obtain the strain value. A set of box-type grips was used to hold the geogrid test specimen. The test specimen with three parallel ribs was mounted to the grips and then the two outer ribs were cut. This mounting method provided uniform

Table 1—Selected Physical Properties of the HDPE Geogrids

Property	Test Standard	Value
Unit weight (g/m²)	ASTM D 5261	826
Density (kg/m³)	ASTM D 1505	949
Aperture size (mm) MD XD	Direct measurement	121 ± 2 5 to 15 (oval)

Note: MD = machine direction; XD = cross-machine direction

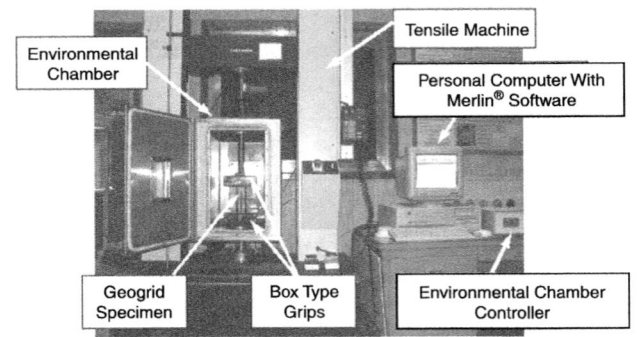

Figure 2—Pictorial view of
the test apparatus for the
SIM tests.

loading to the central single rib. The test temperature was controlled by the environmental chamber. The accuracy of the temperature in the chamber was ± 0.5°C.

TENSILE STRENGTH

The tensile test of the geogrid was performed using a procedure similar to that described in ASTM D 6637. The reference temperature of these tests was 20 ± 2°C. The load/deformation curve is shown in *Figure 3*. The loci of failures invariably occurred near the middle of the specimen. The reported mean tensile strength is the average of six replicates. The average ultimate tensile strength (UTS) was 1.92 kN (± 0.085 kN). The UTS value was used to calculate different applied loads for the accelerated creep tests.

SIM PROCEDURES AND DATA ANALYSIS

The SIM tests on the HDPE geogrid were performed according to ASTM D 6992. *Table 2* shows the values that were applied to the three test parameters to investigate their effects on the creep data. The test specimen was first brought to equilibrium at 20 ± 1°C for 2 h. The test was started by loading the specimen at a strain rate of 10% of gauge length to reach the desired load. The times to achieve each new temperature were approximately 2 and 3 min for 7 and 10°C temperature steps, respectively. The overshoot of the

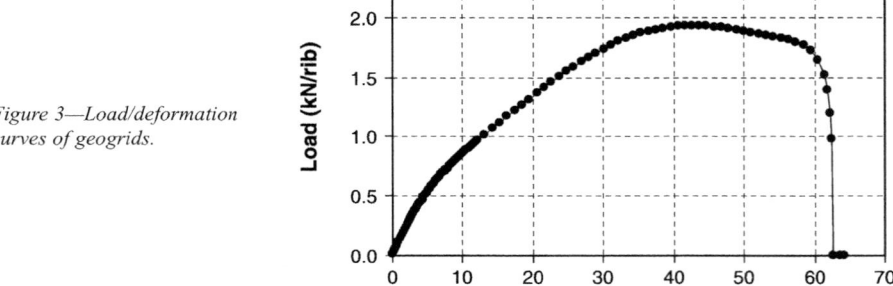

Figure 3—Load/deformation
curves of geogrids.

Table 2—Experiments Conducted using Stepped Isothermal Method (SIM)

Parameters	Conditions		
	Temperature Increment (°C)	Dwell Time (s)	Applied Stress (% UTS)
Temperature increment (°C)	7 and 10	10^4	30
Dwell time (s)	7	2500, 5000, 10000	30
Applied stress (% UTS)	7	10^4	10, 20, 30, 40

target temperature was prevented by a cooling device. The creep deformation was automatically recorded.

The procedure for generation of the creep master curve is described in ASTM D 6992. The most important step in the data analysis procedure is the implementation of the virtual time, t_i' for $i = 1,2,...n$, as shown in *Figure* 4a. Thornton et al.[9,10] indicated that t_i' should be used for rescaling the creep curve at each temperature step except for i=1, the first curve. If a new specimen is used at each temperature step, the beginning of the creep curve should start at an earlier time than the onset time for a new temperature, t_i. The way to determine t_i' is by iteratively varying duration of t_i' until the initial slope of the creep curve seamlessly intersects the final strain for the previous curve.[15] In this case, there is no discontinuities between creep curve segments. The total creep deformation equals the sum of the contributions from individual creep segments, assuming that BSP is applicable to the test data.

Table 3 shows the difference between the virtual time and the segment onset time at each temperature step for the HDPE geogrid together with the PET geogrid as comparison. The HDPE geogrid exhibits a much larger time difference, $t_i - t_i'$, than does the PET geogrid. This is probably because the HDPE geogrid is more susceptible to creep deformation than PET is at the test temperature. In addition, the high thermal expansion of the HDPE geogrid certainly introduces error to the beginning of the creep curve. Therefore, BSP is not applicable and the initial portion of each creep should not be included in the

Table 3—Comparison between the Virtual Time and Starting Time for the HDPE and PET Geogrids

Temperature Step i	PET Geogrid			HDPE Geogrid		
	Segment Starting Time t_i (s)	Virtual Starting Time t_i' (s)	t_i-t_i' (s)	Segment Starting Time, t_i (s)	Virtual Starting Time, t_i' (s)	t_i-t_i' (s)
1	0	0	0	0	0	0
2	10080	9850	230	10101	7890	2211
3	20100	19850	250	20061	18052	2009
4	30120	29850	270	30081	28280	1801
5	40080	39850	230	40101	38250	1851
6	N/A	N/A	N/A	50061	48030	2031
7	N/A	N/A	N/A	60201	57030	3171

Note: Reference temperature (i.e., Step 1): 20°C; temperature increments: 14°C for the PET geogrid and 7°C for the HDPE geogrid.

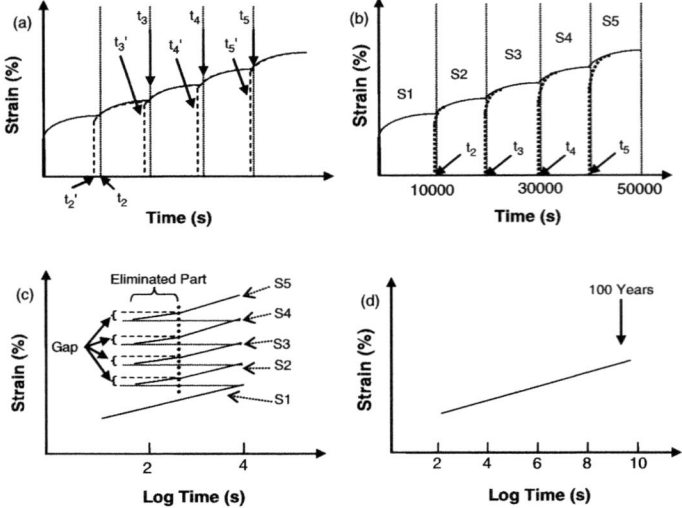

Figure 4—Procedure to generate a creep master curve from test data of SIM and HDPE geogrid.

data analysis. The new procedure to analyze the resulting creep data is illustrated in *Figures* 4b to 4d, and is called the modified SIM (MSIM) in this chapter. *Figure* 4b shows a series of creep strain curves versus time under a constant applied load at different isothermal steps. The creep curves at elevated temperature steps are rescaled to the reference temperature (e.g., 20°C) along the log scale of time axis by eliminating the onset time of each temperature step (i.e., $t_{1, 2, 3,}$ etc.) as shown in *Figure* 4c. After elimination of the initial segments, the creep master curves for creep strain is generated by horizontal shifts of the creep curves at elevated temperatures, as shown in *Figure* 4d. In this procedure, vertical shifts are required due to the elimination of initial parts of each

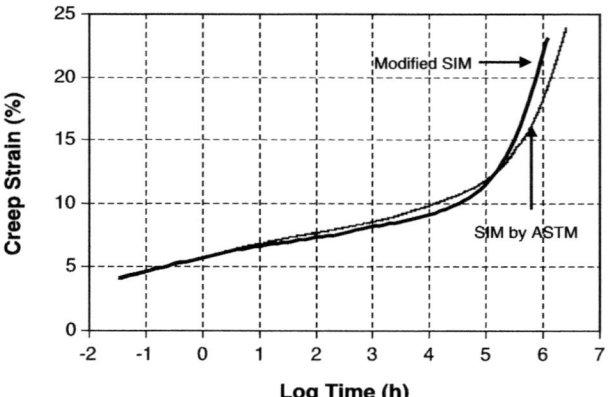

Figure 5—Creep master curves obtained from modified SIM and standard SIM procedures.

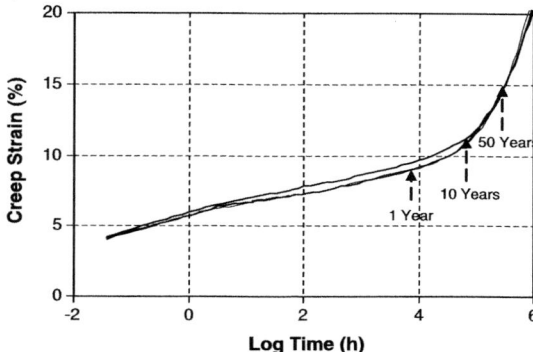

Figure 6—Creep master curves of three replicate tests based on 30% UTS with a junction.

creep curve at elevated temperatures. On the molecular scale, the vertical shifting can be explained by the influence of changes in the crystalline structure of polyethylene on its relaxation modulus with temperature.[16,17]

Figure 5 shows two master curves obtained from SIM and MSIM procedures at 30% UTS using 7°C temperature increment and 10^4 s dwell time. The two master curves are similar; however, their activation energies were found to be very different and will be presented in a later part of this chapter.

RESULTS OF SIM

Repeatability of MSIM

The repeatability (i.e., variability of the test within a single laboratory) of SIM was investigated by performing three replicate tests to obtain statistical significance. A stress ratio, the ratio of the applied load versus UTS, of 30% was used. The resulting three master curves are shown in *Figure 6*. The creep strain values at times of 10^4 h (~ 1 year), 10^5 h (~ 11 years), and $10^{5.6}$ h (~50 years) were arbitrarily selected to provide a quantitative comparison. The average creep strain and the standard deviations are summarized in *Table* 4. Data indicate a good agreement among the three SIM tests.

Table 4—Statistical Significance for SIM at 30% UTS

Time (h)	Creep Strain (%)				Standard Deviation (%)
	Test 1	Test 2	Test 3	Average	
10^4 (~1 year)	9.09	9.70	9.24	9.35	0.32
10^5 (~11 years)	11.52	11.90	11.96	11.79	0.24
$10^{5.6}$ (~50 years)	16.72	16.47	16.49	16.56	0.14

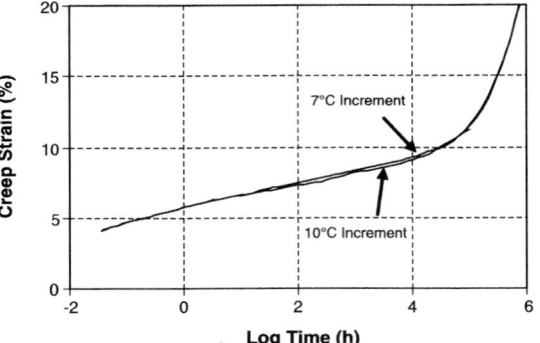

Figure 7—Effect of temperature increment on SIM.

Effect of Temperature Increment

Tests using temperature increments of 7 and 10°C were conducted. The 7°C is the recommended value in ASTM D 6992. The 10°C was selected so that there would not be a large increase in creep strain when the temperature increased. Test specimens were subjected to a constant load stress ratio of 30% UTS and a dwell time of 10,000 seconds. Both treatments were terminated at a maximum temperature of 70°C. As shown in *Figure 7*, the creep properties using two different temperature increments are very similar. Thus, it was found that the creep properties are not sensitive to variation in the magnitude of the temperature increment as long as the same testing temperature range is applied on the specimen.

Effect of Dwell Time

Different dwell times of 2500, 5000, and 10000 s were evaluated using temperature increments of 7°C and an applied load of 30% UTS. An increment of 10000 s is recommended in ASTM, while 2500 s and 5000 s were selected here to observe the effect of dwell time on the early part of the creep curve. The resulting three master curves are shown in *Figure 8*. The maximum temperatures were 83, 76, and 69°C for dwell times of 2500, 5000, and 10000 s, respectively. The creep master curves obtained from these three

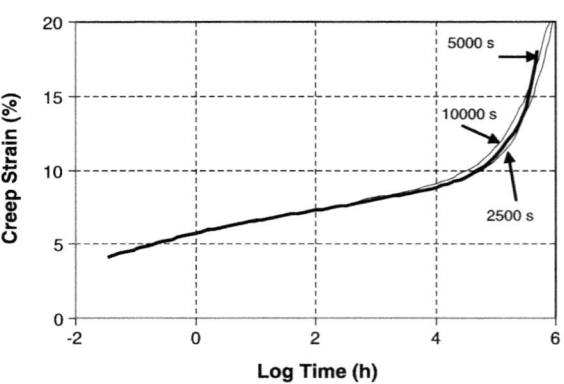

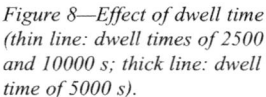

Figure 8—Effect of dwell time (thin line: dwell times of 2500 and 10000 s; thick line: dwell time of 5000 s).

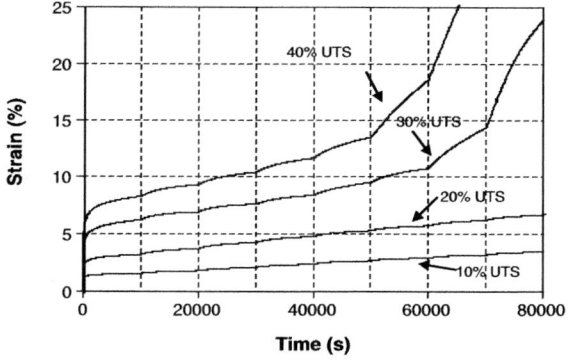

Figure 9—Creep test data from SIM at different percentages of UTS.

different dwell times are very close. Hence, dwell times within the tested range did not affect the creep result for the study HDPE geogrid. On the other hand, Thornton et al.[10] have performed SIM on PET filaments using dwell times of 1000, 10000, and 100000 s. They found a large difference between 1000 and 10000 s. Therefore, dwell time of 10000 s is appropriate to cover both the HDPE and the PET geogrids.

Effect of Different Percentages of UTS

Four tests were performed using stress ratios of 10, 20, 30, and 40% of the UTS. The tests were performed using temperature increments of 7°C and dwell times of 10^4 s. *Figure* 9 shows a series of creep curves at four different percentages of UTS, and *Figure* 10 shows their corresponding creep master curves using the MSIM procedure. While the creep master curves at the 10 and 20% stress ratios exhibit a linear behavior, a non-linear behavior was observed at the 30 and 40% stress ratios. Sherby-Dorn plots[12] were constructed to illustrate different creep stages of the four creep tests, as shown in *Figure* 11. At 10 and 20% UTS, the strain rate [linearly] continuously decreases as the creep strain increases, indicating the geogrid is undergoing a primary creep deformation. In contrast, the secondary creep stage is observed, as indicated by the constant creep strain rate after the primary creep stage at 30 and 40% UTS. The onset of the secondary creep occurs at

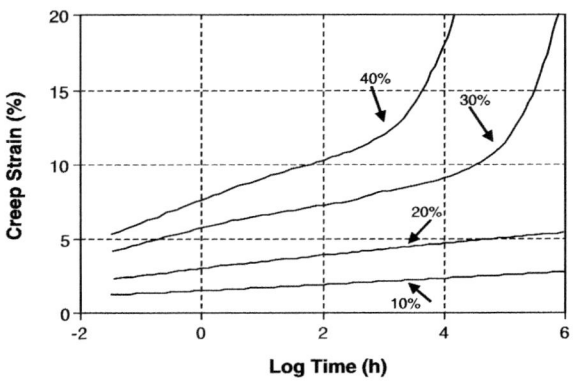

Figure 10—Creep master curve at different percentages of UTS.

214

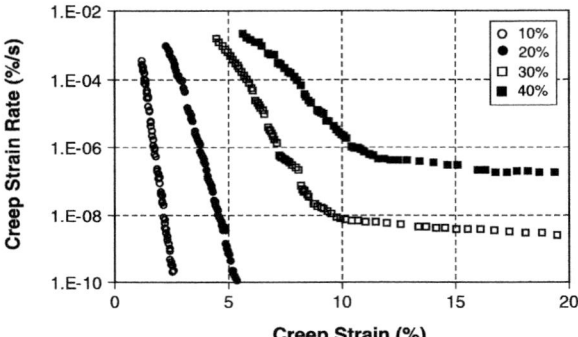

Figure 11—Sherby-Dorn plot for creep results in Figure 8.

a strain exceeding 10% at 30 and 40% stress ratios. The same behavior over 10% of creep strain was also reported by Lothspeich and Thornton[14] on their HDPE geogrid sample.

Figure 12 shows four isochronous curves at 1, 10, 50, and 100 years, which were generated from the creep master curves in *Figure* 10. The isochronous plot (i.e., percentage of UTS versus strain) exhibits nonlinear behavior and can be fitted with a logarithmic equation having the form

$$\sigma = A \ln(\varepsilon) - B \tag{1}$$

where: σ = the applied load (kN), ε = strain (%), and A and B = material constants

Ingold et al.[18] suggested a 10% strain limit strain, corresponding to a load limit of 0.48 kN (23.5%), for a reinforcement wall with 100-year design life. Den-Hoedt[19] also recommended a load level less than 25% UTS for polyethylene (PE) geotextile used in reinforcement applications. In this study, the 10% strain corresponds to a load limit of 24% of UTS for a 100-year design life.

TTS PROCEDURES AND DESIGNS

The TTS tests were performed based on procedures described in Farrag and Shirazi.[20] The test specimen was brought to equilibrium at the test temperature for 2 h and was then

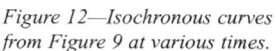

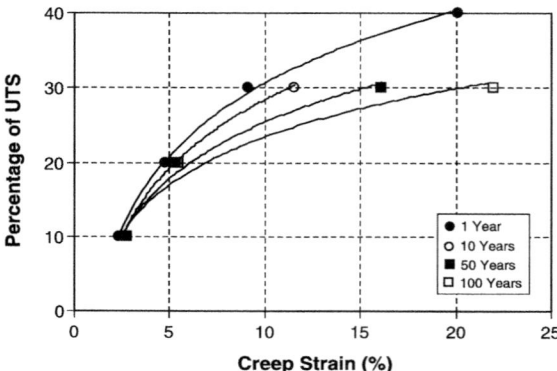

Figure 12—Isochronous curves from Figure 9 at various times.

loaded at a strain rate of 10% of the gage length per minute. The test duration was 10⁴s at each temperature. A new test specimen was used for each test temperature. The test temperatures were 20, 27, 34, 41, 48, 55, and 62°C. Two applied loads at 20 and 30% UTS were tested. *Figures* 13a and 13b show the TTS test data at 20 and 30% UTS, respectively. The creep master curve at each applied load was generated at a temperature of 20°C by shifting the creep curves along the log-time axis.

COMPARING CREEP DATA BETWEEN TTS AND SIM

Figure 14 shows two sets of creep master curves at 20 and 30% UTS obtained from TTS and SIM procedures. The two master curves in each set are relatively similar. The discrepancies in creep strains at 50 years are 0.9% and 0.12% for 20 and 30% UTS, respectively. The results suggest that the strain history at each temperature step does not have a significant effect on the creep behavior of the HDPE geogrid.

CREEP MECHANISM

The creep mechanism of SIM and TTS is evaluated using the activation energy, which is obtained by plotting the shift factor (a_T) against reciprocal test temperatures according to Arrhenius equation, as expressed in equation (2):

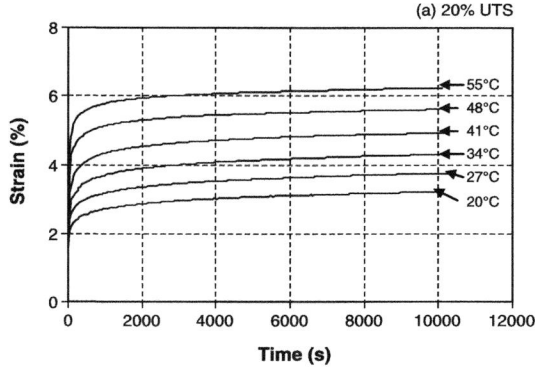

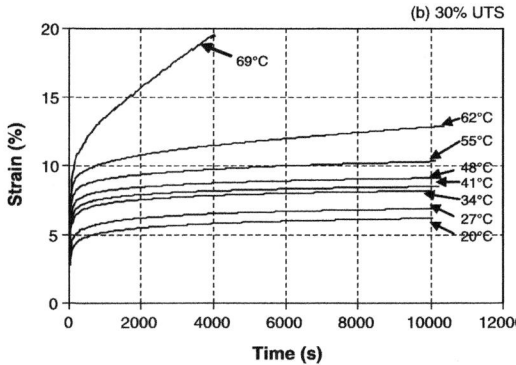

Figure 13—Creep test data from TTS at 20% (a) and 30% (b) UTS.

216

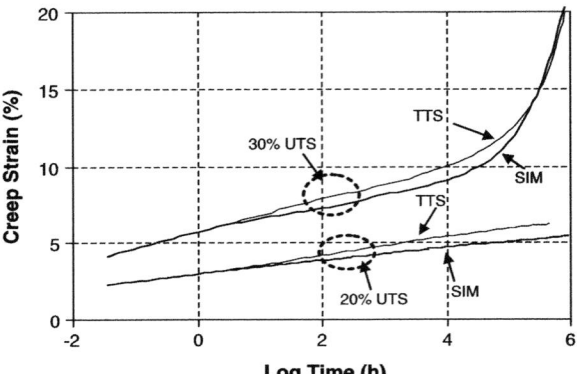

Figure 14—Comparing between SIM and TTS (thin line: TTS, thick line: SIM).

$$\log a_T = \frac{Q}{2.303R}(\frac{1}{T} - \frac{1}{T_{ref}}) \quad\quad (2)$$

where: Q = activation energy, R = gas constant, T = temperature, T_{ref} = reference temperature

Table 5 shows the activation energy values of six creep tests based on MSIM and TTS procedures with a dwell time of 10^4 s. At the 20% UTS, the energy values of SIM and TTS are very close, while a lower value was obtained from MSIM at 30% UTS. In addition, the activation energy decreases as applied load increases from 20 to 40% UTS. Sherby and Dorn[12] accounted for this behavior by contending that stress decreases the thermal energy required for polymer flow.

Comparing the activation energy between MSIM and SIM from the same sets of test data, SIM yields much higher energy values. This is because MSIM requires less horizontal shifting than does SIM due to the large vertical shifting caused by the elimination of initial portion of each creep curve.

However, creep tests that were performed using a dwell time of 10^4 s exhibit much higher activation energies than published values. Govaert et al.[21] found an activation energy value of 118 kJ/mol for PE fibers. Cembrola and Stein[22] obtained an energy value

Table 5—Activation Energies Based on Horizontal Shift Factor

Load (% UTS)	This Work		Hsuan and Yeo[24]	Lothspeich and Thornton[14]	Farrag[23]
	MSIM (kJ/mol)	TTS (kJ/mol)	SIM (kJ/mol)	SIM (kJ/mol)	TTS (kJ/mol)
	Dwell time = 10^4 s				Dwell time = 10^3 h
10	190	—	302		91 (19% UTS)
20	200	217	257		102 (23% UTS)
30	175 ± 6	202	227 ± 10	230 (37% UTS)	96 (31% UTS)
40	140	—	198		

of 85 ± 30 kJ/mol by studying the stress relaxation of an oriented HDPE. In contrast, similar activation energies were obtained by Farrag and Shirazi[20] and Farrag.[23] They performed creep tests using TTS with test durations of 103 h at each temperature. The activation energies calculated from the shift factors at different applied loads are also presented in *Table 5*.

It seems that the duration of the test at each temperature segment plays an important role in the mechanism of the accelerated creep tests. Caution must be carried out when short-term SIM and TTS procedures are used to evaluate the creep properties of new geosynthetics.

CONCLUSIONS

For the SIM test procedure, the temperature increments of $7°C$ and dwell time of 10^4s that are recommended in ASTM D 6992 were found to be applicable for HDPE geogrids. At stress ratios of 10 and 20% of UTS, only primary creep was observed in the HDPE geogrid; the secondary creep stage was observed at stress ratios of 30 and 40%.
The creep master curves obtained from MSIM and TTS are very similar, while their activation energies are slightly different. In addition, the activation energies of these short-term accelerated creep tests are much higher than those from the long-term creep tests, suggesting that different mechanisms may be involved.

ACKNOWLEDGMENTS

The research project was sponsored by the Center for Polymer Reinforced Structures of Geosynthetic Institute (GSI).

References

(1) Carroll, R.G. and Chouery-Curtis, V., "Geogrid Reinforcement in Landfill Closures," *Geotextiles and Geomembranes*, 10 (5-6), 471-486 (1991).
(2) Koerner, R.M., *Designing with Geosynthetics*, 4th ed., Prentice Hall, Englewood Cliffs, NJ, 1998.
(3) Fannin, R.J., "Long-Term Variations of Force and Strain in a Steep Geogrid-Reinforced Soil Slope," *Geosynthetics Intern.*, 8 (1), 81-96 (2001).
(4) Nielsen, L.E., *Mechanical Properties of Polymers and Composites*, Vol. 1 and 2, Marcel Dekker Inc., NY, 1974.
(5) Greenwood, J.H., Kempton, G.T., Watts, G.R.A., and Bush, D.I., "Twelve-Year Creep Tests on Geosynthetic Reinforcements," *2nd European Geosynthetics Conference*, Italy, 333-336, 2000.
(6) Ferry, J.D., *Viscoelastic Properties of Polymers*, 3rd ed., Wiley, NY, 1980.
(7) Painter, P.C. and Coleman, M.M., *Fundamentals of Polymer Science*, 2nd ed., CRC press, Boca Raton, FL, 1997.
(8) Moore, G.R. and Kline, D.E., *Properties and Processing of Polymers for Engineers*, Prentice-Hall Inc., Englewood Cliffs, NJ, p. 209, 1984.
(9) Thornton, J.S., Paulson, J.N., and Sandri, D., "Conventional and Stepped Isothermal Methods for Characterizing Long-Term Creep Strength of Polyester Geogrids," *6th Intern. Conference on Geosynthetics*, Atlanta, GA, 691-698, 1998.
(10) Thornton, J.S., Allen, S.R., Thomas, R.W., and Sandri, D., "The Stepped Isothermal Method for Time-Temperature Superposition and its Application to Creep Data on Polyester Yarn," *6th Intern. Conference on Geosynthetics*, Atlanta, GA, 699-706, 1998.
(11) Greenwood, J.H. and Voskmp, W., "Predicting the Long-Term Strength of a Geogrid Using the Stepped Isothermal Method," *2nd European Geosynthetics Conference*, Italy, 329-331, 2000.

(12) Sherby, O.D. and Dorn, J.E., "Anelastic Creep of Polymethyl Methacrylate," *J. Mechanics Physics Solids*, Vol. 6, 145-162, 1958.

(13) Baker, L.T. and Thornton, J.S., "Comparison of Results Using the Stepped Isothermal and Conventional Creep Tests on a Woven Polypropylene Geotextile," *Geosynthetics Conference 2001*, Portland, OR, 729-740, 2001.

(14) Lothspeich, S.E. and Thornton, J.S., "Comparison of Different Long-Term Reduction Factors for Geosynthetic Reinforcement Materials," *2nd European Geosynthetics Conference*, Italy, 341-346, 2000.

(15) Allen, S.R., "The Use of an Accelerate Test Procedure to Determine the Creep Reduction Factors of a Geosynthetic Drain," *Geotechnical Special Publication*, Geo-Frontiers Conference, p. 3297-3309, 2005.

(16) Tobolsky, A.V., *Properties and Structure of Polymers*, Wiley, NY, 1960.

(17) Popelar, C.H., Kenner, V.H., and Wooster, J.P., "An Accelerated Method for Establishing the Long-Term Performance of Polyethylene Gas Pipe Materials," *Polymer Engineering and Science*, 31 (24), 1693-1700 (1991).

(18) Ingold, T.S., Montanelli, F., and Rimoldi, P., "Extrapolation Techniques for Long-Term Strengths of Polymeric Geogrids," 5th Intern. Conference on Geotextiles, *Geomembrances and Related Products*, Singapore, 1117-1120, 1994.

(19) Den-Hoedt, G., "Creep and Relaxation of Geotextile Fabrics," *Geotextiles and Geomembranes*, 4 (2), 83-92 (1986).

(20) Farrag, K. and Shirazi, H., "Development of an Accelerated Creep Testing Procedure for Geosynthetics-Part I: Testing," *Geotechnical Test. J.*, 20 (4), 414-422 (1997).

(21) Govaert, L.E., Bastiaansen, C.W.M., and Leblans, P.J.R., "Stress-Strain Analysis of Oriented Polyethylene," *Polymer*, 34 (3), 534-540 (1993).

(22) Cembrola, R.J. and Stein, R.S., "Crystal Orientation Relaxation Studies of Polyethylene," *J. Polym. Sci.: Polymer Phys. Ed.*, Vol. 18, 1065-1085 (1980).

(23) Farrag, K., "Development of an Accelerated Creep Testing Procedure for Geosynthetics-Part II: Analysis," *Geotechnical Test. J.*, 21 (1), 38-44 (1998).

(24) Hsuan, Y. and Yeo, S.-S., "Comparing the Creep Behavior of High Density Polyethylene Geogrid using Two Acceleration Methods," *Geotechnical Special Publication*, Issue 130-142, Geo-Frontiers Conference, p. 2887-2901, 2005.

Chapter 14

Methodology Study of Qualification of Electric Wire and Cable Used in Nuclear Power Plant

Masayuki Ito

Advanced Research Institute for Science and Engineering, Waseda University, 245-26 Nakamachi, Kodaira, Tokyo 187-0042, Japan

Electric wire and cable used in the containment vessel of a nuclear power plant are exposed to weak radiation for the lifetime of the plant, which is expected to be more than 40 years. In addition, the materials in the vessel are predicted to be exposed to a higher thermal-radiation environment under a Loss of Coolant Accident (LOCA). LOCA is the basis for the design of power plants. The radiation dose and the temperature depend on the type of nuclear power plant. The maximum temperature and dose estimated in LOCA conditions is 150°C and 2 MGy, respectively, in pressurized water reactor. The properties of the wire and cable are also degraded by various stresses including steam and chemical spray added under LOCA conditions. This chapter demonstrates the importance of the synergistic relationship between thermal and nuclear radiation exposure in the heat- and radiation-induced degradation of elastomers used in electric wire and cable.

This chapter describes an apparatus demonstrating LOCA conditions and gives examples of the test results. A procedure for the test with a particular emphasis on a valid accelerated irradiation condition and the synergistic relationship between heat and radiation on the degradation of elastomers is suggested. The suggestion proposes an ordering of the deteriorating factors added during LOCA conditions to establish an efficacious qualification procedure.

INTRODUCTION

Electric wire and cable used in the containment vessels of nuclear power plants are exposed to low dose radiation during their expected 40-year lifetime. Practical tests for qualification of these materials require a short irradiation time. The materials in the containment vessel are also aged by the relatively modest heat during its working period. The temperature is estimated to be 40°C to 50°C depending on the type of power plant and the position of the container vessel. Assuming that the result of radiation-induced degradation at room temperature defines the rate constant of the decay of a certain property of the material by radiation alone and there is no synergistic relationship between heat and radiation on the degradation of the polymer, sequential exposure of radiation and heat has almost the same effect as simultaneous addition of the two chemical stresses.

A previous article[1] compared two methods of time-accelerated aging and showed that two methods were available for time-accelerated irradiation.

In addition, test methods need to consider the condition of loss of coolant accident (LOCA). The cables in the containment vessel are exposed to high dose-rate radiation, high temperature steam, and chemical (or water) spray simultaneously during LOCA. The maximum temperature during LOCA is expected to be 171°C for a boiling water reactor (BWR) and 150°C for a pressurized water reactor (PWR); the duration is expected to be less than 10 h.[2,3] After LOCA, the materials in the containment vessel are aged under a relatively mild thermal-radiation environment from about 100 days to one year.

The objective of this review is to study the methodology for qualification of electric wire and cable for use in nuclear power plant containment vessels and to suggest methods to improve the technical basis for establishing standard tests.

NORMAL CONDITION

Time Accelerated Irradiation Condition

From a practical point of view, accelerated aging methods for electric wires and cables require high dose-rate irradiation to shorten the exposure time. It has been pointed out that the exposure of polymers in air results in the oxidation of only the surface when the dose rate is high.[4,5] Dissolved oxygen in the polymer is consumed by the reaction with free radicals generated by irradiation. Therefore, when the rate of consumption of oxygen is faster than the rate of diffusion from the surrounding atmosphere to the inside of the polymer, oxidation occurs only in the regions near the surface of the sample.

Clough and Gillen et al. reported the occurrence of heterogeneous degradation, and have summarized those results.[6] They developed a modulus profiling technique[7,8] based on "Time-Temperature-Dose rate superposition."[9] Seguchi et al. studied the "thickness of oxidized layer-oxygen concentration-dose rate relationship."[10,11]

In a previous article,[1] I reported that polymer degradation obeyed the "law of reciprocity"; that is, a short irradiation exposure at high dose gave the same change in properties as a long exposure at low dose. Nine kinds of EPDM, which have different compounding formulas, were used. Specimens were exposed to different doses of Co-60 γ radiation; the maximum dose was 2 MGy. The reference condition to be compared with two short-time test conditions is irradiation of 0.33 kGy/h at room temperature. Two methods were studied as the time-accelerate irradiation conditions:

(1) Irradiation of 4.2 kGy/h in 0.5 MPa oxygen at room temperature.
(2) Irradiation of 5.0 kGy/h in air at 70°C.

After irradiation the mechanical properties of samples were measured at room temperature. The stress-strain relationship at a small strain gives us Young's modulus, which is proportional to network concentration of an elastomer. Since Young's modulus is one of the fundamental physical properties, it is important in the time-accelerated test method to compare the relationship between the dose and Young's modulus obtained by each method. For practical convenience, modulus at 50% strain or modulus at 100% strain is used instead of Young's modulus in the rubber chemistry. The changes in 100% modulus suggest that irradiation in 0.5 MPa oxygen slightly increases molecular chain scission reactions and irradiation at 70°C slightly increases crosslinking, compared with the results obtained under the reference conditions. The deviation was about ± 0.25 for 100% modulus and ± 0.5 for ultimate elongation throughout all doses, where the value obtained at the reference condition is referred to as 1.0. Thus, it was found out that the two methods mentioned above are available as time-accelerated irradiation conditions. From a practical viewpoint, irradiation at 70°C in air is preferable to irradiation in pressurized oxygen.

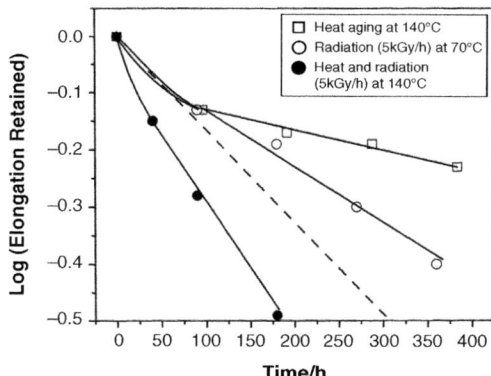

Figure 1—Decrease in elongation retained of EPDM-7 by heat and/or radiation. Dotted line shows the expected values based on an additive relationship between decrease rate of elongation retained by heat and that by radiation, i.e., the gradient of dotted line is Kh + Kr shown in equation (3).

Synergistic Relationship between Heat and Radiation

If there is additivity of changes in a certain property of the polymer by heat, Kh, and by radiation, Kr, the rate constant of simultaneous addition of heat and radiation, Kh+r, is shown in equation (1):

$$Kh+r = Kh + Kr \qquad (1)$$

In this case, a sequential addition of heat and radiation to the polymer can be used for qualification testing. The Arrhenius law might be used to determine the temperature for time-accelerated thermal aging of the polymer, or we could use an empirical relationship, i.e., an increase of 10°C typically increases the rate of oxidative chemical reaction by two times. Based on the empirical low, the damage to the polymers after 40 years at 40°C corresponds to that after 56.3 days at 120°C. For qualification under IEEE Std. 323-1974, 500 kGy is taken as the dose exposed during 40 years.[2] This dose can be applied by using the short-time condition shown in the previous article[1]: dose rate 5.0 kGy/h at a temperature of 70°C for 100 h. If there is additivity between heat and radiation in the degradation of polymers, the sequential addition of the heat aging (120°C, 56.2 days) and radiation under the conditions mentioned above can be used for the qualification testing.

This section compares the additive and synergistic relationship between heat and radiation on the degradation of elastomers. *Figure* 1 shows the elongation retained for EPDM-7 aged by heat and/or radiation.

The decay curves are expressed by equation (2):

$$e/e_0 = \exp(-k\,t) \qquad (2)$$

where e_0 is the ultimate elongation of non-aged sample, e is that of the sample degraded by heat and/or radiation at time t, and k is the rate constant of the decrease in ultimate elongation. The rate constant of decrease in ultimate elongation by heat or radiation was defined as kh or kr, respectively. If there is no synergistic relationship between heat and radiation on the rate constant of decrease in ultimate elongation, the rate constant of the combined addition of heat and radiation, kh+r, should be expressed by equation (3):

$$kh+r = kh + kr \tag{3}$$

The dotted line in *Figure* 1 refers to kh + kr. However, the result obtained was expressed by equation (4):

$$kh+r > kh + kr \tag{4}$$

This result shows the existence of a synergistic relationship between heat and radiation on the rate of decrease of ultimate elongation.

The relationship between logarithms of elongation retained versus aging time is not always linear and it is difficult to define the rate constant of decrease in elongation. Clough and Gillen showed the synergistic relationship between heat and radiation on the degradation of polyvinylchloride, but they did not define the rate constant of decrease in ultimate elongation.[12]

The synergistic relationship between heat and radiation was also found for the rate constant for increase of C=O concentration. Ethylene-propylene copolymer (EP07P, JSR) was mixed with 3.0 phr of dicumyl peroxide by using a mixing roll, and the sample was heat pressed to about 0.1 mm for the measurement by IR spectroscopy. The absorbance of the irradiated samples was measured at 1720 cm^{-1}, and the concentration of C=O was calculated by using $\varepsilon = 300$ (L/cm). *Figure* 2 shows the increase in the concentration of C=O in thin film of ethylene-propylene pure vulcanized (EPR) with the period of aging time. The rate constant of the increase of C=O concentration by heat or radiation is defined as kh(C=O) or kr(C=O), respectively. The dotted line in *Figure* 2 refers to the increase in C=O concentration which is the sum of kh(C=O) and kr(C=O). The combined addition of heat and radiation resulted in a higher value than that of dotted line, showing again the synergistic relationship between heat and radiation. The synergistic relationship described above is a primitive concept, because the radical generated at a certain temperature reacts with the chemical substances near the radical at the temperature; therefore, the reaction by irradiation is intrinsically dependent on the temperature. However, the changes of the properties of the sample induced by irradiation at a certain temperature may be considered to be the effect of irradiation if the reaction by heat alone is negligibly small at that temperature.

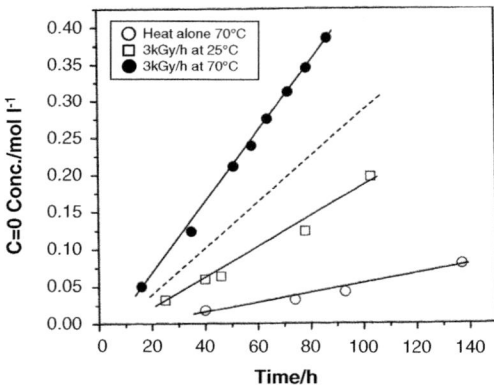

Figure 2—Increase of C=O concentration by heat and/or radiation. Dotted line shows the expected values based on an additive relationship between the rate of increase in C=O by heat (Kh (C=O)) and that by radiation (Kr (C=O). The gradient of dotted line is Kh (C=O) + Kr (C=O).

Chemical stress relaxation has been used for evaluating the heat resistant properties of elastomers, and the method was applied to obtain the efficiency of scission and crosslinking of elastomer by irradiation.[13] The method was also used to quantify the synergistic relationship between heat and radiation on the degradation of elastomers. The rate constant of continuous chemical stress relaxation, which was reflected scission reaction under heat and radiation, Kh+r, is expressed by equation (5):

Table 1—Coefficient of Synergism and Dose Rate Exponent of Various Elastomers

Sample	α	E'/ kJ mol^{-1}
EPDM-1		19.7
EPDM-2	0.97	8.4
Hypalone-1........................		29.0
Hypalone-2		52.1
Chloroprene	0.85	29.8, 40.3
Tetrafluoroethylene-Propylene elastomer.......	0.9	11.8

Note: the symbols are shown in equation (5).

$$Kh+r = Kh + Cc\ I^{\alpha}exp\ (-E' / RT) \qquad (5)$$

Here, Kh is the rate constant of stress relaxation by heat alone, I is dose rate, E' is "coefficient of synergism," R is gas constant, T is Temperature (K), and Cc is the constant which refers to the sensitivity of the specimen to radiation, and α is dose rate exponent.[14] *Table* 1 shows coefficient of synergism and dose rate exponent of various elastomers measured by chemical stress relaxation. Hypalone is chloro-sulfonated polyethylene. The value of E' differed from sample to sample. Clough and Gillen proposed the chemical reaction mechanism for the synergistic relationship.[12] The detailed mechanism which decides the value of E' is not yet clear.

The acceptance test calls for the sequential addition of radiation and heat for experimental convenience, but the recent development of an air oven with light inorganic heat insulation facilitates irradiation of the samples at elevated temperatures. In this case, dose may be reduced by the consideration of degree of synergism. As an example of the qualification condition, we chose the 56.2-day thermal aging at 120°C, which is one of the short time tests corresponding to the aging for 40 years at 40°C, as described above. The estimated dose applied during 40 years of the normal condition of the nuclear power plant is taken to be 500 kGy. If there is no synergistic relationship between heat and radiation, the dose should be 500 kGy.[2,3] However, there is the term of temperature on dose rate in equation (5), showing the magnitude of syner-

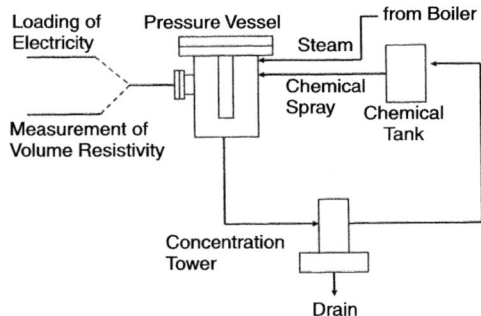

Figure 3—Simplified plant layout of the system for LOCA trial in JAERI (1979–1987). The loading of electricity or measurement of insulation resistance can be carried out during the exposure of radiation. The cylinder welded to the lid of pressure vessel is opened outside, and the radiation source is contained here.

gism, E'. The term $-E' / RT$ is similar in form to the Arrhenius equation, and equation (6) or (7) is given.

$$\log k_1 - \log k_2 = - E' / R \left(1/T_1 - 1/T_2 \right) \tag{6}$$

$$\log \left(k_2 / k_1 \right) = E' / R \left(1/T_1 - 1/T_2 \right) \tag{7}$$

The effect of dose rate multiplies by k_2/k_1 when the irradiation temperature rises from T_1 to T_2.

When T_1, T_2, and E' is 313 K (40°C), 396 K (120°C), and 10 kJ/mol respectively, (k2/k1) is 2.2. This means the dose can reduce from 500 kGy to 227 kGy, considering the synergistic relationship between heat and radiation.

LOCA CONDITION

An Apparatus to Simulate LOCA Condition

When cables and wires are installed in nuclear power plants, the cables are exposed to radiation at a high dose rate, steam at high temperature, and chemical (or water) spray should a LOCA occur. For reactor safety, the cables should be functional even if they are subjected to a LOCA at the end of their service life.

Figure 3 shows the outline of the apparatus[15] to simulate the various LOCA environments including the condition described in IEEE Std. 323-1974.2 The chemical spray

Table 2—Specifications of Apparatus

Size of pressure vessel	700 (I.D.) x 1300
Maximum pressure	2.5 MPa
Maximum temperature	250°C
Size of mandrel	400 (I.D.) x 600
Maximum steam supply	10,000 kg/h
Rate of chemical spray	0-5 l/min.
Co-60 source	30 kCi and 200 kCi
Electricity loading	600V and 90 A
Number of cable loaded	Nine cables
Dose rate	2.5 and 10 kGy/h (designed)
Dose rate uniformity	1.1 (designed)
Heating rate in pressure vessel	20°C to 150°C within 10 s
Uniformity ratio of dose rate	
Vertical	1.04–1.14 (for 200 kCi)
	1.04–1.15 (for 30 kCi)
Radial	1.03–1.06 (for 200 Ci)
	1.07 –1.15 (for 30 Ci)
Air equivalent dose rate at cable position	
200 kCi	10.9 kGy/h (without spray)
200 kCi	9.5 kGy/h (with spray)
30 kCi	2.1 kGy/h (without spray)

Note: the cylinder into which Co-60 γ radiation source can be put from the exterior is welded to the center of the lid of a pressure vessel.

contains 17.3 g of boric acid, 15.9 g of sodium thiosulfate, and 13 g of sodium hydroxide for 600 ml water, but the solution is adjusted by the addition of sodium hydroxide to pH 10.5. The electric cable was inserted into the pressure vessel. The electric wires coiled around the holding fixture are installed in the pressure vessel and the end of the electric wires has come out of the pressure vessel through the connecting tubule. Loading of electricity and the periodical measurement of the resistance of the cables can be done by using the control box. *Table* 2 shows the specification of the apparatus. The relationship between the maximum flow rate of steam and the time to achieve the high temperature in pressure vessel are shown in *Figure* 4.[15] The need for the higher steam flow rate to heat the vessel was due to the long run of steam pipe from the gate valve placed outside of Co-60 γ ray irradiation facility to the pressure vessel that is in the facility.

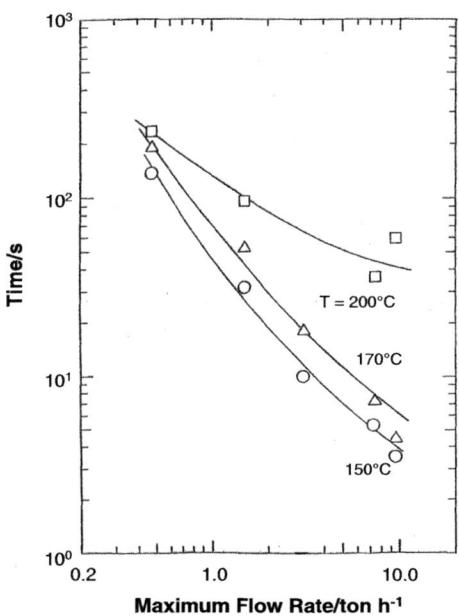

Figure 4—*Maximum steam flow rate and time to achieve the high temperature in pressure vessel. Maximum steam flow rate was measured at the near gate valve which is outside the irradiation facility. □: Relationship between maximum flow rate and required time when pressure vessel reached 200°C. Δ: Relationship between maximum flow rate and required time when pressure vessel reached 170°C. ○: Relationship between maximum flow rate and required time when pressure vessel reached 150°C.*

An Example of the Experiment for Electric Cable Under Simulated LOCA Condition

Figure 5 shows a profile of the LOCA condition examined.[16] In the vessel, electric cables were exposed to saturated steam, chemical spray (2.3l/min), and radiation with electrical loading (600V) under the oxidative condition supplied by 0.05 MPa oxygen. The exposure at 120°C refers to the time accelerated after the LOCA condition; the total dose was 1.5 MGy (9.3 kGy/h) and 1.3 MGy (0.55 kGy/h) for one-week and three-months tests, respectively. The changes of the resistance of the electric cable during the LOCA test are shown in the lower part of *Figure* 5. As the electrical resistance has temperature dependence, the resistance observed after the LOCA experiment measured at room temperature was higher than that measured during LOCA at 120°C. The three-months test brought lower resistance than the result of the one-week test. This is due to the accumulation of higher concentration of oxidative products in the sample during long term thermal aging by steam and air. Thus, the qualification of electric cable has been tested by the system simultaneously, but as a qualification experiment it has the problem of exposure to all of the considerable excitation factors at the same time. Therefore, the next section discusses how to simplify the condition of the test.

226

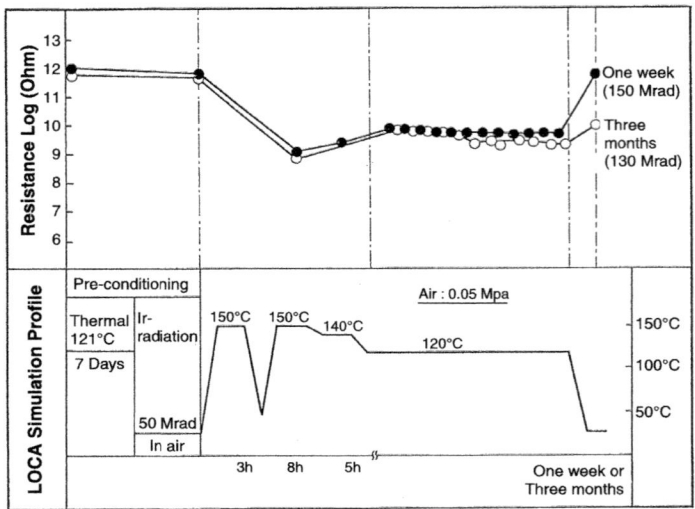

Figure 5—Example of a LOCA profile and changes of resistance of electric cable.

Methodology Study of the Qualification for LOCA

The possible factors that result in deterioration of polymers under a LOCA condition are radiation, heat by steam with air, chemical spray, and loading of electricity. The classification of these deteriorating factors in an efficacious order might be useful for the establishment of a qualification test. Radiation and steam with air significantly damage the polymers. On the other hand, the loading of electricity during one year of LOCA and the post-LOCA period probably do not harm the cable; and therefore the qualification test can omit the electrical loading, which simplifies the apparatus for the test.

The chemical spray system brings complexity to the apparatus because the chemicals and the condensed water in the pressure vessel were mixed in the pressure vessel that must be evaporated by the concentrating tower. Yagi et al. compared the effect of chemical spray and water spray on the water absorption of the irradiated polymers.[17] They measured the water absorption of four kinds of ethylene-propylene elastomer, two kinds of chlorosulfonated polyethylene, and two kinds of chloroprene in water and the solution for chemical spray at about 100°C, and found out that the water absorption was higher in water than in the chemical solution. Since the osmotic pressure of the chemical solution is higher than that of water, the equilibrium water content in the samples was higher in water than in the chemical solution. The chemicals in the spray might not harm the polymers, and while it does contain sodium hydroxide, pH is only 10.5. Therefore, we might use water spray instead of the chemical spray in the qualification.

The simultaneous addition of radiation and heat with steam in the air-containing environment mostly damaged the electric wire and cable during LOCA. IEEE Std. 323 & 383 showed 1.5 MGy as the considerable dose for PWR.[2,3] The recommendation added radiation and steam sequentially for experimental convenience. The electric cable might be oxidized through most of the LOCA period by irradiation, because oxygen penetrates in the sample at higher temperature during irradiation. IEEE-627[18] showed the importance of the account of the degradation mechanism to decide the test condition. Considering these, one of the recommended irradiation conditions is 5.0 kGy/h at 70°C in which the most of the cable might be oxidized (the duration is 300 h). IEEE Std. 323 and 383 showed the various types of steam exposure profile for the test of cable used in a nuclear power plant.[2,3] The profile described in *Figure* 5 is one of the appendices for PWR in IEEE Std. 323, and we can use it for the steam exposure condition for the sequential test. One week of duration may be enough, because one week's thermal aging at 120°C of the polymers corresponds to the damage given during 224 days at 70°C (the average temperature of considerable post-LOCA period), assuming that a temperature increase of 10°C increases oxidative chemical reaction rate by a factor of two.

In the first step of the profile, the temperature rises from ambient to 150°C. This is an important procedure in the test, because one of the possible features of degradation by irradiation is generation of small cracks on the cable surface. The break point is very sensitive to the quick pressurize by steam. The penetration of steam or water enormously decreases the electric resistance of the cable. Giving a quick pressurizing shock twice increases the reliability of the qualification. The lining in the pressure vessel by heat resistant elastomer reduces the amount of steam flow for rapid increase in the pressure and the temperature of the vessel; this is one of the advantages in the sequential method.

Reference

(1) Ito, M., *Nucl. Instr. and Meth. in Phys. Res.*, B236, 229 (2005).
(2) IEEE Std. 323-1974, "IEEE Standard for Type Test of Class IE Electric Cables, Field Splices and Connections for Nuclear Power Generating Stations," IEEE, 1974.
(3) IEEE Std. 383-1974, "IEEE Standard for Type Test of Class IE Electric Cables, Field Splices and Connections for Nuclear Power Generating Stations," IEEE, 1974.
(4) Dole, M., *Report of Symposium IV Chemistry and Physics of Radiation Dosimetry*, Army Chemical Center, , p. 120, 1952.
(5) Charlesby, A., *Proc. Roy. Soc.*, A215,187 (1952).
(6) Clough, R.L., Gillen, K.T., and Quintana, C.A., *J. Polym. Sci., Polym. Chem. Ed.*, 23, 359 (1985).
(7) Clough, R.L., Gillen, K.T., and Quintana, C.A., *Polym. Degrad. Stab.*, 17, 31 (1987).
(8) Clough, R.L. and Gillen, K.T., *Polym. Eng. and Sci.*, 29, 29 (1989).
(9) Clough, R.L. and Gillen, K.T., *Polym. Degrad. Stab.*, 24, 137 (1989).
(10) Seguchi, T., Hasimoto, S., Kawakami, W., and Kuriyama, I., *Report of Japan Atomic Energy Research Institute*, JAERI-M 7315 (1997).
(11) Seguchi, T. and Yamamoto, Y., *Report of Japan Atomic Energy Research Institute*, JAERI 1299 (1986).
(12) Clough, R.L. and Gillen, K.T., *Radat. Phy. Chem.*, 18, 661 (1981).
(13) Ito, M., *Rubber Industry Japan*, 62, 87 (1989).
(14) Ito, M., *Radat. Phy. Chem.*, 18, 653 (1981).
(15) Tanaka, S., Nakase, Y., Kusama, Y., Ito, M., Okada, S., and Yoshida K., *Report of Japan Atomic Energy Research Institute*, JAERI-M 9361 (1981).
(16) Ito, M., Kusama, Y., Yagi, T., Okada, S., and Yosida, M., *Proc. of the U.S., N.R.C.*, NUREG/CP-0058, 5, 351(1985).
(17) Yagi, T., Kusama, Ito, M., Okada, S., Yoshikawa, M., and Yoshida, K., *Report of Japan Atomic Energy Research Institute*, JAERI-M 83-072 (1983).
(18) IEEE Std. 627-1980, "IEEE Standard for Design Qualification of Safety System Equipment used in Nuclear Power Generating Stations," IEEE, 1980.

Chapter 15

Innovative Measurement Methodology in Quality Assessment of Coatings

Kenneth Moller,[1] Anna Backman,[1] Bo Carlsson,[1] Magnus Palm,[1] Sahar Al-Malaika,[2] Husam Sheena,[2] Elizabeth Lakin,[2] Dieter Kockott,[3] Bernd Dawid,[4] Wolfgang Kortmann,[4] Karl Bechtold,[5] Jacques Simonin,[5] Michael Hilt,[6] Joachim Domnick,[7] Dirk Michels,[7] Axel Nagel,[8] Manfred Wunsch,[8] and Karin Hvit Wernstal[9]

[1] SP Technical Research Institute of Sweden, Boras, Sweden
[2] Aston University, Birmingham, UK
[3] Atlas MTT BV, Lochem, The Netherlands
[4] BYK-Chemie, Wesel, Germany
[5] Clariant, Huningue, France
[6] DaimlerChrysler, Sindelfingen, Germany
[7] FhG-IPA, Stuttgart, Germany
[8] PPG, Ingersheim, Germany
[9] Volvo Car Corporation, Gothenburg, Sweden

In the European project MANIAC—Innovative Measurement Methodology in Quality Assessment of Coating, a methodology was developed to assess the long term durability of coatings. The project was focused on environmentally friendly waterborne automotive multi-layer coating systems. Although developed for waterborne coatings, the methodology is very generic and can be applied to all kinds of coatings.

The first step in the evaluation of long-term performance of automotive coatings involves exposures to weathering (natural as well as artificial). Traditionally, the durability of coatings is evaluated by examining the changes in performance properties like gloss, color, and scratch resistance. In the MANIAC methodology, the focus is on changes in structures on molecular levels. Microtomed slices of the coatings were examined by FTIR spectroscopy for monitoring oxidation. UV spectroscopy and GC-MS were used to study the migration of additives like HALS and UVA.

In the methodology, the environmental variables (outdoors as well as indoors), e.g., temperature, relative humidity, UV irradiation, were carefully monitored and included in the modeling scheme. The results from accelerated tests are modeled and, together with outdoor climatic data, the different models are used to predict degradation during outdoor exposure.

INTRODUCTION

The overall aim of the EU project MANIAC was to develop and implement a mechanism-based methodology for accelerated artificial weathering and analytical tests that will enable rapid and reliable prediction of quality, durability, and long-term performance of automotive coating systems. The use of this methodology will reduce considerably the time-to-market of new environmentally friendly paint/coating systems. The innovative aspects in the development of the methodology lies in the application of the knowledge of the fundamentals of degradation processes and advanced analytical techniques to address simultaneously the problems of both the intricate nature of coating systems and the complexity of the weathering conditions, and their effects on changes in terms of chemical and physical properties.

Traditionally, the evaluation of the long-term performance of automotive paint systems has been based on natural weathering for quite a long period of time: 4–8 years in combination with accelerated artificial weathering. The aging of paints has been followed by measuring changes in performance properties such as gloss, color, scratch resistance, etc. In the MANIAC approach the outdoor exposures are reduced to only a few years and the importance of artificial weathering is more pronounced. Of greatest importance is, however, the adoption of a mechanism-based evaluation of aging. Detailed characterization of micro-climatic parameters is also included in the methodology.

The coating systems were selected with the objective to compare well-known coating systems used by many automotive manufacturers in Europe with new, future-oriented waterborne systems focused on further emission reduction, shorter application times, and reduced energy consumption without losing performance and durability.

OUTDOOR AND ARTIFICIAL EXPOSURES

To compare the effects of accelerated weathering on different coating systems with their outdoor service performance, coated panels were exposed at a test site in Miami, FL. The selected accelerated test was SAE J 1960 and the exposures were conducted in an Atlas Ci4000 Xenon Weather-Ometer. To determine the influence of temperature, solar irradiation, humidity, and condensation, modified SAE J 1960 tests were conducted at different levels of these climatic parameters.

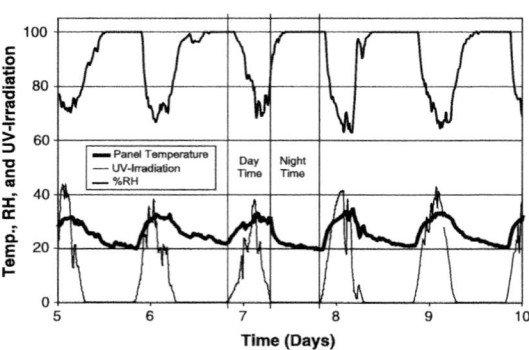

Figure 1—Recorded outdoor climatic data from the Miami test site during five days in October 2004.

To understand aging behaviors of materials, it is important to carefully characterize the local environment or microclimate to which materials are exposed. Moreover, the results from the tests should be evaluated using appropriate mathematical models. From accelerated tests, the parameters in the models can be determined. Knowing the values of these parameters together with outdoor climatic data, the degradation performance out-doors can be predicted. From comparison with measured degradation during outdoor exposure, the reliability of dif-ferent models to predict outdoor performance was analyzed. Therefore, all important environ-mental properties, outdoors as well as indoors, were continu-ously monitored and stored as 15

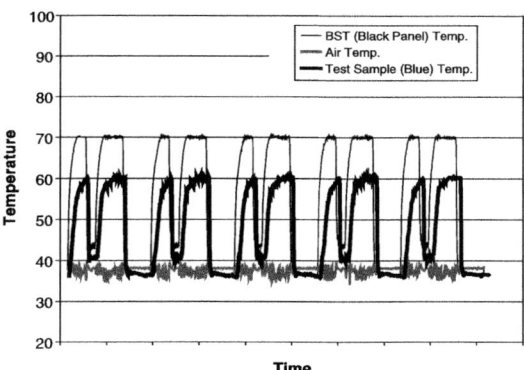

Figure 2—Recorded temperature data from artificial weath-ering according to SAE J 1960.

min mean values. The monitored climatic properties were: ambient air temperature, sam-ple panel temperature, black panel temperature, relative humidity, total solar and UV solar radiant exposure, and pH of rainwater (not measured at regular intervals). *Figure* 1 shows recorded climatic data from the outdoor test site in Miami during five days in October 2004, while *Figure* 2 shows temperature data from artificial weathering accord-ing to SAE J 1960.

PERFORMANCE PROPERTIES

Car owners do not care about changes at a molecular level in the coating. They care about the appearance of the paint in terms of color, gloss, etc. Consequently, it is neces-sary to measure changes in these properties and establish the correlation between chem-ical changes and changes in performance properties. Measured performance properties or conducted tests within the MANIAC project were gloss, color, car wash resistance, chemical resistance, and Distinctness of Image (DOI). Crockmeter and VIEEW scan tests were also conducted. *Figure* 3 shows the gloss retention for samples exposed in Miami.

As can be seen in *Figure* 3, the decrease in gloss was most severe for the waterborne sys-tems 5–10. The most pro-nounced drop in gloss appeared after about one year. A closer examination of these coatings, including microscopy, revealed a considerable growth of mildew (see *Figure* 4). The measured increase in gloss after 12-months exposure was due to the introduction of a new sam-ple cleaning procedure at the

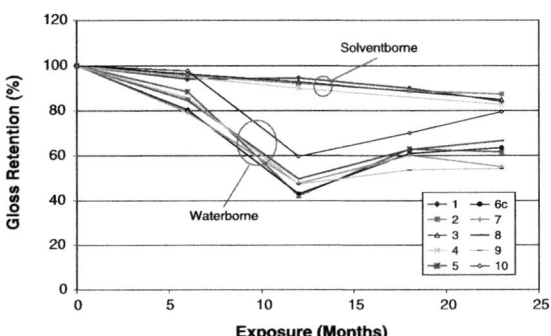

Figure 3—Gloss retention for waterborne as well as solvent-borne coatings exposed in Miami, FL.

232

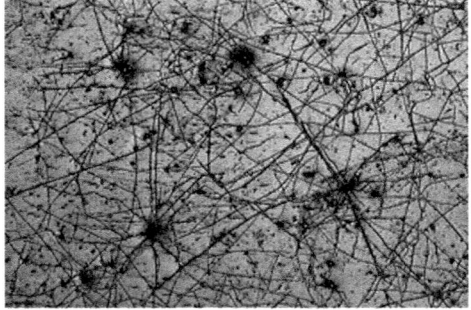

weathering station in Miami, i.e., the loss in gloss was partly caused by scattering of light caused by the mildew, which was removed by the cleaning procedure. *Figures* 3 and 4 exemplify problems associated with gloss retention as a measure of aging. The solventborne coatings 1–4 were not, or were very little, affected by mildew.

Figure 4—Mildew on waterborne coating.

CHANGES IN MOLECULAR PROPERTIES CAUSED BY AGING

Most of the work carried out within the MANIAC project concerns development and use of analytical methods at a molecular level to follow degradation of automotive coating systems. The work done has been focused on migration of additives and chemical changes taking place during aging. As is seen in *Figure* 5, which schematically describes the analytical approach, the analytical procedures to measure additive concentrations and chemical degradation are very interrelated. Almost all instruments and techniques used in the project are commercially available and well established. The only exceptions are devices for migration studies and oxygen 18 ($^{18}O_2$) exposures. Excellent reviews of how these techniques can be used in evaluating aging effects of automotive coatings are given in chapters 9, 10, and 11 of reference 1.

Of great and central importance was microtomy as indicated in *Figure* 5. *Figure* 6 shows a microtomed slice using an in-plane microtome. The thickness of slices used was 5 µm.

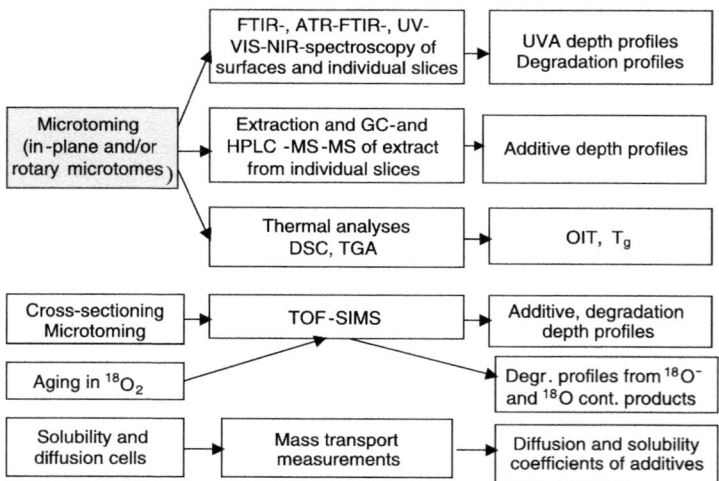

Figure 5—The analytical techniques used in the MANIAC project.

The deterioration of polymer coatings at a macroscopic level is reflected at the microscopic level by early chemical changes. Fourier transform infrared spectroscopy (FTIR) has been shown in earlier studies[2-7] to be an excellent technique that is both chemically specific and highly sensitive.

One important objective of the MANIAC project was to develop and/or use techniques that give generic information. In the IR-region, 3600-2500 cm^{-1} changes caused by

Figure 6—In-plane microtoming.

aging are associated with –OH, –COOH, and –NH groups. The bands between 3000 and 2800 cm^{-1}, which are attributed to the C–H stretching bands, are not considerably altered in intensity due to aging. These bands may, therefore, be used as a reference or an "internal standard". The quotient, Q, between the area of the "aging" bands and the area of the reference bands can thus be regarded as a generic measure of degradation caused by aging, i.e., $Q=A_{-OH, -NH}/B_{Ref.}$ (see *Figure 7*). However, since an unaged coating usually contains groups absorbing in the 3600–3200 cm^{-1} IR-region, the difference between the quotient after and before aging must be used. This difference, $\Delta Q=Q_{aged}-Q_{initial}$, is often referred to as an Oxidation Index or Photo Oxidation Index (PI-index), i.e., $PI=\Delta Q$. Ford Motor Company has successfully used the technique described above.[5,6]

Figure 8 shows PI values plotted versus depth according to the definition in *Figure 7* for a clearcoat containing a UV-absorber (UVA) and exposed for 4000 hours according to SAE J 1960.

An exponential decaying behavior is reasonable, since the UV-radiation at specific wavelengths drops exponentially if the clearcoat contains an evenly distributed UV-absorber. One very severe degradation or failure mode in an automotive coating system is delamination, i.e., the adhesion between two layers is lost. *Figure 9* shows PI-index profiles for a coating system containing no UVA in the clearcoat. Obviously, the lack of UVA in the clearcoat has resulted in a substantial increase in oxidation of the basecoat, which might cause delamination.

As mentioned, the gloss retention of waterborne coatings exposed in Miami was severely affected by mildew. No

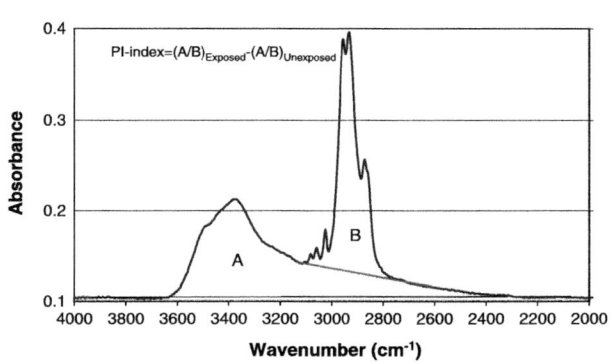

Figure 7—Definition of PI-index.

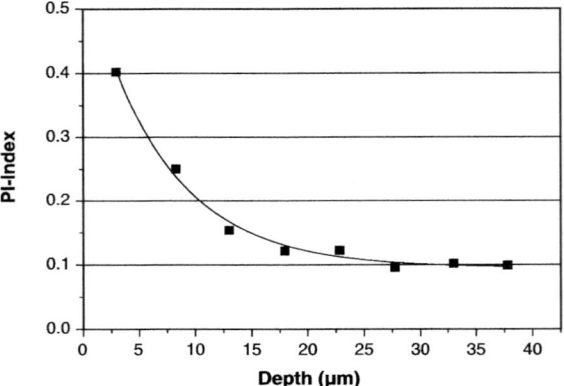

Figure 8—PI-index versus depth for waterborne clearcoat containing UVA and exposed for 4000 hours according to SAE J 1960.

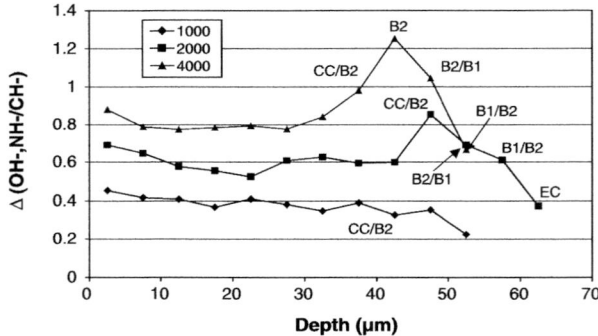

Figure 9—PI-index versus depth for waterborne clearcoat containing no UVA and exposed according to SAE J 1960.

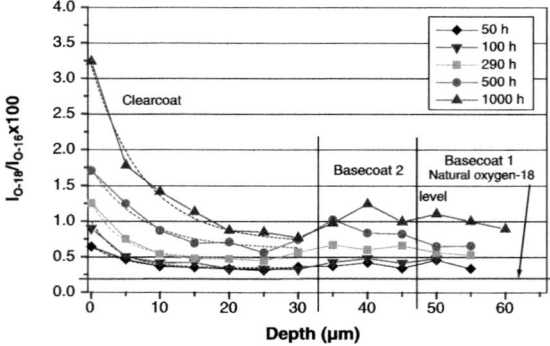

Figure 10—Ratio between oxygen-18 and oxygen-16 TOF-SIMS signals for waterborne clearcoat containing UVA as a function of depths after different exposure times.

corresponding influence of mildew on the PI-index was found, which demonstrated the power of a mechanism-based approach.

TOF-SIMS, OXYGEN 18, AND EARLY DETECTION OF AGING

Time-of-Flight Secondary Ion Mass Spectrometry (TOF-SIMS) is a very sensitive and powerful analytical tool for analysis of solid materials. In TOF-SIMS analysis, O^- and oxygen containing ions are easily detected. To suppress oxygen present in the material prior to artificial accelerated aging, $^{16}O_2$ in the exposure atmosphere can be replaced by $^{18}O_2$. After that, all ions in a mass spectrum containing ^{18}O most likely originate from the artificial aging.[8] The yield of O^--ions is very high, which makes $^{18}O_2$ an excellent tool for the observation of early stages of oxidative degradation.[9] For obvious reasons, the $^{18}O_2$-exposure technique has a great potential when studying oxidative degradation of polymers with oxygen in the backbone, such as automotive coatings.[8,10]

Figure 10 shows the photooxidation of a waterborne paint system containing standard levels of UV-absorber (UVA) and Hindered Amine Light Stabilizer (HALS), while *Figure* 11 shows the photooxidation of the same system, but without UVA. *Figure* 10 shows that already after 50 h of exposure, effects of oxidation can be detected, especially close to the outermost surface. *Figure* 11 shows the response of the paint system without UVA to the exposure. As is seen, the clearcoat is more or less homogeneously oxidized. The clearcoat also gives a much poorer protection of the basecoat than UVA containing clearcoats.

ANALYSES OF ADDITIVE

Of utmost importance for the long-term performance of coatings is additives, especially UVA and HALS.[11-16] Accordingly, it is very important to measure the total amount, as well as concentration depth profiles, of these additives. *Figure* 5 shows the techniques that have been used in the MANIAC project including gas and liquid chromatography combined with mass spectrometry (GC-MS and LC-MS), and UV-spectroscopy using an integrating sphere. *Figure* 12 shows concentration depth profiles of UVA in a waterborne coating system after different exposure times. The depth profiles were obtained by extraction of individual microtomed slices followed by GC analysis of the extracts. By adding the values from the individual slices, the total loss of UVA as a function of time can be estimated (see *Figure* 13). Similar results can be obtained by using HPLC[16] or UV-spectroscopy[5,6,14,15] (in the clearcoat only).

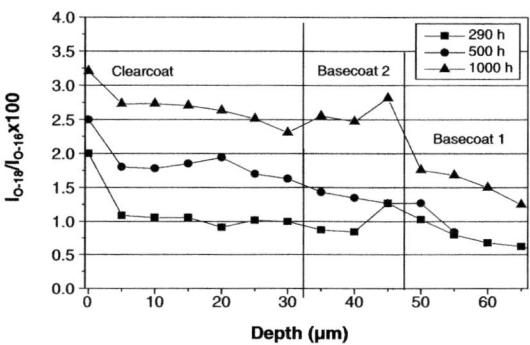

Figure 11—Ratio between oxygen-18 and oxygen-16 TOF-SIMS counts for waterborne clearcoat containing no UVA as a function of depths after different exposure times.

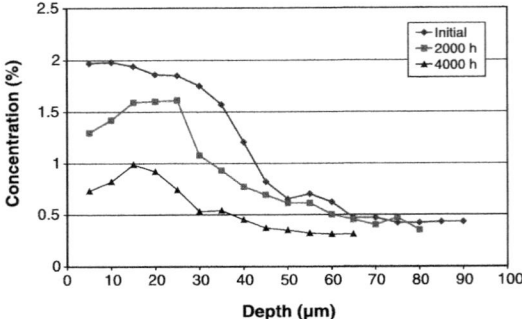

Figure 12—UVA depth profiles for unexposed and exposed waterborne coatings.

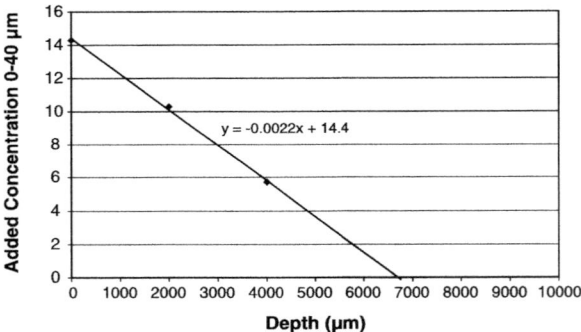

Figure 13—Total UVA loss versus exposure times.

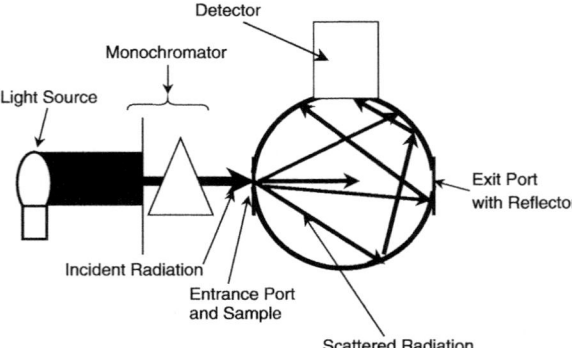

Figure 14—Schematic drawing of a UV-VIS-spectrophotometer equipped with an integrating sphere attachment.

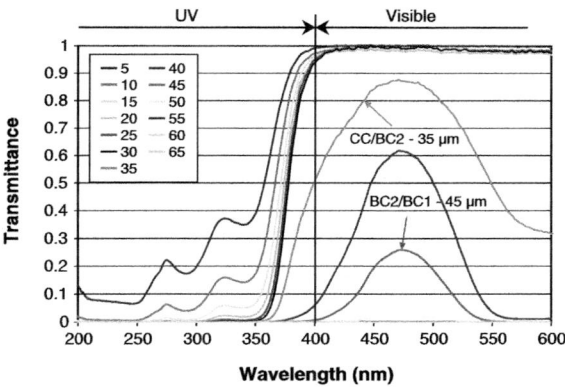

Figure 15—Transmittance as a function of wavelength at different depths.

In UV-spectroscopy on turbid solid materials, such as microtomed slices from automotive coatings, serious problems caused by light scattering occur, which have limited its use. If problems with scattering can be avoided, UV-spectroscopy has great advantages in quantitative analysis of UV absorbing additives in polymers due to high sensitivity and lack of interference from the polymer matrix. Treating microtomed clearcoat films with oil can effectively reduce the problems arising from scattering. However, the films now contain the oil, which may influence other subsequent analyses. An alternative method in eliminating the negative effects of light scattering is to use an integrating sphere attachment, which collects all transmitted light (see *Figure* 14). The reflected part of the incident radiation can be obtained as well. By adding the transmittance and reflectance spectra, essentially all effects of scattering and reflection are removed.[17] As can be seen, the transmittance in the visible region is now essentially equal to unity for clearcoat slices, i.e., the baseline for the added spectra is very close to unity. A 100% transmittance baseline makes it possible to accurately determine the UV-protective action of UVA as function of depths and wavelength as shown in *Figure* 15.

The total transmittance, $T_{tot}(\lambda)$, at a specific depth of a coating system is obtained by multiplying the transmittance spectra of the individual slices. *Figure* 15 shows such total transmittance spectra. Slice No. 6, at a depth of 30 µm, is the deepest lying slice of pure

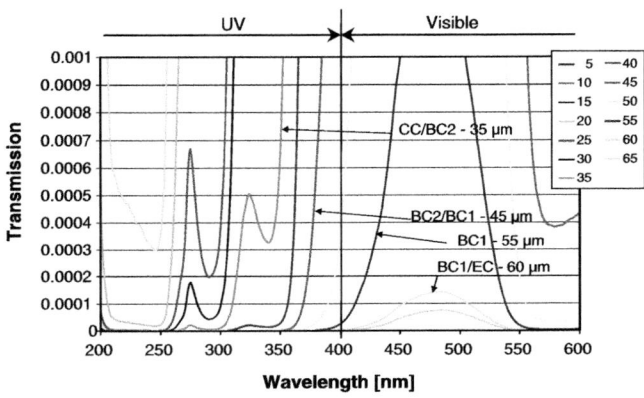

Figure 16—Magnification of Figure 15.

clearcoat. Thus, the added spectra at 30 μm represent the total protective action of the clearcoat. *Figure* 15 shows also that the protective action of this UVA-containing clearcoat is very high below 360 nm. However, above 360 nm the protection decreases drastically, with almost no protection at 400 nm.

The electrocoat layer in an automotive coating system consists often of an epoxy polymer which is sensitive to UV radiation even up to the visible region at 400 nm. It is, therefore, important to be able to measure the UV stress level at the electrocoat. The use of an integrating sphere makes it possible to get such information. *Figure* 16, which is a magnification of *Figure* 15, shows the UV transmittance of the coating system at the interface between the basecoat and the electrocoat. As can be seen, the UV transmittance is very low, only about 0.02% for the coating system under study.

MODELING AND ASSESSMENT OF LONG-TERM PERFORMANCE OF COATINGS

The following approach to accelerated life testing has proved to be successful in the durability assessment of a number of optical materials, especially from the solar energy field, and should be applicable also for accelerated life testing of the kind of automotive clearcoats that have been studied in the MANIAC project.[18]

Assume that for an automotive coating the change in time of a property, e.g., PI-index, due to aging, may be approximated by the following expression:

$$PI(t) = \sum_{i=1}^{m} \Delta PI_i \tag{1}$$

where

$$\Delta PI_i = [A(I_{UV}(t))^\alpha + B(ToW)]e^{E/RT} \Delta t_i \tag{2}$$

where I_{UV} is the intensity of UV irradiance, α is a material-dependent constant, E is an activation energy, and ToW is time of wetness responsible for hydrolysis reactions in the coatings. A and B are additional parameters that have to be determined. The parameters above can be determined using accelerated tests. When all the parameters in equation (2) have been determined, through accelerated tests, outdoor climatic data can be used to predict the progress of *PI*-index. Δt_i is the i:th (15 min) time interval during outdoor exposure. The service life can be determined, if a failure criterion has been defined with respect to *PI*-index. However, service life predictions involve many pitfalls and are not always reliable. The outcome is to a very great extent dependent on the choice of the mathematical model. The chosen model in this case should be regarded as an empirical one that does not truly represent the complex nature of polymer degradation. It is, for example, assumed that the photooxidation and the hydrolysis reactions have the same activation energy.

An interesting alternative might be to leave some of the parameters in equation (2) undetermined, e.g., A and B. These parameters are instead determined by a regression fit using short-term outdoor *PI*-index data. *Figure* 17 shows the result of such a fit for two

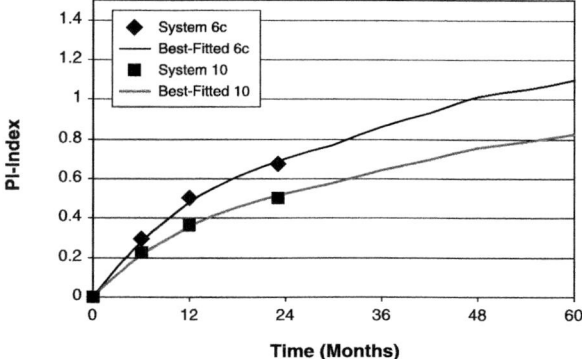

Figure 17—Measured and calculated PI-indexes for two identical clearcoats. The only difference is the pigmentation of the underlying basecoat. System 6c is dark blue while system 10 is silver metallic. Squares and diamonds represent measured values, while the lines represent calculated values.

waterborne systems, 6c and 10, using equation (2). The only difference between the systems is the color. The *PI*-index is obtained at a depth of 7.5 μm (second slice). Outdoor data up to 23 months were used. The correlation between predicted and measured *PI*-indexes is very good. The effect of different temperatures, due to different colors of the coatings, is clearly seen. The wavy pattern of the predicted curves is most likely caused by seasonal variation in the Miami climate.

However, the relatively rapid loss of UVA from the clearcoat (*Figures* 12 and 13) and the indicated susceptibility to photooxidation of the basecoat may suggest that delamination could be a possible failure mode.

CONCLUSIONS

A methodology for the quality assessment of automotive coatings has been developed. The methodology is primarily based on characterization of the aging behavior on a molecular level. Aging parameters are determined mainly through accelerated exposures in which the climatic conditions were carefully measured. Outdoor microclimatic data was collected and used to predict the outdoor aging behavior of the coatings. Different aging effects of identical clearcoats on top of differently pigmented basecoats were clearly observed. This is, of course, due to the fact that different colors of coating systems will lead to different temperatures when exposed to solar irradiation.

ACKNOWLEDGMENTS

John Gerlock, Cindy Peters, Mark Nichols, Alexei Kucherov, and Tony Misovski, Ford Motor Company, are acknowledged for inspiring and valuable discussions. The European Commission (Contract No. G6RD-CT-2001-00638) is acknowledged for financial support.

240

References

(1) Bauer, D.R. and Adamsons, K. (Chapter 9), Adamsons, K. (Chapter 10), and Gerlock, J.L., Smith, C.A., Cooper, V.A., Kaberline, S.A., Prater, T.J., Carter III, R.O., Kucherov, A.V., Misovski, T., and Nichols, M.E. (Chapter 11), in *Service Life Prediction – Methodology and Metrologies*, Martin, J.W. and Bauer, D.R. (Eds.), American Chemical Society Symposium Series 805, Oxford University Press, Washington, DC, 2001.

(2) Bauer, D.R. and Dickie, R.A., "Crosslinking Chemistry and Network Structure in Organic Coatings. I. Cure of Melamine Formaldehyde/Acrylic Copolymer Films," *J. Polym. Sci., Polym. Phys.*, 18 1997-2014 (1980).

(3) Adamsons, K., "Chemical Surface Characterization and Depth Profiling of Automotive Coating Systems," *Prog. Polym. Sci.*, 25 (9) 1363-1409 (2000).

(4) Haacke, G., Brinen, J.S., and Larkin, P.J., "Depth Profiling of Acrylic/Melamine Formaldehyde Coatings." *J. Coat. Technol.*, 67, No. 843, 29-34 (1995).

(5) Adamsons, K., Short Course Presentation at the Atlas School or Natural and Accelerated Weathering (ASNAW), April 28-30, Miami, FL, 1999.

(6) Gerlock, J.L., Smith, C.A., Cooper, V.A., Dusbiber, T.G., Weber, W.H., "On the Use of Fourier Transform Infrared Spectroscopy and Ultraviolet Spectroscopy to Assess the Weathering Performance of Isolated Clearcoats from Different Chemical Families," *Polym. Degrad. Stab.*, 62, 225-234 (1998).

(7) Adamsons K., Sticka, K., Swartzfager, D., Walls, D., Wood, B., and Lloyd, K., "Characterization of Multilayered Automotive Paint Systems Including Depth Profiling and Interface Analysis," *Polym. Mater. Sci. Eng.*, 77 482-483 (1996).

(8) Gerlock, J., Prater, T., Kaberline, S., and deVries, J.E., "Assessment of Photooxidation in Multi-Layer Coating Systems by Time-Of-Flight Secondary Ion Mass Spectrometry," *J. Polymer Deg. Stab.*, 47, 405-411 (1995).

(9) Moller, K., Jansson, A., and Sjovall, P., "Analysis of Polymer Oxidation Using $^{18}O_2$ and TOF-SIMS," *J. Polymer Deg. Stab.*, 80, 345-353, 2003.

(10) Moller, K., Backman, A., and Johansson, U., "Analysis of Photooxidation of Automotive Coating Systems Using $^{18}O_2$ and TOF-SIMS," Third MoDeSt Conference in Lyon, August-September, 2004.

(11) Bauer, D.R., "Chemical Criteria for Durable Automotive Topcoats," *J. Coat. Technol.*, 66, No. 835, 57-65 (1994).

(12) Haacke, G., Andrawes, F.F., Cambell, B.H., "Migration of Light Stabilizers in Acryl/Melamine Clearcoats," *J. Coat. Technol.*, 68, No. 855, 57-62 (1996).

(13) Haacke, G., Longordo, E., Brinen, J.S., Andrawes, F.F., Cambell, B.H., "Chemisorption and Physical Adsorption of Light Stabilizers in Pigments and Ultrafine Particles," *J. Coat. Technol.*, 71, No. 888, 87-94 (1999).

(14) Haacke, G., Longordo, E., Brinen, J.S., Andrawes, F.F., Cambell, B.H., *Proc. 25th Int. Waterborne, High-Solids, Powder Coat. Symp.*, 433-443, 1998.

(15) Haacke, G., Longordo, Andrawes, F.F., and Cambell, B.H., "Interaction Between Light Stabilizers and Pigment Particles in Polymeric Coatings," *Prog. Org. Coat.*, 34, 75-83 (1998).

(16) Adamsons, K., Short Course Presentation at the Atlas School of Natural and Accelerated Weathering (ASNAW), April 28-30, Miami, FL, 1999.

(17) Moller, K. and Gevert, T., "A Solid State Analysis of the Desorption/Evaporation of Hindered Phenols from Low Density Polyethylene (LDPE) Using FTIR and UV Spectroscopy Integrating Sphere: The Effect of Molecular Size on the Desorption," *J. Appl. Polym. Sci.*, 61, 1149-1162 (1996).

(18) Jorgensen, G., Brunold, S., Carlsson, B., Heck, M., and Moller, K., *Performance and Durability Assessment: Optical Materials for Solar Thermal Systems*, Koehl, M., Carlsson, B., Jorgensen, G.J., and Czanderna, A.W. (Eds.), Chapter 6, Elsevier, Oxford, 2004.

Chapter 16

A New Test Method for Weatherability Prediction: Radical Shower using a Remote Plasma Reactor

Masahiko Akahori

Nippon Paint Co., Ltd., R & D Div., 4-1-15 Minami-Shinagawa Shinagawa-ku, Tokyo, Japan 140-8675

It is difficult to predict long-term (30 year) durability using conventional acceleration methods such as the Super UV weather-test machine or the Xenon weather-test machine. A new test method for weatherability prediction using a Remote Plasma Reactor, a down-flow type device, has been studied.

We studied the degradation mechanism by focusing on the fact that exposure to radicals strongly influences the durability of coatings. As a means of direct radical irradiation on the coatings, we used a plasma-generating device. The Remote Plasma Reactor has a plasma generating zone (electrode) placed apart from a sample exposure portion, which allows for more easily controlled test conditions than a direct plasma reactor. Consequently, neutral active oxygen radical species can be easily beamed selectively on the coatings without influence by ion, electron, or ultraviolet rays. We can adjust the condition of plasma with pressure, flow rate of gas and radio frequency power.

The accelerated test for the weatherability of enamel type coatings was studied. For the evaluation of the durability, we investigated changes not only in color and gloss, but also surface changes (SEM, and FTIR technique) through the Remote Plasma Reactor.

Results obtained with the Remote Plasma method correlated very well with outdoor exposure tests and the acceleration behavior with test time was especially well controlled. The radical shower method can be applied to other materials such as plastics and rubbers.

INTRODUCTION

In 1966, Kenneth E. Boulding, an American economist, compared the earth to a spaceship. This idea of "The Spaceship Earth" is thus over 40 years old. Our modern society is in a transition from an economic system of mass-production, mass-consumption, and mass-disposal, to a new sustainable economic system.[1] In the Japanese housing field, we have refurbished or repainted old housing, some of which were built 1000 years ago. The situation in Japan has been changing recently, however. Over 60% of houses have been built within the past 20 years, while less than 10% were built over 50 years ago.[2] This means that the service life of houses in Japan is shorter than in the United States of

America and the countries in Europe. To achieve a sustainable system, we should treasure the limited resources of the Earth and maximize their utilization. This attitude also holds true in the automobile market: we do not need to buy a new car every three years. For the past 10 years, Japanese car consumers have been using their old cars instead of buying new cars. It is thus desirable for automobile manufacturers to enhance paint durability.

From the viewpoint of sustainable housing, there is a strong demand to have a lifetime of 30 years in housing exteriors. From the viewpoint of good appearance and protection of wall substrates, remarkable progress has been made in the development of raw materials. There are also several new durable masonry paints now under development. It is highly desirable to be able to predict and guarantee the lifetime of the modern industrial product. The 30-year goal seems difficult to attain, but the use of clear coatings like those used in automobile coating systems may make it possible to have such a long lifetime. We should predict the weatherability or the durability of coatings in order to be able to guarantee their long lifetime.

Although many acceleration methods are widely used, it is difficult to predict the durability over a 30-year period using conventional acceleration methods. For example, there are metal halide lamp methods that emphasize stronger ultraviolet rays, a Super UV (SUV) weather-test machine, or a Xenon weather-test machine. In other words, it is not possible to predict endurance for 20 or 30 years by conventional methods. Other new methods proposed recently are from 1992 and 2000, respectively. The system described in reference 3 uses oxidation promotion via increased oxygen pressure using an autoclave reactor; the system described in reference 4 uses a metal halide lamp and hydrogen peroxide. These systems are on their way to practical application in Japan.

With respect to artificial weather deterioration of the coatings brought on by radical degradation from these surfaces, we reported the endurance prediction method with use of a plasma device as a radical irradiation method in 1994.[5,6] Also, in recent years, an etching by plasma has been actively used in the IT technology fields. The first analysis, from the point of film erosion, was done in 1984.[7] Later, plasma was used as a technique to improve the quality of a surface. For instance, the study on the improvement of quality of a poly-tetra-fluoro-ethylene surface using a hydrogen radical in a Remote Plasma (down-flow type) Reactor was done.[8,9] Furthermore, starting in 1998, there was a move-

Table 1—Content of Pigments in Each Paint by Weight

Pigment	No. 156	No. 328	No. 530
White	96.5	91.9	87.2
Black	1.9	4.2	6.5
Yellow	1.3	3.2	5.5
Red	0.3	0.7	0.8
Total	100.0	100.0	100.0
Volume concentration of white pigment			
Titanium dioxide	3	8	13
Barium sulfate	15	10	5

White pigment: Titanium dioxide, Barium sulfate.
Colored pigment: Carbon black, Iron (III) oxide, Iron (II) oxide.

ment in Europe to utilize a plasma reactor for endurance prediction and a three-year project was started.[10] However, the study was suspended due to difficulty in the selection of conditions by direct plasma method.

In this chapter, the authors have found that a plasma reactor can be used as an effective tool for long-term prediction of coatings durability.[11-14] The direct plasma system exposes a substrate between two electrodes. Therefore, the deterioration of the coatings is caused not only by the radical species, but also by plasma-comprising ions, electrons, highly reactive neutral atoms, and ultraviolet rays. As a result, it is difficult to optimize control test conditions. In contrast, the Remote Plasma Reactor, which has a plasma generating zone (electrode) placed apart from a sample exposure portion, allows for easily controlled test conditions. Consequently, neutral active oxygen radical species can easily be beamed selectively on the test specimen without ions, electrons, or ultraviolet rays. We should control the condition of plasma. We used the oxygen plasma gas for our environment, which is filled with oxygen and nitrogen; in other words, the degradation is concerned with the oxidation of the products. We also used gases other than oxygen, including nitrogen oxide and carbon dioxide.

Inspection by an electronic microscope (3D-SEM) showed that the surface conditions resulting from natural weathering in Okinawa and those produced by a Remote Plasma Reactor appear similar. The chemical change observed at the coatings surface is also similar. It can therefore be presumed that a deterioration mechanism of a Remote Plasma Reactor is similar to outdoor exposure.

One goal of this work is to standardize the Remote Plasma method to ISO standard and JIS standard as a new test method for the weatherability prediction. Secondly, this quick determination method is useful for the evaluation of material durability. In the summer of 2006, we successfully developed the method and Suga Test Instruments Co., Ltd. started to produce equipment for commercial use. The method will allow us to determine the weathering performance of the product and to guarantee the product for a specified time more quickly.

EXPERIMENTAL

Preparation of Model Paints

The waterborne enamel paints were prepared by blending the synthesized acrylic emulsion resin, titanium dioxide, and barium sulfate as model masonry paints. It also included hindered amine light stabilizer (HALS) and the ultraviolet absorber (UVA) at 1.5 wt% based on resin. The compositions of the pigments in these three types of the paints are given in *Table* 1. These paints were sprayed on the sizing boards, which are commonly used in exterior walls in Japan, and were dried at 100°C for 10 min.

Outdoor Exposure Test

The outdoor exposure test was carried out over five years in Okinawa, which is at the same latitude as the subtropics. The specimen material was exposed to ambient air circulation at a rack angle of 20° South. The latitude of Okinawa is around 23° North;

244

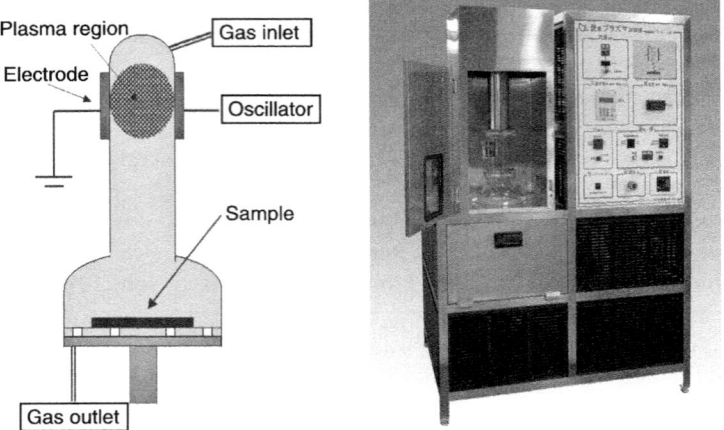

Figure 1—Chamber model of the Remote Plasma device and the commercial machine by Suga Test Instruments Co., Ltd.

for convenience, we used the 20° angle. It is said that the latitude of -5° is most efficient for receiving sunlight.

Super UV Accelerated Test

The Super UV (SUV) machine for accelerated testing is commonly used in Japan. It is different from the Xenon accelerated machine as a result of its irradiation waves. The SUV will damage the substrate quickly because it has stronger ultraviolet rays than the Xenon machine. The SUV test was carried out using a cycle mode. The cycle mode used was 4 h of UV irradiation (100 mW) at 63°C (Black Standard Temperature: a temperature measuring device consisting of an insulated black plastic which absorbs all wavelengths of radiation uniformly, with a thermally sensitive element firmly attached on the center of the unexposed surface) followed by 4 h of darkness and condensation at 35°C, including 10 s of spray between irradiation and darkness.

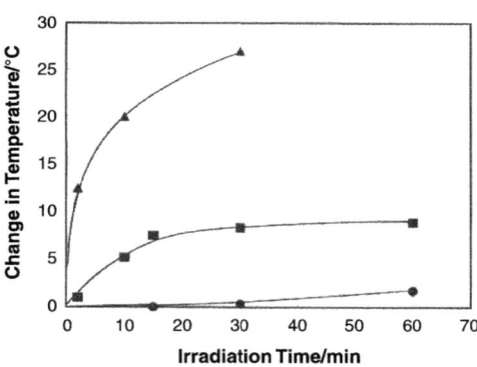

Figure 2—Change in temperature at the same sample surface, 350 mm away from the electrode during O₂ plasma treatment. Conditions: Pressure 133 Pa, oxygen flow rate: 300 ml/min. Radio frequency power:

Remote Plasma Reactor Acceleration Test

The new accelerated test illustrated in *Figure 1* (left) was carried out by the Remote Plasma Reactor (RPR). The vessel consists of a quartz tube, which has a gas inlet and an exhaust. The discharge region is equipped outside with a pair of electrodes that are made of aluminum plates. Gas plasma is produced by a 13.56 MHz generator.

In March 2006, we granted patent rights to Suga Test Instruments Co., Ltd. in Japan. This RPR machine is called the Plasma accelerated test device, Pl-2. This commercial machine is shown in *Figure* 1.

Characterizations and Analysis

Gloss retention and color difference were tested according to JIS Z 2381, K-5600 (corresponding to ISO 7724), and determined by means of the BYK-Gardner micro TRI gloss (60°) and of the Konika-Minolta color difference meter CR-300, respectively.

The surface of the painted boards was observed using a 3D-SEM Erionics Co. ERA-8800FE Apparatus. The IR spectra were recorded by using an FTIR Nihon Electrics Co. FTIR7300 apparatus.

RESULTS AND DISCUSSION

Conditions for New Accelerated Test

Many acceleration methods are widely used. For example, there are metal halide lamp-methods that emphasize stronger ultraviolet rays, a Super-UV weather-test machine, or a Xenon weather-test machine. When we used these weather-test machines, we controlled some conditions like irradiation power, temperature, pressure, and their cycling mode. In addition to these acceleration methods, we carefully studied the plasma state and the irradiation control. This is a key for getting good correlation between the natural exposure and the artificial accelerated methods.

It is reported[9] by Inagaki et al that the ionized species, electrons, highly reactive neutral atoms, molecules, and ultraviolet rays in the plasma region are rapidly decreasing in proportion to the distance from an electrode to a sample on the stage. Those plasma constituents have the influence of raising the surface temperature with which they come in contact. The increased temperature during an irradiation of the board surface will influence the endurance of the coatings. To avoid the increase in the surface temperature, it is important to determine the plasma reactor configuration and processing parameters. We need to adjust the operating parameters with an appropriate selection of plasma configuration. The increased temperature on the board surface as a function of an

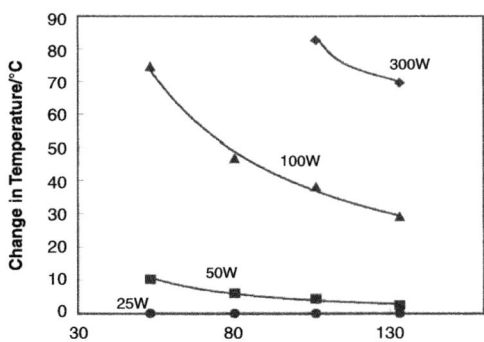

Figure 3—Change in temperature at the same sample surface, 150 mm away from the electrode during O_2 plasma treatment. Irradiation time: 60 min; oxygen flow rate: 200 ml/min. Radio frequency power ●: 25 W, ■: 50 W, ▲: 100 W, ◆: 300 W.

246

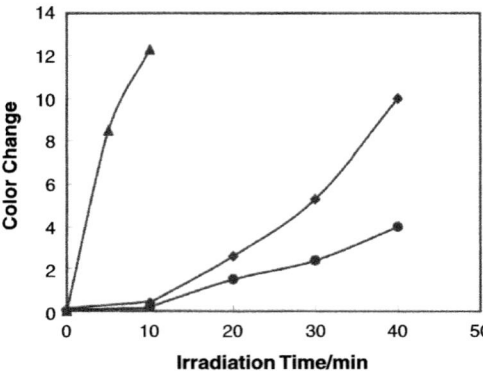

Figure 4—*Color change of the sample by plasma treatment at 50W. Conditions: Pressure: 133 Pa; flow rate: 300 ml/min. Distance from the electrode:* ▲: *10 mm,* ◆: *150 mm,* ●: *350 mm.*

irradiation time at the distance of 350 mm from the electrode with several powers of frequency radiation is shown in *Figure* 2.

The increased temperature on the board surface as a function of the pressure at the distance of 150 mm from the electrode with several powers of frequency radiation is shown in *Figure* 3.

By using a Remote Plasma Reactor, we could selectively irradiate neutral oxygen radicals to the coatings on the board at the temperature below glass transition temperature (T_g). The color changes after irradiation at the different distances from the electrode, and different powers of frequency radiation, are also shown in *Figure* 4.

From these results, the distance of 10 mm from the electrode to the specimen was too small to control ions, electrons, radicals, etc. Consequently, the temperature of the test specimen surface rises, there is mainly heat damage, and there are striking changes in color. Also, with the enamel paints used in this investigation, the L-value changes remarkably with the chalking of the test specimen during irradiation. As a result of the chalking, and the large color difference (e.g., L > 5), there is no need for further discussion. When utilizing the Remote Plasma machine, there is no rain (no water) during irradiation, and the chalk formed remains on the surface. The influence of chalking on the color difference is ambiguous, because a great deal of chalking covers the surface of the specimen. The conditions determined for the RPR are given in *Table* 2.

Plasma Condition Changes with Choice of Gas

We changed the plasma condition using other gases such as carbon dioxide and nitrous oxide instead of oxygen. The point of doing this is that we wanted to see how these gases would affect the degradation of coatings. These are the equations we assume for the decomposition of several gases. Also there are some radical reactions of organic compounds as shown in *Scheme* 1.

In these gas plasmas, gas molecules generate many species of electrons, radicals, and ions of various valence. In the decomposition of oxygen gas, we selectively use the effective radicals. The activated carbon oxide from the decomposition of carbon dioxide is similar to organic compounds. In the case of nitrous oxide gas, it is expected to produce a pair of nitrogen radicals for each

Table 2—Operating Conditions for the Remote Plasma Reactor

Radio frequency power	50 W
Irradiation distance from the electrode	150 mm
Pressure	133 Pa
Flow of oxygen	20–300 ml/min
Irradiation time	0–120 min

Oxygen:	O_2	+ Energy	>>>>>	$2O^{2-}$		
	O_2	+ e^-	>>>>>	O^{3-}	+	e^-
	O_2	+ Energy	>>>>>	$2O^*$	+	$h\nu$
	O_2	+ e^-	>>>>>	O_2^+	+	$2e^-$
	O_2^+	+ e^-	>>>>>	$2O^*$		
Carbon dioxide:	CO_2	+ Energy	>>>>>	C	+	$2O^*$
Nitrous oxide:	$2N_2O$	+ Energy	>>>>>	$2N_2$	+	O_2
			>>>>>	$4N^*$	+	$2O^*$

*Scheme 1—Some decomposition of several gases. *: Radical.*

oxygen radical. We expected more gas volume in the chamber with decomposition. Understanding the plasma state under various conditions is important for the development of suitable test methods. We check the increase in surface temperature to avoid a heat effect. Oxygen gas plasma should be well controlled, or the surface temperature will increase, as shown in *Figure* 2.

The graphs in *Figure* 5 show the results using carbon dioxide and nitrous oxide gas plasma. In the case of carbon dioxide, the temperature does not increase as it does in oxygen plasma even up to 100 watts power. In the case of nitrous oxide, even at 500 watts power, the surface temperature increase is less than 10°C, compared with a temperature increase of over 40°C in the case of carbon dioxide.

Next we studied the color changes of degradation by using several gases in the Remote Plasma device as shown in *Figure* 6. We conducted the remote plasma bombardment experiments with two different gases: one with carbon dioxide and the other with nitrous oxide in comparison with oxygen gas. The materials are the same coatings that we checked in oxygen plasma experiments No. 530 in *Table* 1.

Carbon dioxide plasma has a greater effect than does nitrous oxide plasma. Here, nitrous oxide plasma is less useful than expected. Compared with nitrous oxide plasma, we can recognize the strength of oxygen plasma. The cause of weak attack to the coatings is the effect of carbon black. Carbon dioxide is expected to produce elemental carbon, which is the terminator against the radical propagation and elemental carbon lets

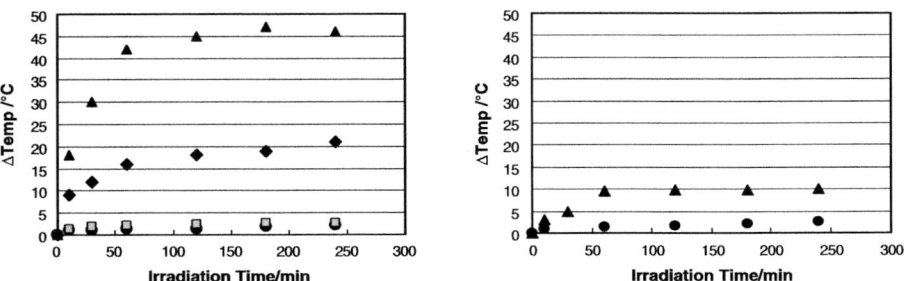

Figure 5—Change in temperature at the same sample surface, 350 mm away from the electrode during carbon dioxide and nitrogen oxide plasma treatment. Conditions: Pressure: 133Pa, flow rate: 500 ml/min. ▲: 300 W, ◆: 200 W, ■: 100 W, ●: 50 W.

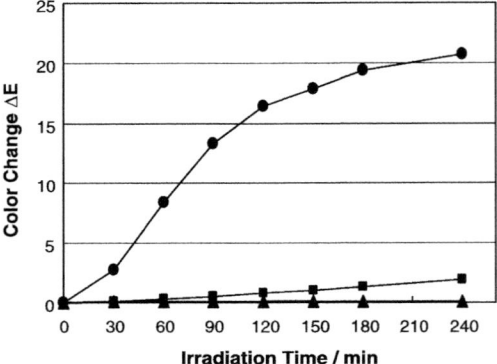

Figure 6— Color changes under acceleration by Remote Plasma using several gases. Conditions: 50 W, 133 Pa, gas flow rate 500 ml/min. ●: Oxygen gas, ■: Carbon oxide, ▲: Nitrous oxide.

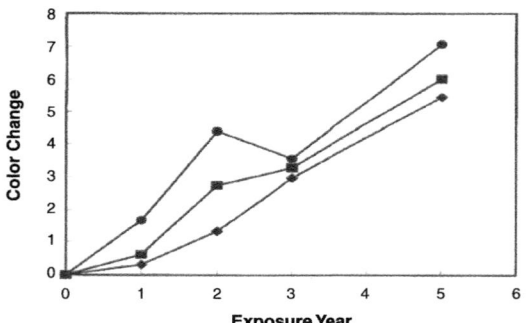

Figure 7— Color change under outdoor exposure in Okinawa. ◆: No. 156, ■: No. 328, ●: No. 530.

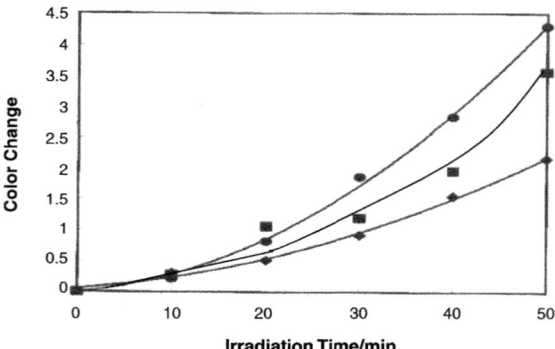

Figure 8—Color change under acceleration by Remote Plasma treatment. Conditions: 50 W, pressure 133 Pa. ◆: No. 156, ■: No. 328, ●: No. 530.

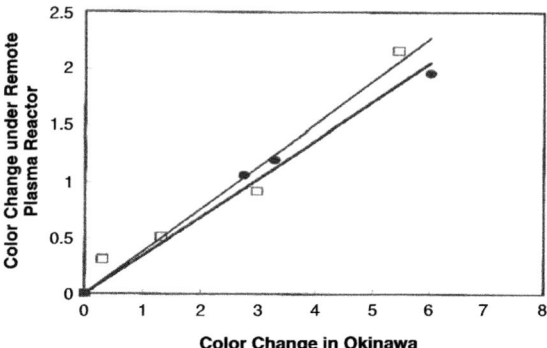

Figure 9—Correlation of color change between outdoor exposure in Okinawa and acceleration by Remote Plasma treatment. Conditions: 50 W, pressure: 133 Pa. □: No. 156, ●: No. 328.

other radicals terminate. The carbon radical itself does not live a long time. In the case of nitrous oxide, the reactions shown in *Scheme* 1 do not occur as frequently, so the number of oxygen radicals generated is just half of the amount compared with oxygen.

Correlation Between Outdoor Exposure and Remote Plasma Acceleration

The color change of outdoor exposure in Okinawa during the five-year period 1994 to 1999 is recorded in *Figure* 7. The increase in color differences of these model coatings follows the same general trends.

Figure 8 shows the color changes of the surface after Remote Plasma Reactor treatment under the conditions in *Table* 2 as a function of irradiation time.

Figure 9 presents the correlations of color change between the outdoor exposure and the accelerations by Remote Plasma device. The color change is plotted with the outdoor exposure on vertical axis, and accelerations on horizontal axis (see *Table* 3). This means it has a good correlation in their durability.

Table 3—Color Change Under Outdoor Exposure and Remote Plasma Treatment

Outdoor exposure in Okinawa	Year	1	2	3	4	5
No. 156	ΔE	0.31	1.32	2.98	—	5.45
No. 328	ΔE	0.62	2.75	3.28	—	6.01
Remote plasma treatment	(min)	10	20	30	40	50
No. 156	ΔE	0.31	0.51	0.92	1.55	2.16
No. 328	ΔE	0.28	1.06	1.21	1.96	3.56

(cf) The condition of Remote Plasma treatment is based on Table 2.

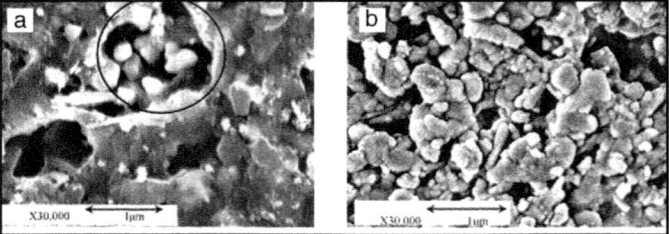

Figure 10—SEM micrograph of coatings A and B after one and three years in Okinawa.

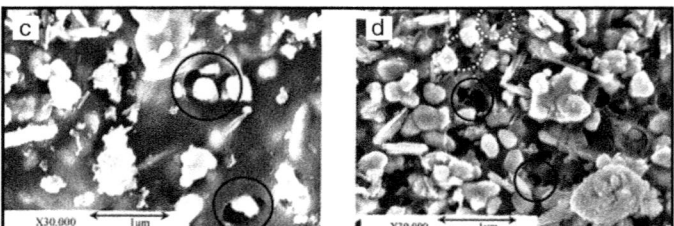

Figure 11—SEM micrograph of coatings C after 15 min. of irradiation in the Remote Plasma Reactor. D: after 394 h in SUV Reactor.

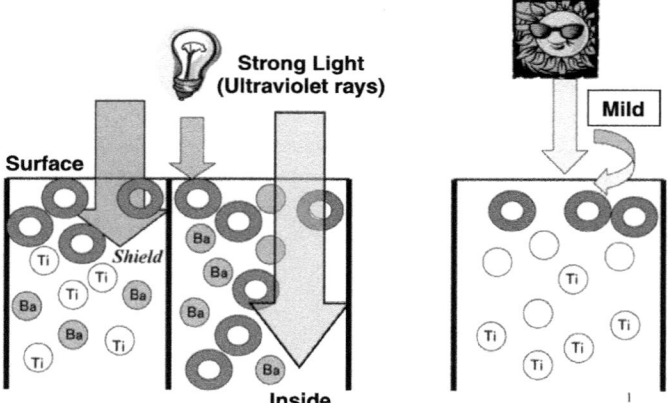

Figure 12—Hypothesis of the shield effect by titanium oxide.

Surface Observations by SEM Technique

The morphological analyses of both the coated samples after the outdoor weathering test and the artificial weathering test are shown in *Figures* 10 (A,B) and 11 (C,D) respectively. As can be seen from *Figure* 11 A and B, the coatings in Okinawa were damaged and chalking appeared on the surfaces. These surface results are similar to those surfaces obtained by the Remote Plasma Reactor (*Figure* 11C).

In these SEM technique observations, at first glance there are similar cavities present (illustrated by solid circles). Both have a particle of titanium dioxide present. On the contrary, a much clearer difference can be observed if we look into the cavity around the particle.

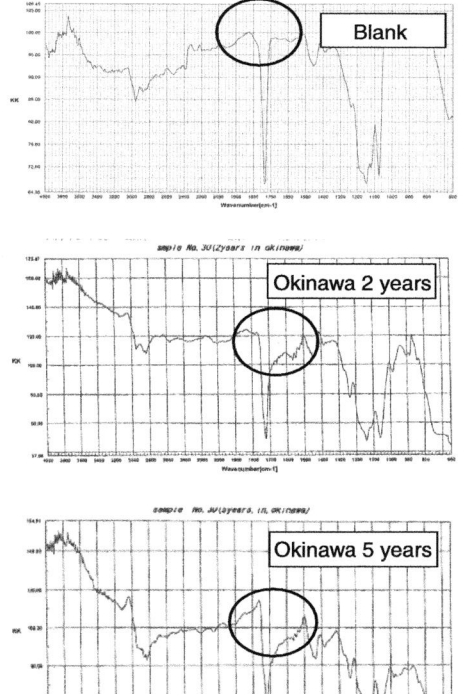

The observation of coatings accelerated by using a SUV machine after the 394 h test (*Figure* 11D), revealed the presence of another, different, cavity (illustrated in broken circle) on the surface. This cavity was produced by a different mechanism of the degradation. This mechanism seems to relate to the photo-catalyst process by titanium dioxide (see *Figure* 12).

This hypothesis of mechanism seems to involve the shield effects of titanium dioxide and its ability to act as a photo-catalyst. The degradation of the surface by sunlight can be seen on the upper right side. It is a mild one. In contrast, the stronger ultraviolet rays cause the degradation on the inside by the activated

Figure 13—Chemical change of FTIR during the exposure in Okinawa.

titanium dioxide. This degradation of the inside is different than the degradation of natural exposure. Currently, this process is under investigation using a depth profile.

Chemical Changes by FTIR Technique

We studied the chemical changes of coatings with the proceeding weathering test. The results are shown in *Figure* 13. The outdoor exposure data in Okinawa steadily rises with weathering time.

Continued weathering increases concentration of the carbonyl groups within the polymer and simultaneously reduces the ester group content. The ratio of each group was defined as the Polarity Index (Y) given below:

$$Y = S (A) / S (B)$$

S (A) : Fraction of Carbonyl group (1650-1730 cm^{-1})

S (B) : Fraction of Ester group (around 1730 cm^{-1})

As can be seen in *Figure* 14 (the Polarity Index), the outdoor exposure data in Okinawa steadily rises with weathering time in contrast with DMW (Daipla Metal

252

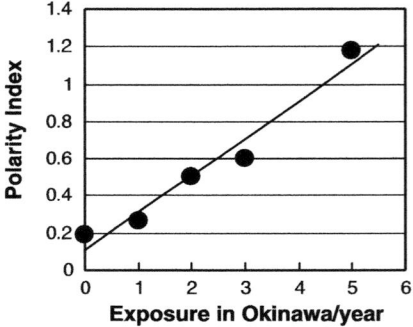

Figure 14—Polarity Index in Okinawa exposure.

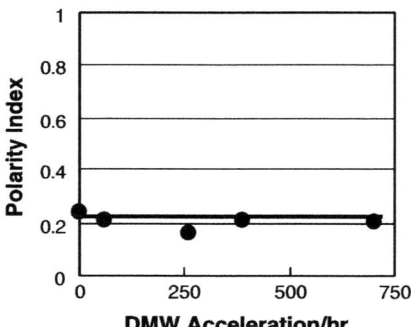

Figure 15—Polarity Index on DMW machine.

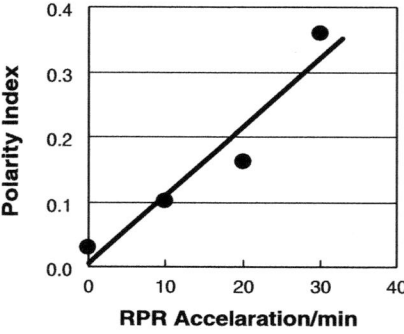

Figure 16—Polarity Index by Remote Plasma machine.

Weather tester, one of the accelerated weathering testers with a metal halide lamp, it has higher irradiance for the ultraviolet wavelength like SUV method than solar radiation). There is no increase in the Polarity Index as in *Figure* 15. This result means that the point of decrease in concentration of the ester group contents, and decomposition of the polymer in coatings, is no greater than the formation of the carbonyl group. In other words, the accelerated method using strong energy of an ultraviolet ray under 300 nm, cannot be correlated with outdoor exposure.

The Polarity Index from the Remote Plasma Reactor given in *Figure* 16 shows the trend as similar to that obtained from outdoor exposure in Okinawa.

SUMMARY AND CONCLUSIONS

The experiments carried out for the weatherability prediction demonstrate that the Remote Plasma Reactor can offer distinct advantages over conventional methods. The results obtained can be summarized as follows:

(1) A Remote Plasma Device for weatherability prediction has been successfully developed. This acceleration machine provides a promising new prediction method.

(2) Although the Remote Plasma methods have a fast acceleration compared with the conventional methods like a Xenon weather-machine, it is important to control and to adjust the proper operating parameters, as well as the appropriate selection of plasma configuration.

(3) For the study of weatherability or durability, we should combine exposure test results with analytical evaluation. Then we can totally predict the weatherability and the durability of the products.

(4) The new method, Remote Plasma, is useful for evaluation of the durability of industrial materials and helpful for the selection of raw materials in polymer design.

ACKNOWLEDGMENTS

This report is partially supported by Japanese Ministry of Economy, Trade and Industry on the project for Research and Development of House Technology Creating Life Values. This five-year project was named "Development of Recycled-Oriented Highly Durable Coating Materials and Coating System."

References

(1) Seike, T. and Akimoto, T., *Sustainable Housing*, Toyo keizai shinnhosha, Tokyo, 2003.
(2) Japanese Ministry of Economy, Trade and Industry, "The Statistics of Japanese Mministry of Internal Affairs and Communications," 2003.
(3) Shiobara, T., Ozaki, H., and Matsunaga, T., JPN Patent H01-28897, 1989.
(4) Mori, K. Okamoto, K., and Tachi, K "Degradation of Coatings Films Coating Titanium Oxide by UV Irradiation in Aqueous Oxidation Agent Solutions," *J. Jpn. Soc. Colour Mater*, 75, 209-226 (2002).
(5) Umino, M. and Higashino, Y., '"A New Acceleration Method for Durability and a Thermo and a Mass Analysis Study of the Deterioration Mechanisms of Epoxy/Block Isocianato Crosslink," *Proc. Conf. J. Jpn. Soc. Colour Mater*, 12 -13 (1995).

(6)   Umino, M. and Higashino, Y., JPN Patent H09-178727A.
(7)   Falla, N., "An Ultra-rapid Method for Predicting Paint Film Durability," *Analytical Proceedings*, 21, 259-261 (1984).
(8)   Hansen, R. and Schonhorn, H., "A New Technique for Preparing Low Surface Energy Polymers for Adhesive Bonding," *J. Polym. Sci., Polym. Lett. Ed.*, 4, 203-209 (1966).
(9)   Yamada, Y., Yamada, T., Tasaka, S., and Inagaki, N., "Surface Modification of Poly (tetrafluoroethylene) by Remote Hydrogen Plasma," *Macromolecules*, 29, 4331-4339 (1996).
(10) Falla, N, "Ultra Rapid Prediction of the Durability of Coatings for Plastics using Plasma Erosion," *Surf. Coat. Int.*, 81, 375-380 (1998).
(11) Akahori, M., Kajino, T., and Umino, M., JPN Patent No. 3934582.
(12) Akahori, M., Kajino, T., and Umino, M. et al., U.S. Patent 7141755.
(13) Akahori, M. and Kajino, T., "A New Technology for Weatherability Prediction," *Techno-Cosmos*, 17, 30-35 (2004).
(14) Akahori, M., "A New Technology for Weatherability Prediction by using a Remote Plasma Reactor," *Surf. Coat. Int. Part B: Coat. Trans.*, 89, B2, 163-168 (2006).

Characterizing and Modeling Color and Appearance

Chapter 17

Metrologies for Characterizing Optical Properties of Clear and Pigmented Coatings

Li-Piin Sung, Xiaohong Gu, Haiqing Hu, Cyril Clerici, and Vincent Delaurent

National Institute of Standards and Technology, Polymeric Materials Group
Building and Fire Research Laboratory, Gaithersburg, MD 20899

Gloss and color performance attributes are commonly used in the coatings industry to assess the appearance of new and weathered coatings. Advances in appearance measurements and models are needed to satisfy continued demands from customers for materials having enhanced appearance qualities that last throughout the life of a product. In this chapter, we present an angular-resolved, surface-optical scattering metrology for characterizing the optical reflectance properties of a coating. Through an analysis of the specular and off-specular optical scattering profiles, a link has been established between surface morphology and subsurface microstructure to optical reflectance properties of a coated material as a function of weathering time. Optical data for coatings having different degrees of pigment dispersion at various UV-exposure times were obtained using laser scanning confocal microscopy and compared to corresponding optical scattering data.

INTRODUCTION

Coatings are designed to achieve a particular appearance effect and to protect consumer products and built structures. The ability to accurately characterize appearance attributes of a coating and to relate these attributes to material and performance properties continues to be of great academic and industrial research interest. Key material properties, such as surface topography, pigment dispersion, and heterogeneity in microstructure, affect the appearance and are believed to impact the service life and mechanical properties of a coated object. Currently, specular gloss measurements using commercially available glossmeters remain the primary measurement tool for assessing the appearance and durability of coated objects. However, it has not been possible to consistently correlate specular gloss measurements with the fundamental micro-physical surface properties, such as surface roughness, pigment dispersion, and subsurface microstructure of a material, that are felt to affect the appearance of a coating. These changes can only be monitored through changes in diffuse scattering intensity, which commercial glossmeters do not measure.

Recent efforts at National Institute of Standards and Technology (NIST)[1-3] in linking the surface morphology and subsurface microstructure to optical reflectance properties of a coated material using a ray scattering model have provided an approach to calculate and understand the optical reflectance for a given surface morphology and subsurface microstructure. The outcome of this research provided (1) an improved understanding of the microstructural basis for coating appearance, and (2) the development of tools for producing computer graphic images (computer rendering) to visualize the color and gloss of surfaces using measured data and models. Computer rendering may potentially provide a powerful tool for identifying coating parameters that primarily determine a coating's appearance. Another important aspect of color and appearance studies is to identify the relative importance of these parameters in affecting the service life of a coated material, a primary goal of the optical scattering research in the Building and Fire Research Laboratory (BFRL) of NIST.

In this chapter, we report on the effect of pigment dispersion on (1) surface morphology and optical properties (appearance), and (2) the weathering durability of a TiO_2 pigmented epoxy coating. Different degrees of pigment dispersion were achieved by varying the mixing conditions. For the purposes of comparison, the time-resolved weathering results of unpigmented clear epoxy coatings are also reported. The microstructure and dispersion of pigmented epoxy coatings were characterized using ultra small angle neutron scattering (USANS). Atomic force microscopy (AFM) and laser scanning confocal microscopy (LSCM) were used to characterize surface morphology of the unexposed and weathered coatings. A full angular-resolved optical scattering profile was used for understanding the physics and impact of surface morphology and subsurface microstructure on the appearance of a coating as a function of weathering time. Commercial gloss measurements were also carried out to study the corresponding optical properties of the coatings, and were compared to surface optical scattering results.

EXPERIMENTAL*

Materials and Sample Preparation

An amine-cured, unfilled (unpigmented) and pigmented epoxy coating was used in this study. The unfilled epoxy was a mixture of a pure diglycidyl ether of bisphenol A (DGEBA) with an epoxy equivalent of 172 g of resin containing 1 g equivalent of epoxide (Der 332 from Dow Chemical) and 1,3-bis(aminomethyl)-cyclohexane (1,3 BAC) at the stoichiometric ratio. Appropriate amounts of toluene were added to the mixtures of epoxy resin and amine curing agent, which were then mechanically stirred. After being degassed in a vacuum oven, the mixture was applied onto the substrates in a CO_2-free, dry air glove box. Then, 150 μm thick films were obtained by casting the mixture onto a glass substrate coated with black coating using a drawdown technique. All samples

---

*Certain trade names and company products are mentioned in the text or identified in an illustration in order to adequately specify the experimental procedure and equipment used. In no case does such an identification imply recommendation or endorsement by the National Institute of Standards and Technology, nor does it imply that the products are necessarily the best available for the purpose.

were cured at room temperature for 24 h in a glove box, followed by heating at 130°C for 2 h in an air-circulating oven.

Pigmented epoxy samples were prepared using the same ratio of epoxy resin and amine curing agent with 15% pigment volume concentration (PVC) commercial pigment grade TiO_2 (Dia. $\approx$ 230 nm). The solvent was n-butyl acetate and 1-methoxy-2-propanol acetate at a 4:1 ratio. Due to the inherent complexity involved in controlling pigment dispersion, samples were prepared by the following procedures: (1) epoxy resin was mixed with/without dispersant in the solvent using a mechanical mixer at 210 rad/s mixing speed for 20 min. six percent (by mass of pigment) of a high molecular weight wetting and dispersing cationic agent was used as dispersion in this case; (2) TiO_2 pigment was added while mixing at 365 rad/s for 60 min; (3) amine curing agent was added at the stoichiometric ratio with respect to epoxy resin and mixed at 105 rad/s for 10 min; (4) the mixture was applied to release paper using a drawdown technique in a CO_2-free, dry air glove box with a wet film thickness of 150 μm; and (5) samples were cured at room temperature for 24 h in a glove box, followed by heating at 130°C for 2 h in an air-circulation oven. The samples were denoted as "Clear" for unpigmented epoxy; and "Disp" and "NonDisp" for the pigmented epoxy coatings with and without dispersant added into the mix, respectively.

Outdoor UV Exposure

Outdoor exposures were conducted on the roof of a NIST laboratory located in Gaithersburg, MD. Specimens were loaded in multiple-window exposure cells and placed in an outdoor environmental chamber at five degrees from the horizontal plane facing south. The bottom of the chamber was made of black-anodized aluminum and the top of the chamber was covered with "borofloat" glass; all four sides of the chamber were perforated and these perforations were covered with a non-moisture absorbing fabric material that acted as a filter to prevent dust particles from entering the chamber. The exposure cell was equipped with a thermocouple and a relative humidity (RH) sensor, which recorded temperature and RH in the chamber at minute intervals throughout the day.

Sample exposures were started at the beginning of each of five different months in 2003 and 2004. The rates of degradation for the epoxy films exposed at each starting date differed; hence, the exposure durations required to achieve the same amount of degradation depended on the starting date. The starting months were October of 2003, and March, July, September, and December of 2004. Specimens are differentiated by their exposure date; thus, specimens have been designated as the October03 group, the March04 group, the July04 group, the September04 group, and the December04 group.

Microstructure, Pigment Dispersion, and Surface Morphology Characterization

Ultra small angle neutron scattering (USANS) was carried out to characterize the microstructure and pigment dispersion of the coated samples. The USANS measurements were performed at the NIST Center for Neutron Research (NCNR) using a perfect crystal diffractometer (PCD) instrument. Details of instrumentation and scattering theory are given in references 4 and 5. The USANS data presented in this chapter were a one-dimensional absolute scattered intensity curve as a function of the scattering wave vector, q =

$4\pi \sin(\theta/2)/\lambda$, where θ is the scattering angle and λ is the wavelength. The estimated extended uncertainties (k=2) in the data are presented in this chapter.

Atomic force microscopy (AFM, Veeco Metrology Dimension 3100) and laser scanning confocal microscopy (LSCM, Zeiss LSM510) were used to characterize surface morphology of the unexposed and weathered coatings. Details of instrument specifications and measurement protocols are given in references 6–9. By combining these two microscopic measurements, a wide range of length scale surface areas could be observed from (1 μm x 1 μm) to (2 mm x 2 mm). The AFM images presented here are height images, while the LSCM images are two-dimensional (2D) intensity projections. The 2D intensity projection images are effectively the sum of all the light backscattered by different planar layers of the coating. The pixel intensity level represents the total amount of backscattered light. Approximately 6 to 10 micrographs were obtained for each sample, with representative images reported here.

From the 3D topographic profiles, the root-mean-square (RMS) surface roughness, S_q, was calculated using a surface tilt correlation and an automatic plane fit.[9] Plane fit is commonly used to remove tilt from images. A single polynomial fit was calculated for the entire image and then subtracted from the image. S_q was computed using the following formula:

$$S_q = \sqrt{\frac{1}{N_x \cdot N_y} \cdot \sum_{i=1}^{N_x} \sum_{j=1}^{N_y} \cdot \left[z(x_i, y_j) - S_c \right]^2}$$

(1)

where

$$S_c = \frac{1}{N_x \cdot N_y} \cdot \sum_{i=1}^{N_x} \sum_{j=1}^{N_y} \cdot z(x_i, y_j)$$

(2)

Here, $z(x_i, y_j)$ is the surface height at position (x_i, y_j), and N_x and N_y are the number of pixels in the X- and Y-directions, respectively.

Optical Properties Measurements

Optical scattering measurements were conducted at several incident angles in the specular and off-specular configurations on surfaces of unexposed and weathered coatings using a custom-built optical scattering instrument in the Building and Fire Research Laboratory at NIST.[9] Specular scattering is dominated by the optical contrast and surface gloss of a coating. The major contributors to off-specular scattering include surface roughness, subsurface scattering from pigments, and other sources of heterogeneity in the polymer matrix. *Figure* 1a presents the optical geometry, where θ_i and θ_s are the incidence and scattering angles measured with respect to the normal of the sample, and θ_{as} is the aspecular angle, the angle measured away from the specular direction. *Figure* 1b shows a representative 2D scattering profile for the incidence angle of 20° and $\theta_{as} = 3°$.

In the remainder of this chapter, optical results are presented as a 2D scattering pattern or as a one-dimensional 1D circularly-averaged scattered intensity curve as a function of the scattering angle away from the specular angle. Estimated uncertainties in the optical scattering intensity data are ± 4% from the mean value computed from measure-

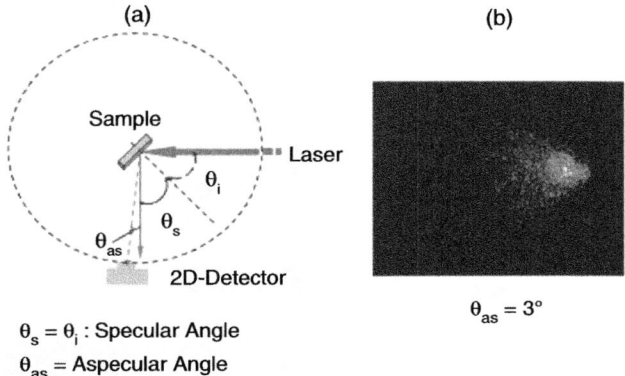

(a)   (b)

Sample

Laser

θ_i

θ_s

θ_{as}

2D-Detector

$\theta_{as} = 3°$

$\theta_s = \theta_i$: Specular Angle

θ_{as} = Aspecular Angle

Figure 1—(a) Optical geometry of the scattering experiment, and (b) 2D scattering profiles obtained from a surface at $\theta_{as} = 3°$. Here, θ_i is the incident angle, θ_s is the specular angle, and θ_{as} is the aspecular angle, which is the angle measured away from the specular direction.

ments at four different locations on a specimen. Specular gloss measurements were carried out using a handheld commercial glossmeter (Minolta, Multi-Gloss model 268) conforming to the ASTM D 523 standard measurement protocol. The data were expressed as gloss retention, which is defined as the percentage change in the gloss of a specimen relative to its initial gloss value. Estimated uncertainties in the gloss retention are one standard deviation from the mean value of eight measurements per sample for each of three replicates or, equivalently, one standard deviation from the grand mean value of 24 measurements.

RESULTS AND DISCUSSION

Exposed Clear Epoxy Coatings

Figure 2 shows the surface morphological LSCM images (top row, image size: 460 μm x 460 μm) and AFM images at the same exposure times (bottom row, image size: 10 μm x 10 μm) of clear epoxy samples for the (a) March04 and (b) July04 groups. The exposure histories, including temperature and humidity, were different for these two groups. For the March04 group, the exposure conditions were initially cold and dry but, later in the exposure, the conditions were warm and humid; the July04 group was immediately exposed to hot and humid conditions. Consequently, the rate and pattern of morphological changes differ for these two groups. Nevertheless, both groups exhibit similar morphological changes as shown in *Figure* 2. At a larger size scale (LSCM images, 460 μm x 460 μm, *Figure* 2a) in the March04 group, the surface started from smooth with a few particle-like features (38 d, *Figure* 2a), then cracks and some surface features appeared (69 d, 77 d, and 84 d). At a smaller size scale (AFM images, 10 μm x 10 μm), some sub-microscale features that were not visible in LSCM images are revealed. The degradation started with the formation of protuberances (38 d, *Figure* 2a), pits (69 d,

262

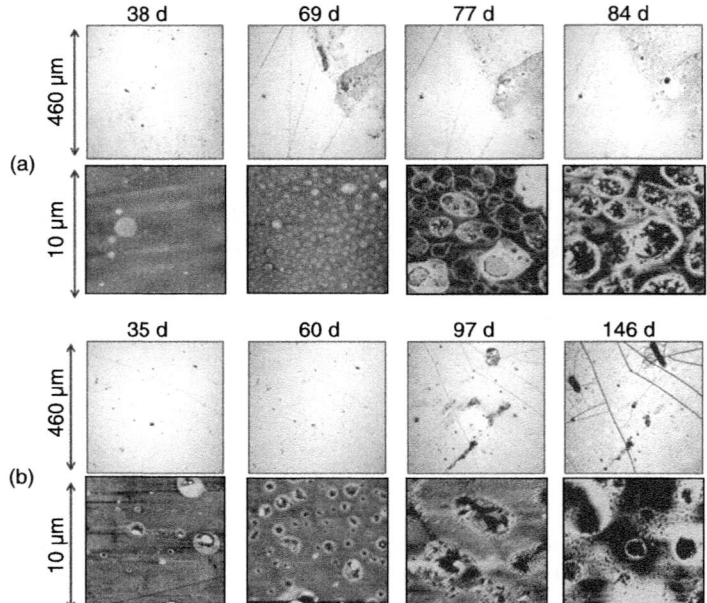

Figure 2—LSCM (top row) and AFM (bottom row) images of (a) March04 and (b) July04 groups epoxy coatings after outdoor exposure in Gaithersburg, MD, at exposure times as labeled. The lateral dimensions for LSCM images are 460 μm and for AFM images are 10 μm.

Figure 2a) and, finally, the aggregation and deepening of the pits (77 d, and 84 d, *Figure* 2a). Compared to the March04 group, the changes in surface morphology of the July04 group (*Figure* 2b) appeared to occur faster due to the warmer, more humid starting exposure conditions. By combining large scale and small scale morphological changes (e.g., large scale changes include the growth and the distribution of cracks, changes in pit density, and surface roughness while small scale changes include the formation, deepening, and aggregation of individual pits), one can observe patterns in the topography of the amine-cured epoxy coatings as they weather.[10]

Insight into topographic changes and how these changes impact the optical properties are provided by an analysis of the surface roughness and the correlation between surface roughness and gloss measurements. *Figures* 3a–3c show the (a) RMS roughness and corresponding (b) 20° and (c) 60° gloss retention as a function of exposure time for four exposure groups. These roughness values, based on 460 μm x 460 μm LSCM images, were calculated using a surface tilt correlation without a numerical filter. Here, gloss retention is defined as the percentage change in the gloss of a specimen relative to its initial gloss value. RMS and gloss retention changes are fastest in the July04 group samples and slowest in the October03 group. On the other hand, it appears that the October03 group has the longest plateau before the gloss significantly drops; while the July04 group has the shortest one, which corresponds to the fast changes in surface roughness in July04 group. As predicted, different groups display different gloss change behaviors as

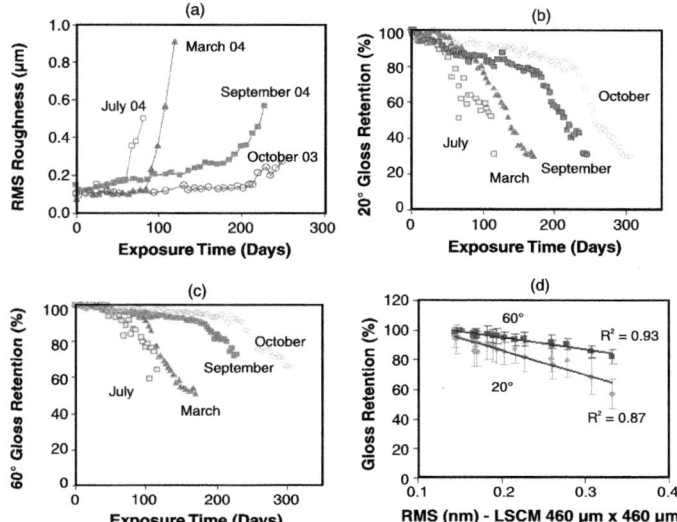

Figure 3—(a) RMS roughness (LSCM 460 μm x 460 μm), (b) 20° and (c) 60°
gloss retention (%) versus exposure time for the October03, March04, July04, and
September04 groups. (d) Correlations between roughness (LSCM 460 μm x 460
μm) and gloss retention (20° and 60°) for the September 04 group. The RMS data
is averaged from at least two samples measured at five different locations. Gloss
data is the average of 8 to 10 measurements on each specimen. Variability for
measured data is less than 10%.

observed in surface roughness changes, indicating that the exposure conditions have a great effect on weathering rate. Normally, the surface topography and gloss values are strongly correlated for a clear coating.[1]

To examine the correlation between these two physical properties for a weathered clear coating, we have also plotted the surface roughness versus the corresponding gloss retention for both 20° and 60° in *Figure* 3d for the September04 group. The estimated correlation coefficient between the gloss retention and RMS roughness is slightly higher for 60° gloss than 20° gloss measurements ($R^2 = 0.93$ versus $R^2 = 0.87$, where R is the correlation coefficient). Note that gloss measurement by a commercial handheld glossmeter only collects a narrow angular range of light scattered around the specular direction. With the increase of RMS roughness, the proportion of the specular scattering in the total light scattering profile decreases.[1] For rougher surfaces, the 60° gloss measurement provides a higher signal to noise ratio than the 20° gloss measurement due to a higher reflectance values for incident angle of 60° than that of 20°. Thus, the 60° gloss measurements provided a better correlation with surface roughness than did the 20° measurements.

To investigate how surface morphology affects the optical properties of a weathered clear coating, an angle-resolved optical scattering technique was used in which optical scattering measurements were conducted at various incident angles in the specular and off-specular configurations. The results are separated into specular and off-specular intensities and compared to the results from the gloss and the roughness measurements.

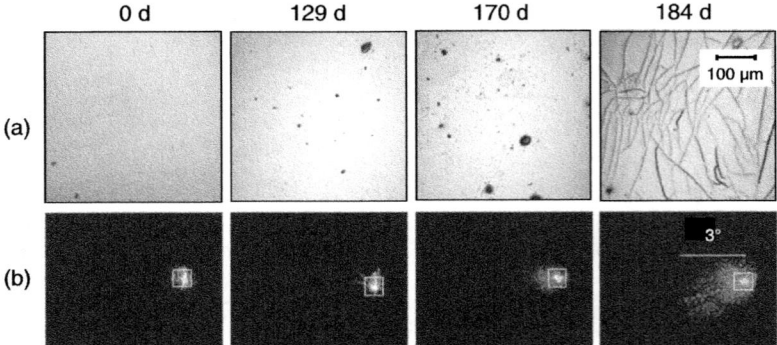

Figure 4—(a) 2D projection LSCM images of a December05 group epoxy coating after outdoor exposure in Gaithersburg, MD, for (from left to right) 0 d, 129 d, 170 d, and 184d, and (b) the corresponding 2D optical scattering (OS) profiles for incident angle of 20° with an aspecular angle of 3°. The box in the OS profiles is the specular region, which is the equivalent angular range measured by a commercial glossmeter.

Figure 4 displays LSCM images of a December05 group specimen after increasing lengths of outdoor exposure along with their corresponding 2D optical scattering profiles. Quantitatively, the intensities of the specular reflectance and off-specular diffusive scattering were plotted against the exposure time (*Figure* 5a). As shown, specular intensity decreases with increased exposure time, which is consistent with the gloss retention changes shown in *Figure* 5b.

On the other hand, changes in the off-specular reflectance intensity are much greater than those for specular reflectance and, indeed, exceed the intensity of the specular reflectance after 179 days of exposure. The general trend of the off-specular scattering is

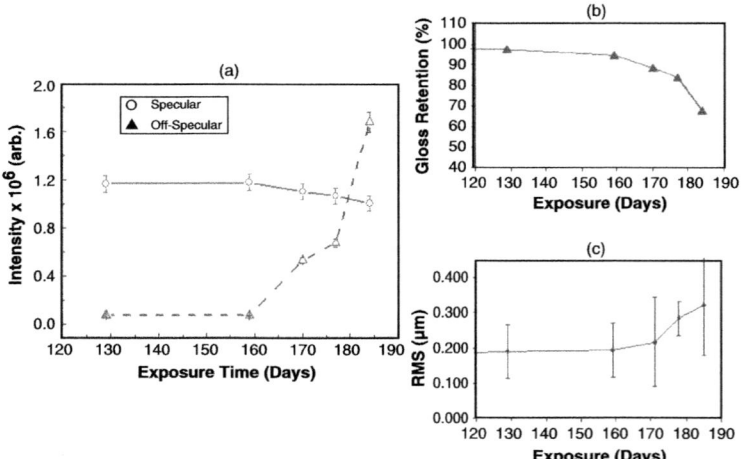

Figure 5—(a) Specular and off-specular intensity obtained from OS profiles; (b) RMS roughness (LSCM) measured at area of 460 μm x 460 μm, and (c) 20° gloss retention versus exposure time for a December05 group exposed to the outdoors.

similar to the change of the RMS roughness (*Figure* 5c). This suggests that the major contribution to off-specular scattering is an increase in surface roughness, assuming that subsurface structures have remained unchanged. These preliminary results indicate that the commercial gloss measurement may not fully capture the weathering response of a rough surface when off-specular scattering becomes significant. Future work includes applying an optical scattering model to the AFM and LSCM profile data, so that the correlation between the nanoscale/microscale topographical changes and macroscale appearance can be predicted.

Unexposed TiO$_2$ Pigmented Epoxy Coatings

The optical property of a clear coating is dominated by the surface roughness and index refraction of the materials. However, the optical properties of a pigmented coating are much more complex due to the contribution of subsurface scattering. We investigated the impact of pigment dispersion on surface and subsurface scattering in unweathered and weathered TiO$_2$ epoxy coatings. The procedure for producing well-dispersed and poorly dispersed pigmented coatings is described in the Experimental section. Characterization of microstructure and degree of dispersion of pigmented TiO$_2$ epoxy coatings was conducted using USANS and the results are shown in *Figure* 6a. In the low q region (i.e., at larger size scale), the scattering intensity of the NonDisp coating is much higher than that of the Disp coating. This implies that larger clusters exist in the NonDisp coating than in the Disp coating. The scattering curve of the Disp coating can be fitted to a form factor of polydisperse spheres with a TiO$_2$ particle size equivalent to a sphere radius of (115.2 ± 5.0) nm with a polydispersity of 19%. However, it is difficult to obtain a meaningful fit to the scattering curve of the NonDisp coating due to the wide range of sizes and shapes in the coating microstructure.

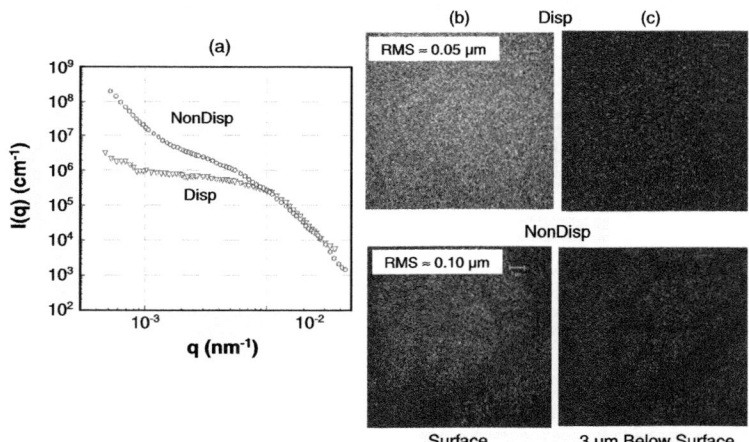

Figure 6—(a) USANS scattering intensity of Disp and NonDisp coatings as a function of scattering wave vector q in a log-log plot. The error bars are smaller than the size of symbols. (b) 2D projection LSCM images and (c) images from 3 μm below the polymer-air surface. All LCSM image sizes are 64 μm x 64 μm. The scale bars in the images are 5 μm.

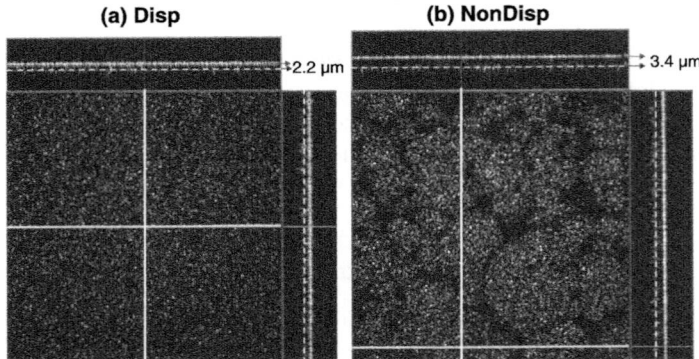

Figure 7—*Orthogonal cross profile of (a) Disp and (b) NonDisp LSCM images. From the coating interfaces between air, polymer, pigment-rich layer, the apparent clear (pigment-poor) layer can be determined as shown in the images. All LCSM image sizes are 64 μm × 64 μm. The uncertainty of the thickness measurement is ± 0.5 μm.*

The surface morphology (*Figure* 6b) and subsurface structure (*Figure* 6c) are presented in the 2D LSCM images for the two coatings. The bright spots originate from the backscattered reflectance from the TiO_2 particles; therefore, by analyzing the intensity distribution of the LSCM images, the pigment distribution at/near the coating surface can be mapped.[9] There is a significant difference in terms of surface topographic features and optical contrast in the 2D LSCM projection images as shown in *Figure* 6b. The differences in subsurface microstructures (3 μm below the surface, *Figure* 6c), however, are greater. As seen in Figure 6c, the subsurface of the NonDisp coating exhibits irregular domains that are not observed in the Disp coating. *Figure* 7 shows more detailed information. In the cross profile of Disp coating (*Figure* 7a), there is a 2.2 μm gap between the polymer-air interface (the solid line) and the pigment-rich layer (the dashed line).

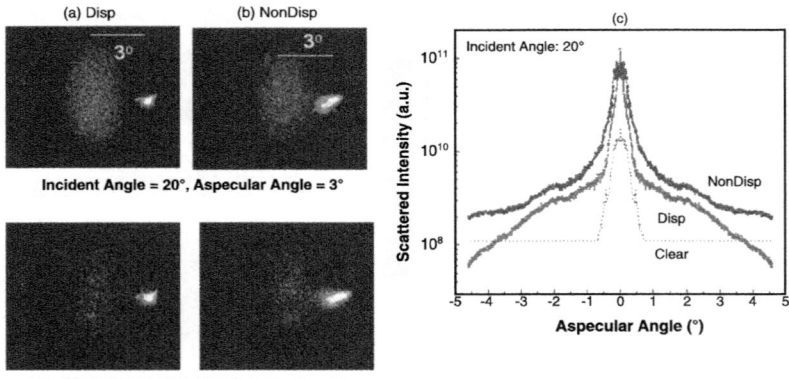

Figure 8—*The 2D optical scattering (OS) profiles of (a) Disp (b) NonDisp coatings at 20° (top) and 45° incident angles with* $\theta_{as} = 3°$, *and (c) 1D scattering curving as a function of angle from aspecular angle at 20° incident angles. The OS scattering curve of Clear coating is also plotted in (c) for reference.*

However, a 3.4 μm gap was observed between the air-polymer and the pigment-rich layer in the NonDisp coating (*Figure* 7b). AFM, LSCM, and Scanning Electron Microscopy (SEM) measurements have been conducted and have independently confirmed the existence of these observations.[11] The existence of a clear layer on top of the pigment layer was first proposed in 1945 by Rutter.[12] Colling and Dunderdale[13] introduced their "clear layer theory" to explain the importance of this layer to gloss and to the loss of gloss as a coating weathers. As we will describe later, we confirm the existence of the clear layer and highlight the significant role that this clear layer plays in the UV degradation process of a TiO_2 pigmented coating.

The corresponding 2D OS intensity profiles at an incident angle of 20° and 45° with $\theta_{as} = 3°$ are displayed in *Figures* 8a and 8b. Note that optical scattering is a result of light scattered from each constitute/component (polymer, pigment) of a coating on the surface and in the subsurface. Additionally, the specular scattering is dominated by the optical contrast (index of refraction) and surface glossiness of a coating, and the major contribution of off-specular scattering is from surface roughness and subsurface scattering from the pigments and heterogeneity in the polymer matrix. By studying optical scattering profiles, valuable information for analyzing microstructure and pigment dispersion near the surfaces can be obtained. A quantitative comparison of optical properties between Disp and NonDisp coatings can be demonstrated by plotting 1D scattering curves as a function of angle from the specular angle. As shown in *Figure* 8c, the optical scattering intensity of the NonDisp displays a higher off-specular (diffuse) scattering than that of Disp. This is due to the higher surface roughness (0.10 μm versus 0.05 μm at size scale 64 μm x 64 μm shown in *Figure* 6b) as well as the irregular subsurface microstructure as observed in *Figure* 6c. Clearly, poor pigment dispersion affects surface morphology and subsurface microstructure, and consequently affects the optical properties of a coating.

Exposed TiO_2 Pigmented Epoxy Coatings

Changes in light scattering from clear and TiO_2 pigmented coatings were studied as a function of exposure time and degree of dispersion. Three epoxy coatings, Clear (as a reference), Disp, and NonDisp, were exposed starting in June 2006; the surface morphology, material properties (such as glass transition temperature and mechanical properties), and optical properties were recorded as a function of exposure time. In all cases, the surface roughness increased, and some surface deterioration (holes, bumps, and cracks) appeared with increasing outdoor exposure time.[11] *Figure* 9 shows surface morphology of the three coatings (Clear, Disp, NonDisp) at different measured size scales and their corresponding 2D optical scattering profiles at 151 d exposure time. The deterioration paths are observed to differ for each coating. As shown in *Figure* 9a, the NonDisp coating exhibited greater surface deterioration than did either the Disp or Clear coating (LSCM images at 5x magnification and the measured area of 1840 μm x 1840 μm). These surface morphology changes directly impact optical properties, and, hence, the appearance of the coatings. As a result, optical scattering profiles reveal a strong increase in diffuse scattering and a specific scattering pattern from cracks as shown in *Figure* 9d for the NonDisp coating.

However, there are no obvious differences at smaller size scale (*Figure* 9c: LSCM images at 150x magnification and the measured size of 64 μm x 64 μm) compared to that

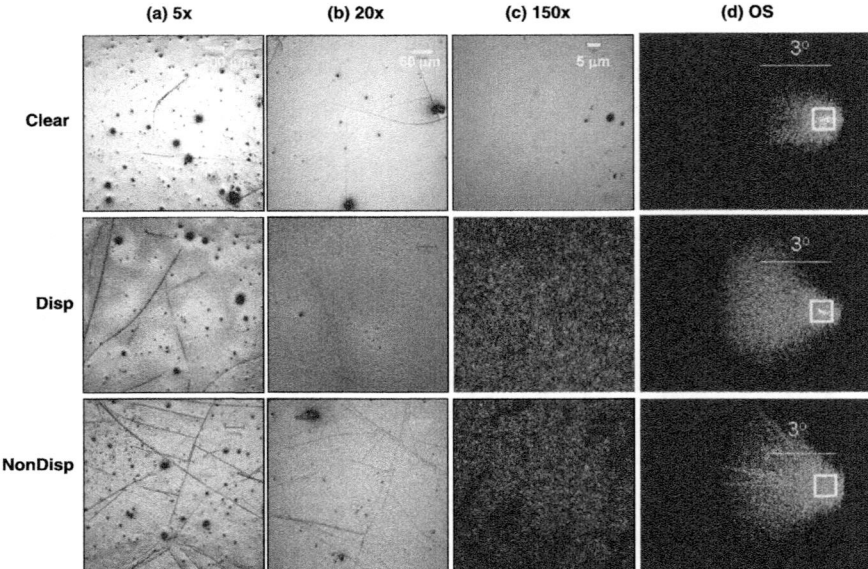

Figure 9—LSCM images at magnification of (a) 5x, (b) 20x, (c) 150x, and (d) OS profiles at 20° incident angle for Clear, Disp, and NonDisp coatings after 151 d outdoor exposure. The size scale bars in (a) to (d) represent 100 μm, 50 μm, 5 μm, and 3°, respectively. The boxed areas in (d) are compatible to the angular range of 20° commercial specular gloss measurements.

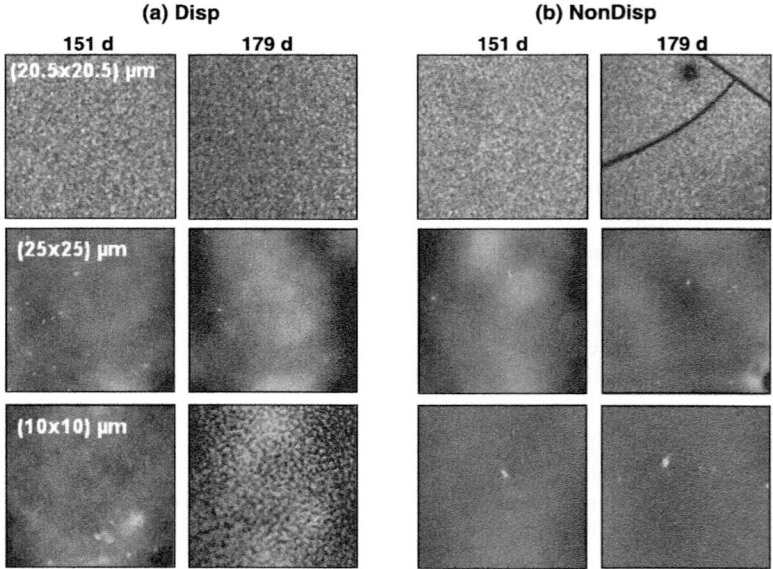

Figure 10—LSCM (top row) and AFM (bottom two rows) images of (a) Disp and (b) NonDisp coatings after outdoor exposure in Gaithersburg, MD, for 151 d and 179 d. The sizes of the measured area are indicated in the micrographs.

of unexposed coatings (as shown in *Figure* 6b). Note that an LSCM is an optical scattering device, which collects the light reflected back from both surface and subsurface; thus, we can not detect the exact location (on or near surface) of the pigments. By combining the AFM measurements, which analyze nanometer-scale surface topographic features, we can determine location of the pigments and the effect of the clear (pigment-poor) layer on the degradation process. *Figure* 10 shows LSCM (top row, 20.5 µm x 20.5 µm) and AFM (two bottom rows, 25 µm x 25 µm and 10 µm x 10 µm) high magnification images for (a) Disp and (b) NonDisp exposed coatings at 151 d and 179 d exposure time. At 151 d exposure time, no significant differences are observed in the LSCM images for the two coatings, but the AFM surface measurements are smoother for the NonDisp than the Disp specimens. At 179 d exposure time, LSCM images shows clear pigment particles distributed throughout the area in the Disp coating, while slightly fuzzy pigment images and cracks are observed in the NonDisp coating. Furthermore, we can clearly distinguish the surface features in the AFM images (10 µm x 10 µm). For the Disp coating, pigments emerge on the surface and exhibit chalking behavior after 179 d of exposure time. On the other hand, there are no pigments appearing on the NonDisp coating, but increased surface roughness and cracks appear. This observation confirms that an apparent clear layer (pigment-poor layer) exists near the coating surface of the NonDisp coating. Due to differences in the thicknesses of these clear layers, we would expect a difference in degradation rate, but not degradation mechanism, for Disp and NonDisp coatings.

Small length scale surface morphological changes, however, might not directly affect changes in appearance (i.e., optical properties) of the coatings. To gain a better understanding of the correlation between surface morphology and optical property of outdoor weathered coatings, we plotted (a) LSCM RMS surface roughness data obtained from a measured area of 1840 µm x 1840 µm, (b) the apparent clear layer thicknesses, and (c) the 20° gloss retention data as a function of the outdoor exposure time for Disp and Nondisp coatings, as shown in *Figure* 11. As expected, the RMS surface roughness increased with increasing exposure time, and the trend in the changes was the same for both coatings up to 225 days. However, there are significant changes observed in the apparent clear layer thickness (*Figure* 11b) and the 20° gloss retention values (*Figure* 11c). There is a large change in the apparent clear layer thickness around 80 d in the NonDisp sample, that caused a corresponding drop in its gloss retention. The big difference in the gloss retention might be due to the different morphological changes among those coatings when exposed to outdoor UV radiation. The Disp coating underwent degradation similar to the contraction model described by Colling and Dunderdale.[13] With the contraction of the polymer matrix, pigments emerge to the surface and chalking eventually occurred (at a later exposure time not presented in this chapter). However, in the NonDisp coating, the top layers of the coatings received the most ultraviolet radiation exposure, so the clear layer was the first to degrade. As a result, the NonDisp coating underwent surface degradation similar to the non-pigmented clear coating[10] first, and then followed the same degradation process as the Disp coating.

Since both Disp and NonDisp coatings had the same PVC of TiO_2 pigments in the entire samples (ca 120 µm thick), the spatial distribution of pigments inside the NonDisp coatings are more agglomerated than in the Disp coating due to the existence of the clear-layer near the surface. This effect has been confirmed by the SEM measurements on the

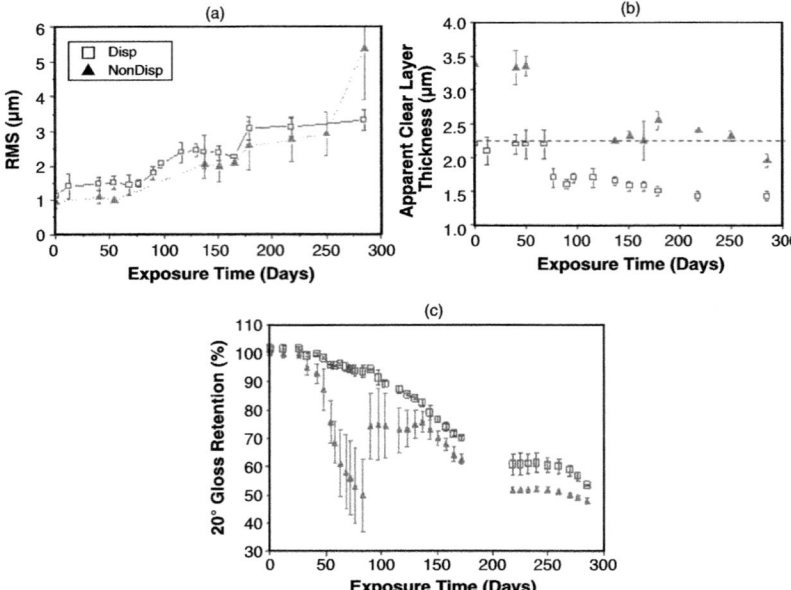

Figure 11—(a) RMS surface roughness measured at area of (1840 x 1840) μm, and (b) apparent clear layer thickness, and (c) 20° gloss retention for Disp and NonDisp coatings as function of outdoor exposure time. Dashed line in (b) indicates the unexposed apparent clear layer of the Disp coating. Data was a mean of four replicates. The error bars represent one standard deviation.

center of the coatings in the cross-sectioned samples. The result shows a higher PVC and a larger cluster size in the center of the NonDisp coating than that in the Disp coating.[11] This result also helps to explain why the gloss value rapidly dropped to 45% (before 100 d) and increased again to almost 70%. As expected, the surface roughening of the apparent clear layer (before 100 d) contributed to the loss of the gloss; then the increase in reflection contribution is due to the higher index of refraction pigment-rich layer near the

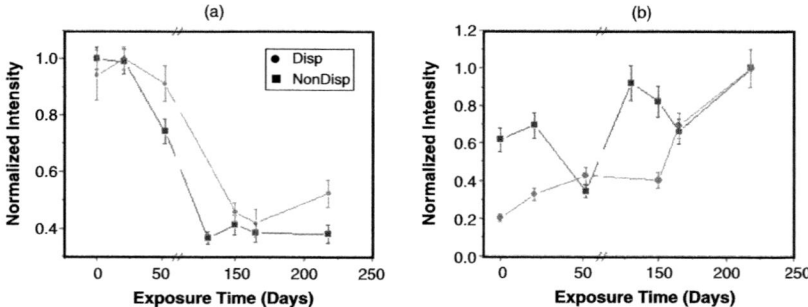

Figure 12—(a) Specular and (b) off-specular scattering intensities versus exposure time for Disp and NonDisp coatings. The error bars represent one standard deviation from measurements from four different locations. There is a break (no measurements) between 52 d and 132 d.

surface when the clear layer became thinner ($\approx 2 \ \mu m \pm 0.5 \ \mu m$). Moreover, the RMS surface roughness data measured using AFM (but not by LSCM) increased up to 100 d then decreased after 100 d. This explains why the gloss value increased after dramatically decreasing around 100 d. This result also implies that the degradation rates and mechanisms are different due to different states of pigment dispersion at this particular set of samples.

To correlate the morphological changes in the weathered coatings to their optical properties, we needed to analyze the full angular-resolved optical scattering profiles in terms of specular and off-specular portions. To obtain the time evolution of the scattering profiles, the specular intensity was integrated within the box area indicated in *Figure* 9d, and the off-specular intensity was calculated from the rest of viewing area of the 2D detector. *Figure* 12 shows the normalized (a) specular and (b) off-specular intensities at different exposure times for both Disp and Nondisp coatings. The specular data are consistent to what we obtained from the 20° gloss data with the exception of the time period between 52 d and 130 d, in which no measurements were conducted. From the off-specular data, we observed a smoothly increasing curve in the Disp coating, but the trend in the NonDisp data appeared to be more complicated and could be related to the evolution of surface/subsurface morphology of the weathered NonDisp coatings. This result also reveals the importance of tracking both specular and off-specular intensities as the coating weathers to fully understand the overall optical properties. Note that the detecting angular range of a commercial specular glossmeter is equivalent to the box areas illustrated in *Figure* 9d. Therefore, the commercial gloss measurements are not always adequate to capture the overall changes in optical properties of a pigmented coating as exposed to UV radiation. Extensive microscopic, optical scattering, and other analyses are ongoing to better correlate the pigment dispersion to performance properties. Our future work will include collaborating with computer-rendering researchers to identify relevant measurements in morphological and optical data using outdoor and accelerated aging to feed in modeling and visual simulation. The final product based on optical measurements and aging models from NIST will be applied to obtain sample visualizations and service life prediction of coatings applied to structures and products using a variety of rendering systems and techniques.

SUMMARY

We have demonstrated that optical scattering is a powerful tool for characterizing surface morphology and subsurface structure due to different pigment dispersion in a coating system. The effect of pigment dispersion on the weathering durability of a TiO_2-pigmented epoxy coating was also investigated. Unlike commercial specular gloss measurements, the full angular-resolved optical scattering profile has revealed both specular and aspecular scattering information, which can be directly linked to the surface morphology and subsurface microstructure measured using LSCM and AFM. The preliminary results have shown a great promise in elucidating the degradation process, and for developing a method that can be used to predict coating appearance during weathering.

In the future, we will use the following approaches to develop a methodology for predicting the appearance properties from weathered coatings: (1) conducting quantitative measurements on the surface morphology and collecting full angular optical scattering profiles in earlier stages of the degradation; (2) analyzing the trends and scaling behavior in the morphological and optical scattering data for different degradation times; (3) modeling optical properties from predicted surface morphological data and comparing them to the measured optical scattering data; and (4) collaborating with computer-rendering programmers and researchers to generate virtual weathered data.

ACKNOWLEDGMENT

We acknowledge the support of the National Institute of Standards and Technology, U.S. Department of Commerce in providing the neutron research facilities used in this work. The authors also thank Dr. Paul Butler of NIST for his help in conducting USANS experiments and insightful discussion in the data analyses.

Reference

(1) McKnight, M.E., Marx, E., Nadal, M., Vorburger, T.V., Barnes, P.Y., and Galler, M.A., "Measurements and Predictions of Light Scattering by Clear Epoxy Coatings," *Applied Optics*, 40, No. 13, 2159-2168 (2001).

(2) Hunt, F.Y., Marx, E., Meyer, G.W., Vorburger, T.V., Walker, P.A., and Westlund, H.B. , "A First Step Towards Photorealistic Rendering of Coated Surfaces and Computer Based Standards of Appearance," in *Service Life Prediction: Methodology and Metrologies*, Martin, J.W. and Bauer, D.R. (Eds.), ACS Symposium Series 805, Oxford University Press, 2002.

(3) Sung, L., Nadal, M.E., McKnight, M.E., Marx, E., and Laurenti, B, "Effect of Aluminum Flake Orientation on Coating Optical Properties," *J. Coat. Technol.*, 74, No. 932, p. 55-63 (2002).

(4) Barker, J.G., Glinka, C.J., Moyer, J.J., Kim, M.H., Drews, A.R., and Agamalian, M., "Design and Performance of a Thermal-Neutron Double-Crystal Diffractometer for USANS at NIST," *J. Appl. Cryst.*, 38, 1004 -101 (2005).

(5) Higgins, J.S. and Benoit, H.C., *Polymers and Neutron Scattering*, Clarendon Press, Oxford, UK, 1994.

(6) VanLandingham, M.R., Nguyen, T., Byrd, W.E., and Martin, J.W., "On the Use of the Atomic Force Microscope to Monitor Physical Degradation of Polymeric Coating Surfaces," *J. Coat. Technol.*, 73, No. 923, p. 43 (2001).

(7) Corle, T.R. and Kino, G.S., "Confocal Scanning Optical Microscopy and Related Imaging Systems," p. 37-39, Academic Press, 1996.

(8) Sung, L., Jasmin, J., Gu, X., Nguyen, T., and Martin, J.W., "Use of Laser Scanning Confocal Microscopy for Characterizing Changes in Film Thickness and Local Surface Morphology of UV Exposed Polymer Coatings," *J. Coat. Technol. Res.*, 1, No. 4, 267 (2004).

(9) Faucheu, J., Sung, L., Martin, J.W., Wood, K.A., "Relating Gloss Loss to Topographical Features of a PVDF Coating," *J. Coat. Technol. Res.*, 3, No. 1, 29 (2006).

(10) Gu X., Sung, L., Kidah, B., Oudina, M., Delaurent, V., Hu, H., Stanley, D., Byrd, W.E., Jean, J.Y.C., Nguyen, T., and Martin, J.W., "Multiscale Physical Characterization of a UV-degraded Polymeric Coating System," *Proc. FSCT FutureCoat! Conf.*, Metro Toronto Convention Centre, Toronto, Ont, Canada, October 3-5, 2007.

(11) Clerici, C., Gu, X., Stutzman, P., Forster, A.M., Sung, L., Ho, D.L., Nguyen, T., and Martin, J.W. "Effect of Pigment Dispersion on Durability of a TiO_2 Pigmented Epoxy Coating During Outdoor Exposure," *Proc. 4th International Symposium on Service Life Prediction of Coatings*, Key Largo, FL, 2006.

(12) Rutter, E.G., *J. Oil Colour Chem. Assoc.*, 28, 187 (1945).

(13) Colling, J.H. and Dunderdale, J., "The Durability of Paint Films Containing Titanium Dioxide— Contraction, Erosion and Clear Layer Theories," *Prog. Org. Coat.*, 9, 47 (1981).

Chapter 18

Computer Graphic Tools for Automotive Paint Engineering

Gary W. Meyer

University of Minnesota, Dept. of Computer Science and Engineering,
4-192 EE/CS Bldg., 200 Union St. SE, Minneapolis, MN 55455.

Recent developments in computer graphics software and hardware have made it possible to write computer graphics programs that can be used to solve automotive paint engineering problems. New surface reflection models have been created for simulating the appearance of automotive paint, and the hardware available today on graphics cards allows these reflection models to be evaluated in real-time. Three interactive computer graphic programs have been written that demonstrate the potential of using the new software and hardware to find answers to engineering problems involving automotive surface coatings. The first program allows the color appearance of new metallic and pearlescent automotive paints to be designed, the second application demonstrates how video projectors can be used to make an object appear as if it has been covered with automotive paint, and the third program is a virtual reality simulation of spraying automotive paint onto a surface.

INTRODUCTION

Computer graphics software and hardware has advanced to the point where innovative computer graphics programs can be written to address automotive paint engineering problems. New computer graphic reflection models have been developed that allow traditional automotive paint measurements to be employed when making renderings of automotive surface coatings. Computer graphic cards now have special purpose hardware that permits these reflection models to be evaluated at interactive rates. The new vehicle paint reflection models and special purpose graphics hardware let new applications be written to accomplish tasks related to automotive paint engineering. These tasks include designing the color of new automotive paints, prototyping the appearance of new paints by using video projectors, and creating virtual automotive spray paint simulations.

The technical advance that will make this new generation of automotive paint engineering tools possible involves software for modeling light reflection from vehicle paint and hardware for evaluating that reflection model at interactive rates. The automotive paint model makes use of standard color appearance measurements for vehicle paint including ASTM gloss and multi-angle Lab color measurements. The programmable shaders now available on graphics cards allow this complicated reflection model to be

---

To view color renditions of the figures in this chapter, see http://www.springer.com/978-0-387-84875-4.

Figure 1—Entrance and exit apertures for the virtual gloss meter.

evaluated in real-time. This permits an object to be rotated and moved on the screen so that its color appearance can be inspected. The color appearance of the paint can be changed interactively by modifying the parameters of the reflection model. The new reflection models and shader technology are discussed in the next section of this chapter.

The new automotive paint engineering tools facilitated by these hardware and software advances are just beginning to emerge. An interactive computer graphics program to design new vehicle paint colors is now possible because of the advanced surface reflection models and the ability to evaluate these models in real-time at each screen position. Video projectors have become brighter and cheaper, and, when driven by an advanced video card, they can be directed at a three-dimensional object to make the object look as if it was painted a new color. The new graphics cards can also be used to produce a video signal for a head-mounted display and offer an immersive simulation of applying automotive paint by using a spray gun. These emerging automotive paint engineering applications are described in separate sections of this chapter.

BACKGROUND

Reflection Model

To simulate the look of automotive paint, a reflectance model has been developed that is based on standard appearance measurements.[1] The goal of this work was to find the correspondence between the parameters of existing computer graphic reflection models and two appearance measurements that are used in industry. The two aspects of surface reflection that are captured by the model include how light reflects from the top layer of the paint (gloss) and from the metal flake below the surface (aspecular measurements). This section describes the individual components of the model and how they are combined to create a complete model.

To develop the portion of the model that accounts for surface gloss, a virtual gloss meter was constructed. A gloss meter is a simple instrument that is used to determine the ratio between the amount of light incident on a surface and the quantity of light reflected from that surface.[2] The entrance and exit aper-

Figure 2—Numerical integration of a 3D computer graphics reflection model across the exit aperture of the virtual gloss meter. Integration is repeated for each discretized region of the entrance aperture.

tures of the simulated device are shown in *Figure 1*. To use the virtual gloss meter, the entrance aperture is discretized, an incident light direction is selected, a computer graphics reflection model is evaluated, and the resulting light distribution is numerically integrated across the exit aperture (see *Figure 2*). The process is repeated for all incoming directions defined by the discretized entrance aper-

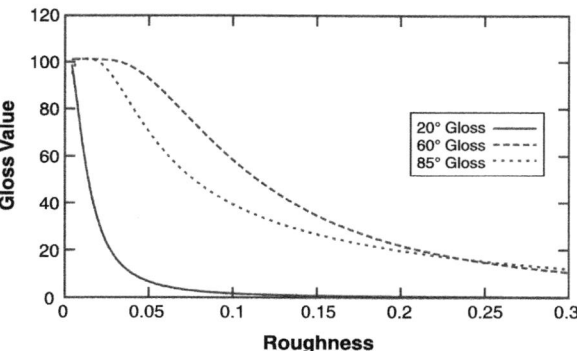

Figure 3—Gloss versus roughness (standard deviation of the surface slope) for the Ward isotropic reflection model.[3]

ture, and a ratio between incident and reflected light is calculated. This ratio determines the gloss value for the parameters used to evaluate the reflection model. Duplicating the procedure for a range of parametric values results in a correspondence between gloss and the model's parameters (see *Figure 3*).

Light that is not reflected from the top of the paint interacts with the colorants and the metal flakes that are located below the surface. The metal flakes determine the shape of the reflectance function that is produced by this subsurface interaction, and the colorants alter the spectral distribution of the light that is reflected away from the surface. The change in color, caused by the colorants, is a function of the angle of reflection. Experimental studies have shown that the rate of variation in both the color and the intensity of the reflection for a metallic automotive paint can be characterized by a low order polynomial.[4] The change in CIE Lab coordinates with aspecular angle (an angle measured using the specular reflection direction as zero) can be fit using a second order polynomial. Because it is easy to compute and lends itself to a good user interface, this simple function is used to model the part of automotive paint reflection that is due to subsurface interaction. (While a second order polynomial is sufficient to model a wide variety of vehicle paint in use today, a higher order function may be required to capture the range of variation possible in newer paints containing effect pigments and flakes.)

The complete reflection model for automotive paint includes both the light reflected from the exterior face of the paint and the light reflected from the metal flakes and the colorant below the surface of the paint. A linear combination of the gloss model and the aspecular

Figure 4—Linear combination of gloss reflection model and metallic paint reflection model.

model is used to create the final reflection model for the vehicle paint. *Figure* 4 shows how the reflection lobes for the two parts of the model are combined to create the complete reflection function.

Real-Time Shading

The simulation of color and appearance has been greatly facilitated by the recent development of real-time shading technology. A shader is a small special purpose piece of computer code that is executed as the color of each pixel in a computer graphic picture is determined. Because shaders are programmable, they make it possible to implement a wide range of surface reflection models including the automotive paint model described in the preceding section. Shaders include reflection model variables that can be altered to change the spectral and the spatial distribution of the light reflected from the material that is being rendered. For the automotive reflection model, these variables include the aspecular measurements and the gloss of the paint. Changing the value of the reflection model parameters at interactive rates makes it possible to consider several alternative color appearances and, in this manner, to design the color of the surface coating.

There are a number of technical issues that must be kept in mind when working with real-time shaders. The computer code for shaders is written using special purpose programming languages such as NVIDEA Cg, DirectX 9.0, and OpenGL 2.0. Once the shader has been composed it is turned into a set of low level machine instructions by using a compiler. These machine instructions must be downloaded onto the graphics card where they will be executed by a special purpose graphics processor. The process of downloading the compiled code takes time and limits how quickly a shading model can be updated. There is also a restriction on how many shading instructions can be placed on the card and on the number of those instructions that can be executed as each pixel color is determined. These limitations must be kept in mind when designing shader code.

Additional Work

In addition to the simulation techniques that are employed in this paper, there have been other attempts to use computer graphics to reproduce the color appearance of automotive paint. Full BRDF measurements of the finish have been made, and this data has been used to generate a realistic image of a car.[5] A three-dimensional model has been created of the titanium-coated mica plates within the paint, a BRDF has been determined by simulating light interaction with this model, and the resulting BRDF has been employed to create a picture.[6] An analytical model has been developed that accounts for the effect of each constituent material within the paint,[7] and an inversion technique has been produced that allows this model to be fit to an existing BRDF.[8] The use of digital images to acquire the reflectance properties of vehicle paint has also been explored.[9]

AUTO PAINT DESIGN PROGRAM

An interactive computer interface has been developed to allow stylists to design new automotive paint colors.[10] The interface is based on the reflectance model for automotive paint that was described in the Background section of this chapter. The user makes modifications to the aspecular reflectance values that define the color appearance of the paint,

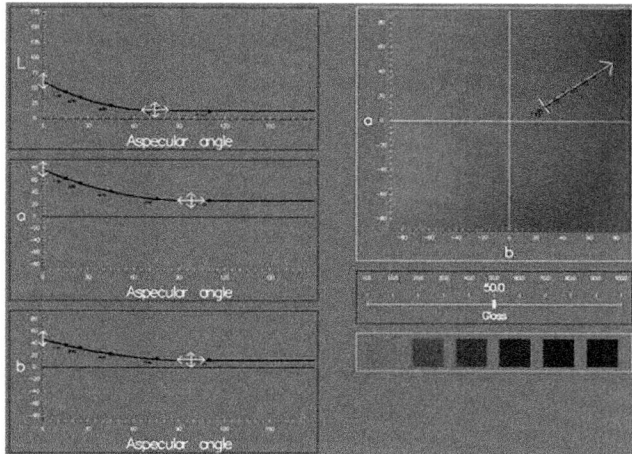

Figure 5—Automotive paint design program interface for selecting face and flop colors.

and the result of making those changes is immediately displayed in a computer graphic simulation. Adjustments can be made directly to the reflectance curve for the paint, or alterations can be accomplished indirectly by using several novel interfaces. Once the design has been completed, the desired aspecular reflectance values for the paint can be written to a file. Given the expected reflectance properties of the coating, paint manufacturers can determine the formula necessary to produce the paint.

The interface for the styling program provides several novel ways to adjust the color appearance of an automotive paint. A simple slider is available to change the 60 degree ASTM standard gloss of the paint from 0 to 100. The aspecular reflectance curve for the paint is drawn as part of the interface, and the reflectance values at the standard measurement angles can be directly manipulated. As adjustments are made to the reflectance data, the program fits a second order curve to the points and redraws the curve. A second order fit is maintained because experiments have shown that the reflectance curve of metallic automotive paints usually has this characteristic. To provide a more natural interface for designers, interactive controls are also provided to allow the average hue, saturation, and brightness of the data values to be adjusted and to permit the face and flop color of the paint to be directly specified (see *Figure 5*).

As modifications are made to the reflectance properties of the paint, a new

Figure 6—Computer graphic image produced by automotive paint design program.

Table 1—Comparison Between the Target Aspecular Measurements for the Designed Colors and the Actual Aspecular Measurements of the Manufactured Colors

	Angle	Designed			Manufactured		
		L	a	b	L	a	b
Green	15	35.8	-29.8	-0.59	35.7	-27.4	-5.8
	25	21.3	-20.2	-3.4	23.3	-20.6	-6.2
	45	7.6	-8.7	-7.7	8.8	-9.8	-7.2
	75	4.2	-4.8	-7.0	3.9	-2.6	-6.4
	110	3.5	-4.1	-6.4	3.0	-1.2	-5.5
Red	15	43.8	55.3	46.2	45.0	53.5	43.8
	25	33.6	49.7	39.7	36.5	47.7	39.1
	45	19.4	40.3	28.6	22.1	36.6	27.9
	75	13.3	30.5	17.2	13.9	28.7	18.1
	110	13.3	25.8	11.7	11.6	26.2	15.2
Blue	15	79.8	-48.4	-41.0	90.1	-48.2	-39.8
	25	58.6	-37.0	-35.7	61.1	-35.3	-32.8
	45	27.8	-19.2	-27.4	28.3	-16.4	-24.6
	75	10.9	-5.1	-20.9	13.1	-6.3	-21.5
	110	10.9	-3.8	-20.2	8.7	-4.5	-19.4

computer graphic image is generated to show the change in the paint's color appearance (see *Figure* 6). This update occurs at interactive rates because programmable shaders are available on modern graphics hardware. The new reflectance function, used to determine the color of each pixel in the image, is downloaded as a shader to the graphics card. The effect of environmental lighting is also taken into account by combining the new reflectance function with a set of pre-filtered environment maps. To facilitate this lighting calculation, the reflectance function is fit to a finite set of special basis lobes.[11] These basis functions have been found, by experiment, to be an optimal set for representing the reflectance functions produced by the paint design program. Representing the reflectance function in this way limits the number of necessary pre-filtered environment maps and accelerates the environmental lighting calculation.

Design exercises were conducted to demonstrate that the paint design program could be employed to create new automotive colors.[12] Stylists used the interface to design new colors, often starting with an existing paint finish and making adjustments to it. The desired measurements for the new paint were then downloaded into formulation software and the necessary paint mixture was determined (see *Table* 1). The results were considered adequate for initial styling exercises.

APPEARANCE SIMULATION USING PROJECTORS

The automotive paint reflection model discussed in the Background section of this chapter has been used to develop two new types of *shader lamps*. A shader lamp involves the use of a video projector to create the illusion that an object has a color appearance different from its normal surface finish.[13] The concept of shader lamps was advanced by employing the automotive surface reflection model to simulate the appearance of a real material: vehicle paint. A new type of shader lamp, involving the color appearance of a flexible sheet of material, was also introduced. In addition, traditional shader lamps were extended by adding illuminating light sources and environmental reflections to the simulation. The use of the automotive paint reflection model with shader lamps is described in this section.

A new type of shader lamp was developed that allows the user to manipulate a flexible sheet of material and examine how light reflects from the surface of the sheet as its shape changes.[14] The approach involved a piece of plastic acetate that was tracked by a magnetic positioning system while it was held in front of a video projector. Two edges of the sheet (bonded to back-projection screen) were

Figure 7—Back projection onto flexible sheet to simulate metallic appearance.

attached to rigid plastic sticks and, given the tracked position and orientation of the sticks, the shape of the plastic sheet was approximated by a Hermite surface. Given the position of an imaginary light source and the viewer's eyes (both of which can also be tracked), the surface normal was determined at each point on the sheet and the automotive reflection model was evaluated. Given the results of this shading calculation, a computer driven video projector was used to produce the required color at each point on the surface of the sheet (see *Figure* 7). The video projector employed a spherical projection lens to spread the light out across the plastic sheet and to provide a large depth of field.

Improvements were also made to the traditional shader lamp paradigm where video projectors are used to make solid three-dimensional objects assume different color appearances.[15] One technical advance involved painting the objects with a dark gray projection screen material to improve the rendition of metallic colors. In addition to the video projector that was used to change the color of the object, other video projectors were employed to imitate light sources and to illuminate the area surrounding the objects (see *Figure* 8). Environment mapping techniques were also employed to incorporate the reflection of sur-

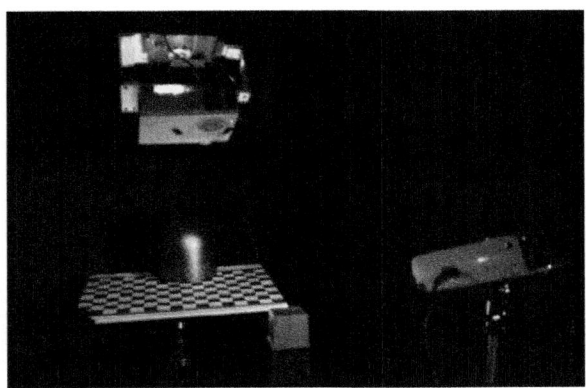

Figure 8—Shader lamp setup used to make cylindrical object assume metallic color appearance. Upper projector projects color while right projector provides general illumination.

rounding objects into the reproduction. To validate the quality of the simulation, comparisons were made between real objects painted with automotive lacquer and the simulations created using shader lamps. The automotive paint design program described in the previous section was connected to the shader lamp simulation to provide an interactive interface that could be used to change the color of the object.

VIRTUAL SPRAY PAINT PROGRAM

A virtual reality-based spray paint simulation program is also under development. This software is intended to teach automotive technicians the correct method to use when applying vehicle paint. A student wears a tracked head-mounted display that generates a view for each eye of the object to be sprayed, and they hold a tracked pointing device in their hands that determines the position and orientation of the spray paint gun. The advantage of this virtual approach includes the fact that the training is administered under controlled conditions, the performance of the student can be carefully monitored, and there is no waste of paint or exposure to harmful fumes. The research was based on an existing spray paint simulation program that taught painters how to spray standard household paint onto flat sheet metal.

Several improvements have already been made to extend the capability of the existing spray paint software and make it more appropriate for use in automotive applications. The most important enhancement has been the introduction of the automotive paint reflection model described in the Background section of this chapter. The addition of this model makes it possible to separately apply the base metallic coat and the final clearcoat. The paint also appears to have metallic and pearlescent properties as the user looks at the surface from different directions. Another advance has been the development of software that makes it possible to apply paint to a curved surface instead of being restricted to flat sheet metal. Improving the reflection model and adding curved surfaces allows the user to spray complex automotive shapes such as the car hood shown in *Figure* 9.

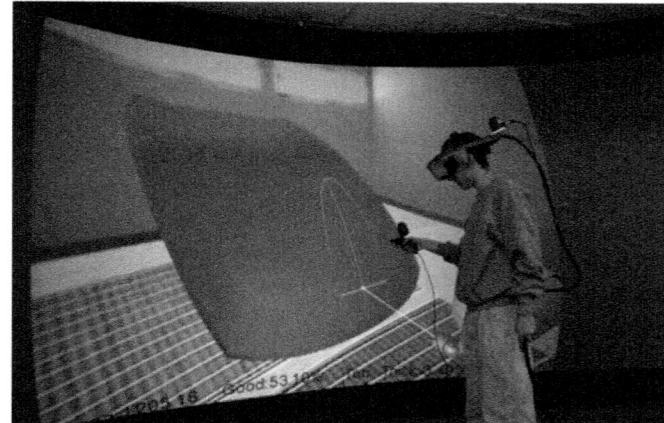

Figure 9—Virtual spray paint system simulates application of automotive paint to complex three-dimensional shape.

A number of features must still be added to the virtual spray paint program. The simulation of paint flaws, such as orange peel and paint runs, will improve the realism of the software. The addition of these imperfections will also make it possible to create an educational module that teaches students how to identify these problems and shows them how they are produced due to poor spray painting technique. Another training capability to be added involves the correct set up of the spray gun to apply the paint. Proper adjustment of the gun is critical to achieve satisfactory results. Finally, mixing or shading the paint to achieve a proper match is a critical skill that an automotive spray painter must master. It would be valuable to extend the simulation program to include such training.

CONCLUSIONS

Even though they were not developed for this purpose, the paint design, shader lamp, and virtual spray paint applications described in this chapter demonstrate the potential of using the computer graphics hardware and software to solve automotive paint engineering problems. The paint design program illustrates how a new paint color with specific reflectance properties can be created. This capability could be used to fix a paint mismatch problem between two components of a car (for example, the body and the bumper cover) by designing a special "touch-up" paint for one of the parts. The shader lamp application shows how video projectors can be employed to make a simple cylindrical shape assume a different color. This work could be extended so that a new color scheme is tested by projecting it onto an actual car before the car is painted. Finally, the virtual spray paint program was created to improve the quality of the paint jobs done by automotive repair technicians. An application similar to this could be used to test the motions that will be followed by a robot as it paints a car on the assembly line.

The examples in this chapter provide only a small indication of how computer graphics will be used in the future to help visualize and solve automotive paint engineering problems. Important questions still remain regarding our ability to make judgments about reality from simulations, and critical tests with human subjects must be performed to resolve this key issue. However, the desire to apply computer graphics to real world surface coatings problems will continue to grow as graphics processing power increases, displays become brighter, images gain resolution, and hardware costs decrease. The day will come when paint engineers regularly use desktop workstations to answer color appearance questions just as mechanical engineers have used computer graphics to solve geometric design problems for over 30 years.

ACKNOWLEDGMENTS

This paper summarizes research that was done jointly over a number of years with Jonathan Konieczny, Clement Shimizu, and Harold Westlund. The work was performed at the University of Minnesota Digital Technology Center. The author would like to thank DuPont Automotive Products for providing paint samples and reflectance measurements. This research was partially funded by the National Institute of Standards and Technology and by the National Science Foundation under grant number EEC-0438693.

References

(1) Westlund, H. and Meyer, G., "Applying Appearance Standards to Light Reflection Models," *Proc. SIGGRAPH '01*, 501-510, 2001.

(2) Hunter, R. and Judd, D., "Development of a Method of Classifying Paints According to Gloss," *ASTM Bulletin*, 1939.

(3) Ward, G., "Measuring and Modeling Anisotropic Reflection," *Proc. SIGGRAPH '92*, 265–272, 1992.

(4) Alman, D., "Directional Color Measurement of Metallic Flake Finishes," *Proc. ISCC Williamsburg Conference on Appearance*, 53–56, 1987.

(5) Takagi, A., Takaoka, H., Oshima, T., and Ogata, Y., "Accurate Rendering Technique Based on Colorimetric Conception," *Proc. SIGGRAPH '90*, 263-272, 1990.

(6) Gondek, J., Meyer, G., and Newman, J., "Wavelength Dependent Reflectance Functions," *Proc. SIGGRAPH '94*, 213-220, 1994.

(7) Ershov, S., Kolchin, K., and Myszkowski, K., "Rendering Pearlescent Appearance Based on Paint-Composition Modelling," *Computer Graphics Forum*, 20, C221-C238 (2001).

(8) Ershov, S., Durikovic, R., Kolchin, K., and Myszkowski, K., "Reverse Engineering Approach to Appearance-Based Design of Metallic and Pearlescent Paints," *Visual Computer*, 20, 586-600 (2004).

(9) Gunther, J., Chen, T., Goesele, M., Wald, I., and Seidel, H.P., "Efficient Acquisition and Realistic Rendering of Car Paint," in *Vision, Modeling, and Visualization*, Adademische Verlagsgesellschaft Aka GmbH, Berlin, 487-494, 2005.

(10) Shimizu, C., Meyer, G., and Wingard J., "Interactive Goniochromatic Color Design," *Eleventh Color Imaging Conference*, 16-22, 2003.

(11) Shimizu, C. and Meyer, G., "Computer Aided Color Appearance Design using Environment Map Based Lighting," *Eurographics Workshop on Computational Aesthetics in Graphics, Visualization, and Imaging*, 223-230, 2005.

(12) Meyer, G., Shimizu, C., Eggly, A., Fischer, D., King, J., and Rodriguez, A., "Computer Aided Design of Automotive Finishes," *Proc. 10th Congress of the International Colour Association*, 685-688, 2005.

(13) Raskar, R., Welch, G., Low, K., and Bandyopadhyay, D., "Shader Lamps: Animating Real Objects with Image-Based Illumination," *Proc. 12th Eurographics Workshop on Rendering Techniques*, 89-102, 2001.

(14) Konieczny, J., Shimizu, C., Meyer, G., and Colucci, D., "A Handheld Flexible Display System," *Proc. IEEE Visualization 2005*, 591-597, 2005.

(15) Konieczny, J. and Meyer, G., "Material and Color Design Using Projectors," *CGIV 2006: Third European Conference on Colour in Graphics, Imaging, and Vision*, 438-442, 2006.

Chapter 19

Computer Graphics Techniques for Capturing and Rendering the Appearance of Aging Materials

Holly Rushmeier

Yale University, Dept. of Computer Science, P.O. Box 208285
New Haven, CT 06520

Computer graphics photorealistic rendering techniques are capable of rendering images that predict the appearance of yet to be manufactured objects. A challenge in computer graphics realism is creating the digital models of shape, materials, and lighting that are required for such rendering. Models for materials that have been aged by weathering or usage are difficult to produce. A recent trend in computer graphics is to attempt to capture aged materials in a form that allows them to be applied to arbitrary new shapes. We present the techniques used for capture and some sample results. Current techniques can be applied to various types of visual simulation, and we outline some future potential applications of rendering aged materials.

INTRODUCTION

Computer graphics is used in a wide variety of familiar applications, ranging from abstract representations of statistical and scientific data to synthetic characters in feature films and games. One important sub-area of computer graphics is the generation of realistic images. A key component in generating realistic images is creating digital models of the appearance attributes of the materials used in the scene being simulated. A variety of methods for modeling the attributes of pristine homogeneous materials have been developed in computer graphics over the past 30 years. However, many applications require simulating materials that have undergone spectral, spatial, and directional changes due to weathering and usage.

In applications, realistic images may be either required to be *plausible* or *predictive*. To be *plausible*, images must appear to an observer to be indistinguishable from a photograph of a physical scene, although the observer never needs to make a critical decision based on the image. Many applications such as computer games or synthetic props in film require only plausibility. To be *predictive*, images must appear the same to an observer as an image acquired of a physical scene. Unlike plausible images, predictive images are used by observers to make decisions, such as selecting a design or training for a task that requires target visibility. Methods for generating predictive images are based on physical models derived from work in other disciplines, and are subject to validation

To view color renditions of the figures in this chapter, see http://www.springer.com/978-0-387-84875-4.

by physical and/or psychophysical experiment. While the requirements for a plausible image are lower, a difficult question to answer is whether a method can be relied upon to always produce a plausible result. Heuristics that produce plausible results in some cases may require a lot of manual tuning to obtain the desired output for a particular application. As a result, computer graphics researchers often focus on predictive techniques that have proven reliability. Generalizing predictive techniques to create reliable plausible results is easier than starting from pure heuristics that are not founded on physical models.

Most existing models for weathered materials in computer graphics are purely heuristic. Over the past 10 years various attempts have been made to generate models from first principles and from measured data. Existing methods are generally too slow or hard to control. Modeling aged materials continues to be a challenge in computer graphics research. Unlike other challenges that can be met with faster processors and increased memory, modeling materials requires new techniques and algorithms built on theory and data from other disciplines including chemistry and materials science.

Computer graphics has a history of borrowing from and contributing to other disciplines. An example of this interchange is radiative heat transfer and computer graphics lighting simulations. In the 1980s, techniques such as radiosity[1-2] from heat transfer were adapted to accurately compute visible light transfer in scenes for predictive rendering.[3] Subsequent work in computer graphics refined and improved the computational techniques and the results have been included in heat transfer texts.[4] Collaboration between chemists and material scientists and researchers in computer graphics clearly has the potential to improve graphics systems, and may also result in new methods for designing and evaluating the appearance of new materials.

We begin with a brief review of the elements of rendering images and capturing input that have become common in computer graphics. Next we give an overview of current methods that attempt to model aging materials. We conclude with an outlook for possible new work in this area, and the potential for new applications that could be enabled by improved material appearance models.

BACKGROUND—RENDERING AND CAPTURE

Material models are required to render realistic images. Specifically, in computer graphics the term rendering refers to the process of using numerical descriptions of three-dimensional scenes and a virtual camera and producing a two-dimensional image. The description of a three-dimensional scene may be produced entirely by a user interacting with a computer using mathematically based modeling software. However, over the past 10 years with the increasing availability of digital cameras, the use of captured data along with completely human specified models has become increasingly common.

Figure 1 illustrates the rendering process. A virtual camera specifies a viewer position and the viewing frustum. The image plane is perpendicular to the view direction, and is discretized into an array of pixel locations. The image is specified by defining an emitted (illuminated display) or reflected (for print images) color at each pixel location. Typically for computer images, values of red, green, and blue are specified, with the specifics of the element spectra taken into account for consistent appearance cross devices.[5] The view point and pixel locations define a set of rays in the three-dimensional scene. A realistic image is formed by setting the pixel values according to the visible light that would arrive at the view point from the scene. The light arriving from a particular object visible along a ray depends on three things: shape, incident illumination, and properties of the object material.

The shape of an object can be defined in a variety of ways including triangle meshes, tensor splines, and subdivision surfaces. The specification of shape, computer aided geometric design, is a broad area in itself that is described in numerous texts.[6-7]

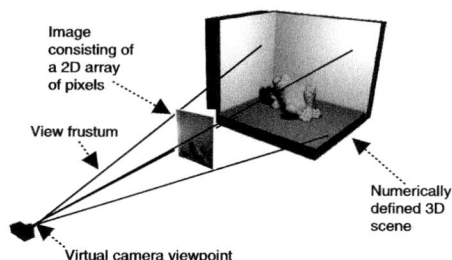

Figure 1—A 2-D image is defined by specifying a virtual camera and the lighting, shape, and materials in a 3-D scene.

Computing the incident illumination is also an extended topic. Light may arrive at a point directly from a light source such as a lamp or the sun. Light may also arrive after one or more scattering effects in the environment. Accounting for all scattering events is referred to as "global illumination" in computer graphics. Methods for computing global illumination include the radiosity methods mentioned earlier, as well as many variations of ray tracing. Global illumination methods are described in detail in texts such as those by Larsen and Shakespeare[8] and Dutre[9] et al. Despite being a complex problem, current global illumination methods have been validated relative to ground truth as being reliable for being predictive, given accurate models of shape and materials.[10] Global illumination systems such as the freely available Radiance software[11] or commercial products such as Lightworks[12] are used in design applications requiring predictive results as well as for creating appealing and plausible imagery.

Material models in computer graphics have mainly focused on specifying the materials bidirectional reflectance distribution function (BRDF), or more generally the bidirectional scattering surface reflectance distribution function (BSSRDF).[13] The BRDF gives the reflected radiance as a function of wavelength, incident direction, and exitant direction. Both phenomenological models[14,15] that are constructed to fit measured data and first principles models that use surface roughness models and material index of refraction[16-17] have been developed and validated. For materials that are not spatially homogeneous, many systems depend on procedural methods that vary the BRDF parameters on a surface.[18]

BRDF specifications require some sort of measured data—either reflected radiances or surface roughness and index of refraction. Since microscopic surface roughness is difficult to measure, attention in graphics has focused on measuring reflected radiance. For spatial variations, procedural textures can be difficult to tune to achieve the appearance of specific materials. Interest has grown therefore in using digital cameras to capture spatially varying BRDFs.

A simple digital camera image is not adequate to model a material. As in the synthetic image, an image from a digital camera includes the effect of shape and incident illumination as well as material. The basic strategy for capturing materials with a digital camera is to control the lighting, measure the shape, and then process the image to estimate the mate-

286

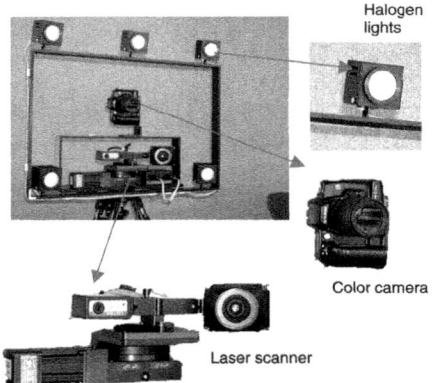

Halogen lights

Color camera

Laser scanner

Figure 2—Object shape and appearance are captured by a hardware set-up including computer controlled range scanner, color camera, and lights.

rial appearance attributes. Since most methods for this processing in computer graphics do not include error estimates in this processing, we refer to this as "capture," rather than as "measurement." Several techniques have been developed with digital camera and controlled lighting for computer graphics to capture reflected radiance from simple known convex shapes.[14,19] These methods account for factors such as camera response nonlinearities and the spectrum of the illuminating source used, as well as novel geometric arrangements of objects and mirrors to reduce the number of images required.

For materials of interest on existing objects, more complex techniques are needed. *Figure* 2 shows one set-up for capturing the appearance of existing objects. The apparatus consists of a laser triangulation scanner (ShapeGrabber SG1002),[20] digital camera (Olympus C8080WZ),[21] and small computer controlled light sources (in this case, halogen bulbs in custom housings with custom control). The laser triangulation scanner is one of many optical devices developed over the past 20 years for measuring shape.[22] An emitter and sensor are mounted a known distance apart. Knowing the angle of the emitted laser spot and the angle from the sensor at which the reflection of the laser from the object is observed allows calculation of the distance from the object to the scanner. Scanning the object from a series of views produces a series of range images (images in which a depth, rather than a color, is recorded at each pixel) which can subsequently be

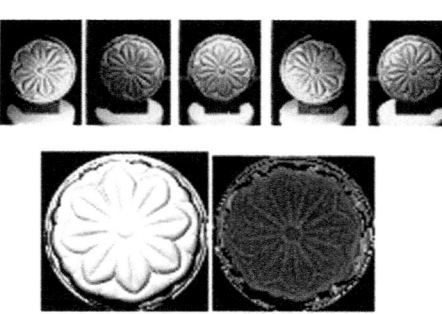

Figure 3—Hardware set-up captures images under five different lighting conditions (top row), which are processed with captured geometry (lower row left) and color corrected for the light source color to produce a map of the diffuse reflectance of the object surface.

geometrically registered and merged to form a three-dimensional model of the object shape.[23]

The parameters of the color digital camera (position, orientation, focal length, and distortion correction) and position of the light sources can be calibrated in terms of the three-dimensional scanning system.[24-25] With these calibrations the projection of the captured image on the shape and the direction of the incident light are known. The image can be processed with this information to estimate either just the diffuse albedo[26] or, with assumptions about the similarity of the material viewed through different pixels, the BRDF at each surface point.[27] *Figure* 3 shows five images and geometry captured by the system in *Figure* 2, and the texture map of diffuse reflectance obtained after processing for directional lighting and accounting for the non-white spectrum of the incident light.

Analogous to the problem of aligning and merging all of the geometric range images, the individual processed textures need to be registered and combined.[23] The processed texture values are stored in images that are associated with the 3-D geometry via texture mapping. That is, each 3-D vertex on the object is mapped to

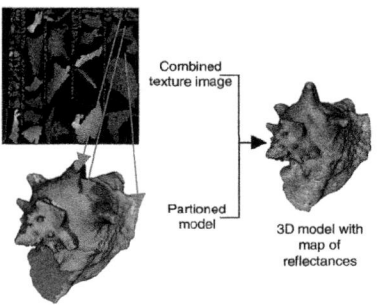

Figure 4—A 3-D object is partitioned into nearly flat regions to map texture images to the surface.

a 2-D location in the texture image. To facilitate this mapping, the 3-D object is partitioned into approximately flat regions, and the textures for each region are stored in one combined large image. The result of this partitioning is illustrated for a captured complex natural object, a seashell, as shown in *Figure* 4.

CURRENT CAPTURE TECHNIQUES FOR AGING MATERIALS

Various methods have been developed for limited simulation of material aging. A complete summary is given by Lu et al.[28] Methods include modeling the mechanics of adhesion and cohesion of paint and substrates to simulate cracking and peeling,[29] and flow over objects resulting in the deposition of dirt and corrosive agents.[30] Most simulations methods are computationally slow, and can result in noticeable visual artifacts. Simulations have only been validated by general comparisons of synthetic results with photographs of the same phenomena that show similarities such as crack density. Because simulation has not proved efficient, most weathering effects in applications such as film and games are currently produced by hand painting of textures, or by artistically altering and manually applying photographs as texture maps. To reduce the labor involved in these applications and to enable predictive applications, researchers have turned to capturing physical aging effects.

The problem of capturing aging effects differs from the capture of full objects. Rather than capturing an image of reflectances that maps to a particular object, such as the mapping in *Figure* 4, a set of spatially varying reflectances is sought that can be mapped to a new object that is significantly different in shape from the original capture object.

Capturing the appearance of aging materials requires accounting for spectral, directional, and positional changes over time. Obtaining data for such a high dimensional problem in a transferable form is difficult. Four recent methods for capturing the effects each work by simplifying one or more dimensions. All of the methods can be applied to produce plausible results in some cases, but all are restricted to the extent they can produce predictive images.

Assumptions or restrictions in the different methods include:

(1) The effect of object shape on temporal appearance variations can be ignored.

(2) The end effect only, rather than temporal evolution of appearance, is of interest.

(3) Processes that can be reproduced in a reasonable length of time (weeks or less) are considered.

(4) Directional variations can be inferred from spectral variations on the material surface.

(5) The sequence of temporal effects acting on a sample can be inferred.

Assumptions (1) and (3) apply to the results by Gu et al.[31] who performed a series of laboratory experiments to capture the temporal variation of appearance on flat 1" x 1" samples of materials. Appearance was captured using an array of cameras and lights supported by a dome structure to capture the full BRDF across the sample. A library of data for 26 materials was captured, and has been made available online.[32] To apply the results, assumptions need to be made such as the effect of the object geometry and edge effects around the area where the object was exposed to the agent producing the weathering. An advantage of the data collected is that subtle variations such as changes in glossiness are captured using the full dome system. The output of this method is a temporal texture. That is, for each pixel in the texture rather than just having a BRDF, there is a time series of BRDF values. Gu et al.[31] also introduced effective mechanisms for storing this data in compact form.

Assumptions (1), (4), and (5) apply to the technique introduced by Wang et al.[33] Wang et al. begin with the assumption that a sample of material is found which has different areas that have been exposed to weathering effects for different periods of time. An example is a rusted plate with some original metal finish still exposed, some areas with flecks of rust, and some thoroughly rusted areas. The sample is scanned to determine the BRDF at equally spaced spatial positions (in general, pixel locations on an image of the sample). The BRDF samples are characterized by a seven parameter model. Three parameters each give the diffuse and specular reflectance in the red, green, and blue channels, and the seventh parameter gives the width of the specular lobe. A high dimensional manifold is formed by connecting each of the samples to each of its nearest eight neighbors. A human observer is asked to identify the most and least weathered areas of the sample. The samples identified are labeled as the end points of the weathering process on the manifold. All other points are assigned a degree of weathering based on their distance along the manifold between the greatest and least weathered samples. On the original spatial sample then, each sample can be assigned a degree of weathering. To capture the effect of spatial texture, the degree of weathering associated with each position is equal to the average degree of weathering of samples in a small surrounding region. Degree of weathering is roughly interpreted as the length of time an area has been exposed to a weathering process. The output is a space in which a texture can be extracted for any time in the assumed time history of the aging of a particular material.

Assumptions (2) and (4) apply to the approach by Mertens.[34] In this work, only the end appearance, or the "look and feel," of an aged object is of interest. This has the advantage that the effect of the object shape at all length scales is accounted for, but the specifics of what caused the weathering and over what time period are not known. Mertens et al.[34] start with an object that has been captured with texture. Machine learning techniques (specifically canonical correlation analysis) are used to correlate texture variations with geometric measures such as curvature, surface orientation, and relative height of the position on the object. The result is a texture correlated with geometric quantities.

Finally, assumptions (3) and (4) apply to the method and data developed by Lu et al.[28] This work is similar to Gu et al.,[31] in that aging effects are produced in the laboratory. Unlike Gu et al.,[31] though, the effects of shape on the effects are captured, and directional effects are not. Lu et al. produced accelerated effects such as rusting of an ironing surfacing compound with exposure to vinegar fumes, molding of cheese, and cracking of a thick paste coating of plaster and paint. The shape of the objects with related texture was captured using the set-up shown in *Figure* 2. By being able to geometrically register the shapes captured at different times, it was possible to move the objects during the experiment, rather than keeping them still relative to the capture device. This extended the range of effects that were practical to capture to phenomena that required weeks to develop. Experiments were repeated on objects to study the reproducibility of the effects. Similar to Mertens et al.,[34] the texture variations were correlated to geometric parameters, although in Lu et al. geometric harmonic analysis was used to organize the data, and temporal texture series (similar to the series captured by Gu et al.[31] for flat objects) rather than static textures were used in the correlations. The same aging process was physically applied to different shapes, so that the predictive capabilities of the model derived from an experiment could be assessed by comparison of synthesized and physical results. The output of this work is temporal textures for particular aging effects, correlated with object geometry.

For all of these methods, the models are used on new synthetic objects by texture transfer methods inspired by the Markov random field methods introduced by Efros and Leung.[35] These transfer methods were extended to texture synthesis on non-flat surfaces in work such as Wei and Levoy.[36] In these methods a texture is transferred from an original sample while enforcing consistency and constraints. The process begins by selecting a location on the source object. For textures such as those modeled by Mertens et al.[34] or Lu et al.[28] a location in the original texture with the same value of the relevant geometric parameter (or parameters) is found, respecting the constraint on the texture. Either a pixel, or small patch of pixels, is copied from the source texture to the new target. The method proceeds to a new untextured location on the target. Once again, a location on the source is found with the same relevant geometric parameter value. However, on this and all subsequent

Figure 5—An example using data from Lu et al. of using a captured map of aging material to render a synthetic object.

transfers, consistency is enforced. That is, not only must the pixel or patch copied come from an area with the same geometric parameter, it must also be bounded by pixels with similar values to the pixels already transferred to the target object bounding the target location. An example of data captured by Lu et al.[28] applied to a synthetic object via texture transfer is shown in *Figure 5*.

FUTURE OUTLOOK

Captured textures of aging materials have been shown to be capable of producing rich visual effects on synthetic objects that add to the realism of synthetic images. Current methods are clearly very limited in the types of effects they can produce. Only very preliminary validation experiments have been conducted for assessing how reliable the methods are in predicting visual impression. More data sets, and suitable models for extending the application of the data, are needed for captured effects to be used in mainstream graphics applications.

In the literature from different fields there are many reports of long-term aging effects on materials. Examples include a two-year field test of painted frames,[37] and the 50-year stone wall test at NIST.[38] Since these tests, however, were not conducted for the sort of detailed simulation now possible, data reported for such work includes graphs of summary parameters rather than detailed colored imagery and/or geometric measurements. As future accelerated aging techniques are developed, it may require only simple modifications to acquire image data under documented conditions that could subsequently be used in visual simulations. Image data is required to track the spatial variations such as cracking that occur during aging. Shape and illumination data are needed to disambiguate changes in images resulting from shading, and changes resulting from changes in the material. A shape description may simply be that the sample is flat or spherical. For other shapes a CAD file for manufactured samples could be provided, or the sample could be scanned with a device such as the ShapeGrabber. Illumination data can be provided by including a standard object, such as a white diffuse sphere, into the image captured of the sample.

Any release of data or modification of procedures involves costs. Why should data be released to benefit another industry such as training, film, or computer gaming? One possibility is that such data could be packaged as a product for sale. Examples of such data are stock photographs sold as texture libraries for graphics applications, or product-specific content provided (for paints or furniture) through software rendering systems.

A more speculative benefit of capturing and releasing data is that the availability of synthetic material aging models could in the long term be used as a tool by material designers and manufacturers. One scenario would be the availability of differing products or procedures for preventing or retarding weathering effects. Realistic imaging could be applied to demonstrate the effect of choosing a set of products in a particular geographic and architectural setting. Results for several different treatments could be compared. The uncertainty in results could be illustrated by rendering results for several different cases of anticipated weathering conditions.

Another scenario is the design of materials that are meant to change in appearance to uniquely adapt to an environment. Faux finishes and rapid forming patinas are popular for applying to small decorative objects to give them a unique look. Materials to produce similar effects could be designed for larger structures. While the local effects of age and geometry could be predicted by experiment and modeling, computer graphics image generation could be used to predict the large scale visual effect on different geometries in different environments.

ACKNOWLEDGMENTS

Discussions with members of the Yale Computer Graphics Group—Julie Dorsey, Athinodoros Georghiades, Andreas Glaser, Jianye Lu, Bing Wang, Hongzhi Wu, Chen Xu, and Songhua Xu—contributed to this paper. This work was supported by a grant from the National Science Foundation, # CCF-0528204.

References

(1) Goral, C.M., Torrance, K.E., Greenberg, D.P., and Battaille, B., "Modelling Interreflections Between Diffuse Surfaces," *Computer Graphics* (*SIGGRAPH 84*), Vol. 18, p. 213-222 (1984).

(2) Nishita, T. and Nakamae, E., "Continuous Tone Representation of Three-Dimensional Objects Taking Account of Shadows and Interreflection," *Computer Graphics (SIGGRAPH 85)*, Vol. 19, p. 23-30 (1985).

(3) Meyer, G.W., Rushmeier, H.E., Cohen, M.F., Greenberg, D.P., and Torrance, K.E., "An Experimental Evaluation of Computer Graphics Imagery," *ACM Trans. on Graph*, 5, 1, p. 30-50 (January 1986).

(4) Siegel and Howell, *Thermal Radiation Heat Transfer*, 4th Edition, Taylor and Francis, 2001.

(5) Stone, M., *A Field Guide to Digital Color*, AK Peters Ltd., Wellesley, MA, 2004.

(6) Farin, G.E., *Curves and Surfaces for Computer-Aided Geometric Design: A Practical Code*, Academic Press, Inc., 1996.

(7) Warren, J.D. and Weimer, H., *Subdivision Methods for Geometric Design: A Constructive Approach*, Morgan Kaufmann Publishers Inc., 2001.

(8) Larson, G.W. and Shakespeare, R., *Rendering with Radiance: The Art and Science of Lighting Visualization*, Morgan Kaufmann, San Francisco, CA, 1998.

(9) Dutre, P., Bala, K., and Bekaert, P., *Advanced Global Illumination*, 2nd ed., AK Peters, Boston, MA, 2006.

(10) Drago, F. and Myszkowski, K., "Validation Proposal for Global Illumination and Rendering Techniques," *Computers & Graphics*, Vol. 25, 3, p. 511-518 (June 2001).

(11) Radiance Synthetic Imaging System, http://radsite.lbl.gov/radiance/HOME.html, accessed May 2007.

(12) Lightworks Advanced Rendering, http://www.lightwork.com/products/advrender.htm, accessed May 2007.

(13) Nicodemus, F.E., Richmond, J.C., Hsia, J.J., Ginsber, I.W., and Limperis, T., "Geometrical Considerations and Nomenclature for Reflectance," *NBS Monograph*, 160, U.S. Dept. of Commerce, 1977.

(14) Ward, G.J., "Measuring and Modeling Anisotropic Reflection," in *Proc. 19th Annual Conference on Computer Graphics and Interactive Techniques*, ACM Press, p. 265–272, 1992.

(15) Lafortune, E.P.F., Foo, S-C., Torrance, K.E., and Greenberg, D.P., "Non-Linear Approximation of Reflectance Functions," in *Proc. 24th Annual Conference on Computer Graphics and Interactive Techniques*, ACM Press/Addison-Wesley Publishing Co., p. 117–126, 1997.

(16) Cook, R.L. and Torrance, K.E., "A Reflectance Model for Computer Graphics," *ACM Trans. on Graph.*, 1(1):7–24, (January 1982).

(17) He, X.D., Torrance, K.E., Sillion, F.X., and Greenberg, D.P., "A Comprehensive Physical Model for Light Reflection," in *Proc. 18th Annual Conference on Computer Graphics and Interactive Techniques*, ACM Press, p. 175–186, 1991.

(18) Ebert, D.S., Musgrave, F.K., Peachey, D., Perlin, K., and Worley, S., *Texturing and Modeling: a Procedural Approac*h, 3rd. Ed., Morgan Kaufmann Publishers Inc., 2002.

(19) Marschner, S.R., Westin, S.H., Lafortune, E.P.F., Torrance, K.E., and Greenberg, D.P., "Image-Based BRDF Measurement Including Human Skin," *Proc. 10th Eurographics Rendering Workshop*, Granada, Spain, pp. 139-152, June 21–23, 1999.

(20) ShapeGrabber, http://www.shapegrabber.com, accessed May 2007.

(21) Olympus, http://www.olympusamerica.com, accessed May 2007.

(22) Blais, F.F., "Review of 20 Years of Range Sensor Development," *J. Electronic Imaging*, 13(1): 231.240 (2004).

(23) Bernardini, F. and Rushmeier, H., "The 3D Model Acquisition Pipeline," *Computer Graphics Forum*, Vol. 21, No. 2, pp. 149-172 (2002).

(24) Farouk, M., El-Rifai, I., El-Tayar, S., El-Shishiny, H., Hosny, M., El-Rayes, M., Gomes, J., Giordano, F., Rushmeier, H., Bernardini, F., and K. Magerlein, "Scanning and Processing 3D Objects for Web Display," in *4th International Conference on 3D Digital Imaging and Modeling (3DIM '03)*, Banff, Alberta, p. 310-318, October 2003.

(25) Bernardini, F., Giordano, F., Gomes, J., and Rushmeier, H., U.S. Patent #7,084,386 (Light source calibration) 2006.

(26) Rushmeier, H. and Bernardini, F., "Computing Consistant Normals and Colors from Photometric Data," *Proc. Second Intenational Conference on 3-D Digital Imaging and Modeling*, p. 99-108, 1999.

(27) Lensch, H.P.A., Kautz, J., Goesele, M., Heidrich, W., and Seidel, H-P., "Image-based Reconstruction of Spatial Appearance and Geometric Detail," *ACM Trans. Graph.*, Vol. 22, 2, p. 234–257 (2003).

(28) Lu, J., Georghiades, A.S., Glaser, A., Wu, H., Wei, L-Y., Guo, B., Dorsey, J., and Rushmeier, H., "Context-Aware Textures," to appear in *ACM Trans. Graph.*, 2007.

(29) Paquette, E., Poulin, P., and Drettakis. G., "The Simulation of Paint Cracking and Peeling," in *Graphics Interface 2002*, p. 59–68, 2002.

(30) Dorsey, J., Kohling Pedersen, H., and Hanrahan. P., "Flow and Changes in Appearance," in *Proc. 23rd Annual Conference on Computer Graphics and Interactive Techniques*, ACM Press, p. 411–420, 1996.

(31) Gu, J., Tu, C., Ramamoorthi, R., Belhumeur, P., Matusik, W., and Nayar, S., "Time-Varying Surface Appearance: Acquisition, Modeling and Rendering," *ACM Trans. Graph.*, 25, 3, 762-771 (2006).

(32) Columbia, STAF: Database of Time-Varying Surface Appearance, http://www1.cs.columbia. edu/CAVE/databases/staf/staf.php, Columbia, accessed May 2007.

(33) Wang, J., Tong, X., Lin, S., Pan, M., Wang, C., Bao, H., Guo, B., and Shum, H. "Appearance Manifolds for Modeling Time-Variant Appearance of Materials," *ACM Trans. Graph.*, 25, 3, p. 754-761 (2006).

(34) Mertens, T., Kautz, J., Chen, J., Bekaert, P., and Durand, F., "Texture Transfer Using Geometry Correlation," *Eurographics Symposium on Rendering*, p. 273-284 (2006)

(35) Efros, A.A. and Leung, T.K., "Texture Synthesis by Non-Parametric Sampling," in *Proc. International Conference on Computer Vision—Volume 2*, ICCV. IEEE Computer Society, Washington, D.C., 1033-8 , Sept. 20-25, 1999.

(36) Wei, L. and Levoy, M., "Texture Synthesis Over Arbitrary Manifold Surfaces," in *Proc. 28th Annual Conference on Computer Graphics and Interactive Techniques (SIGGRAPH '01)*, ACM Press, New York, NY, p. 355-360 (2001).

(37) Hunt, M.O., O'Malley, A.J., Feist, W.C., McCabe, G.P. et al., "Weathering of Painted Wood Construction: Facade Restoration," *Forest Products Journal*, 53, 4, p. 51 (2003).

(38) Stutzman, P.E. and Clifton, J.R., "Stone Exposure Test Wall at NIST," in *Proc. Degradation of Natural Stone*, Labuz, J.F. (Ed)., American Society of Civil Engineers Annual Meeting, Minneapolis, MN, 1997.

Mechanistic Measurements

Chapter 20

The Influence of Water on the Weathering of Automotive Paint Systems

Tony Misovski,[1] Mark E. Nichols,[1] and Henry K. Hardcastle[2]

[1]Ford Research and Advance Engineering, Dearborn, MI USA
[2] Atlas Material Testing Technology LLC, Chicago, IL

The role of water in the long-term weathering performance of complete automotive paint systems was investigated by various means. Experiments using accelerated outdoor exposure (Fresnel-type exposure) were able to reproduce the chemical degradation gradient only when paint systems were completely saturated with water once per day. Saturation with water was shown to require approximately one hour of contact with liquid water or 100% humidity at moderate temperatures. Appearance and degradation gradient differences between panels exposed in Arizona and Florida were ascribed to the washing away of photooxidation products by rainwater in Florida. The implications of these results on designing improved accelerated tests were demonstrated.

INTRODUCTION

A modern automotive paint system performs two main functions: corrosion protection and appearance enhancement. Corrosion protection is mainly provided by the underlying layers in the paint system: the phosphate, electrocoat, and primer layers. The colorful, high gloss appearance that consumers have come to expect from their vehicles is provided by the basecoat and clearcoat, the top two layers in the paint system. Because of their proximity to the surface, these two layers are exposed repeatedly to environmental pressures such as heat, water, sunlight, and atmospheric pollutants. Each of these can cause the paint to chemically degrade. If severe enough, this chemical degradation can lead to physical failure by either gloss loss, environmental etching, cracking, or delamination.

Much research has been published in the last 10 years that examines the mechanisms of photooxidation in automotive coatings and the means to measure such degradation. Sensitive analytical techniques including infrared spectroscopy, electron spin resonance spectroscopy, ultraviolet spectroscopy, and mechanical testing have shown that those coating systems whose chemical composition and mechanical behavior change slowly tend to perform acceptably during exposure in Florida, while coatings whose chemical composition changes rapidly and/or whose mechanical properties deteriorate quickly tend to perform poorly during Florida exposure.[1-6]

While photooxidation appears to dominate the chemical degradation process outdoors, the role of water cannot be overlooked. Because many of the chemical changes in a coating induced by photooxidation and hydrolysis are similar, it can be difficult to separate the two processes. However, extensive IR spectroscopy by workers at NIST has elucidated the details of the mechanisms by which some coatings undergo hydrolysis.[7,8] They have shown that under warm and wet conditions the rate of hydrolysis can be comparable to the rate of photooxidation. However, under more mild conditions typical of outdoor exposure, the rate of hydrolysis still appears to be relatively low, such that at the surface of the coating photooxidation dominates, while deeper in the coating, where light cannot penetrate, hydrolysis dominates.[1]

In addition to chemically degrading a coating, water physically interacts with the coating in a variety of ways. This has been explored by Perera and coworkers who showed that wet/dry cycles induce stresses in coatings that have the potential to physically degrade the coating.[9] These stresses have recently been modeled and shown to be on the order of a few MPa for clearcoats in automotive paint systems.[10] In addition to stressing the coating, liquid water provides a mechanism for material removal such that low molecular weight species can be removed from the coating, either by solubalizing the species or physically washing them away. For example, water has been shown to remove residual acid catalyst in acrylic/melamine coatings, leading to a reduction in the hydrolysis rate after the acid is removed.[11] The washing away of material also appears to be the main mechanism by which film is lost during exposure, which is typically manifest as a reduction in the gloss of a coating.

In summary, the evidence to date suggests that water plays a key, but secondary, role in the chemical degradation of coatings and, perhaps, the dominant role in some of the surface appearance changes that take place during outdoor exposure. However, the specifics of much of this role are not well understood or quantified. When trying to accelerate the degradation process, the interactions between light intensity, water, and temperature become much more complex, and thus, the need to understand the details becomes more pressing. In this paper, we investigate the role water plays in accelerated outdoor Fresnel-type exposure as well as the differences between natural exposure in Florida and Arizona where the atmospheric moisture content and time of wetness are significantly different. In addition, we examine the rate of water uptake in complete automotive paint systems in order to better understand how the duration and magnitude of wet/dry cycles must scale with light intensity in any accelerated weathering protocol.

EXPERIMENTAL

Materials

The water uptake experiments were performed on a paint system consisting of an aluminum substrate, cathodic electrocoat (25 μm), polyester primer (20 μm), and acrylic/melamine based basecoat (15 μm) and clearcoat (40 μm). The basecoat was either black or white in color, and samples were made from both solventborne and waterborne basecoats. The same coating system was used for both the accelerated Fresnel-type exposure and the natural Florida and Arizona exposures. Specimens for water uptake experiments were made by cutting 2.54 cm × 2.54 cm pieces from the larger panels. Due to processing, the backs of the panels were e-coated and contained some overspray. These paint layers were removed by sanding the backs of the panels to bare aluminum. Sanding to bare aluminum not only removed the overspray, but also removed the oxide layer protecting the aluminum so the panels were allowed to sit in a humidity box until the oxide layer redeveloped. Samples exposed in the natural Florida and Arizona exposures were held at 5° from the horizontal facing south.

Microtomy

Paint systems were sectioned using a slab microtome (Leica) parallel to the surface of the samples. In doing so, 5 μm thick slices were removed from the top to the bottom of the paint system. Each slice was then subjected to various analyses, including infrared spectroscopy. Details of this technique are given elsewhere.[12]

Infrared Spectroscopy

Infrared spectroscopy was performed on slices from the microtomy experiments to assess the amount of degradation in the paint system at various depths into the system. Fourier transform infrared (FTIR) spectra were obtained using a Mattson FTIR 5000 at 4 cm^{-1} resolution. Each 5 μm thick slice was placed between salt plates and the spectra was taken in transmission mode. The amount of photooxidation was generically quantified using the $\Delta[(-OH,-NH)/-CH]$ method, which ratios the absorbance in the 3800–2000 cm^{-1} region to that in the 3100–2800 cm^{-1} region.[1]

For the complete paint systems, photoacoustic (PAS) infrared spectroscopy was used to quantify the chemical changes taking place during weathering. All PAS data were obtained on a Mattson Cygnus 100 rapid scan FTIR system equipped with a water cooled source, a variable source aperture at 50%, and an MTEC 2000 cell. All spectra were transformed using 4000 data points and one order of zero filling to give spectral resolution of 8 cm^{-1}, and a digital resolution of 4 cm^{-1}. A scan velocity of 3.6 kHz and an assumed coating thermal diffusivity of 1×10^{-3} cm^2/sec yielded a thermal diffusion length at 4000 cm^{-1} of 6 μm and at 2000 cm^{-1} of 8 μm. This procedure led to a sampling depth of between 12 and 16 μm into the coating. The amount of photooxidation was again quantified using the $\Delta[(-OH,-NH)/-CH]$ method.

Water Absorption/Desorption Measurements

Water absorption/desorption measurements were conducted with a Mettler AT20 balance. The balance had a maximum capacity of 22 g and an accuracy of ±0.002 mg. Measurements were taken every 30 seconds after the specimen was placed on the balance. Data was taken via a computer connection to the balance. The inherent stability of the balance was measured prior to collecting data. Data was logged from the empty balance every minute for 24 h then plotted. No noticeable drift associated from temperature or humidity was noted if the balance was allowed to equilibrate in its environment for several hours before the experiment was started. To prevent noise while testing outdoors, the balance was put on a table that had a hinged cover that would isolate the balance.

Pieces of 3 mm thick rubber were placed under each leg of the balance to help absorb vibrations. The balance specifications state that the admissible ambient conditions are 40°C and a relative humidity of 85%. Tests at 95% relative humidity showed no adverse effects in relation to drift or stability of the measurements.

A humidity box was built out of aluminum to enclose the balance for controlled water absorption experiments. All joints were sealed and a hinged door was installed on the front of the cabinet. The box was heat taped on all sides and insulated with thick Styrofoam insulation. The balance was placed inside the box along with a hot plate and two muffin fans. The hot plate was used to regulate the temperature in the box, and the muffin fans were used to ensure the air was properly circulated. Humidity was generated by several deep dishes of water that had sponges semi-submerged. Humidity was checked with a chilled mirror detector (Omega Instruments). The percentage of humidity and the temperature were held at a constant value over the course of 24 hours.

Before specimens were tested for water absorption, they were first dried in a vacuum oven. The samples were removed and promptly placed in a glass jar containing DriRite. The jar was placed in the humidity box for several hours to allow the sample temperature to equilibrate to the box temperature. Once samples reached testing conditions the humidity box door was opened and the balance was zeroed. The sample was then removed from the jar and placed on the balance. The balance doors were then closed along with the humidity box door. Next, the exterior insulation was affixed to the door and the data logging was started no more than 20 s after placing the sample in the humidity box. The sample mass was then logged for a minimum of 1 h at 20 s intervals.

Accelerated Outdoor Exposure

Outdoor accelerated weathering testing was conducted using a Fresnel-type system in Arizona. Details of the machine design and operation are outlined in ASTM G 90-05.[13] A schematic of the specimen position in the machine is shown in *Figure* 1. ASTM G 90-05 outlines three typical water spray cycles. Cycle 1 prescribes an 8-min water spray every hour during the day, and three 8-min water sprays at night. Cycle 2 prescribes no

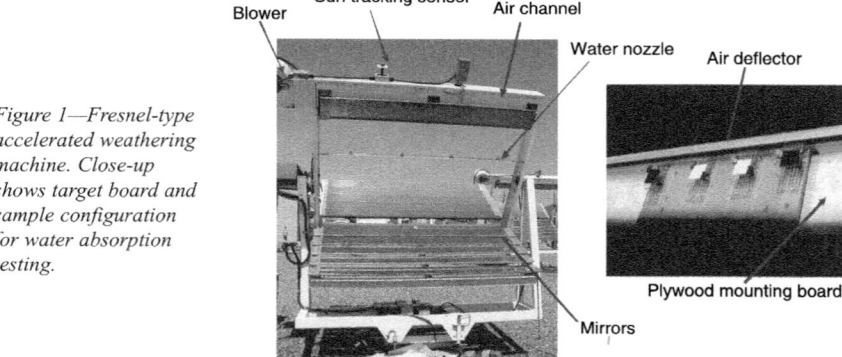

Figure 1—Fresnel-type accelerated weathering machine. Close-up shows target board and sample configuration for water absorption testing.

water spray during the day, and no water spray during the night. Cycle 3 prescribes no water spray during the day, and four 3-min sprays every hour during the night. For the experiments detailed here, the machine was run in the daytime spray mode where air blew over the specimens during testing and water was sprayed intermittently on the irradiated specimens. Additionally, specimens were tested in an "inverted" mode where the specimen rack is rotated downward out of the concentrated irradiance and water was sprayed on the specimens while air was blown over the specimens.

Samples in Fresnel-type exposures are usually mounted to the target board with screws. For our experiments the mounting procedure was modified. A piece of wire mesh was cut into a strip about 100 cm × 20 cm. The mesh was bent into the shape of a squared off "U". The ends of the mesh were then fastened to the target board with screws so that the panels sat about 25 mm above the target board.

Specimens were first dried in the field by simply placing them on the bench in the sunlight for 10 min. The machine was then inverted and the specimens mounted to the mesh by placing two office style binder clips on the edges of the paint panel, being careful to only cover about an 3 mm of the panel. The machine was then rotated right-side-up and adjusted until the sun was focused on the samples. After this, the machine was then run in its standard operating condition with samples receiving concentrated sunlight and blown cooling air. After water exposure the drive was then disengaged and the machine inverted. Samples were then removed and patted dry and weighed on the balance.

For inverted exposures, samples were allowed to dry in the sun for 10 min. The machine was then inverted facing down (so samples were parallel to the ground and paint layers faced up) and locked in place. The samples were then attached with clips to the mesh and the water spray was turned on. The machine blower was active during all spraying cycles.

For soaking experiments, water was obtained directly from the spray nozzles of the machine to ensure that the water and temperature were the same in all experiments. The samples were dried using the procedure above and then placed in a container with water obtained directly from the machine. Samples were removed, patted dry, and weighed at one minute intervals until equilibrium mass was attained.

Separate experiments were conducted to measure the effect of various exposure variables using the accelerated Fresnel-type machines on acrylic melamine automotive coating systems. The factors evaluated in the DOE (design of experiment) included: exposure temperature, number of mirrors (intensity), water spray frequency, nighttime water soaking, chemical pretreatment prior to exposure, high intensity UV exposure before accelerated exposure, mechanical abrasion prior to exposure, soak-freeze-thaw prior to exposure, and thermal aging prior to exposure. The response variable in all cases was the change in the (–OH,–NH)/–CH value as determined by PAS-IR. Further details of the experimental design and factors can be found elsewhere.[14]

Fracture Energy

The fracture energy, a quantitative measurement of a material's brittleness and a key indicator of a materials propensity to crack, was measured for each of the three clearcoats in the complete paint systems using techniques that have been described in detail elsewhere.[15] Briefly, 8 mm wide strips were cut from the paint panel with a hand shear. These strips of the paint panel were then pulled in tension at 20 mm/min in a mechanical testing machine (Instron model 5565). The strain at which the clearcoat cracked was then recorded. Multiple samples were tested for each panel at each weathering time. The fracture energy, G_c, was then calculated using

$$G_c = 0.5\pi h \varepsilon^2 \overline{E}_f \, g(\alpha, \beta) \tag{1}$$

where ε is the strain at cracking, h is the clearcoat thickness, $\overline{E}_f$ is the biaxial modulus of the clearcoat, and $g(\alpha, \beta)$ is a constant related to the mismatch in moduli between the coating and substrate. The modulus of the coating was previously determined from tensile testing on free films of similar clearcoats.

RESULTS

Accelerated Fresnel-Type Exposure DOE

The main effects plot from the accelerated outdoor weathering DOE array is shown in *Figure 2*. For each run in the DOE the output variable monitored was the change in the (–OH, –NH)/–CH ratio.[1] As can be seen, the variable that had the biggest impact on the photooxidation rate was the nighttime soak, where the panels were removed from the Fresnel device and placed in a soak tank with 40°C water overnight. In this DOE, the nighttime soaked specimens exhibited significantly higher photooxidation rates than specimens that were not soaked overnight. Similar effects have also been reported for gloss reading taken on the same panels.[14]

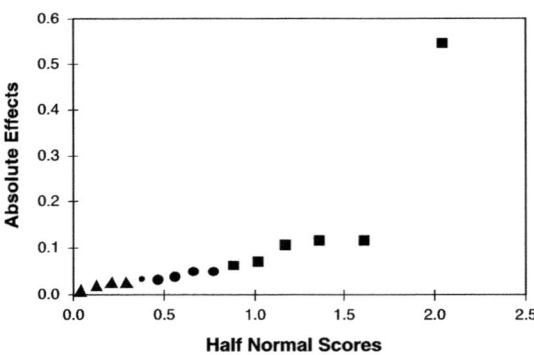

Figure 2—*Main effects plot for Fresnel-type testing with various environmental factors. Response variable is the change in the (–OH, –NH)/–CH value as determined by PAS-FTIR. Data points on graph from left to right indicate the following factors: daytime and nighttime spray, temperature and irradiance, irradiance and nighttime soak, temperature and nighttime soak, thermal pretreatment, temperature and daytime spray, chemical pretreatment, irradiance and daytime spray, mechanical pretreatment, high UV pretreatment, temperature-irradiance-daytime spray, irradiance, temperature, daytime spray, and nighttime soak. Note that the only significant effect is that of soaking panels over night in water.*

Water Uptake

The extent of water uptake in paint systems exposed during accelerated Fresnel-type exposure is shown in *Figure 3*. The data is presented as mg of water uptake. The same test specimens were used for each experiment; thus, mg of water uptake was a direct measure of the amount of water absorbed and can be used to compare the results of different test conditions. Results are presented for both waterborne and solventborne basecoat/clearcoat systems. For all paint systems tested, the mass uptake during spraying on the Fresnel-type machine with the daytime spray configuration (water spray with concentrated irradiance and blown cooling air) was minimal compared to the mass uptake after soaking for the same time period. Even spraying the panels in the "inverted" configuration led to only 50% of the water absorption that occurred during soaking.

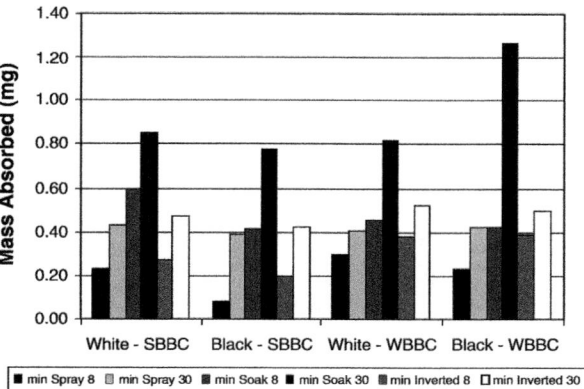

Figure 3—Water uptake during Fresnel-type testing for waterborne and solventborne paint systems. All uptakes are in mg. Testing occurred after either 8 or 30 min after either: spraying, inverted spraying (Fresnel-type machine upside down during spraying), or during soak.

The time required to fully saturate a waterborne basecoat paint system in 100% humidity at 30°C is shown in *Figure* 4. The horizontal line represents the mass uptake of the same specimen immersed in liquid water at 30°C for 12 h, well past the time to reach equilibrium. Similar results were obtained for solventborne paint systems. The rate of uptake in liquid water is similar to that obtained in 100% humidity, but reliable data was difficult to obtain due to the rapid release of water from the surface of the coatings once the specimens were removed from the water for weighing.

The mass of a paint system in a natural Florida outdoor exposure as function of time is shown in *Figure* 5. The high mass periods were the times when the panel was wet and the lowest mass regions were when the panel was dry. For each 24-h period, the panels were covered with standing water for at least 7 h. This occurred during the evening when moisture from the humid Florida atmosphere condensed on the panels. The water was driven off quickly as soon as the sun rose and the panel temperature rose above the air temperature. The initial rise in mass is due to the sample being taken from a dessicator and placed into the humid Florida environment. The dashed lines indicate when no data was taken. While the absolute mass gain is small due to the coupon size, the percentage uptake is on the order of 1.5% by mass.

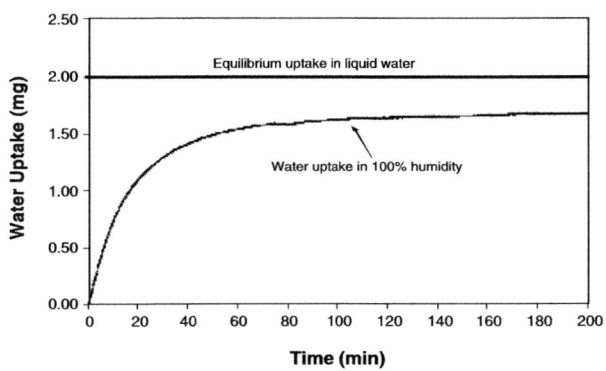

Figure 4—Water uptake in mg for paint systems exposed to 100% humidity at 30°C.

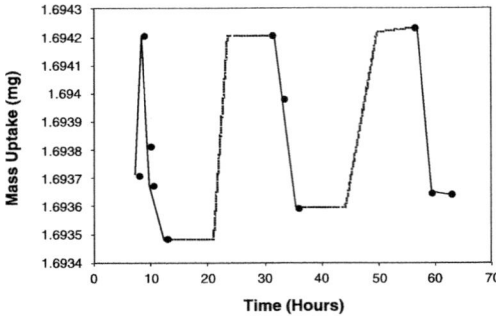

Figure 5—Mass of paint system exposed in Florida for several days. High mass periods indicate liquid water on the surface of the samples. Dashed lines indicate estimated data. Data points indicate actual time and mass data collected. Hours are real military time of day in south Florida.

The extent of photooxidation in paint systems exposed naturally in both Florida and Arizona was investigated to determine the influence of water during the natural weathering process. *Figure* 6 shows the photooxidation depth profile of the solventborne basecoat/clearcoat paint system weathered in Florida and Arizona. The graph shows the amount of photooxidation throughout the depth of the system as obtained by microtomy and transmission FTIR. The Florida exposed system showed a gradient in the amount of photooxidation through the clearcoat, with the surface being more photooxidized than the bulk. The amount of photooxidation at the surface of the system exposed in Arizona was significantly greater than that of the Florida exposed system, but leveled off to approximately the same amount as in the Florida exposed specimen approximately 6 μm below the surface of the clearcoat. The data points on the graph representing the top of the coating were not measured by transmission FTIR on microtomed slices, but by scraping off the top one to two microns with a razor, collecting the scrapings, and mixing with KBr powder to produce a pellet for transmission IR analysis. The fracture energies of the clearcoats in the systems exposed in Florida and Arizona are shown in *Figure* 7. In each case the fracture energy is shown after exposure for four years, and in the case of the Arizona exposed system, after four years plus a light surface polish, which was estimated to have removed less than 5 μm of the clearcoat surface. The frac-

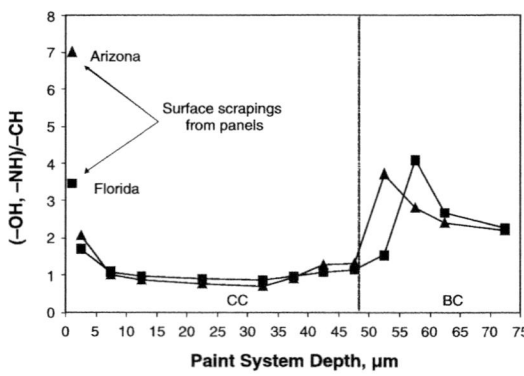

Figure 6—(–OH,–NH)/–CH values as a function of depth into the coating system for paint systems exposed in both Florida and Arizona for four years. Note higher surface degradation in Arizona exposed system.

ture energy of the Florida system is higher (less brittle) than that of the Arizona exposed system. After polishing the Arizona exposed system to remove the top 5 μm of clearcoat, the fracture energy of that system was similar to that of the Florida exposed system.

The influence of water also can be seen in *Figure* 8, where the degradation versus depth profile for four paint systems is shown for: an unexposed paint system, a system exposed in Florida, a system exposed in a Fresnel-type exposure with daytime spraying (ASTM G 90 cycle

1), and a system exposed in an ASTM G 90 exposure where the specimens were removed each night and soaked overnight in a soak tank with 40°C water. The gradients in degradation are different in most cases, but the Florida and nighttime soak samples are very similar except in the basecoat region, which can be ascribed to differences in basecoat formulation.

DISCUSSION

Previous work has shown that water plays a secondary role in the chemical degradation of automotive coatings when exposed to natural weathering.[6] Results presented here and by other workers have shown that the physical effects of water are significant and cannot be ignored when examining potential physical failures, such as gloss loss, cracking, and delamination, during outdoor exposure. Of particular importance is how to properly introduce water into accelerated testing, either accelerated outdoor testing or chamber testing, such that the correct physical failure mechanisms are reproduced. These failures must not only look the same to the minimally trained technician in the field, but occur at a predictable and scalable time during the accelerated test.

Accelerated Outdoor Exposure

Figures 4 and 5 clearly show that to fully saturate a paint system with water the paint system must be exposed to liquid water or high humidity for a minimum amount of time, ~ 1 h. Shorter water exposure times will not saturate the material. This saturation appears to be important for two reasons.

First, the results from the extensive Fresnel-type DOE testing (*Figure* 2) show that nighttime soaking has a significant effect on the photooxidation of the clearcoat during accelerated outdoor weathering. Other variables (temperature, chemical pretreatment, intensity, etc.) show no or little effect in this designed experiment. During the 40°C nighttime soak the panels had sufficient time to become saturated with water. Saturation and later desorption are hypothesized to aid in the transport of other small molecules within the coating system which may affect the degradation rate. Indeed the effect of nighttime soaking was positive, meaning that the photooxidation of the top 10 microns of the clearcoat was accelerated compared to non-soaked panels. Transport of photolabile molecules from deeper in the coating system to the surface may enhance the photooxidation of the surface. In addition, swelling of the coating by the water will lower the T_g of the coating, again allowing for easier mass transport within the coating system. Whatever the mechanism, it is clear that nighttime soaking enhances the rate of photooxidation in Fresnel-type testing.

Second, paint systems soaked overnight during Fresnel-type exposure show a gradient in degradation similar to that seen in Florida exposed panels. The chemical composition changes as a function of depth into the coating system are shown in *Figure* 8 for a paint exposed to accelerated outdoor exposure (ASTM G 90 cycle 1 with daytime spray and ASTM G90 cycle 2 with nighttime soaking), an unexposed panel, and the same paint system exposed naturally in south Florida. Because all three panels were exposed to similar outdoor solar radiant UV doses the level of degradation should have been similar. The changes in the (–OH,–NH)/–CH value demonstrate that the system exposed with the nighttime soak exposure exhibits increased degradation at both the surface of the clearcoat and into the depth of the clearcoat as compared to the system exposed to ASTM G 90 cycle 1 daytime spraying. This gradient matches the gradient in the Florida exposed system (differences in the –OH,–NH value in the basecoat are due to initial basecoat color/chemistry difference in the Florida and G 90 exposed panels and should be ignored). We conclude that daytime spraying in ASTM G 90 does not introduce sufficient water (less than half that of soaking) into the paint system such that surface and in-depth degradation is not as pronounced as that in Florida exposed paint systems, and we attribute this enhanced degradation to both hydrolysis of the coating binder and possibly enhanced photooxidation due to increased mobility and a lowering of the T_g due to plasticization by the 40°C water during overnight soaking.

Florida vs. Arizona Natural Exposure

For natural outdoor exposure, the presence of liquid water alters the physical appearance of panels. The panels whose photooxidation and fracture energy behavior are shown in *Figures* 7 and 8 were exposed naturally in either Florida or Arizona for four years. The 20° gloss of the panel exposed for four years in Florida was 45, while the gloss of the panel exposed in Arizona for four years was 75. The relative UV dose was approximately the same in both locations (slightly higher in Arizona) yet the appearance of the panels is dramatically different. The most obvious difference between these exposures was the amount of liquid water to which these panels were exposed. Panels exposed in south Florida become wet at least once a day due to the formation of dew and annual rainfall amounts in excess of one meter. Annual total time of wetness in south Florida for panels exposed at 5° south often exceeds 4200 h. Panels exposed in Arizona rarely had any dew formation and the total annual rainfall is typically less than 25 cm while many months had only trace amounts of precipitation. Annual total time of wetness for panels in Arizona rarely exceeds 370 h. The relative UV radiant exposure was less than 15% different between the two locations.

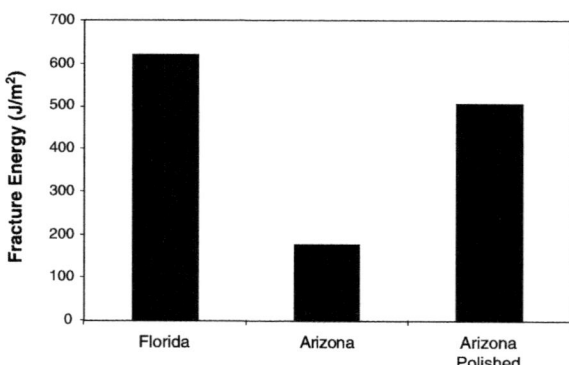

Like differences in appearance, differences were also apparent in the physical properties of the coating systems. The Florida exposed system was much tougher (less brittle) than the Arizona exposed system. After polishing away the highly degraded surface of the Arizona exposed system, the fracture energy recovered near to the level of the Florida exposed system. Thus, the uppermost degrad-

Figure 7—Fracture energy of coating system exposed in Florida and Arizona for four years. Bar on right is the same Arizona exposed system after having its top ~2 μm removed by polishing.

ed layer of the Arizona exposed panel dominated the fracture behavior of the system. Once this degraded layer was removed the mechanical performance was similar to that of Florida as was the level of surface degradation.

The Arizona/Florida panels demonstrated that chemical and physical performance did not necessarily follow each other. In this case, the Florida exposed panels clearly had inferior appearance yet were less susceptible to cracking due to their higher fracture energy. The Arizona exposed panels maintained a high level of gloss but were at a higher risk for cracking failure due to their high levels of chemical degradation and embrittlement at the surface. This dichotomy is explained by the washing away and removal of the highly degraded surface in the Florida exposed panels due to the presence of liquid water. During Arizona exposure, the surface of the coating degraded but no mechanism existed to wash away the highly degraded/embrittled surface.

Implications for Improved Accelerated Testing Cycles

In designing the correct cycles for an accelerated weathering test multiple factors must be considered. To improve upon the current state of the art two elements must be altered to improve accuracy and maintain or increase acceleration: (1) the duration/nature of the wet cycle that is required to saturate a panel and (2) the manner in which the wet/dry cycles scale with the UV dose. Clearly the data presented in *Figures* 4 and 5 show that samples must be kept wet for considerably longer than prescribed by ASTM G 90 or SAE J1960 to mimic Florida conditions and become saturated. How much longer is required is currently unknown and will play a role in determining the amount of acceleration that can be achieved as well as the scaling factors.

The two most common accelerated weathering tests for automotive coatings are SAE J1960 Jun89 and ASTM G 90. Neither currently meet the minimum water saturation time criterion demonstrated by the experiments discussed in this manuscript. SAE J1960 calls for 20 min of spray on the panel's front side during irradiation, and a one hour spray on the back side of the panels during the dark period. The back spray provides little wetting to the front of the panel and while the humidity in the chamber is raised, it is not sufficient to saturate the panels. ASTM G 90 uses eight minute sprays every hour during the day. However, the air circulation around the machines, the heat of the samples, and the low humidity of the atmosphere prevent the samples from taking up much more than 20% of saturation

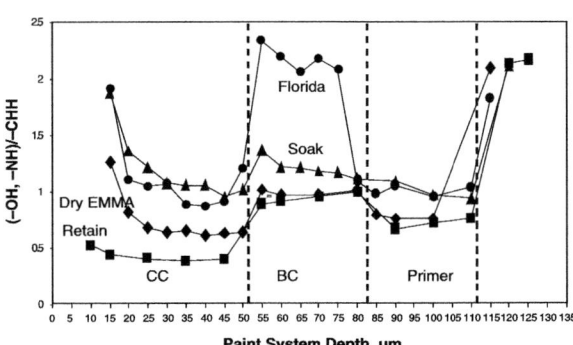

Figure 8—(–OH,–NH)/–CH values as a function of depth into the coating system for paint systems: unexposed, exposed in Florida, Fresnel-type with daytime spray, and Fresnel-type with nighttime soaking. All exposed systems are approximately equal dose ~four year Florida equivalent.

(*Figure* 3). By adding the water soak to the ASTM G 90 test, the gloss behavior may more closely match Florida exposure and the through depth degradation may more closely match Florida exposure as well. However, this may be an impractical solution due to the labor involved in unmounting and remounting the samples every day. Redesign of the machines and wetting cycles may allow for significant wetting without removing the samples from the Fresnel-type device.

ASTM G 90 cycle 1 mandates an eight-minute water spray every hour during the day and three eight-minute water sprays at night. As shown previously, the daytime spray of cycle 1 does not sufficiently wet the specimens. ASTM G 90 cycle 3 mandates four three-minute sprays every hour during the night. Even if the cycle 3 spray did sufficiently saturate the specimens, cycle 3 would not result in the same proportion (scaling) of UV radiant exposure to wetting events observed in natural south Florida exposures.

For example, a specimen on summertime exposure in south Florida may receive about 1 MJ/m^2 total ultraviolet radiation from 295 to 385 nm in a day followed by a time of wetness each night. Consider the natural south Florida exposure variable ratio of 1 MJ/m^2 total UV to one saturating wet cycle in a 24-h period. The current ASTM G 90 wetting cycle 3 will not allow this proportion to be simulated on the Fresnel-type exposure. It is observed on the Fresnel-type exposures that a sample may receive about 5 MJ/m^2 total ultraviolet radiant exposure in a day and using cycle 3 must wait until night for water spray. The 5 MJ/m^2 total UV to one wet cycle at night ratio in ASTM G 90 is very different than the 1 MJ/m^2 total UV to one wet cycle at night ratio observed in natural south Florida exposure.

SAE J1960 Jun89 calls for the use of a Xenon arc weathering chamber outfitted with quartz inner and borosilicate glass outer filters. To more closely approximate sunlight, some users modify this test method and use borosilicate inner and outer filters. The use of the boro/boro filters decreases the risk of false positives and negatives as the quartz/boro filter combination contains unnaturally short wavelength light that is not present in the terrestrial UV spectrum. Clearly, an improved filter that matches sunlight more closely is needed, and several workers have reported on such efforts.[16]

The average annual south Florida dose is approximately 2700 kJ/m^2 @340nm. The dose during each three hour J1960 cycle is approximately equal to half of the average daily dose received by a panel on exposure in south Florida. The panel also undergoes one short wet-dry cycle (front spray) and one humidity cycle (back spray). However, during every 24-hour period, panels exposed in Florida become wet at night for an extended time. Including rainfall events, each panel in Florida may experience on average more than one wet-dry cycle per day. Thus, on a dose basis each three hour J1960 cycle ($\sim$ 4 kJ/m^2 @340 nm) is approximately equal to half a day of exposure in south Florida and on a wet-dry cycle basis each three hour J1960 cycle is approximately one day. The current wet-dry cycle is inadequate as it does not completely saturate the panels. While this scaling is not exact, it is at least within a factor of two with respect to Florida parameters.

To increase acceleration, the intensity of the light must be increased. If the light source correctly matches sunlight, acceleration may be achieved without distorting the chemistry.[17] However, increasing the intensity will necessitate reducing the water soak times to maintain the approximate scaling between dose/cycle and wet-dry cycles/day. The two parameters appear to work at cross purposes. The compromise between dose scaling and wet/dry cycle scaling must be understood to improve these cycles. One strategy for overcoming this issue may be to heat the water moderately to improve the rate of diffusion of water into the paint system. This may allow for shorter times to saturation. However, care must be taken not to significantly increase the rate of hydrolysis with respect to the rate of photooxidation. More investigation into this strategy must be undertaken.

CONCLUSIONS

Both the chemical and physical impact of liquid water on the degradation of automotive coating systems has been explored. The presence of liquid water in accelerated testing is critical to reproducing the appearance of paint systems exposed outdoors and to reproducing the chemical changes that take place through a coating system when it is exposed to outdoor weathering in wet environments such as south Florida. The rate of water uptake appears controlled by diffusion into the paint system. Current accelerated weathering protocols do not allow for sufficient time during their nominal wet cycles for paint systems to become saturated. Improving accelerated testing protocols may only be possible by coupling improved, higher light intensities with improved environmental cycles such that both the physical and chemical changes that occur outdoors are more accurately reproduced in an accelerated weathering device.

ACKNOWLEDGMENTS

The authors would like to thank Dr. Alexi Kucherov for his microtomy work on some of the specimens, Ms. Cindy Peters for her thoughtful discussions of the data, and Joe Farley and the technicians at Atlas Weathering Services Group for exposure testing.

References

(1) Gerlock, J.L., Smith, C.A., Nichols, M.E., Tardiff, J.L., Kaberline, S.L., Prater, T.J., Carter, R.O. III, Dusbiber, T.G., Cooper, V.A., and Misovski, T., *Proc. 2nd Conference on Service Life Prediction of Organic Coatings*, Monterey, CA, ACS, Washington, D.C., November 1999.
(2) Gerlock, J.L, Smith, C.A., Nunez, E.M., Cooper, V.A., Liscombe, P., Cummings, D.R., and Dusibiber, T.G., in "Polymer Durability," Clough, R.L., Billingham, N.C., and Gillen, K.T. (Eds.), *ACS Advances in Chemistry Series 249*, Washington, D.C., p 335, 1996.
(3) Gerlock, J.L., Smith, C.A., Cooper, V.A., Dusbiber, T.G., and Webber, W.H., *Polym. Deg. and Stab.*, 62, 225, 1998.
(4) Gerlock, J.L., Prater, T.J., Kaberline, S.L., and deVries, J.E., *Polym. Deg. and Stab.*, 47, 405 (1995).
(5) Hill, L.W., Korzeniowski, H.M., Ojunga-Andrew, M., and Wilson, R.C., *Prog. Org. Coat.*, 24, 147, (1994).
(6) Wernstahl, K. M., *Polym. Deg. Stab.*, 54, 57, (1996).
(7) Nguyen, T., Martin, J., Byrd, E., and Embree, N., *Polym. Deg. and Stab.*, 77 (1), 1-16 (2002).
(8) Nguyen, T., Martin, J., Byrd, E., and Embree, N., "Relating Laboratory and Outdoor Exposure of Coatings: II. Effects of Relative Humidity on Photodegradation and the Apparent Quantum Yield of Acrylic-Melamine Coating," *J. Coat. Technol.*, 74, No. 932, 65-80 (2002).
(9) Perera, D.Y. and Eynde, D.V., "Moisture and Temperature Induced Stresses (Hygrothermal Stresses) in Organic Coatings," *J. Coat. Technol.*, 59, No. 748, 55 (1987).
(10) Nichols, M.E., and Darr, C.A., "Effect of Weathering on the Stress Distribution and Mechanical Performance of Automotive Paint Systems," *J. Coat. Technol.*, 70, No. 885, 141-149 (1998).
(11) Bauer, D.R., Mielewski, D.F., and Gerlock J. L., *Polym. Deg. and Stab.*, 38, 57, (1992).

(12) Gerlock, J.L., Kucherov, A.V. and Nichols, M.E., "On the Combined Use of UVA, HALS, Photooxidation, and Fracture Energy Measurements to Anticipate the Long-Term Weathering Performance of Clearcoat/Basecoat Automotive Paint Systems," *J. Coat. Technol.*, 73, No. 918, 45-44 (2001).

(13) ASTM G90-05 Practice for Performing Accelerated Outdoor Weathering of Nonmetallic Materials Using Concentrated Natural Sunlight. *2005 Annual Book of ASTM Standards*, vol. 14.02, American Society for Testing and Materials, West Conshohocken, PA, 2005.

(14) Hardcastle, H.K., in *Natural and Artificial Aging of Polymers*, Reichert, T. (Ed.), Gesellschaft fur Umweltsimuation e.V. GUS, Pfinztal, Germany, 2004.

(15) Nichols, M.E. and Tardiff, J.L., *Proc. 2nd Conference on Service Life Prediction of Organic Coatings*, Monterey, CA, ACS, Washington D.C., November, 1999.

(16) US Patent # US6906857 B2.

(17) Gerlock. J.L., Peters, C.A., Kucherov, A.V., Misovski, T., Seubert, C.M., Carter, R.O. III, and Nichols, M.E., "Testing Accelerated Weathering Tests for Appropriate Weathering Chemistry: Ozone Filtered Xenon Arc," *J. Coat. Technol.*, 75, No. 936, 35-45 (2003).

Chapter 21

Chemiluminescence Detection: Principles, Chances and Limitations for the Shortening of Weathering Tests

Volker Wachtendorf,* Anja Geburtig, and Peter Trubiroha

Federal Institute for Materials Research and Testing, Berlin, Germany

One way to realize a higher throughput in weathering consists in applying more sensitive means of detection. Within the group of luminescence techniques, which are generally characterized by high sensitivity, chemiluminescence (CL) is investigated as a possible candidate. Its usefulness in evaluating weathering effects and the limitations in correlating the effects of weathering tests at early stages to those at later stages commonly used for established macroscopic detection will be discussed. Also, principles in the correlation between shortening weathering tests and field exposure are highlighted.

INTRODUCTION

Improvements in the stabilization of polymeric materials mean that the duration of weathering tests increases. At the same time, however, an increasing number of new developments in ever-shorter development cycles brings about greater demands to shorten the test duration. Apart from accelerating the degradation rate by enhancing exposure parameters, another approach to increasing the test throughput consists in using more sensitive means of detecting degradation effects in earlier stages of the degradation process.

The degradation process begins on a molecular level before it starts to spread over higher proportions of the polymer. The effect of degradation on the material builds up until macroscopic properties begin to change, either showing an abrupt threshold behavior or a more gradual response to increasing degradation on a molecular level. From a theoretical point of view, a detection of the earlier stages of the degradation process can be expected by switching from the established, more macroscopic properties in the evaluation of effects to more microscopic ones.

While both physical processes and chemical reactions contribute to the degradation of materials, chemical reactions usually are more dominant since their effects tend to be irreversible. Within these chemical reactions, oxidation reactions with oxygen usually are of

---

* Corresponding author: Federal Institute for Materials Research and Testing, BAM VI.3, Unter den Eichen 87, 12205 Berlin, Germany. E-mail: volker.wachtendorf@bam.de.

highest importance. For saturated polymer bonds, the autooxidation process[1] usually is accepted as the main pathway to oxidative degradation. One of the termination reactions of this radical chain process can lead to an electronically excited state that can return to the ground state under emission of a chemiluminescence (CL) photon. In this way, the CL becomes directly linked to the ongoing oxidation process. It allows one to monitor the ongoing degradation process on a molecular level by simply counting the photons emitted in the macroscopic world at the extremely high sensitivity of luminescence techniques, such as CL.

While the first application of CL must be attributed to evolution from biolumines-cence, i.e., very efficient CL emission from enzymatically controlled oxidation reactions mostly for communication purposes, its exploitation for analyzing the oxidation of poly-mers by humans started with Ashby,[2] and Schard and Russell[3] in the 1960s. The method-ology received a boost by the introduction of sensitive and less expensive photomultipli-ers that made a wider range of CL reactions experimentally accessible. However, for the study of polymer degradation, the technique still remained exotic. Most investigations focused on polypropylene[4] although in the 1980s studies of the degree of curing of epoxy resins also received some attention by the aviation industry.[5] The oxidation behavior of many other polymers was covered[6] although actual applications in the monitoring of the production process of relevant properties such as the degree of curing or the monitoring of the aging process remained scarce. In recent years, only a few publications dealt with the application of CL to monitor weathering exposure stages of polymers.[7-9]

EXPERIMENTAL

Artificial weathering was carried out using a commercially available weathering device, Global UV Test (Weiss Umwelttechnik, Germany), which applies spectral irradi-ance type A2 according to Table 1 of ISO 4892-3:2006 with a combination of fluores-cent UV lamps. Radiation is cut off at 290 nm, subjecting the samples to 45 W/m^2 in the wavelength range from 290 to 400 nm. The Global UV Test device allows for running the test at defined, reproducible weathering conditions. Temperature can be held constant within $\pm$ 1 K and relative humidity within $\pm$ 5% RH.

A special BAM-developed weathering test that combines climatic stress with stress by acid precipitation, the so-called Acid Dew and Fog (ADF) test,[10,11] was used. The 24-h weathering cycle used for the ADF test is characterized by continuous UV radiation. It starts with a short-term spraying of the simulated acid dew or fog (mixture of H_2SO_4, HNO_3, and HCl in the weight ratio 1 : 0.3 : 0.17 diluted to pH 2.5) onto the surface of the test specimen, followed by a 14-h dry period of different climatic conditions (9 h at 35°C/75% RH and 5 h at 60°C/40% RH) followed by a 4-h rain period with demineral-ized water at 35°C, continued by a final dry period of 6 h at 60°C/40% RH.

Chemiluminescence (CL)—The setup consisted of a BAM-built heating sample cham-ber[12] attached to a photomultiplier and commercial equipment for single counting detec-tion of emitted photons.

Heating was done electrically, controlled by a Lake Shore (Westerville, OH, USA) 340-temperature controller. A self-built gas-tight sample chamber was used that had a gas inlet and outlet, with the sample in its middle, and a quartz window on its top connecting it to a side-on low-noise photomultiplier tube (HAMAMATSU, Herrsching, Germany, Type R1527P select). The signal of the photomultiplier tube was processed by a Perkin Elmer/EG&G/ORTEC (Oak Ridge, TN, USA) single photon-counting device consisting of pre-amplifier VT120C, discriminator 935, and counter 994.

The ORTEC counter and the Lake Shore temperature controller were connected via an IEEE bus to a PC that uses a self-programmed Turbo Pascal software to read out and to control.

Measurements were conducted under oxygen at a flow rate of about 20 ml/min.

For the spectrally resolved CL experiments, a less sophisticated sample cell was connected to the photomultiplier via a monochromator, which was used in scanning mode to obtain the respective spectra at a certain duration of the CL experiment.[13] Using cubic spline interpolation, measured emissions at respective monochromator wavelengths at different durations of the CL experiment were recalculated for a whole respective spectrum of wavelengths to begin at a common starting point. These measurements were conducted under air at a low flow rate (the cell was not constructed gas-tight and the resulting air flow rate was extremely low).

RESULTS AND DISCUSSION

Principles of CL Detection

The link between the ongoing chemical oxidation reaction that leads to degradation under weathering exposure and observed CL emission is via the radical chain character of the autooxidation process, which can be regarded as the underlying mechanism in the case of saturated bonds.

The autooxidation cycle sketches the reaction of a polymer R–H with oxygen (*Figure* 1).

Polymer radicals R• generated by the decomposition of hydroperoxides readily react with oxygen to form alkylperoxy radicals. This step repeats quickly in the propagation cycle. Among several possible termination steps in which free radicals can react with each other with a certain probability after a certain number of cycles in the propagation step, the depicted termination step renders an electronically excited carbonyl group RO*.

This can relax to ground state by emission of a CL photon; in this way, CL becomes a cycle counter of the ongoing oxidation reaction. The ration of terminations to propagation cycles is specific to a polymer and to the reaction conditions (especially the temperature).

The basic relationship between observed CL emission and ongoing chemical reaction can be described by the following equation:

$$I_{CL} = G\Phi r$$

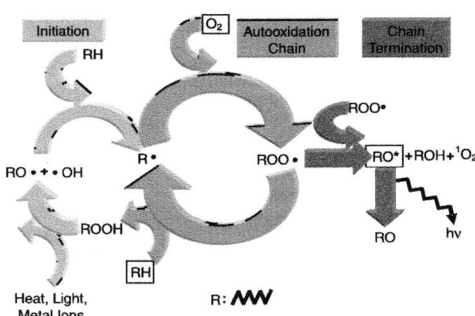

Figure 1—Autooxidation scheme for the oxidation of saturated polymers with oxygen.

where:

I_{CL} = observed CL emission
G = the instrument's detection efficiency for each photon emitted
Φ = the quantum efficiency of the process $(0 \leq \Phi \leq 1)$
r = the rate of the CL reaction

If not only the instrument's detection efficiency, G, is regarded as constant but also the quantum efficiency of the CL process, then the observed CL emission is proportional to the reaction rate. However, in terms of the quantum efficiency, this is only an approximation since the quantum efficiency is only accessible via arduous calibration measurements and its value is specific for a particular material. It can vary by orders of magnitude, from systems with very high quantum efficiencies occurring in bioluminescence near 1 down to the very low quantum efficiencies usually found for the oxidation of polymers of about 10^{-3} to 10^{-8}.

With the CL emission being proportional to the oxidation reaction rate at first approximation, the CL emission can be seen as a measure of the oxidability of a sample or as a measure of its residual stabilization. High emissions and short times to reach the emission maximum mean easy oxidability and low residual stabilization.

A critical parameter to be carefully chosen is the investigation temperature at which the CL experiment is run. This is shown in the example of three isotherm CL experiments using the same kind of LDPE samples (*Figure 2*). On one hand, increasing the CL investigation temperature results in increasing CL emissions and decreasing times to reach the maximum; both would be desirable in conducting the CL experiment. On the other hand, high temperatures usually are far beyond the application conditions to be described and therefore could trigger the activation of possibly "unnatural" reaction paths with potentially unrealistic effects.

While this relationship may be as simple for a homogeneous reaction, the reaction of a solid polymer with gaseous oxygen obviously cannot be regarded as a homogeneous reaction and actually is treated by more recent works[14-16] as much more heterogeneous than just in terms of the two phases in which the two reactants exist.

In the case of a solid polymer reacting with gaseous oxygen, all physical transport phenomena have to be taken into consideration. Depending upon the phase the oxidation reaction is in (induction period, autoacceleration, etc.), the partial oxygen pressure, the density of the polymer, etc., extremes of dominance of the whole process can occur between reaction control and diffusion control; whereas, in between these extreme situations both contributions will play a role.

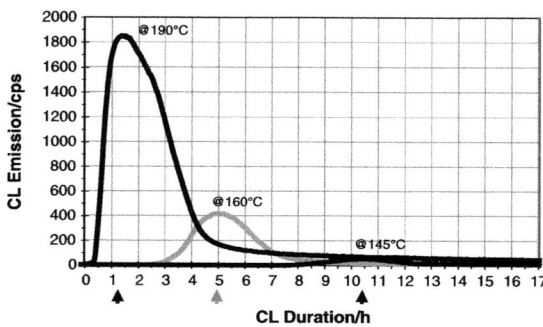

Figure 2—Principal course of CL curves for polyolefins using the example of an LDPE. A variation in the CL investigation temperature results in increasing CL emission and decreasing times to reach the maximum with increasing temperatures.

In the case of several components in the system, the CL emission can be written as follows:

$$I_{CL} = G \sum_{i=1}^{n} \Phi_i x_i r_i$$

The CL emission can be regarded as the weighted sum of the reaction rates of the species contributing to CL. The weighing factor is the product of the fraction x_i and the individual quantum efficiency Φ_i. As the quantum efficiency can vary by orders of magnitude, this can mean that a component with high quantum efficiency contributes higher to overall CL emission than other components even though its relative concentration might be much lower than that of the other components.

This consideration, however, would still mean that all components are independent species in infinitely dissolved solution, a prerequisite that definitely is not met in the case of several components densely packed in a solid. In this case, just as for any non-infinitely dissolved species, energy transfer can occur which will become a dominant pathway if the species to which the energy transfer occurred, B, has a higher quantum efficiency than the original species, A:

$$A^* + B \Leftrightarrow A + B^*$$

This again could mean that a component with low relative concentration might become dominant for the emission.

The resulting CL emission can be described as the sum over all oxidizable species in terms of the superposition of the effects of chemical reaction rate changes and physical transport properties. For this reason, at least for technical polymers, it is hardly possible to find an analytical description of a single species. Thus, it does not seem to be suitable for its emission curves to be used to calculate an absolute lifetime. However, as it is extremely sensitive, it is extremely useful for any mechanistic study of the results of all systematic parameter changes related to oxidation for all sorts of investigations into relative stabilities.

Apart from the chemical reaction in an indirect way, the physical processes can influence the observed CL emissions. This indirect influence mostly comes via transport processes of the reactant oxygen into the solid phase of the polymer. Gas permeability and diffusion, its change-over temperature and crosslinking or fragmentation of the polymeric matrix, as well as migration of stabilizer or crystallinity changes in the polymer in the course of aging, are to be considered.

WAYS TO ABSTRACT FURTHER INFORMATION FROM THE CL SIGNAL

Spectral Resolution

One possible way to extract individual contributions of the emitting species to the integral CL signal could be by means of wavelength dispersion of the CL signal and try-

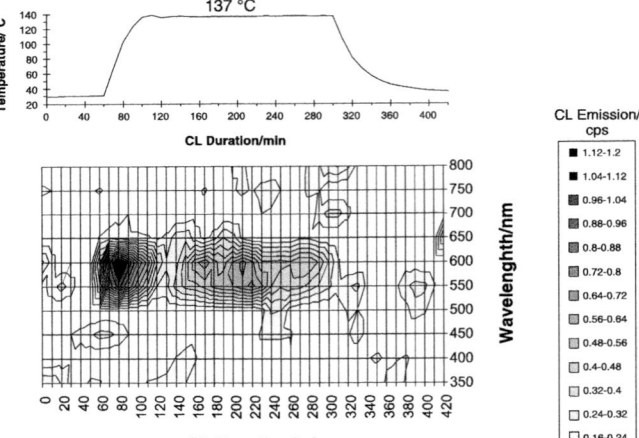

Figure 3—Spectrally resolved CL of the conjugated polymer poly(phenyl-p-phenylene vinylene) showing an about constant CL emission maximum. Heat-up and isotherm 137°C under air.

ing to deconvolute into known spectra of contributing components. However, in the case of condensed matter energy transfer, this means that emissions will be derived primarily from species with the highest quantum efficiency of emission.

This can be demonstrated for CL measurements on the oxidation of a conjugated polymer poly(phenyl-p-phenylene vinylene), P-PPV, oxidized under air[17] (*Figure* 3). Here, the unoxidized polymer has a particularly good quantum efficiency, which is related to its conjugation. In the oxidation process the conjugation is gradually lost and the emission efficiency of the oxidized polymer is decreased. While the oxidizing polymer actually generated the electronically excited species that gives rise to CL, it is the spectrum of the unoxidized polymer that is observed due to energy transfer to the better emitting species. A comparison of the obtained CL spectrum and a photoluminescence spectrum of unoxidized P-PPV shows good agreement. This means that even though in the course of the CL experiment the chemical environment of the emitting oxidized species changes, the spectrum stays constant at the same maximum wavelength. Also, the spectral halfwidth is much broader (about 200 nm for PE compared to about 50 nm for P-PPV) in this case, which could be explained by the large number of similar but spectrally not identical species that give rise to emission as opposed to the case of P-PPV, in which it is assumed that only the unoxidized species emits.

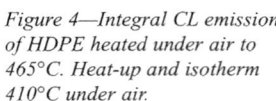

Figure 4—Integral CL emission of HDPE heated under air to 465°C. Heat-up and isotherm 410°C under air.

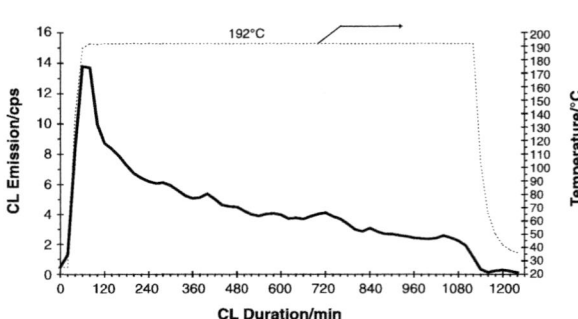

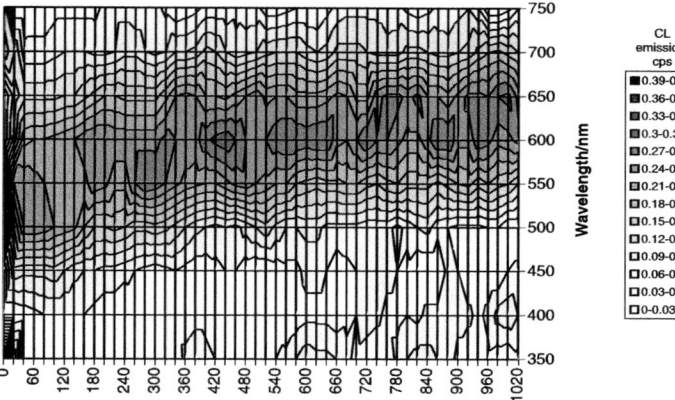

Figure 5—Spectrally resolved CL of HDPE heated under air to 465°C within first 60 min showing spectral red-shift of wavelength of maximum emission from initial 500 nm up to 620 nm at the end of CL investigation; for the temperature program used, see Figure 4.

A different situation exists for an intrinsically poorly emitting polymer like polyethylene. In this case, the carbonyl species formed in the oxidation process can be assumed to exhibit a better quantum efficiency than the unoxidized polymer (*Figures* 4–5). For this reason, a change of the chemical neighborhood (e.g., the degree of oxidation of the pending alky residues of the carbonyl group) means a shift in the emission maximum over the course of the CL experiment in this case.

The CL investigation temperature used in this example is very high and will lead to the destruction of the stabilizer system.

Imaging Chemiluminescence[18]

Through imaging onto an array of photon detectors (mostly CCD), heterogeneities within the oxidation of the sample surface can be investigated. Since this surface could be prepared as the cross-section of a material, this can be used to investigate oxidation profiles.[19]

Combination with Other Techniques

By combining the information from CL with the information gained from other techniques, an interpretation of CL measurements becomes possible. Interestingly, techniques such as oxidation uptake[20] and carbon dioxide evolution[21] have emerged only over the last couple of years, while the combination or coupling with established techniques such as IR[22] and DSC[23] has been in use for a longer time.

BASICS OF CL FOR SHORTENING WEATHERING TESTS

Acceleration of weathering tests can be achieved by an exaggeration of parameters far beyond field conditions (*Figure 6*).

However, this exaggeration of exposure parameters (in terms of high temperatures or shorter irradiation wavelengths than within global irradiation) means that reaction paths are activated that may be completely irrelevant for the exposure under field conditions

316

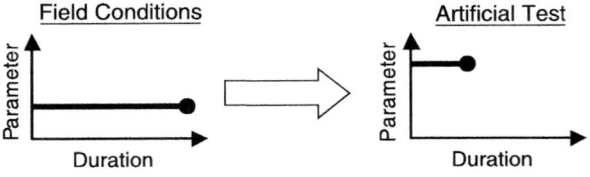

Figure 6—Acceleration of weathering tests by means of exaggeration of exposure parameters.

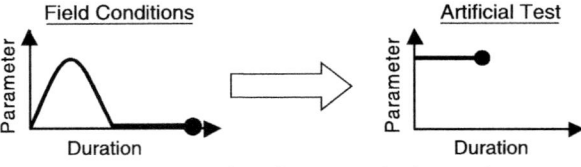

Figure 7—Acceleration of weathering tests by the time-lapse approach.

that was to be simulated. As a result, properties could be completely out of proportion on an absolute scale and the relative ranking of different systems might come out completely inversed.[24] Therefore, this kind of testing can be characterized as quick and useless. Throwing dice would be cheaper, quicker, and possibly even would bear higher probabilities for correct rankings than this exaggeration strategy.

The most common approach for achieving acceleration of weathering tests is the time-lapse strategy (*Figure 7*), in which more effective phases of exposure parameters are applied more frequently. For instance, instead of simulating the diurnal increase and decrease of UV irradiation in the fashion of global irradiation, the noon level is kept constant for 24 h. While there could be processes that need a dark period in general, this strategy maintains a correlation to field conditions. Similar approaches can be used for temperature or humidity, although for these, it might be most effective to frequently change from high to low values that could result in mechanical tensions in the material, instead of maintaining constant phases on high values.

A third approach for shortening weathering tests consists in a more sensitive method of detection (*Figure 8*). In many cases, weathering tests have to be carried out until different performance properties can be detected on the samples. Therefore, the sensitivity threshold in the determination of weathering effects can determine the duration of the test.

In a weathering test,[25] the stability of a relevant property under the action of a weathering exposure is examined. The aging process manifests itself in the change of certain properties of the material under investigation, which usually means a degradation of functional properties. Chemical reactions (most importantly, oxidation reactions) and physical processes (such as transport reactions) both contribute to the aging process. Typically, however, the chemical reactions are the dominant ones, since they tend to be irreversible. Initiated mostly by UV radiation, the degradation process usually starts with single molecules on the surface, and then spreads over the whole surface, finally reaching ever-deeper layers of the bulk of the material (*Figure 9*). If

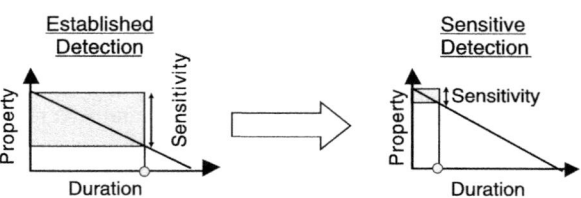

Figure 8—Acceleration of weathering tests by means of more sensitive detection.

the concentration of degradation products on the surface has accumulated beyond certain threshold values, even macroscopical physical properties of the material may change.

It is not sufficient to optimize the initial performance of a property, as the sensitivity to degradation under a weathering exposure changes from one material to the next. This manifests itself in different degradation rates. Initial performance rankings may change completely during a weathering exposure (*Figure* 10).

This is why weathering tests are necessary. If there were a good correlation between the property value after weathering to the value before weathering, it would be possible to predict the final value from the initial one.

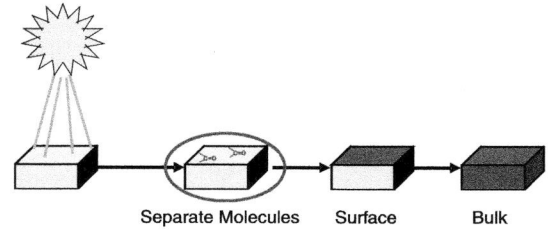

Duration of Weathering

Figure 9—Schematic illustration of the weathering process starting at separate molecules, spreading over the surface, and finally reaching ever deeper layers of the bulk of the material.

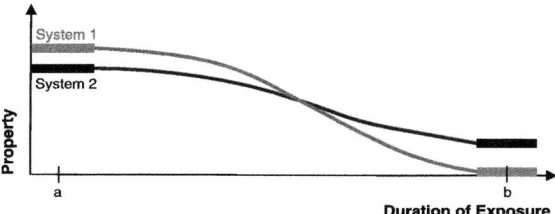

Figure 10—Different degradation rate behavior of two materials resulting in final values of a property at duration B that are not correlated to the initial values before weathering A.

The weathering test is only as good as the detection of its effects on the samples. While very few techniques allow for the detection of weathering effects as degradation rates (a rare example is chemiluminescence), the next closest to detecting the degradation rate itself is the detection of degradation products on the surface by techniques like ATR infrared spectroscopy. These products are the integrated rate curves, i.e. concentration curves (*Figure* 11).

Many established detection techniques are not able to distinguish the degradation products themselves but detect a physical property (such as gloss or mechanical strength), which needs degradation products to be accumulated until this physical property is affected. A transfer function mediates the change in concentration as a change of a physical property (*Figure* 12). This extra transfer function implies a loss in sensitivity and selectivity.

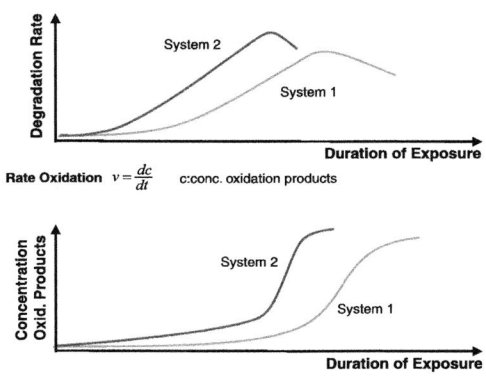

Rate Oxidation $v = \frac{dc}{dt}$ c:conc. oxidation products

Concentration Oxidation Products $c = c_0 + \int v \, dt$

Figure 11—Relationship between degradation rate and oxidation products.

318

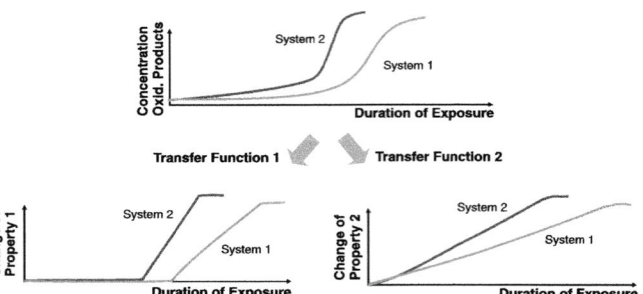

Figure 12—Different behavior of two different properties as a function of exposure duration of same respective materials.

Since each individual property can have its own specific sensitivity to weathering exposure, each property needs its own specific weathering stability test (*Figure* 12). Both the onset of degradation and the condition (gradual or abrupt) of the property decline can be different.

Demonstrating that the same samples exposed to the same weathering can show quite different results of weathering effects in terms of different properties, *Figure* 13 shows that not only the course of the degradation but also the ranking of weathering effects can be a function of the property used.

Prerequisites for the Correlation of Results Out of Accelerated Tests and the Field

The parameters of an artificial test are chosen in order to optimize the correlation of exposure results to those of an outdoor test. This correlation can achieve satisfactory results only if a number of conditions are met.

For any acceleration, the results obtained for the shorter time of the artificial test must balance with the results of the real-time outdoor test. It must be demonstrated that the rankings of property changes at the earlier detection time correlate with the rankings found for the later time of the outdoor test.

Only for a group of materials that shows a homogeneous course of degradation (*Figure* 14), with no overlap of degradation curves versus time, can it be expected that ranking for an earlier detection will be the same as for the later time at which established

Figure 13—Experimentally derived example of aging behavior of the same set of materials expressed differently in the properties of gloss and yellowing: two coating systems being artificially weathered. In terms of gloss (a), system 1 shows fewer effects than system 2, while in terms of yellowing (b) the sequence changes and system 2 shows less yellowing than system 1.

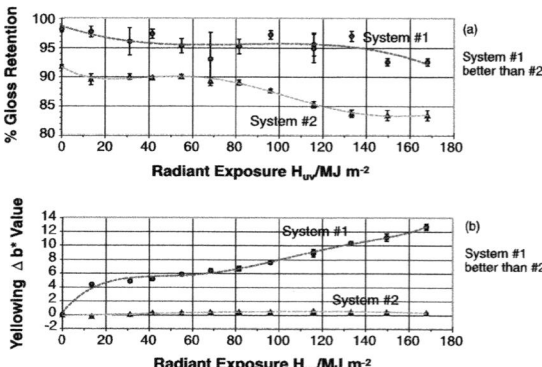

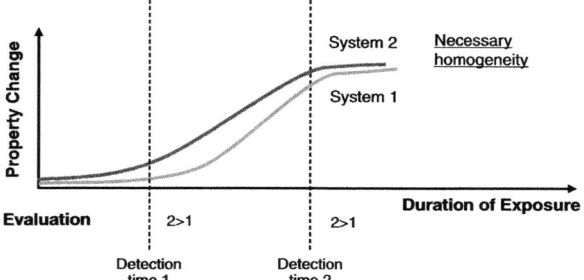

Figure 14—Group of materials showing homogeneous course of degradation leading to consistent ranking between earlier detection time 1 of the accelerated test and detection time 2 of the real time outdoor test.

detection was able to distinguish differences. For a group that shows inhomogeneous degradation behavior (*Figure* 15), with resulting overlap of degradation curves, consistent ranking is not possible. This behavior is specific for both the materials and the exposure. While a more homogenous degradation situation is likely to occur for good or bad performances, the inhomogeneous situation is very likely during the middle part of performances.

This necessary scaling is further complicated if different properties are detected in the artificial and the outdoor test. In this case, the homogeneous degradation behavior of materials as well as homogenous behavior of the two properties is required.

Scaling of the complete exposure can be regarded as the result of the superposition of individual exposures (irradiation, temperature, humidity, chemicals, etc.). Even if the correlation for one of these individual parameters is known, there still is no analytical interpretation of the combined effects of all these parameters. This analytical description is complicated by the fact that the mutual interaction of these parameters (synergistic and antagonistic effects) is very strong. It is not possible therefore to describe the combined effect simply as an addition of individual independent exposure terms and add up the interaction terms.

The most important of these exposure parameters is UV irradiance. It currently is the subject of an investigation, in which ranges of irradiance on the photodegradation effect depend only on the radiant exposure. The so-called reciprocity law[26] holds true if different ways leading up to the same integral of irradiance over time also cause the same exposure effect (*Figure* 16).

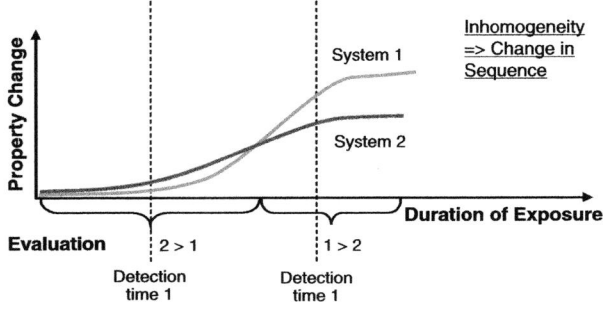

Figure 15—Group of materials showing inhomogeneous course of degradation leading to different rankings at earlier and later points of degradation.

320

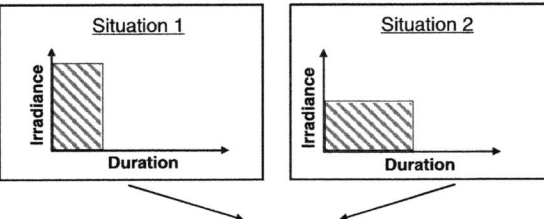

Reciprocity: Same integral of irradiance = Same weathering effect

Figure 16—Validity of the reciprocity law between radiant exposure and photodegradation effects.

While reciprocity can be expected to hold in many cases and artificial testing and outdoor exposure, hence, could be directly related in their effects, the additional effects of the other parameters still cannot be calculated. Therefore, it cannot be predicted what the additional effect of, for instance, a temperature increase of 5°C might cause. This would be the case even if the exact exponential relationship between temperature and degradation effect were known. However, this knowledge only relates to the isolated effect of temperature on the weathering effect for a constant set of all other parameters. It is hoped that the developing numerical environmental simulation will be able to fill this gap in the not too distant future.

Examples of CL Detection for Shortening Weathering Tests

For a CL investigation of weathering effects, weathering exposure is used to cause a certain oxidative stage in the material. The sample then is removed from the weathering exposure and is subsequently investigated in a CL experiment, which can be regarded as a titration of the residual stabilization in the material (*Figure* 17). If this procedure is repeated for all other samples, the CL allows a ranking to be done in terms of stabilization left after the weathering exposure up to this stage. The results of different samples at this stage can be compared to the results at a later weathering stage as the ranking might change in the course of weathering.

COMPARISONS OF CHARACTERISTIC DURATIONS—In the first example presented, the CL ranking is carried out on the basis of characteristic points in the CL over the duration of the experiment. Characteristic points are, for instance, the duration of the induction time or the duration to reach the respective emission maximum. The advantage in this approach is that it avoids having to account for different quantum efficiencies, as would be necessary if absolute CL emissions were compared. The disadvantage is that, in order to be realistically stabilized, the induction times and times to reach the CL maximum are still very long even for exposed samples, and

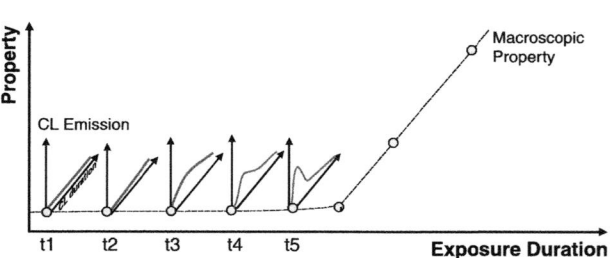

Figure 17—Principal course of CL investigations of increasing degradation during a weathering exposure test. CL allows probing the residual stabilization of the sample after a particular exposure duration like a titration.

temperatures have to be elevated considerably over field conditions. As pointed out previously, this carries the danger of loose correlation of the ranking under the actual application conditions, especially as modern UV stabilizers quickly degrade at temperatures exceeding about 100°C.

The first example for model LDPE agricultural films stabilized with a low concentration of HAS of 0.2% (*Figure* 18) shows the systematic influence of a variation in the duration of a weathering exposure from unexposed to 42 days artificial ADF test on the subsequent CL emission curves.

The example shows a systematic decrease in the duration to reach the CL maximum with increasing weathering exposure. In absolute CL emissions there also is a systematic increase in CL emissions with increasing exposure duration up to 21 days, but there is a decrease in emissions for 42 days again, which is thought to be caused by a change in quantum efficiency due to the weathering exposure-caused degradation process.

The influence of an increased stabilizer concentration on the resulting CL emissions curves is demonstrated when compared to the next example (*Figure* 19), in which an agricultural film that was stabilized at about six-fold higher HAS concentration was studied. Here, the time to reach the CL emission maximum for the unexposed sample has increased to 85 h of the CL experiment compared to 12 h in the case of the 0.2% stabilized film discussed previously. The evaluation in terms of induction times should lead to shorter CL investigation durations, but in the case of PE samples the transition at the end of the induction time is very gradual, making the determination of the induction time difficult.

In this example, the spread differentiation of different exposure durations in the respective CL curves is much better than for the lower stabilized example shown in *Figure* 18. Furthermore, the difference in absolute CL emissions from an almost constant value between unexposed to 21 days exposure up to the much higher value after 42 days exposure suggests that after 21 days exposure the effective concentration of stabilizer left in the polymer is no longer sufficient for the autooxidation to enter the accelerated stage. The lower CL emissions for the field-exposed sample could be a result of the CL quenching chemicals used as pesticides and fungicides within the greenhouse.

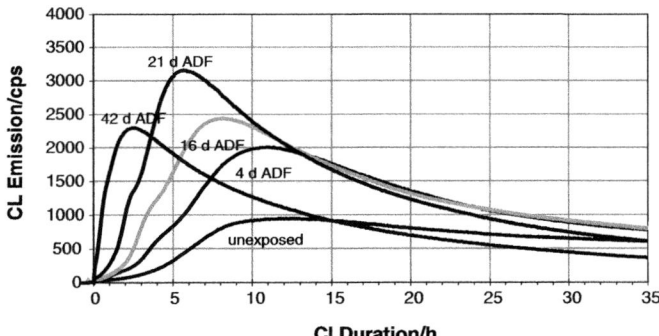

Figure 18—CL investigation of the influence of artificial weathering duration (BAM developed Acid Dew and Fog test ADF-J 1,5) for a model LDPE stabilized with 0.2% Hostavin N30 HALS under air, isotherm at 172°C.

322

Figure 19—CL investiga-
tion of the effect of artificial
weathering (BAM devel-
oped Acid Dew and Fog
test, ADF_J 1,5) on the
oxidative stability of a commercially available LDPE-
EVA film containing 1.2%
Hostavin N30 HALS com-
pared to the CL curve of a
sample of the same material
that failed after 20 months
outdoor exposure under
field condition. CL isotherm
at 172°C under air.

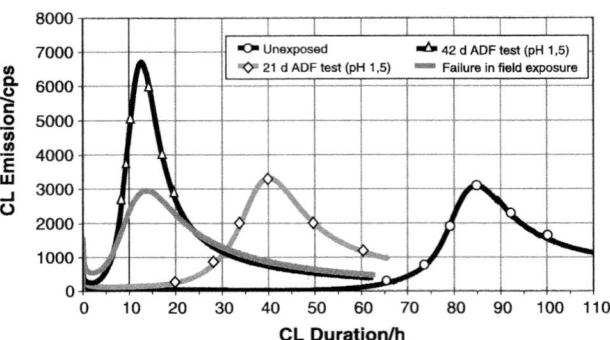

This example also demonstrates that the duration of a CL experiment can be long if the amount of stabilizer is high (85 h for the time to reach the maximum of the unexposed sample). In this particular case, acceleration could have been achieved by using pure oxygen instead of air as the oxidant.

The CL evaluation allowed a ranking of stabilizer systems after 21 to 42 days of weathering exposure, while evaluation by means of tension tests needed at least 98 days of artificial weathering.

It should be mentioned, though, that the CL investigation temperature used in this example is very high, primarily to allow a shortening of induction times despite full concentrations of stabilizers (designed to allow several years of use under outdoor conditions). This means that the stabilizer system will break down quickly and, under these conditions, the CL predominantly investigates the oxidation behavior of the polymer itself. The conclusion about the efficiency of the stabilizer is an indirect one: the amount of polymer that still can be oxidized at this stage in the weathering exposure experiment depends on the exposure history and the effectiveness of the stabilizer in this time.

COMPARISONS ON THE BASIC OF ABSOLUTE CL EMISSIONS—To avoid the potential negative effect that can result from the use of high investigation temperatures for the CL experiment in terms of the correlation to actual service conditions, and still allowing for acceptable durations for the CL experiment, in the following example another approach is followed. Here, the high sensitivity of CL is used to make comparisons during the induction time of the material. Furthermore, to compare different chemistries, a normalization of exposed CL emissions to unexposed CL emissions of respective materials is used.

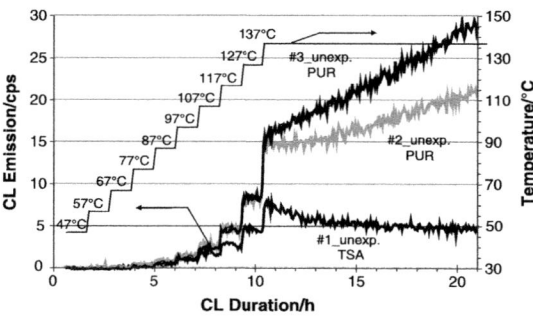

Figure 20—CL emission of unexposed coating systems (#1: TSA; #2: PUR, #3: PUR) showing differentiation between TSA and PURs due to different quantum efficiencies.

In the example demonstrated, three different automotive coat-

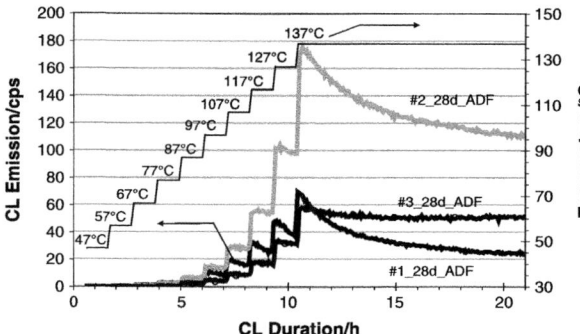

Figure 21—CL emission of automotive coatings that had been exposed for 28 days in the artificial ADF weathering test. For a comparison in terms of stability, the respective quantum efficiencies have to be considered. Mind the changed scaling of Y-axis in comparison to unexposed systems (#1: TSA; #2, #3: PUR).

ing systems were evaluated using CL after outdoor exposure for 98 days in Jacksonville, FL, in comparison to 28 days of exposure to the artificial ADF weathering test. To rank the three coatings, an attempt was made to use absolute CL emission before exceeding the induction time. In this case, the respective unexposed systems were standardized at a temperature of 137°C to account for the different quantum efficiencies of different materials, as described earlier.[8,27]

While a complete coating system was studied, the CL evaluation was predominantly sensitive to the uppermost clearcoat layer. This mostly is due to absorption of emission in deeper layers of the material and could be demonstrated by a comparison of the CL of the whole coating system to the CL of free films of the clearcoat.

Figure 20 shows the result of the CL investigation on the three unexposed coating systems. The two polyurethane systems not only show higher CL emission on all temperature levels but also show a different behavior within the highest temperature level of 137°C. The decrease with increasing duration on this temperature level for the thermosetting acrylic #3 contrasts to the increase for the two polyurethanes, which is likely due to different oxygen diffusion behavior.

The CL evaluation of artificially exposed systems (*Figure* 21) shows emissions that are about six-fold higher than for the unexposed systems. For a comparison in terms of stability, however, the different quantum efficiencies have to be considered.

Normalization of the CL emission maximum at 137°C of the respective unexposed system results in a change of rankings for system #3 (*Figure* 22).

To allow a better comparison of quantum efficiency corrected CL emissions, the respective initial CL emissions upon reaching a new

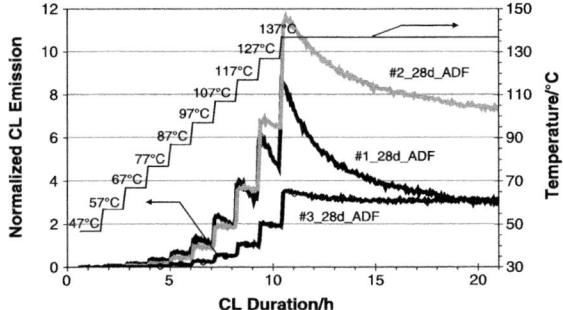

Figure 22—CL emission that was corrected to different quantum efficiency by normalizing to emission of respective unexposed systems at 410 K. (#1: TSA; #2, #3: PUR).

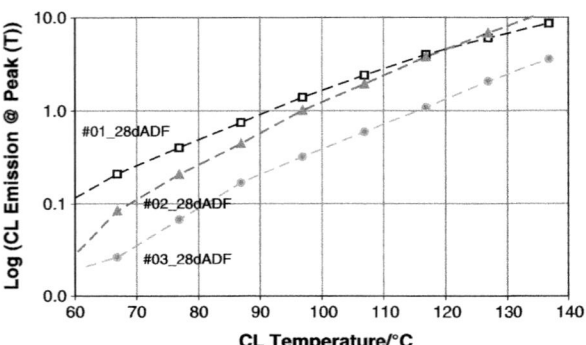

Figure 23—Normalized CL emissions after 28 days of the artificial ADF weathering test as in Figure 22, but logarithmic CL emission of peak value at each respective temperature step is shown for representational reasons (#1: TSA; # 2,#3: PUR).

temperature level are plotted versus this temperature (*Figure* 23). The evaluation shows that for a temperature range between 60°C and 115°C, a stability sequence #3>#2>#1 is valid, whereas for temperatures between 1150°C and 140°C, the stability ranking is #3>#1>#2. As temperatures on real car surfaces can reach up to 90°C, the region up to this temperature and below it are considered most relevant.

The same procedure was used to obtain the representation shown in *Figure* 24 for the outdoor exposure of the same coating systems for 98 days in summer in Jacksonville, FL.

In comparison to the results of the CL evaluation of the artificial ADF weathering test (*Figure* 23), the CL evaluation of the outdoor exposure in Jacksonville (*Figure* 24) shows good agreement. Two differences should be mentioned: first, the CL emissions of all three coatings are lower than for the case of artificial weathering. This could either mean that the artificial exposure was more degrading or that quenching from components that only were present in outdoor exposure occurred. It seems to confirm the former interpretation that the spread of CL emissions between the three coatings also is lower for the outdoor exposure compared to the artificial exposure. The second difference is the additional overlap of the curves for system #2 and system #3 for temperatures below 75°C, which most probably is caused by the lower spreading of the three curves.

More comprehensive studies of the weathering effects of automotive coatings using the CL evaluation[28] showed that the duration of the artificial weathering could be reduced by at least a factor of two compared to macroscopic evaluation techniques like visual

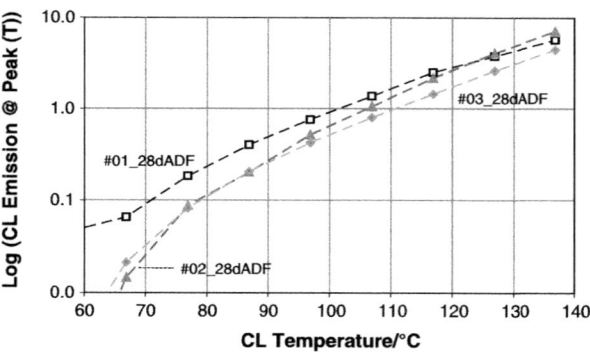

Figure 24—CL evaluation of three automotive coatings that were exposed for 98 days in outdoor exposure in the summer of 2002 in Jacksonville, FL. (#1: TSA, # 2,#3: PUR). The analogue normalization procedure was as described for artificial weathering.

assessment or haze and gloss measurements. This implies however that the demands to rank correlation are reduced to screening group correlations of good, medium, and low performances.

SUMMARY

Chemiluminescence's high sensitivity and close link to oxidation as a driving force of the degradation process make it a high priority candidate for use as a more sensitive detector to allow for the shortening of weathering tests. Its emission is a measure of the oxidability of a material, but also is influenced by the transport properties of the oxidation process. However, as its response is a weighted sum of the emissions of the individual components in the system with unknown values of the respective weights, its interpretation for a technical system is the major drawback. These difficulties only can be overcome by the parallel use of complementary techniques, such as oxygen uptake or IR measurements.

Regarding information content of CL experiments, it was found that, without an analytical physical-chemical model, CL cannot estimate absolute service lifetimes. It is useful, though, for characterizing a momentary oxidation or stabilization state relative to other samples. CL carried out after a weathering exposure probes the current stabilization. CL detection is not suited for mass throughput of samples as the CL experiment takes hours.

A more general question that not only applies to CL-enabled accelerated weathering tests, but all accelerated tests, is that of the scalability of early weathering stages to the stages to be described under actual field conditions. In first approximation, it seems that this scalability only can be assumed for systems that show very good or very low performance, while for the large group of systems that show mid-range performance, frequent ranking inversions are to be expected. Also, the CL signal must be related to the macroscopic property that usually is the primary focus. For a set of samples to be compared at a certain stage of aging, homogeneous ageng behavior in the group is necessary or otherwise ranking inversions will occur in the course of the exposure.

References

(1) Bolland, J.L. and Gee, G., "Kinetic Studies in the Chemistry of Rubber and Related Materials. 2. The Kinetics of Oxidation of Unconjugated Olefins," *Trans. Faraday Society*, 42, 3-4, pp. 236-243 (1946).

(2) Ashby, G.E., "Oxyluminescence from Polypropylene," *J. Polym. Sci.*, 50, pp. 99-106 (1961).

(3) Schard, M.P. and Russell, C.A., "Oxyluminescence of Polymers. I. General Behavior of Polymers," *J. Applied Polym. Sci.*, 8, 2, pp. 985–995 (1964).

(4) Billingham, N.C., O'Keefe, E.S., and Then, E.T.H., "Chemiluminescence from Oxidation of Polypropylene," *Polym. Mater. Sci. Eng.*, 58, S.431-435 (1988).

(5) Wolf, C.J., Fanter, D.L., and Grayson, M.A., "Chemoluminescence of Thermosetting Resins," In *Chemorheology of Thermsetting Polymers*, May, C.A. (Ed.), ACS Symposium Series 227, American Chemical Society, Washington, D.C., pp. 121-139, 1983.

(6) George, G.A., "Use of Chemiluminescence to Study the Kinetics of Oxidation of Solid Polymers," In *Developments in Polymer Degradation*, vol. 3, Grassie N. (Ed.), London, Applied Science Publishers, pp. 173-206, 1981.

(7) Dudler, V., Bolle, Th., and Rytz, G., "Use of Chemiluminescence to the Study of Photostability of Automotive Coatings," *Polym. Degrad. Stab.*, 60, pp. 351-365 (1988).

(8) Wachtendorf, V., and Schulz, U., "Shortening of Weathering Tests for Automotive Coatings by Chemiluminescence Detection," *Proc. XXVII FATIPEC Congress,* Aix-en-Provence, France, pp. 931-940, April 19-21, 2004.

(9) Fratricova, M., Simon, P., Schwarzer, P., and Wilde, H.-W., "Residual Stability of Polyurethane Automotive Coatings Measured by Chemiluminescence and Equivalence of Xenotest and Solisi Ageing Tests," *Polym. Degrad. Stab.,* 91, pp. 94-100 (2006).

(10) Schulz, U., Trubiroha, P., Schernau, U., and Baumgart, H., "The Effects of Acid Rain on the Appearance of Automotive Paint Systems Studied Outdoors and in a New Artificial Weathering Test," *Prog. Org. Coat.,* 40, pp. 151–165 (2000).

(11) Verband Deutsche Ingenieure, guideline VDI 3958, part 12, "Environmental simulation – Effects of Acid Precipitation on Polymers. Test Methods," (German/English). VDI/DIN-Handbuch Reinhaltung der Luft, Band 1a, 2004.

(12) Wachtendorf, V., Jansen, K., Schulz, U., and Tjandraatmadja, G., "Combining Artificial Weathering with Chemiluminescence for Lifetime Predictions of Polymeric Materials," In *Proc. 9th Intern. Conf. on Durability of Building Materials and Components,* Brisbane, Australia, 17–21, 2002.

(13) Wachendorf, V., "Untersuchung Thermooxidativer Veränderungen an Polymeren durch Chemilumineszenz," dissertation, published by Dr. Köster, Berlin, pp. 40-43, 1997.

(14) Billingham, N.C. and George, G.A., "Chemiluminescence from Oxidation of Polypropylene: Some Comments on a Kinetic Approach," *J. Polym. Sci., Part B, Polym. Phys.,* 28, pp. 257-265 (1990).

(15) Celina, M. and George, G.A., "A Heterogeneous Model for the Thermal Oxidation of Solid Polypropylene from Chemiluminescence Analysis," *Polym. Degrad. Stab.,* 40, 3, pp. 323-335 (1993).

(16) Celina M., George G.A., and Billingham, N.C., "Physical Spreading of Oxidation in Solid Polypropylene as Studied by Chemiluminescence," *Polym. Degrad. Stab.,* 42, pp. 335 (1993).

(17) Wachendorf, V., "Untersuchung Thermooxidativer Veränderungen an Polymeren durch Chemilumineszenz," dissertation, published by Dr. Köster, Berlin, pp. 87-88, 1997.

(18) Ahlblad, G., Reitberger, T., Terselius, B., and Stenberg, B., "Imaging Chemiluminescence Technique Applied to Thermo-Oxidation of Polymers, Possibilities and Limitations," *Angew. Macromol. Chem.,* 262, pp.1-7 (1988).

(19) Gustav A., Reitberger, T., Terselius, B., and Stenberg, B., 'Thermal Oxidation of Hydroxyl-Terminated Polybutadiene Rubber II. Oxidation Depth Profiles Studied by Imaging Chemiluminescence," *Polym. Degrad. Stab.,* 65, pp. 185-191 (1999).

(20) Gijsman, P. and Hamskog, M., 'Simultaneous Oxygen Uptake and Imaging Chemiluminescence Measurements," *Polym. Degrad. Stab.,* 91, pp. 423-428 (2006).

(21) Jin, C.Q., Christensen, P.A., Egerton, T.A., Lawson, E.J., and White, J.R., "Rapid Measurement of Polymer Photo-Degradation by FTIR Spectrometry of Evolved Carbon Dioxide," *Polym. Degrad. Stab.,* 91, 5, pp. 1086-1096 (2006).

(22) Blakey I. and George G.A, "Simultaneous FTIR Emission Spectroscopy and Chemiluminescence of Oxidizing Polypropylene: Evidence for Alternate Chemiluminescence Mechanisms," *Macromolecules,* 34, pp. 1873 (2001).

(23) Fearon, P.K., Bigger, S.W., and Billingham, N.C., "DSC Combined with Chemiluminescence for Studying Polymer Oxidation," *J. Therm. Analy. Calorimetry,* 76, 1, pp. 75–83 (2004).

(24) Pilcher, G.R., Van de Streek, G.N., Chess, J.A., and Cocuzzi, D.A., "Accelerated Weathering: Science, Pseudo-Science or Superstition?," in *Service Life Prediction of Organic Coatings,* ACS Symposium Series 722, American Chemical Society, Washington, D.C., pp. 130-148, 1999.

(25) Wypych, G., *Handbook of Material Weathering,* 3rd ed., ChemTec Publishing, Toronto, 2003.

(26) Martin, J.W., Chin, J.W., and Nguyen, T., "Reciprocity Law Experiments in Polymeric Photodegradation: A Critical Review," *Prog. Org. Coat.,* 47, pp. 292–311 (2003).

(27) Wachtendorf, V. and Geburtig, A., "Using Chemiluminescence as Sensitive Detector of Early Stages Weathering Effects on Car Coatings to Shorten Weathering Tests," *Proc. 2nd European Weathering Symposium* (ISBN 3-9808382-9-3), Gothenburg, Sweden, June 2005, p. 63-74. Re-print: in *Galvanotechnik,* 96, 10, pp. 2470-2474 (2005).

(28) Krüger, S., Wachtendorf, V., Rauth, W. Klimmasch, T., and Krüger, P., "Screening the Weathering Stability of Automotive Coatings by Chemiluminescence," *Proc. XXVIIIth FATIPEC Congress,* ISBN: 963-9319-55-4, Budapest, June 12-14, 2006.

Chapter 22

Enhancement of Photoprotection and Mechanical Properties of Polymers by Deposition of Thin Coatings

Agnès Rivaton,[1] Jean-Luc Gardette,[1] Sandrine Morlat-Therias,[1] Bénédicte Mailhot,[1] Eric Tomasella,[2] Oscar Awitor,[3] Kyriakos Komvopoulos,[4] and Paola Fabbri[5]

[1] Laboratoire de Photochimie Moléculaire et Macromoléculaire, UMR CNRS 6505, Université Blaise Pascal (Clermont-Ferrand), 63177 Aubière Cedex, France

[2] Laboratoire des Matériaux Inorganiques, UMR CNRS 6002, Université Blaise Pascal (Clermont-Ferrand), 63177 Aubière Cedex, France

[3] Laboratoire LASMEA, UMR 6602, Université Blaise Pascal (Clermont-Ferrand), 63177 Aubière Cedex, France

[4] Department of Mechanical Engineering, University of California, Berkeley, CA 94720

[5] Dipartimento di Ingegneria dei Materiali e dell'Ambiente Universita di Modena e Reggio Emilia-Sede di Modena Via Vignolese 905/a 41100 Modena (Italia)

INTRODUCTION

Exposing polymers to environmental atmosphere changes their external appearance and their properties and modifies their surface. Protection from photoaging of polymers with aromatic structures such as bisphenol-A polycarbonate (PC), poly(ethylene terephthalate) (PET), or poly(ethylene naphthalate) (PEN) is a difficult challenge as their monomer units strongly absorb UV light. UV radiations induce chemical reactions such as the rupture of covalent bonds initiating photolytic (without intervention of oxygen) and photooxidation (fixation of oxygen) reactions. This causes, in turn, breakdown of the materials, with yellowing and embrittlement as major consequences.

To prevent these damages, a very effective method to protect the polymer is the deposition of ceramic coatings that are transparent in the visible light range. A ceramic coating physically screens the incident radiation, which reduces the undesirable effects of photochemical processes. In addition, the photooxidation rate decreases because the coating acts as an oxygen barrier. Low temperature crystalline deposition is of particular interest for polymers that cannot tolerate high temperatures.

The efficiency of photoprotection by coatings, associated with an enhancement of the surface properties of the coated polymer, will be illustrated by three examples described in this chapter. The first two examples include a ceramic coating which enhances the nanomechanical properties of PC, and a self-cleaning ceramic coating deposited on PET. A third example will show the potentiality of coatings obtained by sol-gel process and deposited on poly(ethylene oxide) (PEO).

$$\text{--}\bigcirc\text{--O--}\underset{\underset{O}{\|}}{C}\text{--O--}\bigcirc\text{--}\underset{\underset{CH_3}{|}}{\overset{\overset{CH_3}{|}}{C}}\text{--} \qquad \text{--}\bigcirc\text{--}\underset{\underset{O}{\|}}{C}\text{--O--}CH_2\text{--}CH_2\text{--O--}\underset{\underset{O}{\|}}{C}\text{--} \qquad \text{--}CH_2\text{--}CH_2\text{--O--}$$

PC   PET   PEO

The effectiveness of ceramic coatings has been demonstrated in recent studies dealing with various polymers coated with zinc oxide (ZnO) and titanium dioxide coatings.[1-3] Even if the efficiency of the ZnO layer is incontestable, one of the limitations of the use of ZnO is its photocatalytic activity[4] that leads to photocatalytic oxidation of the polymer at the interface. It is therefore advantageous to first coat the polymer with a photocatalyt-ically inactive ceramic layer, such as aluminum oxide (Al_2O_3). However, because Al_2O_3 does not provide a screening effect on damaging UV radiation, it is necessary to synthe-size a two-layer coating consisting of an Al_2O_3 underlayer and a ZnO top layer.[5]

Enhancing the surface nanomechanical properties of the coating could increase the range of engineering applications of the layered ceramic/polymer systems. Al_2O_3 pos-sesses better mechanical properties compared to ZnO, with lower scratch and wear sen-sitivities. To improve the quality of the surface, Al_2O_3 has also been deposited as an upper layer, sandwiching the ZnO layer. Thus, PC/Al_2O_3-ZnO-Al_2O_3 assemblies were success-fully obtained.[5]

Self-cleaning property is required to have surfaces without alteration of their superfi-cial properties in outdoor exposure. Reactive radio-frequency magnetron sputtering using a titanium target in an argon-oxygen mixture is a very valuable process to get crystalline anatase on unheated polymer. On one hand, our results indicate that TiO_2 has a very good photocatalytic activity as shown by its ability to photodegrade Rhodamine B dye solu-tion. On the other hand, our results indicate that transparent, thin bi-layered ceramic coat-ings Al_2O_3/TiO_2 deposited on PET films efficiently protect the polymer from photodegra-dation. The combination of these two properties allows for photostable and transparent polymer/ceramic layered media with superficial self-cleaning property.[6]

The use of the above mentioned technique is limited by high costs, small item dimen-sions, and simple geometrical forms requirements. An interesting alternative that permits optimizing the surface protection and the mechanical properties of the polymer is to deposit nanostructured organic-inorganic hybrid coatings (ceramers). These hybrid mate-rials can be prepared by sol-gel process, which allows the incorporation of organic struc-tures in a three-dimensional inorganic network. While transparency and clarity are main-tained, our results show that enhanced resistance can be obtained towards different envi-ronmental factors, such as scratching and abrasion.[7] The possibility of applying this kind of protective coating even on objects of complex shape could draw wide attention for several industrial applications.

ENHANCEMENT OF PHOTOPROTECTION AND NANOMECHANICAL PROPERTIES OF PC

Experimental

SPECIMENS: PC films (50 µm-thick) were supplied by Technifilm (ref. Makrofol D.E. 6-2) or Goodfellow. They are amorphous, transparent, and antioxidant free. They were ultrasonically cleaned in ethanol. Before sputtering they were treated by CO_2 cold plasma in the sputtering chamber. The PC surface energy was deduced from the study of the wettability using the two-liquid method.[8] The chemical bonds at the PC surface were analyzed by XPS using SIA 200 RIBER CAMECA UHV with non-monochromated Mg Kα source.

DEPOSITION AND CHARACTERIZATION OF ZNO AND AL$_2$O$_3$ THIN FILMS: The deposition of ZnO and Al$_2$O$_3$ thin films was carried out in an Alcatel SCM 450 sputtering unit equipped with 13.56 MHz radio-frequency generator. ZnO and Al$_2$O$_3$ targets (purity: 99.9%; diameter: 100 mm) were fixed on cooled magnetron cathodes. The substrate was situated at 90 mm from the target and the sputtering chamber was evacuated below 10^{-4} Pa pressure before admitting gases at low pressure for plasma treatments or deposition. The thickness of the sputtered coatings was measured by ellipsometry and interferometry. The structure and the microstructure of the deposits were examined by X-ray diffraction (XRD) and scanning electron microscopy, respectively. The film composition was analyzed by Rutherford Backscattering Spectroscopy (RBS). Stresses were evaluated by the bending beam method.[9] The optical properties of the thin films were characterized by ellipsometry, interferometry, and study of the UV-visible spectra by the Swanepoel method[10] in accordance with their thicknesses.

The properties of ZnO deposits depend on sputtering conditions and on the thickness of the layer.[3,11] The deposition conditions which give the best results for the photoprotection are: argon (95%)-oxygen (5%) plasma, 1 Pa pressure, and 0.89 W/cm^2 power density. Films crystallize in the hexagonal würtzite phase with a preferential orientation. Their microstructure is columnar and compact. The deposits grow with c-axis and columns perpendicular to the substrate surface. Their compactness increases with thickness and also with the deposition power. Their density is about 4.6 g/cm^3 for thicknesses in the 50–100 nm range. The zinc oxide deposits have an oxygen excess: the O/Zn atomic ratio is about 1.15. High compressive stresses (about 1 GPa) are present.

The Al$_2$O$_3$ thin films were deposited in conditions which give a dense microstructure and a low stress level, i.e., pure argon plasma, 1 Pa pressure, and 1.27 W/cm^2 power density.[12] The deposits are amorphous and consist of columns with domed tops. These alumina layers contain a light oxygen excess (the O/Al atomic ratio is about 1.57) and a low argon concentration (about 0.6 atomic %). Compressive stresses are also present in these deposits but their level is lower than in ZnO deposits (about 140 MPa).

IRRADIATIONS AND SPECTROSCOPIC STUDIES: Irradiations were carried out in a SEPAP 12.24 unit at a temperature of 60°C with medium-pressure mercury lamps. This medium-accelerated photoaging device has been described previously.[13] The evolutions of UV-Vis and infrared spectra were recorded respectively on a Shimadzu UV-2101PC equipped with an integrating sphere and on a Nicolet Magna-IR 760 FTIR spectrophotometer.

Because the light absorption by PC extends up to 330 nm, this polymer absorbs the UV-light present in the terrestrial solar radiation, which produces photochemical degradation of the polymer. Photodegradation of PC results in significant changes in the UV-visible and IR spectra.[13–15] The rates of degradation can be characterized by the measurement of the concentration of the stable photo-products that are produced during irradiation, indicated by the changes in the UV-Vis and IR spectra.

• The photo-yellowing of PC can be evaluated by measuring the increase of absorbance at 400 nm. The discoloration is a consequence of both photolysis (photo-Fries) and photooxidation.

• Measuring the increase of absorbance in the hydroxyl region of the infrared spectrum (3470 cm^{-1}) can be used to characterize accurately the progress of the photooxidation reactions in PC.

NANOMECHANICAL TESTING: The nanomechanical behavior of both coated and uncoated PC was evaluated with a surface force apparatus consisting of an atomic force microscope (Nanoscope II, Digital Instruments) retrofitted with a force transducer (Triboscope, Hysitron, Inc.), which allows loading and unloading to be performed in a controlled fashion. Indentation experiments were performed with diamond tips of nominal radius of curvature equal to ~117 nm (Berkovich) and ~850 nm (conospherical) using triangular force functions with loading and unloading times both equal to 5 s. Tuning of the electrostatic force constant of the transducer and tip shape calibration were carried out before testing. The tip calibration procedure was based on the method of Pharr et al.[16] To determine the tip-shape function, indentations of varying contact depth were produced on fused quartz of hardness and elastic modulus equal to ~10 and ~73 GPa, respectively. The hardness and reduced elastic modulus of the various samples were calculated using the contact depth at maximum load and the slope of the unloading portion of the force-displacement curve determined at maximum load, respectively.

Sliding tests were performed with the ~850 nm radius conospherical diamond tip. Wear tracks of 2 µm in length were produced by traversing the tip on the specimen surface at a constant speed of 0.2 µm/s, while increasing the normal load from 3 to 250 µN in a linear fashion. The coefficient of friction was obtained as the ratio of the lateral (friction) force to the instantaneous normal force. To avoid deformation effects from neighboring wear tracks, the distance between sequential wear tracks was kept larger than 2 µm. Additional information about the testing procedure used in the nanoindentation and nanoscale sliding experiments can be found in a previous publication.[17]

Results and Discussion

Figure 1 shows the UV-visible spectra of PC coated with ZnO and Al$_2$O$_3$ layers. Al$_2$O$_3$ deposits are perfectly transparent in visible and near-UV domains. ZnO coatings also have a good transparency in visible UV but absorb UV radiations as it is expected for our application. This UV absorption increases with the thickness of the ZnO layer and reaches very high value above 200 nm.

TREATMENTS AND CHARACTERIZATION OF THE PC SURFACE: The adhesion of ZnO and Al_2O_3 onto PC is high if the sputtering parameters previously indicated are used. In this case, a mechanical adhesion test like "peeling test" is not able to give reliable results because the rupture of the assembly is cohesive (in the materials) instead of adhesive (at the interface). Nevertheless, it is essential to obtain the highest possible adhesion. This objective can be reached by an increase of the surface energies of the two materials which are brought face to face. The surface energy of the deposit can be modified by acting on the sputter

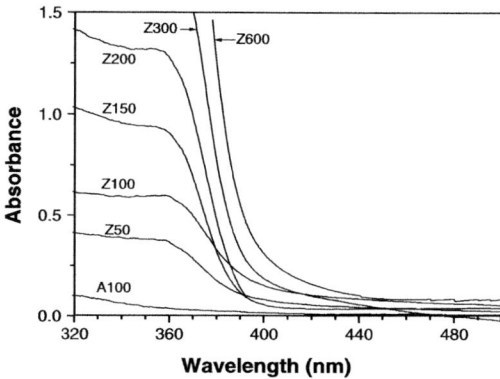

Figure 1—*UV-Vis absorbance spectra of ZnO and Al_2O_3 coatings deposited on PC (The ZnO and Al_2O_3 coatings are denoted by Z and A, respectively, while the numbers indicate the corresponding layer thickness in nanometers).*[18]

ing conditions. Appropriate treatments of PC can eliminate its surface pollution and induce changes in its chemical composition. With this aim, PC surfaces can be treated by a CO_2 cold plasma of short duration (from 10 to 30 s), treatment that usually gives a moderate oxidation of the polymers. Moreover, risks of post-deposition contamination are eliminated because such treatment is carried out in the sputtering chamber.

The variation of the surface energy γ_s of PC due to the treatment and its two components is presented in *Table* 1. The dispersive component γ_s^D slightly increases under the treatment effect. On the contrary, the polar component γ_s^p, which is insignificant for reference PC, becomes very important (38 mJ/m²). This indicates the formation of a great deal of polar groups i.e., there is functionalization of the surface.[18]

X-ray photoelectron spectroscopy (XPS) analyses were performed in order to characterize the surface modifications of PC. Important modifications were observed after treatment by a CO_2 plasma. A new component at 288.6 eV appeared in the C1s peak. This bond was identified as free carbonyl (C=O). This attribution was confirmed by the increase of the O=C component against the O–C component in the O1s peak. In addition, the Sat component of the C1s peak disappeared. This indicates that aromatic rings are broken, which can permit the fixation of oxygen. Moreover, a small broadening of the initial peak components was observed, which corresponds to the appearance of secondary bonds.

Table 1—Surface Energy (γ_s) of PC: γ_s^p = **Polar Component;** γ_s^D = **Dispersive Component**

	γ_s^p (mJ/m²)	γ_s^D (mJ/m²)	γ_s (mJ/m²)
PC (reference)	4	38	42
PC (CO_2 treated)	33	57	90

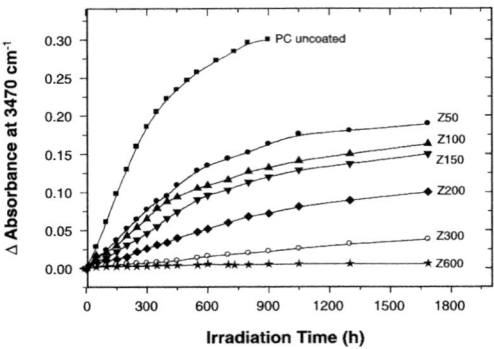

Figure 2—Increase of the IR absorbance of the band located at 3470 cm^{-1} versus the irradiation time for PC covered with ZnO coatings of different thicknesses.[3]

In conclusion, the presence of C=O bonds is in good accordance with the increase of the polar component of the surface energy as previously seen. This functionalization of the PC contributes to better adhesion because it allows for chemical bonds with a deposit ceramic layer.[18]

ZnO Coatings on PC: *Figure* 2 shows the increase of absorbance at 3470 cm^{-1} versus irradiation time. This increase is important for the uncoated PC and characterizes its photooxidation. When the thickness of the ZnO layer increases, a decrease of the PC photodegradation is observed. The same conclusion is drawn when the absorbance is measured at 400 nm. ZnO coatings can efficiently protect PC from photoaging. ZnO exhibits a strong absorption in the region 300–400 nm which depends on the thickness of the layer. The screening effect therefore reduces the PC photodegradation produced by the direct absorption of sunlight radiation. In addition, we have shown that the grain size and the density increase with the coating thickness. The coatings are denser and have lower microvoids. Consequently, they are more impermeable to oxygen and so the oxidation is limited.

Multilayer Coatings on PC: *Figure 3* shows the increase of absorbance at 3470 cm^{-1} and at 400 nm versus irradiation time for different polymer/ceramic assemblies.

The analysis of the curves shows that the photooxidation of PC is not inhibited by Al$_2$O$_3$ (monolayer of 100 nm). The deposit causes a decrease in the rate of oxidation of PC. The same tendency is observed for the yellowing (400 nm), but at a lower extent. These results can be explained according to both photolytic and photooxidative reactions.[13–15] The UV-Vis spectrum clearly shows that the layer of Al$_2$O$_3$ cannot protect the polymer against incident radiation. Consequently, photo-Fries rearrangement of PC, which does not involve oxygen, is not inhibited by the presence of the Al$_2$O$_3$ deposit. However, the subsequent oxidative reactions, initiated by photo-Fries products, are inhibited as Al$_2$O$_3$ acts as a barrier to oxygen. In other words, Al$_2$O$_3$ does not prevent the photolytic process (without intervention of oxygen) but reduces the extent of oxidative reactions as this coating is impermeable to oxygen.

With a ZnO layer, as explained above, the photolytic process of degradation of PC is inhibited and the increase of absorbance observed is only due to the photooxidation process. The degradation rate of PC/ZnO is logically lower than that of PC/Al$_2$O$_3$.

With the aim of giving evidence of the photocatalytic effect, PC was covered by Al$_2$O$_3$ and then by ZnO. In order to observe some degradation after a reasonable duration of exposure, we limited the thickness of the ZnO and Al$_2$O$_3$ coatings to 50 nm each. To have acute evidence of the photocatalytic activity of ZnO at the interface with the polymer, another experiment was realized in which Al_2O_3 was deposited above ZnO. The rate of formation of hydroxyl products (3470 cm^{-1}) is also reported in *Figure 3*. Analysis of the curves shows that both bi-layer coatings show a reduced degradation compared to a 100 nm monolayer of ZnO or Al_2O_3. The Al_2O_3-ZnO (50–50 nm) coating presents a higher photoprotective efficiency than ZnO (100 nm) coating. This effect can be attributed to the suppression by Al_2O_3 of the photocatalytic effect of ZnO, which causes the degradation of PC at the interface of PC/ZnO. This is confirmed by the results obtained with the PC/ZnO-Al_2O_3 coating, which have an interface of PC/ZnO. This system is less stable than the PC/Al_2O_3-ZnO coating.

In conclusion, PC/Al_2O_3-ZnO coatings present the highest photoprotective efficiency because this assembly combines the absorption ability of ZnO in the UV-Vis range with the oxygen barrier property of Al_2O_3 without the photocatalytic effect.

With the three layer coating of

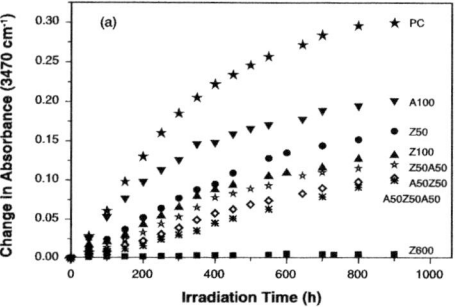

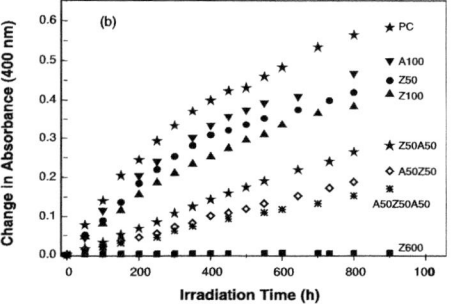

Figure 3—Change in absorbance at (a) 3470 cm^{-1} and (b) 400 nm of PC coated with single- and multi-layer coatings of ZnO and Al_2O_3 of different thickness.[5]

Al_2O_3-ZnO-Al_2O_3, the photoprotection is again increased. The gain is explained by a reduction of oxygen diffusion due to the outer layer of Al_2O_3. Despite the fact that the

improvement is not proportional to the increase of the coating thickness, the enhancement of the surface nanomechanical properties obtained with the three-layer coating is significant, as shown in the next paragraph.

SURFACE MECHANICAL BEHAVIOR: Nanomechanical experiments have been carried out at the samples' surface before photooxidation and after 200 h of irradiation.[5] We compared surface mechanical properties of the uncoated PC to PC covered by 50 nm of ZnO or by 50 nm of Al_2O_3 or combinations of both.

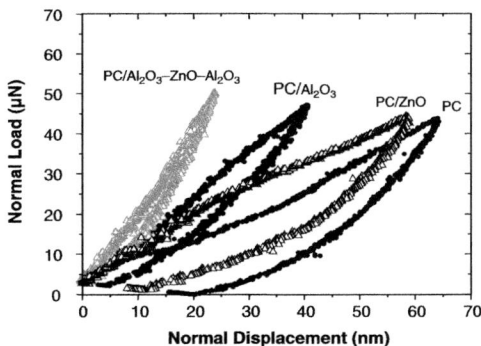

Figure 4—Force versus displacement curves of PC, PC/ZnO (50 nm), PC/Al_2O_3 (50 nm), and PC/Al_2O_3-ZnO-Al_2O_3 (50-50-50 nm).[9]

Table 2—Coefficient of Friction Obtained with a Normal Load of 50 µN

Sample	COF
PC	0.34 ± 0.02
PC/ZnO (50 nm)	0.18 ± 0.03
PC/Al$_2$O$_3$ (50 nm)	0.13 ± 0.01
PC/Al$_2$O$_3$-ZnO-Al$_2$O$_3$ (50–50–50 nm)	0.07 ± 0.03

Indentation Tests—For each sample, load/displacement curves were recorded. *Figure* 4 compares the force versus displacement curves of PC, PC/ZnO, PC/Al$_2$O$_3$ and PC/Al$_2$O$_3$-ZnO-Al$_2$O$_3$. The hardening effect of the films can be directly observed by comparing their indentation responses. It is clearly seen from the figure that Al$_2$O$_3$-ZnO-Al$_2$O$_3$ multilayer film gives the strongest protection to the PC substrate (the shallowest normal depth) while the ZnO film offers the weakest one (the deepest normal depth). No obvious difference was found between the indentation curves before and after 200 h of irradiation for the coated PC samples.

Scratch Tests: A scratch was performed on the different samples with an increasing load up to 250 µN. The results given in *Table* 2 show that the PC/Al$_2$O$_3$-ZnO-Al$_2$O$_3$ sample displays the smallest coefficient of friction (COF), and uncoated PC the highest. This classification results from the different hardening effects discussed previously. After irradiation, for each coated PC sample, similar values of COF were obtained. It can be then concluded that after 200 h of irradiation, the chemical processes due to the photodegradation of the polymer do not modify significantly the interactions between the polymer and the ceramic. At that stage of degradation, the photochemical reactions do not modify significantly the adhesion between the polymer and Al$_2$O$_3$ or ZnO.

Conclusions

The efficacy of relatively thin ceramic coatings to reduce photodegradation and to enhance the surface nanomechanical properties of PC was evaluated in light of IR and UV-Vis spectroscopy results and nanomechanical property measurements obtained with a surface force apparatus. It is shown that a thin ZnO coating is effective in reducing the photodegradation rate of PC because of the UV screening effect. However, the efficiency of the thin ZnO coating is limited by the photocatalytic activity at the interface with the PC substrate. The deposition of a thin Al$_2$O$_3$ layer between the ZnO layer and the PC substrate increases the resistance to photodegradation by suppressing the photocatalized degradation of the polymer at the interface. In addition, the Al$_2$O$_3$ and ZnO layers act as barriers to oxygen and limit the oxidative degradation caused by photo-induced aging. A higher photooxidation resistance is obtained with the three-layer coating (Al$_2$O$_3$/ZnO/Al$_2$O$_3$) than with the two-layer coatings (Al$_2$O$_3$/ZnO and ZnOAl$_2$O$_3$) because of the increase of the impermeability to oxygen with the coating thickness. The significant improvement of the surface resistance to plastic deformation and the decrease of the coefficient of friction is another benefit with the three-layer coating. Prolonged irradiation of PC resulted in the increase of the reduced elastic modulus and hardness in spite of chain scissions, presumably due to reorganization of the chain molecules under the simultaneous irradiation and temperature effects, in conjunction with the formation of new oxidized functionalities associated with hydrogen bonding. The ceramic coatings not only protected PC from photodegradation but also enhanced the nanomechanical properties and lowered the coefficient of friction. The three-layer coatings exhibited the lowest coefficient of friction over the entire load range investigated. In contrast to the PC, the similar nanomechanical properties obtained before and after extensive irradiation indicate that photo-chemical reactions at the polymer interface are not detrimental to the adhesion of the Al_2O_3 and ZnO layers to the PC substrate. The results of this study demonstrate that thin ceramic coatings are effective inhibitors of photo-chemical reactions and deformation processes occurring in polymers, which lead to accelerated photodegradation, high friction, and excessive wear.

PHOTOPROTECTION OF PET AND SUPERFICIAL PHOTOCATALYTIC ACTIVITY

Experimental

SPECIMENS: PET films (25 μm thick) were supplied by Terphane (Rhône Poulenc). These polymers do not contain antioxidants. They were ultrasonically cleaned in ethanol before coating. The deposition of TiO_2 and Al_2O_3 coatings was carried out in an Alcatel SCM 450 sputtering unit equipped with an r.f. generator operating at 13.56 MHz. Bulk Ti and Al_2O_3 targets (purity: 99.9%; diameter: 100 mm) fixed on cooled magnetron-effect cathodes were used as source materials. The substrates were situated at a distance of 90 mm from the targets. The sputtering chamber was evacuated below a pressure of 10^{-4} Pa before admitting the sputtering gas. The TiO_2 film sputter deposition took place in a reactive Ar-O_2 gas mixture, while Al_2O_3 deposition took place in a pure argon atmosphere of 1 Pa. The magnetron source was operated at 200 W.

Surface topography characterization was performed using a JEOL 880 scanning electron microscope (SEM) and a Nano Scop 3a (Digital Instrument) atomic force microscope (AFM). The structure of the TiO_2 thin film was determined by XRD patterns using a Philips X'PERT PRO diffractometer with Cu K_α radiation.

IRRADIATIONS: Photodegradation of Rhodamine B in aqueous solution was used to assess the photocatalytic activity of the film. The sample (10 cm²) was immersed into 40 ml of Rhodamine B with a concentration of 10^{-5} mol.L^{-1} and placed in a cylindrical glass reactor. The film was irradiated with a polychromatic fluorescent UV lamp (Philips TDL 8 W 300 mm, wavelength range 315–450 nm) at a distance of 10 cm directly above the film. The photocatalytic decomposition of Rhodamine B was monitored by the decrease of the absorbance at wavelength of 550 nm by a UV-Visible spectrometer (Jenway 6400).

Results and Discussion

PHOTOCATALYTIC ACTIVITY:

Films Deposition and TiO_2 Film Characteristics—The typical deposition conditions are summarized in *Table 3*.

The crystalline phase and crystallite size of thin film obtained were analyzed by X-ray diffraction. The film presented lines characteristic of the anatase phase. No characteristic peak of titanium was observed. The crystalline size, calculated using the Scherrer equation, was determined to be close to 15 nm.

Table 3—Thin Films Operating Conditions

Target	Pressure (Pa)	Power (W.cm^{-2})	Plasma Composition
Ti...................................1		2.5	Ar (88%)-O$_2$ (12%)
Al$_2$O$_3$..............................1		2.5	Ar

Surface Morphologies and Roughness—The surface morphologies and roughness of thin TiO$_2$ films were obtained by AFM. The sample had a large specific surface area because of the sharp crystallites and furrows. The sample roughness was 15 nm.

The SEM image of the sputtered TiO$_2$ thin film indicated that the film was composed of TiO$_2$ particles with diameters of 50–200 nm and furrows.

Figure 5 shows AFM two- and three-dimensional images of the sputtered TiO$_2$ thin film. It can be seen that the film has a granular microstructure and is composed of TiO$_2$ particles of about 120 nm diameter. In addition to particle diameter, AFM image analysis also gives the value of surface roughness. The mean value of the surface height Ra and the root-mean-square (RMS) average roughness profile of the surface height, within the given area RMS, respectively 6.6 nm and 8.2 nm, indicate that the surface is not smooth. Roughness causes a large specific surface area of the thin film, which increases the photocatalytic efficiency.

PHOTOCATALYTIC ACTIVITY MEASUREMENT: *Figure* 6 illustrates the degradation of Rhodamine B dye solution in the presence of TiO$_2$ thin film. C$_o$ and C$_t$ are the initial concentrations and the reaction concentrations of Rhodamine B, respectively. Obviously, Rhodamine B can not be degraded by long wavelengths of irradiation alone. However, when the sample is added to the dye solution irradiated with UV light, the Rhodamine B concentration starts to decrease. The UV radiation duration was fixed at 6 h. No change in the concentration of the dye solution was observed in the absence of the photocatalyst, confirming the high resistance of Rhodamine B to direct photodegradation.[19]

After 6 h exposure, TiO$_2$ thin film has degraded 50% of the dye. TiO$_2$ photocatalytic activity lies in its specific surface area. Larger TiO$_2$ specific area implies greater photo-

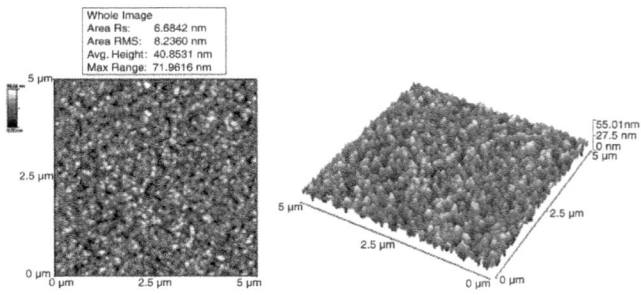

Figure 5—AFM two- and three-dimensional images of the sputtered TiO$_2$ thin film.[6]

catalytic activity while maintaining constant sample size. Thus, this degradation agrees with AFM morphology studies of specific surface area.

PET / CERAMIC PHOTOOXIDATION RESISTANCE: Because the light absorption by PET extends up to 330 nm, this polymer is directly accessible to UV-light present in terrestrial solar radiation.

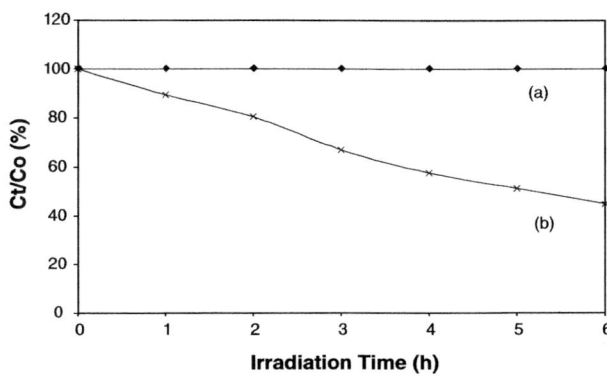

Figure 6—Photodegradation of Rhodamine B dye solution, (a) without TiO$_2$; (b) in presence of TiO$_2$.

The resulting photodegradation at $\lambda > 300$ nm results in significant changes in the UV-Vis and infrared spectra. In the hydroxyl absorption region of the infrared spectra (3800-3000 cm^{-1}), an increase of the absorbance occurs with the exposure time and was attributed to the formation of alcohols, acids, and hydroperoxides.[20-22]

The degradation rate is characterized by the increase of the absorbance in the hydroxyl region. The irradiation time effect on the absorbance in the hydroxyl region of specimens is illustrated in *Figure 7*.

The absorbance curves of single-layer coatings shown in *Figure 7* indicate that the Al$_2$O$_3$ coating was much less effective in preventing photodegradation than the TiO$_2$ coating. The initial photooxidation rate for the 100-nm thick Al$_2$O$_3$ coating is similar to that of the uncoated PET. For irradiation time less than 250 h, the rate of formation of hydroxyl products is not significantly affected by the presence of the Al$_2$O$_3$ coating. However, a decrease in the oxidation rate of PET was obtained after longer exposure. This result can be attributed to both photolytic and photooxidative reactions. This finding may be explained if one considers the complex mechanism of PET photodegradation in which preponderant photolytic processes and minor photooxidative reactions interfere.[21,22] The UV-Vis spectrum of the Al$_2$O$_3$ coating shown in *Figure 1*

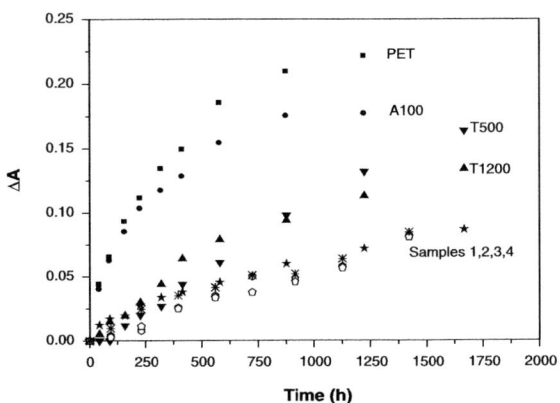

Figure 7—Change in absorbance in the hydroxyl region of PET uncoated, coated with single- (A100, T500, T1100), and multi-layer coatings of TiO$_2$ and Al$_2$O$_3$ of different thicknesses. (Samples 1,2,3,4).[6]

Table 4—Thickness (nm) of Al_2O_3 and TiO_2 Deposited on PET

Sample	Al_2O_3	TiO_2
s1	60	1150
s2	114	550
s3	114	1100
s4	500	1200

indicates that this material cannot protect the polymer against incident radiation. Consequently, the Al_2O_3 layer did not inhibit photolytic reactions, which do not depend on the presence of oxygen. The subsequent oxidative reactions, however, initiated by photolytic products were inhibited because Al_2O_3 is a barrier to oxygen diffusion. Therefore, while Al_2O_3 does not suppress the photolytic process, it is effective in reducing the extent of oxidative reactions because of its impermeability to oxygen.

To explore other means of enhancing further the photooxidation resistance of PET, the feasibility of bi-layer coating on PET was tested. $PET/Al_2O_3/TiO_2$ samples were successfully obtained with different thicknesses of coatings as reported in *Table 4*.

IR spectra were obtained from samples of PET with two-layer Al_2O_3/TiO_2 coatings, and the corresponding absorbance changes in the hydroxyl region were contrasted with those of the single-layer coatings. The two-layer coatings exhibit lower degradation rates than the single-layer coatings. This can be attributed to the added Al_2O_3 layer that prevents the photocatalytic effect, which is partly responsible for the degradation of PET at the interface with the TiO_2 layer. Hence, the Al_2O_3/TiO_2 coating provides greater protection against photodegradation because it combines the absorption characteristics of TiO_2 in the UV-Vis range and the impermeability of Al_2O_3 to oxygen without inducing a photocatalytic effect at the PET interface. As the photodegradation rates of samples 1–4 are observed to be roughly equivalent, it can be concluded that a layer of 60 nm of Al_2O_3 is sufficient to suppress the photocatalytic activity of TiO_2 at the interface with the polymer.

Conclusions

Reactive rf magnetron sputtering using a titanium target in an argon-oxygen mixture is reported to be a very valuable process to deposit crystalline anatase on a low heat tolerance substrate such as polymer. This coating was used to photodegrade Rhodamine B in water solution under polychromatic light, assessing, therefore, the good photocatalytic activity of the coating.

The efficacy of relatively thin ceramic coatings to reduce photodegradation of PET was evaluated in light of IR spectroscopy results. It was shown that a thin TiO_2 coating is effective in reducing the photodegradation rate of PET because of the UV screening effect. However, the efficiency of the thin TiO_2 coating is limited by the photocatalytic activity at the interface with the PET substrate. The deposition of a thin Al_2O_3 layer between the TiO_2 layer and the PET substrate increases the resistance to photodegradation by suppressing the photocatalyzed degradation of the polymer at the interface. In addition, the Al_2O_3 layer acts as a barrier to oxygen diffusion and limits the degradation caused by photooxidative aging.

The combination of these two sets of experiments allows production of photostable and transparent polymer/ceramic layered media with superficial photocatalytic activity. These layered media can be used in outdoor exposure without alteration of their superficial aspect due to their self-cleaning properties. This work could also be an excellent opportunity to protect steels against corrosion.

ORGANIC-INORGANIC HYBRID FOR PC

Experimental

MATERIALS: α,ω-hydroxy-terminated poly(ethylene oxide) (PEO, purchased from Fluka and with a number average molecular weight of 600 g·mol^{-1}), 3-isocyanatopropyltriethoxysilane (ICPTES, Fluka), tetraethoxysilane (TEOS, Aldrich), hydrochloric acid at 37% concentration (Carlo Erba), and ethanol (EtOH, Carlo Erba) were high purity reagents and were used without further purification.

Preparation of a,ω-triethoxysilane terminated poly(ethylene oxide)—α,ω-triethoxysilane terminated polymer chains (PEOSi) were prepared by bulk reaction of α,ω-hydroxy-terminated polymers with ICPTES (molar ratio of 1:2). The reaction was usually carried out in a 50 ml glass flask equipped with a calcium chloride trap and under magnetic stirring, at 120°C, for 3 h as already reported in a previous paper.[23] The molecular structure of the final product can be schematically represented as follows: (EtO)$_3$Si-**PEO**-Si(OEt)$_3$.

Preparation of organic-inorganic hybrids—Mixtures of TEOS and α,ω-triethoxysilane terminated polymer were dissolved in EtOH at a concentration of about 10% wt/vol, then water (for the hydrolysis reaction) and hydrochloric acid (as catalyst) were added at the following molar ratios with respect to ethoxide groups of the functionalized polymer and TEOS: EtO-:H$_2$O:HCl=1:1:0.05.

A typical preparation for PEOSi/TEOS hybrids was as follows: PEOSi (1.20 g) and TEOS (0.80 g) were added to 20 ml of EtOH in a screw-thread glass vial and mixed until a homogeneous solution was obtained. Then water and HCl (37% wt solution) were added under vigorous stirring at room temperature for about 10 min. The closed vial was placed in air circulating oven at a temperature of 70°C for 30 min in order to allow only a partial progress of the sol-gel reaction. The clear solution was then cast into a closed polytetrafluoroethylene dish and the solvent was slowly evaporated at room temperature for about one week to get self-consistent samples for further analysis. Alternatively, the solution was deposited onto clean PC films by spin-coating—obtaining a typical coating thickness of about 0.3–0.4 μm.

Hybrid materials based on PEO were characterized by a final organic/inorganic weight ratio of 7/3 and 3/7, assuming the completion of the sol-gel reactions (see following scheme).

Coating	Organic/Inorganic Ratio (wt/wt)	Coating Thickness (μm)
PEOSi/SiO$_2$ 7/3	7/3	0.3–0.4
PEOSi/SiO$_2$ 3/7	3/7	0.3–0.4

Hydrolysis reactions:

$$Si(OEt)_4 + x\ H_2O \rightarrow Si(OEt)_{4-x}(OH)_x + x\ EtOH$$
$$(EtO)_3Si\text{-}PEO\text{-}Si(OEt)_3 + (y+z)\ H_2O \rightarrow (EtO)_{3-y}(HO)_ySi\text{-}PEO\text{-}Si(OEt)_{3-z}(OH)_z + (y+z)\ EtOH$$

Condensation reactions:

$-Si-OH + EtO-Si- \rightarrow -Si-O-Si- + EtOH$

$-Si-OH + HO-Si- \rightarrow -Si-O-Si- + H_2O$

Curing Treatments—Partially cured PEO-based hybrids were subjected to different post-curing treatments:

- Conventional heating in an air circulating oven (WTB Binder) at 80°C for a time ranging from 20 to 60 min;

- Microwave irradiation by using an Alter TE10n Applicator (single-mode 2.45 GHz, 350 W) at irradiation times ranging from 3 to 15 s.

Scratch Tests—Scratch tests were carried out on a CSM Micro-Combi Tester by using a Rockwell C diamond scratch indenter (R = 0.2 mm) and progressively increasing the load from 100 mN to 2000 mN at a load rate of 494 mN·min^{-1} and for a scratch length of 1 mm.

Results

To evaluate the enhancement in the scratch resistance of PC slabs when coated with PEOSi/SiO$_2$ hybrids, opportune samples were prepared in a further part of the work; hybrid solutions in ethanol were deposited by spin-coating onto PC slabs and the systems were subjected to the above discussed post-curing treatments.[7]

Scratch tests were carried out by performing a scratch with a progressively increasing load of 100 mN to 2000 mN for a final scratch length of 1 mm. As expected, all the samples showed a progressive increase of penetration depth (PD) due to the increasingly applied load. The results of scratch tests for all samples are reported in *Table 5*, in which

Table 5—Scratch Test: Final Penetration Depth (Scratch Length = 1 mm) of Uncoated PC and PC Coated with Different Hybrids Subjected to Different Post-Curing Treatments

Coating	Post-Curing Treatment		Final Penetration Depth (nm)
	Type	Time (s)	
None	–	–	30.5
PEOSi/SiO$_2$ 7/3	None	–	31.1
	Microwave	3	25.5
		5	24.9
		10	24.9
		15	24.9
	Oven (T=80°C)	1200	27.9
		2400	28.2
		3600	28.5
PEOSi/SiO$_2$ 3/7	None	–	29.8
	Microwave	3	23.1
		5	23.3
		10	23.8
		15	23.3
	Oven (T=80°C)	1200	23.5
		2400	23.2
		3600	23.6

final PD values (i.e., at a scratch length of 1 mm) of PC uncoated and coated with different hybrids and subjected to different post-curing treatments (type and time) are reported.

The final PD values were lower when hybrids coated onto PC were post-cured by oven heating or by microwave irradiation. These data indicate that organic-inorganic hybrids are effective coatings for the protection and the improvement of the scratch resistance of polymeric substrates such as PC. In order to obtain the desired protective effect, a post-curing step is necessary to improve the degree of the sol-gel reaction and obtain a high crosslinking density of the hybrid coating, and hence a harder surface. Both oven heating and microwave irradiation

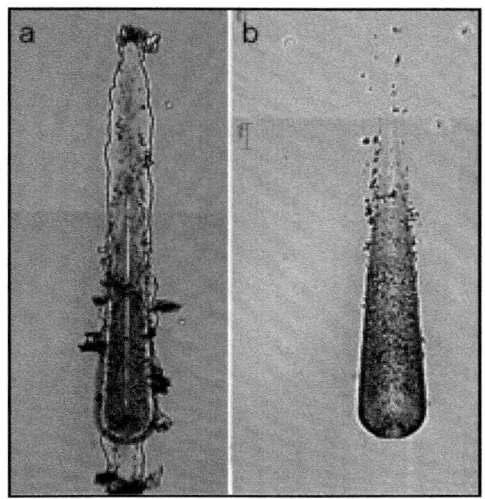

Figure 8—Optical micrographs of scratches performed on PC coated with PEOSi/SiO$_2$ 7/3 (a) and PEOSi/SiO$_2$ 3/7 (b) hybrids, post-cured by 5 s microwave irradiation.

were effective post-curing methods and the scratch resistance (in terms of final penetration depth) was almost independent of the type and the duration of the treatment.

On the other hand, the hybrid composition (organic-inorganic final ratio) seems to have only a slight effect on the scratch resistance; in fact, substrates coated with post-cured hybrids having a higher inorganic content (PEOSi/SiO$_2$ 3/7) showed an average value of final PD (independently of the time of treatment) of 23.4 nm for both oven heating and microwave irradiation while post-cured samples based on PEOSi/SiO$_2$ 7/3 hybrids showed an average value of 25.0 nm (microwave irradiation) and of 28.2 nm (oven heating), respectively. For these samples a harder surface was obtained through the MW irradiation curing treatment, and this result is a further factor supporting the effectiveness of this method versus the traditional thermal one.

These considerations were further supported by the optical micrographs of the scratches present on the surface after the test. Some examples are reported in *Figure* 8 from which the higher scratch resistance related to coatings with the higher inorganic content is well evident. From *Figure* 8b, it is clearly evident that no detachment of the silica-rich coating from the substrate occurred in the region around the scratch profile, which indicates a hard, resistant, and effective surface coating.

As expected, scratch resistance tends to improve by increasing the inorganic phase content. However, the differences between coatings with markedly different organic-inorganic content were not so great and this behavior could be attributed to a preferential surface segregation of silica domains independently on the bulk composition, a phenomenon already observed, reported, and discussed for this type of materials.[24]

Conclusions

Poly(ethylene oxide)/silica hybrids to be applied onto PC substrates were easily prepared by the sol-gel process. In order to increase the degree of reaction and thus to crosslink the thermoset system, different post-curing treatments were carried out by conventional oven heating and microwave irradiation. Both post-curing treatments were efficient enough to almost complete the sol-gel reaction.

Scratch tests carried out on PC substrates coated with poly(ethylene oxide)/silica hybrids evidenced a significant increase of scratch resistance with respect to uncoated PC. Post-curing treatment (both microwave irradiation and oven heating) was absolutely necessary to achieve a sufficiently high conversion degree of the sol-gel reaction (crosslinking degree), and thus the desired anti-scratch effect of the coating. The main objective of the upcoming experiments is to focus on the screen effect of the coatings in order to limit the UV radiations at the surface of the polymer.

CONCLUSION

Research carried out in the last five years in the field of ceramic coatings has considerably enhanced the photoprotection and surface properties of polymers. On one hand, it has been shown that layered ceramic coatings provide photostability and improve the nanomechanical and tribological properties of coated polymers. On the other hand, getting photostable and transparent polymer/ceramic layered media with superficial self-cleaning property is also feasible. The improvement of the photocatalytic activity of the ceramic layer is closely linked to is specific surface. The use of Anodic Aluminum Oxide Templates (AAO), within Al_2O_3 barrier layer characterized by a well ordered hexagonal array of pores with scalable size, could both significantly increase the specific surface and efficiently minimize photocatalytic oxidation of the polymer. AAO pattern process is relatively simple and cheap.

Organic-inorganic hybrid coatings prepared by the sol-gel process could be an interesting alternative having many advantages: mild conditions, low costs, no geometrical forms requirements. These coatings were shown to significantly improve the scratch resistance of coated polymers. The main objective of the upcoming experiments is to focus on the screen effect of these coatings by developing proper formulations using organic and mineral UV absorbers.

ACKNOWLEDGMENTS

The authors would like to recognize A. Moustaghfir and M. Jacquet (Laboratoire des Matériaux Inorganiques, UMR CNRS 6002) for their collaboration.

References

(1) Ben Amor, S., Baud, G., Jacquet, M., and Pichon, N., *Surf. Coat. Technol.*, 102, 63 (1998).
(2) Giancaterina, S., Ben Amor, S., Baud, G., Gardette, J.-L., Jacquet, M., Perrin, C., and Rivaton, A., *Polymer*, 43, 6397 (2002).
(3) Moustaghfir, A., Tomasella, E., Rivaton, A., Mailhot, B., Jacquet, M., Gardette, J.-L., and Cellier, J., *Surf. Coat. Technol.*, 180-181, 642 (2004).

(4) Penot, G., Arnaud, R., and Lemaire, J., *Angew. Makromol. Chem.*, 117, 71 (1983).

(5) Mailhot, B., Rivaton, A., Gardette, J.-L., B Moustaghfir, A., Tomasella, E., Jacquet, M., Ma X.-G., and Komvopoulos, K., *J. Appl Phys.*, 104310, 99 (2006).

(6) Awitor, O., Rivaton, A., Gardette, J.-L., Down, A.J., and Johnson, M.B., *Thin Solid Films,* in press.

(7) Fabbri, P., Leonelli, C., Messori, M., Pilati, F., Toselli, M., Veronesi, P., Morlat-Thérias, S., Rivaton, A., Gardette, J.L., *J. Appl. Polym. Sci.*, 108(3), 1426-1436 (2008).

(8) Schultz, J., Tsutsumi, K., and Donnet, J.B., *J. Coll. Interf. Sci.*, 59, 272 (1977).

(9) Maissel, L. and Lang, R.G., *Handbook of Thin Film Technology*, McGraw-Hill, New York, p 12-29, 1983.

(10) Swanepoel, R., *J. Phys.* E16, 12134 (1983).

(11) Bachari, E.M., Ben Amor, S., Baud, G., and Jacquet, M., *Thin Solid Films*, 348, 165 (1999).

(12) Cueff, R., Baud, G., Besse, J.P., and Jacquet, M., *Thin Solid Films*, 226, 198 (1995).

(13) Rivaton, A., *Polym. Degrad. Stab.*, 49, 163 (1995).

(14) Rivaton, A., Mailhot, B., Soulestin, J., Varghese, H., and Gardette, J.-L., *Polym. Degrad. Stab.*, 75, 17 (2002).

(15) Rivaton, A., Mailhot, B., Soulestin, J., Varghese, H., and Gardette, J.-L., *Eur. Polym. J.*, 38, 1349 (2002).

(16) Pharr, G.M., Oliver, W.C., and Brotzen. F.R., *J. Mater. Res.*, 7, 613 (1993).

(17) Ma, X.-G., Komvopoulos, K., Wan, D., Bogy, D. B., and Kim, Y.-S, *Wear*, 254, 1010 (2003).

(18) Moustaghfir, A., Tomasella, E., Rivaton, A., Mailhot, B., Jacquet, M., Gardette, J.-L., and Bêche, E., *Thin Solid Films*, 662-665, 515 (2006).

(19) Yu, J.C., Tang, Y.H., Ju, J., Chan, H.C., Zhang, L., Yie, Y., Wang, H., Wong, S.P., *J. Photochem. Photobiol. A: Chem.*, 153:211-219 (2002).

(20) Day, M. and Wiles, D.M., *J. Appl. Polym. Sci.*, 16, 175 (1972).

(21) Grossetête, T., Rivaton, A., Gardette, J.L., Hoyle, C.E., Ziemer, M., Fagerburg, D.R., and Clauber, H., *Polymer*, 41:3541 (2000).

(22) Ben Amor, S., Baud, G., Jacquet, M., and Pichon, N., *J. Surf. Coat. Technol.*, 102:63 (1998).

(23) Messori, M., Toselli, M., Pilati, F., Fabbri, E., Fabbri, P., Pasquali, L., and Nannarone, S., *Polymer*, 45: 805-813 (2004).

(24) Brunelle, D.J., in *Technical Information Series*, G. E. R. D. Center, General Electric, p. 32, 2002.

Chapter 23

A Critical Assessment of Techniques for Monitoring Polymer Photodegradation

Jim R. White

School of Chemical Engineering and Advanced Materials, University of Newcastle upon Tyne, Newcastle upon Tyne, NE1 7RU, UK

Methods used to assess the extent of polymer photodegradation after exposure to ultraviolet irradiation (UV) are reviewed and compared. Characterization methods reviewed include Fourier transform infrared analysis (FTIR), differential scanning calorimetry (DSC), X-ray diffraction (XRD), gel permeation chromatography (GPC), residual stress analysis, and mechanical testing. Emphasis is placed on techniques used in the author's laboratory, especially methods that help to link chemical degradation with engineering failure. The use of FTIR to monitor carbon dioxide emission dynamically while the polymer is under UV exposure is also described: this method gives a very rapid assessment of the photosensitivity of the polymer and of the effectiveness of photo-stabilizing (or photo-catalytic) additives. Results are compared with those given by FTIR measurements of carbonyl group development during UV exposure. Computer analysis of molecular weight distributions (MWDs) made by GPC to measure scission and crosslinking rates is discussed. Crystallinity measurements made by DSC and XRD are compared. The use of DSC re-heating runs to investigate the type of molecular damage caused by UV exposure is discussed briefly. Measurement of residual stresses and their relationship with mechanical test failures is discussed.

INTRODUCTION

The conventional approach to the assessment of ultraviolet (UV) photodegradation of a polymer is to expose it for a series of different times, then conduct one or more characterization procedures to determine the effect of photo-exposure.[1,2] The motive is usually to determine the weatherability of the polymer. Exposure can be conducted outdoors, preferably in a region with high UV levels, or in the laboratory, using light sources with UV outputs designed to emulate solar radiation. Some artificial UV sources have a spectral distribution biased towards lower wavelengths—to accelerate degradation—but with the probable disadvantage that the higher energy radiation may provoke chemical degradation mechanisms that are not found under natural conditions. For polymers intended for engineering applications, the characterization may involve mechanical testing such as a tensile test or an impact test to determine the onset of embrittlement,

345

whereby a normally tough polymer fails in a brittle, catastrophic manner when a mechanical load is applied.

Some characterization procedures may attempt to track chemical changes such as the build-up of carbonyl groups that are a consequence of photooxidation. FTIR can be used to perform this fairly routinely, but interpretation is not always totally straightforward since carbonyl groups are both formed and destroyed by photooxidation reactions. These methods are often adapted for "depth profiling," in which measurements are made at various depths from the exposed surface.[3-6] In a thick sample (typically ≥ 1 mm in depth), chemical degradation usually declines with distance from the exposed surface because of attenuation of UV intensity and diminished availability of oxygen.

Other characterization methods are used to determine changes in the structure and morphology of the exposed polymer and the information is used to help to explain the observed changes in engineering properties. For example, the crystallinity of a semi-crystalline polymer is an important characteristic that influences stiffness, strength, and toughness. Crystallinity can be measured by XRD or DSC[7-9]; the sampling procedure for the two methods is usually different, leading to small disagreements between the measurements made by the two methods. Both methods are based on an over-simplified, two-phase crystal-amorphous model that may lead to further discrepancies. Further insight into the changes that take place during photodegradation can be obtained by MWD analysis (using GPC): the small molecule fragments generated by chain scission events become available for secondary crystallization and lead to increases in crystallinity. As with the carbonyl group analyses, the MWD and crystallinity measurements are often made at several distances from the exposed surface to build up a comprehensive picture of the changes that occur during UV exposure.[7,8,10-13] The changes in MWD during exposure can be used to estimate scission and crosslinking rates.

Changes in crystallinity may lead to changes in the residual stress distribution in the exposed polymer artefact. Secondary crystallization occurs predominantly near the exposed surface, causing shrinkage and, consequently, development of tensile stresses. Equivalent densifying effects occur in non-crystalline polymers. The tensile stresses can promote fracture and contribute to the decline of engineering properties. Therefore, it is appropriate to monitor residual stresses during photodegradation.

In the sections that follow, selected methods are discussed that monitor reaction products during photodegradation changes to the polymer at the molecular level, morphological changes, and changes in engineering properties including residual stresses and tensile failure.

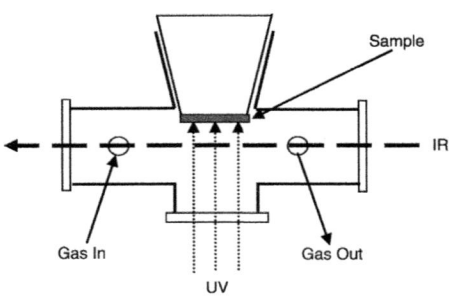

Figure 1—Cell for in-situ CO$_2$ measurement. In a later version of the cell, the sample is relocated to be positioned adjacent to the window that admits the UV illumination.[14-16]

DYNAMIC CO$_2$ MONITORING

Many of the techniques used to monitor polymer photodegradation require exposure times of the order of several days to several weeks in order to observe significant changes with most polymers. The technique described here is capable of yielding meaningful results in tests lasting of the order of three hours and can be used to determine the relative sensitivity of different polymers to photodegradation or to compare the effect of different additives on the susceptibility of a polymer to photodegradation. It is based on monitoring carbon dioxide evolved from the sample when exposed to UV.

Design of In-Situ CO_2 Monitoring Equipment

The method evolved from experiments to monitor paint photodegradation.[14] At the heart of the equipment is a cell in which the sample is mounted and exposed to UV provided by a 150 W xenon tube via a flexible light guide and a calcium fluoride window[15,16] (*Figure* 1). The cell sits in a FTIR instrument and the interrogating IR beam passes through the cell parallel to the sample surface (and perpendicular to the UV illumination). The IR beam passes into and out of the cell via calcium fluoride windows and the absorption of the gaseous reaction products emitted from the sample is measured. The cell has ports to admit the chosen gas phase in which the reaction is to take place, and the cell is flushed for at least an hour before the reaction run commences. At the commencement of the run the ports are closed and no gas is admitted from outside of the cell. The gas phase is monitored for an hour to check that it is stable, then the UV source is switched on. A water filter removes IR, to minimize sample heating, and in most tests conducted to date, solar filters type AM 0 and AM 1.5 (Oriel) were used alone or together, to remove UV irradiation below the terrestrial solar radiation cut off (~290 nm wavelength) and to alter the spectral distribution within the UV range. The progress of photodegradation is followed by monitoring the build up of carbon dioxide within the cell using the IR band centred at 2360 cm^{-1}.

Spectra are normally recorded for a further three hours with the lamp switched on— then the lamp is switched off and spectra recorded for an additional hour with no irradiation falling on the sample. With some polymers, CO_2 emission stops almost as soon

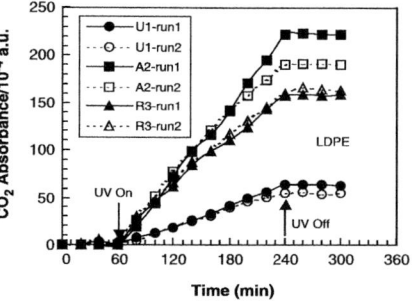

Figure 2a—*Typical in-situ CO_2 runs with a polyolefin (LDPE). Results are given for an unpigmented grade (U1) and pigmented grades of the same polymer containing TiO_2 in the form of anatase (A2) or rutile (R3). These data were obtained using only an AM0 filter, which probably accounts for the differences in measurements compared with those obtained by Jin[15] using a combination of AM0 + AM1.5 filters. These data courtesy of S. Fernando.*

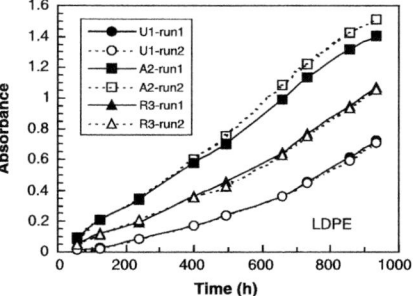

Figure 2b—*Carbonyl absorbances for the same polymer compositions as used for Figure 2a. Results are shown for duplicate samples. These data courtesy of S. Fernando.*

as the UV is switched off, but for others there is continued emission, presumably the result of continuing oxidation promoted by the radicals that had accumulated during photodegradation.

Example Results

A typical example of the application of this equipment is given in *Figure* 2, which shows runs with a low density polyethylene (LDPE) in unpigmented form and pigmented with TiO_2 in the form of anatase or rutile. Duplicate runs show good reproducibility. With the LDPE samples used in the investigation from which *Figure* 2a is taken, CO_2 emission ceased almost as soon as the UV was switched off. In similar experiments with a family of poly(vinyl chloride) (PVC), samples it was observed that CO_2 emission continued for an extended period after the UV was switched off.[16] Carbonyl group absorbances measured on similar LDPE samples exposed to Q-Panel UVA-340 tubes, for which the spectral output is similar to solar radiation in the UV range,[17] are given in *Figure* 2b. Similar ranking is shown: the partitioning of the different data sets developed over much longer times than was required using the CO_2 method.

MOLECULAR WEIGHT DISTRIBUTION (MWD)

The molecular weight of a polymer is often determined using gel permeation chromatography. It is based on measurements of the time taken for the molecules to pass through a column containing a porous gel. This depends on the mass of the molecules (and, to a lesser extent, on their shape). The sample is dissolved in a suitable solvent and allowed to pass through the column; elevated temperature is often required. The time-molecular size relationship for the column is calibrated and the mass passing through in a particular interval of time is determined by measuring the viscosity or the optical absorption of the eluted solution. By making continuous measurements, the data can be converted into a molecular weight distribution (MWD). It is common to use the data to generate molecular weight averages (number average, M_n, or weight average, M_w) but the MWD contains more information. When a polymer is photodegraded it is very common to find that the molecular weight averages are reduced. This is deduced to be the result of chain scission, a common consequence of photooxidation. With many polymers, crosslinking is a competing process. If crosslinking dominates, the molecular weight averages would rise, but in most reported studies (on polyoelfins, polystyrene, polycarbonate, etc.) chain scission dominates and the molecular weight averages fall. The whole of the MWD often shifts towards lower molecular weights, sometimes with little change in the shape of the distribution. However, in polymers that have a tendency to crosslink, a high molecular weight tail sometimes develops and indicates that some molecules are larger than those in the original population, prior to the photodegradation.[17,18] Although the shape of the molecule may influence the elution time through the column, this is considered to be a secondary effect and that the information relates to molecule size.

When both scission and crosslinking occur, the molecular weight averages no longer reflect properly the extent of degradation that has occurred because scission causes a decrease in the averages and crosslinking causes an increase. Shyichuk has developed a method of analyzing MWDs to determine the macromolecule scission and crosslinking concentrations.[19] It involves comparing experimental MWDs with MWDs generated using Monte Carlo computer-aided modification of the starting MWD, assuming scission and crosslinking are both random events (Molecular Weight Distribution Computer Analysis: MWDCA). This procedure has been used to examine the nature of photodegradation in polystyrene and several polyolefins,[11,20,21] and the effect of the inclusion of stabilizer and/of pigment.[22] The remark concerning the effect on molecule shape at the end of the previous paragraph is of relevance, and the MWDCA results are limited by any error that is introduced into the MWD through the shape effect, if present. Furthermore, if preferential reaction sites are present, the assumption that the scission and crosslink events occur at random will no longer apply. It is possible that the Monte Carlo analysis could be modified to take account of this but in the studies conducted to date it has not been considered necessary to do this.

An example of the application of MWDCA is given in *Figure* 3, in which data are shown for samples taken at different depths from the exposed surface of 3 mm thick bars of a low densilty polyethylene (LDPE) and a polypropylene homopolymer (PPHO) after three and six weeks UV exposure, respectively. The scission concentration in LDPE is greatest near the exposed surface and falls in the interior of the bars, due mainly to limited oxygen availability (*Figure* 3a). After six weeks exposure, the scission concentration was fairly close to double that of three weeks at all depths examined. PPHO showed a much lower scission concentration than LDPE after three weeks exposure but, after six weeks, the scission concentration near the exposed surface was much higher than in LDPE. It is evident that in PPHO there was either an incubation time effect or auto-acceleration (or both) in the near-surface region. Near the center of the PPHO bar (at a depth of 1.5 mm), the scission concentration after six weeks exposure was still low, probably because of oxygen starvation, which will be particularly acute when reactions are proceeding more rapidly, consuming oxygen diffusing in from the surface before it can get very far. The

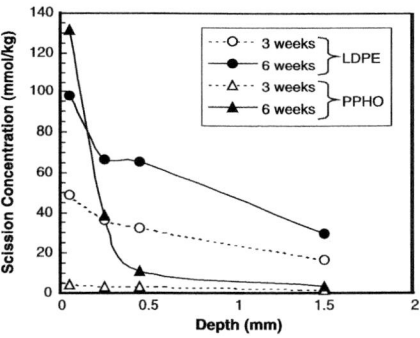

Figure 3a—Scission concentrations at different depths from the exposed surface of bars made from low density polyethylene (LDPE) and a polypropylene homopolymer (PPHO) after three and six weeks of UV exposure.[11]

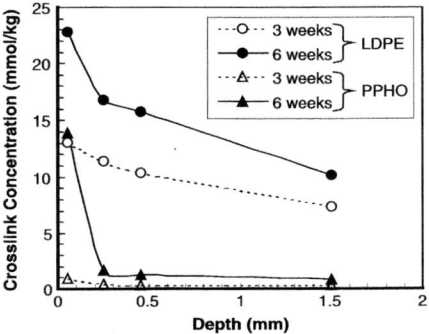

Figure 3b—Crosslink concentrations at different depths from the exposed surface of bars made from low density polyethylene (LDPE) and a polypropylene homopolymer (PPHO) after three and six weeks of UV exposure.[11]

development of the crosslinking concentration in the two polymers was fairly similar to that of scission (*Figure* 3b) but close inspection of *Figures* 3a and b indicates that the scission/crosslink ratio was much higher for PPHO than for LDPE (see also reference 11), which is consistent with the general experience that polyethylene tends to crosslink more readily than polypropylene.

CRYSTALLINITY MEASUREMENT

Most semi-crystalline polymers normally have a lamellar structure in which thin ribbon-like crystals are constructed from molecule segments in the manner shown in *Figure* 4. The molecules fold at the surface (often called the "fold surface") but do not necessarily re-enter at the adjacent site. Molecules pass through the crystal phase and the surrounding amorphous phase, providing strong adhesion between the two phases. Long-range mechanical integrity is provided by "tie molecules" that enter more than one of the crystal lamellae, connecting them across the intervening amorphous material, and by molecular entanglements within the amorphous phase. The crystalline and amorphous phases have very different properties and the mechanical properties depend strongly on the fraction of material in each phase. Thus, it is of importance to have methods to measure the (fractional) crystallinity. This can be done using XRD or DSC.

XRD Crystallinity Measurement

X-ray crystallinity is normally measured using a diffractometer. The scattered X-rays give strong peaks in directions that obey Bragg's law, but in all other directions there is destructive interference. The amorphous phase scatters X-rays into all directions and the crystalline and amorphous diffraction components are normally separated relatively easily. The X-ray intensities in the crystalline and amorphous parts of the scattering envelope are in proportion to the mass fraction of the respective phases. There are geometric factors to be applied (Lorentz factor, etc.[9]).

DSC Crystallinity Measurement

The DSC crystallinity is obtained from the crystal melting endotherm obtained during a DSC heating run. In order to estimate the fraction of material that was in crystal form prior to melting, it is necessary to know the melting enthalpy for a fully crystalline sample. Since very few polymers can be obtained in the form of defect-free crystals with no amorphous phase, this quantity cannot be measured directly and the value deduced from measurements made on semi-crystalline samples is a source of error. It is noted that if the sample was cooled very rapidly when crystallized from the melt state (e.g., material in the skin of an injection molding, frozen by contact with the cold mold wall), there may be some crystallization during the DSC heating run just prior to reaching the crystal melting temperature, as molecules become mobilized at the elevated temperature. This is another potential source of error in the measured values that are meant to represent crystallinity in the state of the material prior to running the DSC test.

Comparison of XRD and DSC Crystallinity Measurements

Both XRD and DSC crystallinity measurements assume simple two-phase morphology. Inspection of *Figure* 4 reveals that this is only an approximation. At the fold surface the molecule segments entering the amorphous phase are constrained to be nearly parallel to one another in the immediate vicinity of the crystal. Therefore, the material immediately adjacent to the crystal surface will have a structure that is intermediate between that in the fully ordered crystal and the random chain structure associated with the amorphous phase—it is sometimes referred to as the "inter-phase." It is expected that there will be some contribution from this material to the crystalline diffraction peaks, producing some broadening but distinct from the amorphous scattering. Similarly, the inter-phase material will have some influence on the melting of the crystal and will have a secondary effect on the size of the melting endotherm. It is not expected that the contributions of the inter-phase to the respective crystallinity measurements will be the same for the two techniques, and it is reasonable to accept values that are not equal. A further difference between the two methods concerns molecule segments freed from entanglements by photo-initiated chain scission. They may require elevated temperature to become sufficiently mobilized to display secondary crystallization; they may therefore provide enhancement of the crystal signal obtained in DSC but not in XRD.

Sampling

For X-ray diffractometer studies, a flat surface is required. This can be obtained conveniently by milling or microtoming away material to expose material at a chosen depth. The X-ray signal will then come from material close to this surface, with depth limited by the X-ray absorption characteristics of the material. Milling should be done using a single point cutter with fly cutting action at high speed. Using an end mill or a side-and-face cutter should be avoided because the severity of the cutting action results in heat generation and the closeness of the tool to the work prevents the escape of heat, and, thus, unacceptable temperature rise occurs. This may cause morphological change.

For DSC studies, the material ideally should be fairly finely divided. Thin slivers can be taken from the surface using a microtome. Alternatively, the material removed by the same high speed milling procedure described above for preparing a surface for X-ray diffractometer study is suitable for DSC measurement. In the studies conducted in the author's laboratory, the depth of cut is normally 0.1 mm. When the original sample is a standard tensile test bar, this generates sufficient material for DSC measurement (and also for GPC measurement). For bars in which there is a steep depth profile, this sampling interval is larger than ideal. It is a compromise between the conflicting needs of a small measure-

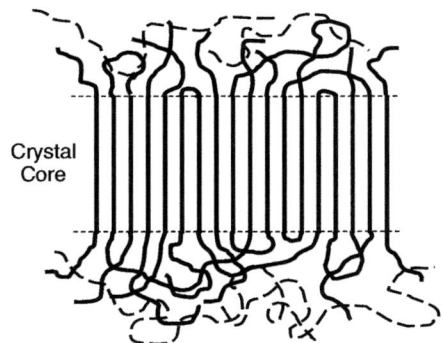

Figure 4—Schematic section of a lamellar crystal. Broken lines represent molecules that reside entirely in amorphous phase.

ment interval and of having sufficient material for reasonably accurate measurement. Thus, the DSC crystallinity measurement represents an average over a depth of 0.1 mm. This contrasts with the XRD measurement that comes from a much narrower depth at the exposed surface.

Results

Example results are shown in *Figure* 5 from a study that used both XRD and DSC measurements on injection molded bars made from a polypropylene copolymer and exposed to UV for different times. Crystallinity measurements by both techniques increased with UV exposure, indicating that secondary crystallization occurred. It is expected that both crystallinity measurements will level off after long exposure times because of the exhaustion of suitable molecule segments for secondary crystallization. Contributions to crystallinity caused by secondary crystallization will diminish at higher exposures because the molecule segments progressively acquire molecular defects that inhibit secondary crystallization.[23] This will not affect XRD measurements and could explain why the XRD crystallinities in *Figure* 5 increase slightly more than those obtained using DSC. More insight into the relative importance of chain scission and the development of molecular defects on the molecules can be obtained by making a second heating run in the DSC.[23] The first heating run provides information on the molded state, as modified by photo-exposure. After melting during the first heating run, the molecules are then crystallized in the DSC equipment under strictly controlled cooling conditions. The second heating run provides information relating to the modified molecules, free from any memory of the original molding operation.

RESIDUAL STRESS ANALYSIS

When polymer moldings cool, residual stresses form when the melt solidifies because differential cooling rates are present, the result of the low thermal conductivity of the material.[24,25] When the surface of the bar drops to below the crystal melting temperature (semi-crystalline polymers) or the glass transition temperature (amorphous polymers), a solid layer forms; the interior is at a higher temperature and liquid-like. As the interior cools and solidifies progressively, it shrinks but the shrinkage is opposed by the solidified outer zone, causing it to go into a state of compression, while tensile stress develops in the interior zones. In the case of a straight bar, the stress distribution can be conveniently measured using the "layer removal procedure" developed by Treuting and Read for metal parts[26] and later exploited with polymers by Broutman[27] and others, as reviewed elsewhere.[24,25] Thin layers are removed from one surface, causing an imbalance in the stress distribution that is relieved by the bar bending to restore moment equilibrium. By measuring the curvature of the bar after each layer removal, the original stress distribution can be derived if the Young's modulus of the material is known.[24–27] The analysis is quite straightforward if the Young's modulus is uniform but more complex if it varies through the depth of the bar, as sometimes happens in moldings that have a penetrant (such as water) that is not uniformly distributed through the depth.[28,29] The Treuting and Read method[26] can be used when bi-axial stresses are present, though this has rarely been exploited in studies of polymers. Examples in which the full bi-axial analysis has been applied are given in references 24 and 30.

Residual stresses cause a molding to distort and may influence the fracture behavior (sometimes beneficially). Much of the research on residual stresses in polymers has been directed at controlling them at the point of processing but this may not always be of critical importance because residual stresses in a molded polymer bar often decay significantly upon ageing at room temperature.[30-32] The use of an elevated temperature increased the changes observed[30] and applying a temperature gradient caused the stresses to become imbalanced and distortion developed.[33] Exposure of one surface of a (nylon) molding to water caused distortion.[34] Changes is stress can be caused by (1) swelling; (2) stress relaxation, which may be accelerated by the plasticizing effect of the absorbed water; or (3) secondary crystallization, also

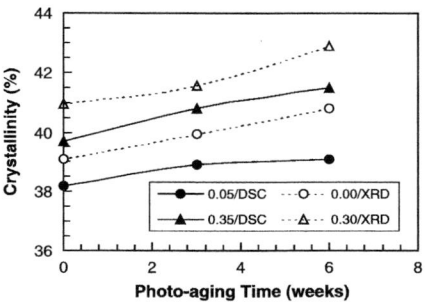

Figure 5—Crystallinity measurements at different depths within a 3 mm thick bar of a polypropylene copolymer after different photo-aging times. DSC measurements were made on samples removed by milling 0.1 mm layers at the surface (0–0.1 mm, denoted by layer mid-point "0.05") and between 0.3–0.4 mm of the surface ("0.35"). XRD measurements were made with a diffractometer using the exposed surface ("0.00") and the surface 0.3 mm beneath the exposed surface, revealed by milling away 0.3 mm ("0.30"). These data were obtained in a study by Craig, White, and Kin.[8]

facilitated by enhanced mobility of molecule segments due to the plasticizing effect of water. Changes in Young's modulus occur when water is taken up and also affect the distortion[34,35]; this must be taken into account when measuring the residual stresses.[28,29]

The residual stress distributions in polymer bars exposed outdoors in a very hot climate have been observed to change very markedly. Several examples have been observed in which the stress near the surface has reversed, becoming tensile.[36,37] This is particularly important because it means that tensile stresses are now located in a region that has become weakened by the molecular degradation that has taken place and is also where there is the greatest chance of finding a flaw (e.g., caused by contact with a foreign body). The presence of residual stresses in this vulnerable region may cause spontaneous failure, without the intervention of external loading. Outdoor exposures are usually applied to one surface so that a temperature gradient is likely to develop during sunny periods and this may cause development of residual stresses, as found in the laboratory tests.[33] In other laboratory experiments it has been established that residual stresses and distortion develop when moldings are exposed to UV even when no temperature gradient is present,[38,39] and it is clear that the temperature gradient effect is not the sole contributor to the observed changes. In semi-crystalline polymers it is believed that the change is residual stress is the result of secondary crystallization that occurs when entanglements in the amorphous phase are released as the result of photooxidative chain scission. This phenomenon has been dealt with in the DSC Crystallinity Measurement section. From measurements of the depth profile of crystallinity changes in polypropylene, residual stress distribution changes were calculated using the density difference between the crystalline and amorphous phases,[40] and showed good agreement with the measured changes in

354

residual stresses distribution.[38,39] In non-crystalline polymers, density changes occur on ageing and this process appears to be accelerated when the polymer is exposed to UV.[41] Therefore, a depth-varying change in density may develop in a molding when exposed to UV, causing changes in residual stress similar to those produced by secondary crystallization in a semi-crystalline polymer. The presence of photo-stabilizers reduces significantly the level of residual stress change during outdoor exposure[42,43] and this supports the idea that densification permitted by chain scission is responsible rather than imbalanced relaxation in a temperature gradient.

Results from a study of residual stress changes in 3 mm thick polystyrene (PS) bars when exposed to UV are given in *Figure* 6.[38] In the as-molded condition, prior to UV exposure, the residual stresses show a fairly typical distribution with modest tensile stresses in a broad zone in the interior of the bar and stronger compressive stresses in a narrower zone near the surface. After four weeks of exposure, a narrow tensile zone had developed immediately at the surface but tensile stresses were still prevalent in the centre with a zone of compressive stress in between. After 14 weeks, the residual stresses had reversed at all locations, becoming tensile in the near-surface regions and compressive in a broad central zone.

MECHANICAL TESTS

It is essential to determine the effect of photodegradation of polymers on their mechanical properties, especially the fracture behavior. This is because a component is normally rendered useless when it breaks, even if its major purpose is not load-bearing.

Impact Testing

The major change in engineering properties caused by the photodegradation of polymers is the reduction in toughness. Although impact tests are designed to assess this property, they are not very popular in photodegradation studies because of the difficulty in interpreting the results. Standard impact tests, such as Charpy and Izod that are based on a pendulum striker, use notched samples and there is an immediate dilemma concerning the introduction of the notch. If it is made after the UV exposure, the notch penetrates the most heavily degraded material and the root of the notch from which the crack grows is located in a less damaged zone and therefore does not probe the most damaged region adequately. On the other hand, if the notch is made before the UV exposure, the notch tip is likely to be located in a region with quite different morphology to the surface, and which is likely to have quite different degradation character-

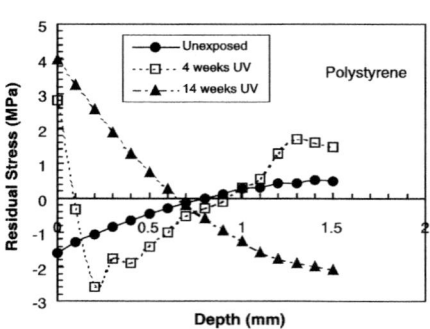

Figure 6—Residual stress distributions as a function of depth from the exposed face of polystyrene bars (3.1 mm thick) in the unexposed state and after exposures of four weeks and 14 weeks. Data from Li Tong; see also reference 37.

istics. Consequently, it is more popular to use tensile tests to assess the progress of pho-todegradation in polymers.

Tensile Testing

The same problem concerning the effect of a notch prevents the use of conventional fracture toughness tests conducted under monotonic loading conditions. Instead it is cus-tomary to employ a standard tensile test and to examine the fracture that occurs at the end of the test. The total area under the stress-strain curve can be considered a measure of toughness and correlates reasonably well with service behavior.

The conventional tensile test in which a specimen bar is held between the grips of a hard-beam machine and one of the beams (crossheads) is driven at a constant speed, is very suited to polymer property assessment. During the early stages of the test, the bar extends in an elastic or quasi-elastic manner, so that the stress applied is proportional or nearly proportional to the strain. In polymers, the departure from a linear relationship is sometimes quite large. In polymers, as with many metals, yield is observed in which there is a sudden reduction in the stress-strain gradient. A neck forms at this point and, with many polymers, the neck stabilizes and the polymer cold-draws. This phenomenon is associated with tough behavior in other loading arrangements and is displayed by poly-carbonate and some other non-crystalline polymers with a good reputation for toughness as well as many semi-crystalline polymers such as polyethylene, polypropylene, and the nylons. Loss of this property following photodegradation is probably the best indicator of all mechanical parameters that the material has become brittle and that a component in the same state of degradation is likely to fail in service. An example is given in *Figure 7*, in which the strain at break is given for samples of a polypropylene copolymer after various UV exposure times.[44] The strain was more than 250% (the limit of the crosshead displacement) for short exposure times and this was maintained after one week exposure, but after three weeks the strain at break fell very considerably and soon afterwards reached catastrophically low values. Samples containing some recycled polymer that had been photodegraded prior to recla-mation showed even more severe degrada-tion than the virgin material after three weeks exposure.

The tensile test can be used to measure the change in Young's modulus and stress at failure caused by photodegradation. Both of these parameters present difficulties in inter-pretation. With many polymers, the surface becomes stiffer on UV exposure, either because of chemi-crystallization or the for-mation of crosslinks. This causes the overall Young's modulus (as measured by the test) to increase, though the test gives an average for the whole bar and the results are domi-

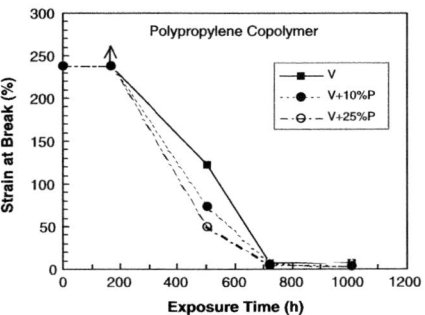

Figure 7—Strain at break of polypropylene copolymer samples after various UV exposure times. "V" indicates virgin polymer while "V+10%P" and "V+25%P" indicate samples made from material containing, respectively, 10% or 25% recycled polymer that had been photodegraded prior to reclaiming.[44]

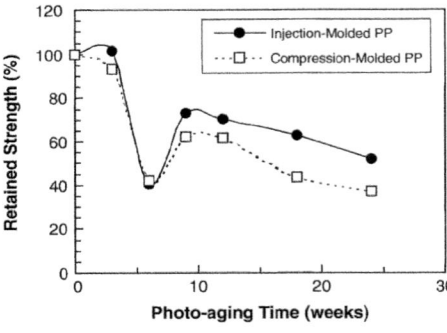

Figure 8—Retained strength versus photo-aging time for (1) injection molded polypropylene and (2) compression molded polypropylene, showing recovery behavior. The standard deviations for four repeated tests were often smaller than the data decals except near the recovery region and are omitted here for the sake of clarity. They can be checked in the original papers.[7,46]

nated by the material in the interior, which remains virtually unchanged. After prolonged UV exposure, the surface becomes embrittled with many polymers and surface cracks appear. This means that the surface zone is no longer load bearing and the test load is supported on a smaller section of the test-bar than that used to calculate the (overall) modulus, which is therefore an underestimate of the Young's modulus in the interior (load-bearing) zone. The stress at failure is of limited value because, even in a heavily degraded sample, failure often does not occur until yield, by which point the sample has reached a load close to the maximum observed in the unexposed state. Under carefully controlled laboratory conditions, the stress at failure can sometimes show a systematic variation that correlates with UV exposure but the interpretation of the results is not always straightforward, partly because of the depth-dependent changes in modulus that are not easily determined and partly because of the influence of the surface cracks that can develop into critical flaws that promote premature fracture. An illustration of the use of failure stress is given in *Figure* 8, where results are given for a series of polypropylene moldings that were exposed for periods up to six months and then tensile tested.[7,45,46] A rapid decay in strength was followed by an apparent "recovery," usually after about six weeks exposure. A brittle surface zone developed after short exposure times and broke easily during the tensile test, providing flaws from which net section fracture occurred. After longer exposure times, the surface zone became so brittle that multiple cracks formed and stress transfer to the relatively undegraded material underneath did not occur so that flaws in the surface zone lost their potency and "recovery" occurred. Under service conditions, the component is likely to have failed before this recovery appears and it is of little practical use, but the results underline the importance of testing samples at closely spaced exposure intervals because it would be possible to miss the property minimum in a series of tests conducted using a large interval, and so obtain an over optimistic assessment of the material.

Fractography

An essential adjunct to the tensile test is the subsequent microscopical examination of the fracture surface and the adjacent region. Scanning electron microscopy is usually preferred because of the large depth of field providing images of rough surfaces superior to those obtained using light microscopy. The fracture surface usually contains characteristic markings that indicate the site at which the fracture initiated. This may be a flaw such as a foreign particle that constitutes a stress concentration, and failure occurs according to the general predictions of fracture mechanics (though quantitative predictions are dif-

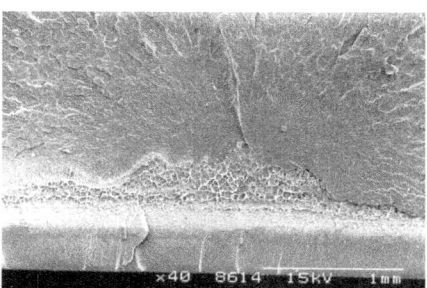

Figure 9b—Polypropylene fracture surface obtained after nine weeks UV exposure. (Image courtesy of M.S. Rabello.)

Figure 9a—Foreign particle that has initiated fracture on the surface of a PP bar exposed outdoors in Jeddah, Saudi Arabia, for three years prior to tensile testing. (Image courtesy of M.M. Qayyum.)

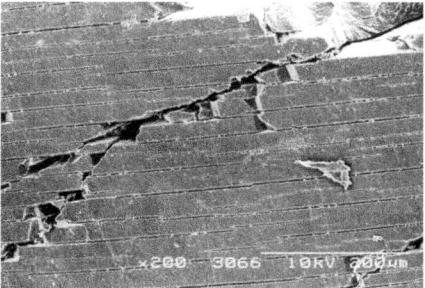

Figure 9c— Fissures on the molded surface of a PVC bar exposed outdoors for four years in Saudi Arabia, then tensile-tested. (See also reference 48.)

Figure 9d—Surface of polypropylene bar after six weeks UV exposure followed by tensile test. Fragments of the embrittled surface layer fell out during testing. (Image courtesy of B.O'Donnell.)

ficult to make because of the viscoelastic nature of the polymer). An example is shown in *Figure* 9a. After UV exposure failure usually initiates in the embrittled surface layer, leaving behind a smooth fracture zone that indicates the depth within which significant damage had occurred (*Figure* 9b) and which offered very little resistance to cracking. Turton and White showed very strong correlation between the smooth zone depth and the region containing advanced chemical degradation, as indicated by molecular weight measurement, in a study of polypropylene.[47] Inspection of the as-molded surface is just as revealing as the fracture surface in the interpretation of the deterioration of properties in photodegradation studies. Multiple fissures often form in the embrittled zone during the tensile test (*Figure* 9c). A dominant flaw often develops and different stages of this can sometimes be identified with the help of a scanning electron microscope.[48] With different polymers and/or different exposure conditions it is sometimes possible to observe that the surface layer has become so fragile that pieces actually fall out and have virtually no adhesion to the underlying (less damaged) material (*Figure* 9d); this is believed to be associated with the phenomenon of recovery, discussed above.

CONCLUSIONS

This chapter shows how the application of a range of techniques can be used to relate changes that occur at the molecular level to engineering property changes when a polymer is exposed to UV. The sub-set of the many techniques available that has been used here is neither unique nor complete. Different methods could be substituted and/or added to provide an equally good or better analysis. Nevertheless, the combination of techniques introduced here can provide a fairly comprehensive picture of the changes that occur.

FTIR analysis of the gas phase adjacent to the sample during UV exposure can be used to indicate that photo-chemical reactions are taking place and to provide a semi-quantitative measurement of the rate. This enables convenient comparison of the susceptibility of different polymers to UV exposure and the effect of different additives and of different exposure conditions (UV intensity, spectral distribution, gas phase composition, etc.).

The molecular characteristic that is most important for the engineering properties is the molecular weight, and this can be measured using GPC. The GPC data can be analyzed to produce MWDs and, from them, scission and crosslinking concentrations can be obtained. These show, among other things, that the shortage of oxygen in the interior of the sample not only reduces the reaction rate but also changes the balance between different reactions, causing the build-up of crosslinks to be relatively higher than the development of scissions compared to observations near the surface.

Conditions favorable to scission exist near the surface of many polymers, including polyolefins and polystyrene, and, in the case of semi-crystalline polymers, this leads to secondary crystallization (chemi-crystallization). Crystallinity and other related morphological characteristics tend to have a very strong influence over mechanical properties. Secondary crystallization occurs preferentially near the surface and this leads to changes in residual stress distribution, with the surface zone becoming tensile. Similar changes occur in polystyrene due to photodegradation-stimulated densification in the amorphous phase. The residual stresses can lead to distortion and will often affect fracture behavior, usually detrimentally because tensile stresses form at the surface in the embrittled zone and produce a much more vulnerable condition than the as-molded state when the tensile stresses are in the interior and low in magnitude.

Tensile tests are normally the most meaningful way to determine the fall off of mechanical properties during photodegradation. The strain to break is very sensitive to the embrittlement of the polymer and is a good indicator of the fall in toughness that normally marks the end of the service life of a component.

References

(1) Davis, A. and Sims, D., *Weathering of Polymers*, Applied Science, London, 1983.
(2) White, J.R. and Turnbull, A., *J. Mater. Sci.*, 29, 584-613 (1994).
(3) Girois, S., Delprat, P., Audouin, L., and Verdu, J., *Polym. Degrad. Stab.*, 56, p. 169-177 (1997).
(4) Gardette, J.-L., Gaumet, S., and Phillipart, J.L., *J. Appl. Polym. Sci.*, 48, p. 1885-1895 (1993).
(5) Hoekstra, H.D., Spoormaker, J.L., Breen, J., Audouin, L., and Verdu, J., *Polym. Degrad. Stab.*, 49, p. 251-262 (1995).

(6) O'Donnell, B. and White, J.R., *Polym. Degrad. Stab.*, 44, p. 211-222 (1994).

(7) Rabello, M.S. and White, J.R., *Polym. Degrad. Stab.*, 56, p. 55-73 (1997).

(8) Craig, I.H., White, J.R., and Kin, P.C., *Polymer*, 46, p. 505-512 (2005).

(9) Campbell, D., Pethrick, R.A., and White, J.R., *Polymer Characterization*, 2nd Ed., Stanley Thornes, Cheltenham, 2000.

(10 Rabello, M.S. and White, J.R., *Polymer*, 38, p. 6379-6387 (1997).

(11) Shyichuk, A.V., White, J.R., Craig, I.H., and Syrotynska, I.D., *Polym. Degrad. Stab.*, 88, p. 415-419 (2005).

(12) Craig, I.H., White, J.R., Shyichuk, A.V., and Syrotynska, I., *Polym. Eng. Sci.*, 45, p. 579-587 (2005).

(13) Craig, I.H. and White, J.R., *Polym. Eng. Sci.*, 45, p. 588-595 (2005).

(14) Christensen, P.A., Dilks, A., Egerton, T.A., and Temperley, J., *J. Mater. Sci.*, 34, p. 5689-5700 (1999).

(15) Jin, C., Christensen, P.A., Egerton, T.A., Lawson, E.J., and White, J.R., *Polym. Degrad. Stab., 91*, p. 1086-1096 (2006).

(16) Jin, C., Christensen, P.A., Egerton, T.A., and White, J.R., *Mater. Sci. Tech.*, 22, p. 908-914 (2006).

(17) O'Donnell, B. and White, J.R., *J. Mater. Sci.*, 29, p. 3955-3963 (1994).

(18) Shyichuk, A.V. and White, J.R., *J. Appl. Polym. Sci.*, 77, p. 3015-3023 (2000).

(19) Shyichuk, A.V. and Lutsjak, V.S., *Eur. Polym. J.*, 31, p. 631-4 (1995).

(20) Shyichuk, A.V., Stavychna, D.Y., and White, J.R., *Polym. Degrad. Stab.*, 72, p. 279-285 (2001).

(21) Shyichuk, A.V., Turton, T.J., White, J.R., and Syrotynska, I.D., *Polym. Degrad. Stab.*, 86, p. 377-383 (2004).

(22) White, J.R., Shyichuk, A.V., Turton T.J., and Syrotynska, I.D., *Polym. Degrad. Stab.*, 91, p. 1755-1760 (2006).

(23) Rabello, M.S. and White, J.R., *Polymer*, 38, p. 6389-6399 (1997).

(24) White, J.R., *Polymer Testing*, 4, p. 165-191 (1984). (Also appeared as Chapter 8 in *Measurement Techniques for Polymeric Solids*, Brown, R.P. and Reed, B.E. (Eds.), Elsevier, Barking, 1984.)

(25) Isayev, A.I., and Crouthamel, D.L., *Polym. Plast., Technol. Eng.*, 22, p. 177-232 (1984).

(26) Treuting, R.G. and Read, W.T. Jr., *J. Appl. Phys.*, 22, p. 130-134 (1951).

(27) So, P. and Broutman, L.J. *Polym. Eng. Sci.*, 16, p. 785- 791 (1976).

(28) White, J.R., *J. Mater. Sci.*, 20, p. 2377-2387 (1985).

(29) Paterson, M.W.A. and White, J.R., *J. Mater. Sci.*, 24, p. 3521-3528 (1989).

(30) Hindle, C.S., White, J.R., Dawson, D., and Thomas, K., *Polym. Eng. Sci.*, 32, p. 157-171 (1992).

(31) Coxon, L.D. and White, J.R., *J. Mater. Sci.*, 14, p. 1114-1120 (1979).

(32) Coxon, L.D. and White, J.R., *Polym. Eng. Sci.*, 20, p. 230-236 (1980).

(33) Thompson, M. and White, J.R., *Polym. Eng. Sci.*, 24, p. 227-241 (1984).

(34) Paterson, M.W.A.and White, J.R., *J. Mater. Sci.*, 27, p. 6229-6240 (1993).

(35) Paterson, M.W.A. and White, J.R., *Polym. Eng. Sci.*, 33, p. 1475-1482 (1993).

(36) Qayyum, M.M. and White, J.R., *J. Mater. Sci.*, 20, p. 2557-2574 (1985).

(37) Qayyum, M.M. and White, J.R., *J. Mater. Sci.*, 21, p. 2391-2402 (1986).

(38) Li Tong and White, J.R., *Plast. Rubb. Compos. Procs. Applics.*, 25, p. 226-236 (1996).

(39) Li Tong and White, J.R., *Polym. Eng. Sci.*, 37, p. 321-328 (1997).

(40) White, J.R., *J. Mater. Sci. Letts.*, 9, p. 100-101 (1980).

(41) Arvidsson, A. and White, J.R., *J. Mater. Sci. Lett.*, 20, p. 2089-2090 (2001).

(42) Qayyum, M.M. and White, J.R., *Polym. Degrad. Stab.*, 39, p. 199-205 (1993).

(43) Qayyum, M.M. and White, J.R., *Polym. Degrad. Stab.*, 41, p. 163-172 (1993).

(44) Craig, I.H. and White, J.R., *J. Mater. Sci.*, 41, p. 993-1006 (2006).

(45) Rabello, M.S. and White, J.R., *Plast. Rubb. Compos. Procs. Applics.*, 25, p. 237-248 (1996).

(46) Rabello, M.S. and White, J.R., *J. Appl. Polym. Sci.*, 64, p. 2505-2517 (1997).

(47) Turton, T.J. and White, J.R., *J. Mater. Sci.*, 36, p. 4617-4624 (2001).

(48) Qayyum, M.M. and White, J.R., *Polymer*, 28, p. 469-476 (1987).

Chapter 24

Polymer Films in Photovoltaic Modules: Analysis and Modeling of Permeation Processes

Michael Köhl, Odón Angeles-Palacios, Daniel Philipp, and Karl-Anders Weiß

Fraunhofer Institute for Solar Energy Systems ISE, Dept. Thermal Systems and Buildings, Heidenhofstrasse 2, 79110 Freiburg, Germany

Polymeric films are commonly used as water-vapor barriers, or as substrate materials for barrier coatings in food packaging, photovoltaic devices, organic light emitting diodes (OLEDs), and vacuum isolation materials. The permeation properties of these films determine the level of water in the polymers and inside the devices. The water can cause corrosion and/or degradation of functional properties of the devices. The temperature of solar devices and the ambient water vapor concentration vary over time according to the solar radiation and the ambient climate. Temperature-dependent permeation and diffusion properties are needed for modeling the water concentration over time in order to predict the long-term behavior and the service life of such devices. This chapter describes the measurement of the temperature-dependent permeation coefficient for polymer films, different laminates of polymer films, and first results of modeling the processes by integrating time series of conditions experienced during outdoor exposure.

INTRODUCTION

Polymeric films are used in very different applications to protect commercial products against ambient substances or mechanical stress. In this chapter we describe the functionality of barrier foils and embedding polymers used in photovoltaic (PV) modules as an example where both properties are relevant.

Materials consisting of back sheets and encapsulants provide a barrier to keep water, atmospheric gases, and pollutants away from the silicon solar cell. Nevertheless, during a service life of more than 25 years with frequently changing meteorological conditions, any migration of different chemical species through the polymer barriers has to be taken into account. The transport mechanisms in polymeric materials are not well understood,[1] and present a very active research field.[2] Methodologies for service life estimation based on successfully established SLP methodologies for other solar energy materials[3] are presently under development.

The determination of the water concentration during the service life of materials under experimental and real environmental conditions is a starting point for understanding

361

many other processes that might occur in materials used in PV modules. Of course, the ingress of other atmospheric gases and pollutants also can play a detrimental role. However, in this chapter we focus on the measurements of water diffusivity in some polymer materials used as encapsulants, namely ethylene vinyl acetate copolymer (EVA), polyvinyl butyral (PVB), and thermoplastic polyurethane (TPU). These measurements are used to estimate the water concentration at equilibrium, and are applied as boundary conditions for simulations under different climates.

THEORETICAL BACKGROUND

Water Concentration Distribution

PV modules are installed in different locations with extremely diverse climatic conditions which vary from place-to-place and from season-to-season. There is evidence that the irreversible changes occurring in materials, and consequently, the service life of PV modules, depend strongly on the local environment.[4] This fact justifies setting up dynamic equations of distributed parameters for modeling the water ingress in PV modules. In our modeling approach the dynamic independent variables are the local weather conditions, namely temperature, relative humidity, global solar radiation, and wind velocity. This set of variables represents the dynamic environment that interacts with the PV module.

Assuming Fickian diffusion of water through the polymers of interest, the time variation of concentration c_i inside each material i is given by

$$\frac{\partial c_i}{\partial t} = D_i \nabla^2 c_i \tag{1}$$

where D_i is the diffusion coefficient of water vapor in material i. In this chapter, we present the calculation of the water solubility that might be used as boundary conditions for solving equation (1). The water concentration in a polymer as a function of the temperature and the relative humidity (i.e., water partial pressure) is estimated using the experimental values of diffusivity and permeability.

Water Content Curves

For estimating the diffusion coefficients of encapsulant materials, the polymer films were considered as membranes; that is, barriers between two phases, one of which is considered as the feed side and the other the permeate side.[5] This corresponds to the geometry of the measurement equipment to be described in the "Experimental Setup" section. As shown below, this membrane approach allows us to relate the diffusivity to the permeation rate for estimating the water content of the tested materials as a

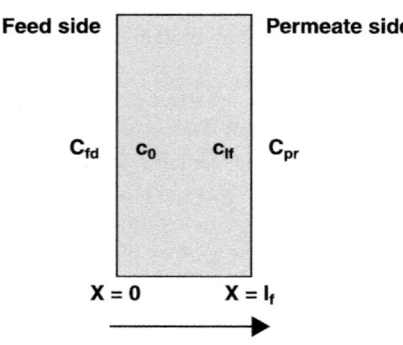

Figure 1—Polymer film considered as a membrane between two gas-filled volumes.

function of temperature and relative humidity. Without losing generality and for the sake of simplicity we consider the one dimensional case shown in *Figure* 1, representing a membrane of thickness l_f. On the feed side of the membrane (left-hand side of *Figure* 1), a gas with concentration C_{fd} is in *equilibrium* with the membrane surface layer with a concentration c_0, while on the permeate side, the gas with concentration C_{pr} is in equilibrium with the concentration in the surface layer of the membrane, c_{lf}.

The mass balance equation for water in the film is given by

$$\frac{\partial c}{\partial t} = -\frac{\partial}{\partial x} j_x \tag{2}$$

where c is the water concentration *in* the film and j_x is the flux in x direction.

We require a relationship between the flux and measurable quantities. Since the water content at saturation in the encapsulant polymers tested in this work is less than 0.1%, we are working with very dilute systems; hence, we assume Fickian diffusion,[6] namely

$$j_x = -D\frac{\partial c}{\partial x} \tag{3}$$

The water concentration distribution in the film is then described by combining equations (2) and (3):

$$\frac{\partial c}{\partial t} = D\frac{\partial^2 c}{\partial x^2} \tag{4}$$

For determining the diffusion coefficient of the film, we consider the steady state diffusion across the film, represented by setting equation (4) equal to zero. For this case the appropriate boundary conditions are

$$v = c_0, x = 0 \tag{5}$$

$$c = c_l, x = l \tag{6}$$

Since the system is in a steady state, c_0 and c_l are independent of time. Hence, the concentration profile through the film is given by

$$c = c_0 + \left(c_{l_f} - c_0\right)\frac{x}{l_f} \tag{7}$$

and the flux across the film is

$$j_x = \frac{D}{l_f}\left(c_0 - c_{l_f}\right) \tag{8}$$

This steady state equation is expressed as a function of film variables only. We use the definition of the partition coefficient,[7] K, to relate the concentration in the membrane, c, in equilibrium with the concentration of the gas outside of the membrane, C:

$$c = KC \tag{9}$$

The partition coefficient is also termed "solubility" by some authors. Using the definition of the partition coefficient for substituting the respective concentrations on both sides of the membrane in equation (8), we have:

$$j_x = \frac{DK}{l_f}\left(C_{fd} - C_{pr}\right) \tag{10}$$

Expressing the concentration outside of the membrane as the ideal gas partial pressures on the respective sides, equation (10) can be rewritten as

$$j_x = \frac{\bar{P}}{l_f}\left(p_{fd} - p_{pr}\right) \tag{11}$$

where $\bar{P}$ is known as the permeation coefficient, equal to $\frac{DK}{RT}$.

For the encapsulation materials considered in this work, we have determined experimental values for the diffusivity and permeation coefficients. With the estimated values of diffusivity, we set up a conventional three-dimensional dynamic partial differential equation like that of equation (4) for the one-dimensional case and calculated water concentration profiles considering the conventional geometry of PV modules. For setting up the boundary conditions for the case of the PV modules, it is necessary to know the water concentration in the surface layer of the film as a function of the partial pressure of water in the environment, which in turn depends on the temperature. We have estimated the water content in the polymer surface layer using equation (9). The partition coefficient was calculated using the temperature dependence of the diffusion and permeation coefficients. According to our measurements, the temperature dependence is an Arrhenius-like function. For the diffusion coefficient we have

$$D = D_0 e^{\frac{E_D}{R}\left(T_0^{-1} - T^{-1}\right)} \tag{12}$$

where D_0 is the diffusivity measured at temperature T_0 and E_D is the activation energy for the diffusion. For the corresponding expression for the permeation, P_0 is the permeation at temperature T_0 and $E_{\bar{P}}$ is the activation energy for the permeation coefficient:

$$\bar{P} = \bar{P}_0 e^{\frac{E_{\bar{P}}}{R}\left(T_0^{-1} - T^{-1}\right)} \tag{13}$$

Using equations (12) and (13) the partition coefficient can be calculated as follows:

$$K = K_0 e^{\frac{\left(E_{\bar{P}} - E_D\right)}{R}\left(T_0^{-1} - T^{-1}\right)} \tag{14}$$

The difference $\left(E_{\bar{P}} - E_D\right)$ in equation (14) has been interpreted as the sorption enthalpy.[1]

The partition coefficient expression was used to estimate the value of the water concentration in the surface layer of the film as a function of the environmental temperature and partial pressure of water, which depends on the relative humidity.

Temperature Distribution

For the temperature distribution in the PV module, a single energy balance was set up for each material. Expressed as a function of the temperature, the energy balance reads

$$\rho_i C_{pi} \frac{\partial T}{\partial t} = k_i \nabla^2 T \tag{15}$$

where ρ_i, C_{pi}, and k_i are the density, heat capacity, and thermal conductivity of material i, respectively.

On the glass cover and the back of the module, a convective boundary condition was considered. For these contributions, we used heat transfer coefficient expressions as functions of the wind velocity v_{wind},[8]:

$$h_{gl} = 4.7 + 3.8 v_{wind} \tag{16}$$

$$h_{bs} = 6.98 + 0.38 v_{wind} \tag{17}$$

h_{gl} and h_{bs} are the heat transfer coefficients for the glass cover and back surface of the module respectively, both in W/m^2, while the wind velocity is given in m/s.

The heat generation due to global solar radiation, q_s, is calculated as follows:

$$q_s = G(\alpha - \eta) \tag{18}$$

where G is the global radiation in W/m^2, α is the solar absorption of the module, and η is the efficiency of the module.

EXPERIMENTAL SETUP

Figure 2 shows a picture of the experimental setup. On the left side of the figure is a climate cabinet, and on the right side a mass spectrometer is shown.

The polymer film to be tested is exposed to the climatic conditions of the cabinet (feed side, *Figure* 1). The permeate side is the lid of a hermetically sealed volume. The sample film is mounted on ultra-high vacuum flanges of up to 250 mm diameter and exposed to defined temperature and relative humidity conditions. The

Figure 2—Experimental set-up: the climatic cabinet and the mass spectrometer.

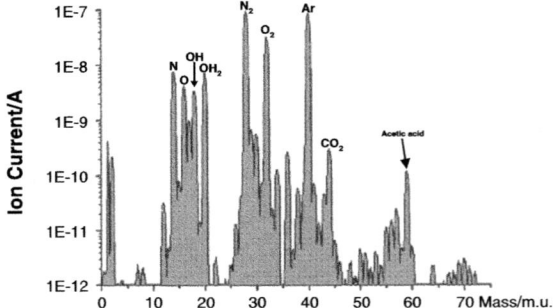

Figure 3—Mass-spectrum of the permeated species from an EVA-sample.

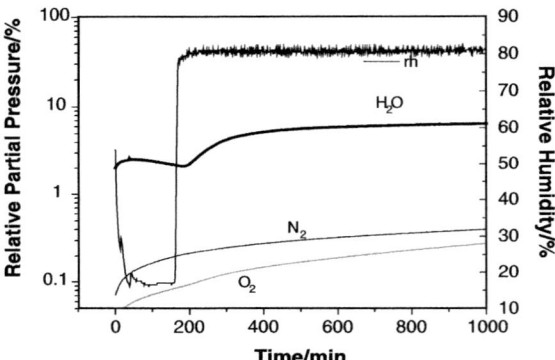

Figure 4—Example for the monitoring of the mass-spectrum and the relative humidity in the climatic cabinet over time.

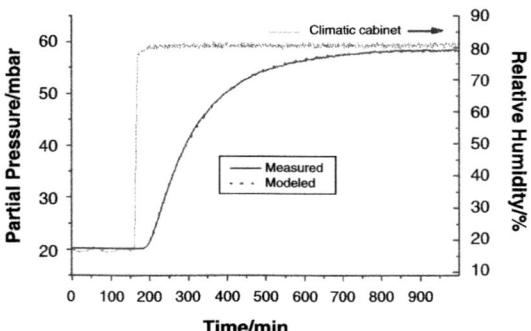

Figure 5—Example for the evaluation of the monitoring of the mass-spectrum and the relative humidity in the climatic cabinet over time.

increase of the humidity content in the volume behind the sample film, which was initially filled with argon under atmospheric pressure, was monitored by the mass spectrometer. Using a mass spectrometer, we obtained a full spectrum of the substances in the test volume. Thus, we were able to track not only the permeation of water vapor, but also the permeation of other substances like O_2 or N_2, which can also be important for the performance and degradation of the materials (*Figure* 3) and to identify substances, which are released from the materials or produced in degradation processes.

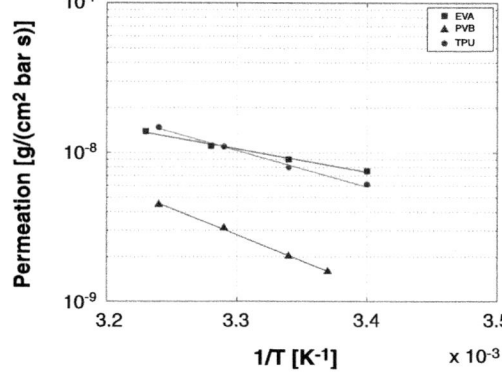

Figure 6—Arrhenius plots on a logarithmic scale of water permeation coefficients for different polymer films used as encapsulant materials in PV modules, measured at different temperatures.

The climate cabinet allows temperature and humidity conditions to vary over a very broad range and therefore facilitates the determination of permeation coefficients for different exposure conditions.

Before the beginning of the experiment, the sample of the polymeric film, a 250 mm diameter circle, and the permeate side volume were pre-conditioned to be completely dry by rinsing with argon at atmospheric pressure. When the experiment starts, the climatic cabinet changes the ambient relative humidity very fast compared to the time-constants of the permeation processes, so that we can assume that the humidity follows a step function (*Figure* 4). Then, water vapor should be absorbed on the film surface and start diffusing through the film. Water molecules that have reached the other side of the film will desorb from the polymer and increase the water concentration in the volume at the permeate side (*Figure* 4). From the time lag and the shape of the answer function to the step function we can calculate the diffusion and permeation coefficients (see *Figure* 5).

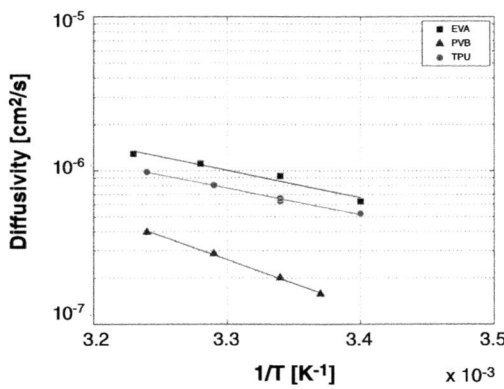

EXPERIMENTAL RESULTS

Figures 6 and 7 show the diffusion coefficients and permeation coefficients of three different polymer films measured at different temperatures. *Table* 1 lists the measured values at a reference temperature of 26°C.

Figure 7—Arrhenius plots on a logarithmic scale of water diffusivity coefficients for different polymer films used as encapsulant materials in PV modules, measured at different temperatures.

Table 1—Measured Values at 26°C[a]

	$D \times 10^7 \left[\dfrac{cm^2}{s}\right]$	$\overline{P} \times 10^9 \left[\dfrac{g}{cm^2 \cdot bar \cdot s}\right]$	$E_D \left[\dfrac{kJ}{mol}\right]$	$E_{\overline{P}} \left[\dfrac{kJ}{mol}\right]$
EVA 9.26	9.04	34.23	29.56
PVB 2.01	2.01	59.08	66.88
TPU 6.31	8.02	33.10	46.57

(a) D stands for Diffusivity, $\overline{P}$ for the Permeability. E_D and $E_{\overline{P}}$ are the Activation Energy for the Diffusion and Permeability, respectively.

Table 2—Permeation Measurements for a Tedlar-PET-Tedlar Back Sheet. The Calculated Activation Energy is 41.6 kJ/mol

T [°C]	31	36	40	50
$\overline{P} \times 10^4 \left[\dfrac{g}{m^2 bar \cdot s}\right]$	0.78	1.08	1.28	2.07

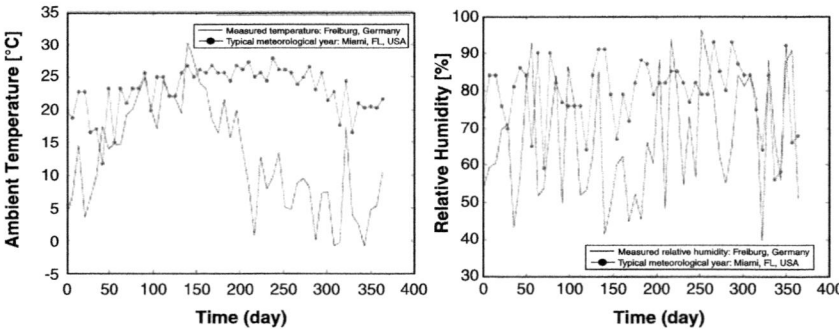

Figure 8—Ambient temperature and relative humidity during one year in Freiburg (Germany) and Miami (USA).

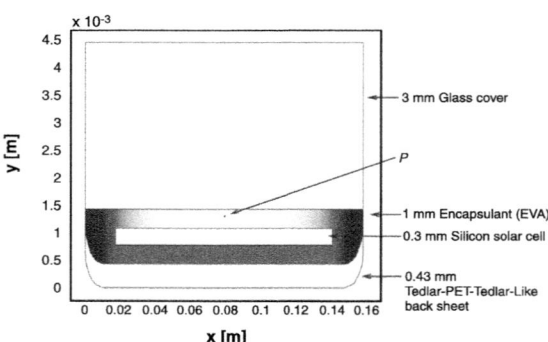

Figure 9—Physical configuration of the PV module used in the simulations.

The experimental setup has also been applied for measuring back sheets. For example, in *Table* 2 permeation coefficients at different temperatures are presented for a Tedlar™-PET-Tedlar back sheet.

With the diffusivity and permeability measured experimentally, the partition coefficient was calculated according to equation (14) and then used in equation (9) to estimate the equilibrium water content of the polymer as a function of the temperature and the relative humidity.

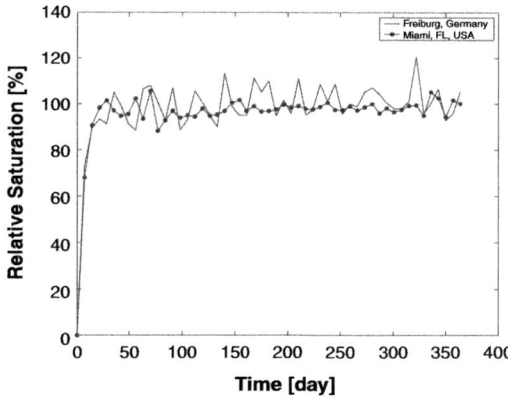

Figure 10—Relative saturation of water in point P of the PV module shown in Figure 9, simulated using weather data of two different regions. An equilibrium of the moisture content is reached at both places relatively fast.

MODELING

The estimated diffusion coefficients and water content curves can be used to simulate the concentration distribution, i.e., to solve equations (1) and (15) with real environmental conditions. Two examples are presented here. The first one uses weather conditions measured in Freiburg, Germany, as dynamic input. These data consist of 5-min averages for each variable. The measured data correspond to March 15, 2003 to March 14, 2004. In the second example, a typical meteorological year of Miami, FL, was taken.[9] In this second example, the weather data are reported as hourly values. To manage the numerical complexity of the model, in both examples the wind velocity was considered to be constant and equal to 1 m/s.

Figure 8 shows the temperature and relative humidity data at the two cities which were used for the calculations.

For the simulation, a 16 × 16 cm² solar module was considered with a 12.5 × 12.5 cm² solar cell. The module consists of EVA encapsulant and a back sheet for which a diffusion coefficient of 5.5 × 10⁻⁷ cm²/s was taken. *Figure* 9 shows the PV module geometry implemented for the simulations. The water concentration in point P of *Figure* 9 was calculated (using equation (1)) and expressed as a percentage of the concentration at saturation. The

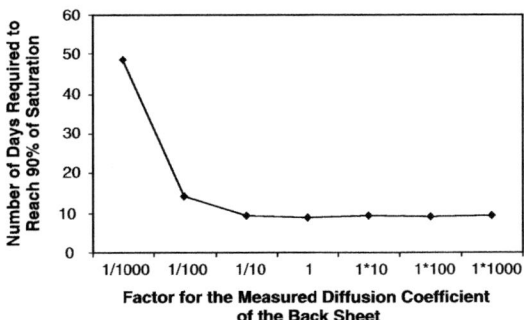

Figure 11—Time required to reach 90% of the saturation for different values of the measured diffusion coefficient $D_o = 5.5 \times 10^{-7}$ cm²/s (at 26°C). These simulations were carried out at constant environmental conditions: air temperature 25°C; relative humidity 80%; global solar radiation 800.0 W/m²; wind velocity 1.0 m/s. Under these conditions the temperature of the module was calculated to be 43°C for an efficiency η of 20%.

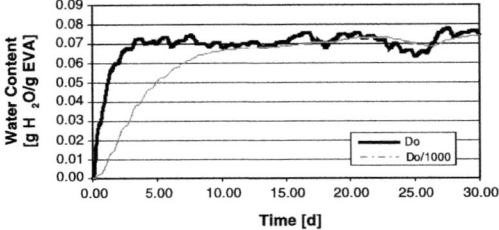

Figure 12—Dynamic simulation of the water content in point P of Figure 9 for weather conditions of Freiburg, Germany (Figure 8). The diagram allows the comparison of the effect of a back sheet with a diffusion coefficient 1000 time lower than the measured value of 5.5 m^2/s for a commonly used laminate.

concentration at saturation was calculated at the present ambient temperature and relative humidity, given by equation (9). *Figure* 10 shows the simulations for one year of exposure time. Simulations like those allow one to determine the magnitude and fraction of time a module is under super-saturation conditions. These conditions may cause the degradation in PV modules.

For determining the influence of the diffusion coefficient of the back sheet, D_o, simulations were carried out for different values of D_o. *Figure* 11 shows the time required to reach 90% of the saturation concentration. For these simulations, a constant value of 0.08 g H_2O/g EVA at the saturation was considered. *Figure* 12 shows dynamic simulations using weather conditions of Freiburg for the first 30 days of the data shown in *Figure* 8. With this diagram two diffusion coefficient values of the back sheet were compared. The results suggest that the major contribution of the humidity ingress comes from the unsealed borders of the mini-module, a real PV-module consisting of many cells should be modelled, and that a very good barrier is needed as back sheet, if the humidity should be kept out of the encapsulation material.

CONCLUSIONS

This chapter shows a new method to measure the temperature-dependent diffusion and permeation coefficients of polymeric materials for water vapor and different other substances in parallel. Experimental results for PV module materials have been presented. Water ingress into PV modules represents an important factor for degradation of materials composing PV modules, although its role has not been clarified. Further investigations are to be done for determining diffusion coefficients and water solubility properties of materials. These efforts help to estimate service life under different climatic conditions by using dynamic simulation of heat- and mass-transfer processes induced by varying border conditions (weather and application) and facilitate the evaluation and usage of new materials.

The approach presented offers a new possibility to support the understanding of functional polymers and barrier foils and their behavior in different ambient conditions. It might also be used for composite films for which effective diffusion coefficients are measured and a gravimetric water uptake measurement is too difficult.

ACKNOWLEDGMENT

Odón Angeles-Palacios thanks Tilmann Kuhn for suggesting the use of correlations for the heat transfer coefficients of equations (16) and (17).

References

(1) Vieth, W.R., *Diffusion In and Through Polymers, Principles and Applications*, Hanser, Germany, 1991.

(2) Kempe, M.D., "Modeling of Rates of Moisture Ingress into Photovoltaic Modules," *Solar Energy Materials and Solar Cells*, (90), 2720-2738 (2006).

(3) Köhl, M., Carlsson, B., Jorgensen, G., and Czanderna, A.W., *Performance and Durabilitiy Assessment of Optical Materials for Solar Thermal Systems*, Elsevier, 2004.

(4) Czanderna, A.W and Pern, F.J., "Encapsulation of PV Modules Using Ethylene Vinyl Acetate Copolymer as Apottant: A Critical Review," *Solar Energy Materials and Solar Cells*, (43), 101-181 (1995).

(5) Mulder, M., *Basic Principles of Membrane Technology*, Kluwer Academic Publishers, The Netherlands, 2000.

(6) *Diffusion in Polymers*, Neogi, P. (Ed.), Marcel Dekker, Inc., 1996.

(7) Cussler, E.L., *Diffusion, Mass Transfer in Fluid Systems*, 2nd Ed., Cambridge University Press, 2003.

(8) ISO Norm: ISO15099.

(9) *User's Manual for TMY2s*, National Renewable Energy Laboratory, 1995.

Chapter 25

3-D Characterization of Multi-Layered Automotive Coating Systems: Stage-1 Analytical Toolbox for Surfaces, Interfaces, and Depth Profiles

Karlis Adamsons

DuPont Company, 3401 Grays Ferry Ave., Philadelphia, PA 19146

Chemical surface/near-surface, interface, and depth profile measurement technologies, as well as traditional microscopy measurement techniques, are very powerful tools in the ever-evolving efforts to study changes in multi-layered automotive coating systems. This chapter is intended as an overview of the analytical toolbox used initially (Stage-1) to obtain a relatively detailed understanding of the chemical composition of coating systems as a function of application environment, formulation, cure conditions, exposure, and storage history. Optical microscopy (OM) and environmental scanning electron microscopy (ESEM) with high resolution digital imaging and elemental analysis (EDS) capabilities are often employed with spectroscopy-based investigations (IR ATR/Trans-mode; UV-Vis) to more completely characterize the chemical changes, species distribution/migration, and identify defect locus/composition. Corresponding sample preparation techniques are noted in each of the case studies presented. Application examples are taken from both plant/field problem-solving cases and service life prediction efforts in characterization of finishes from the automotive OEM and Refinish markets.

INTRODUCTION/HISTORICAL

Durability and service life predictions of polymer-based products have been very important to many businesses worldwide. A broad range of environmental and artificial exposure protocols have been developed over a period of decades, as well as measurement technologies and associated sampling techniques, to drive these efforts in the automotive finishes market (OEM and Refinish) businesses. Tools to track the statistical significance of experimental study results, tools to help improve experimental/project design, and tools to correlate data-based results have evolved steadily. The available information to assist in better material/product design and testing has gone from being largely data-based to now being more knowledge-based. Many investigators have contributed to get to the state-of-the-art in this area, ultimately allowing for development of better materials and products, reducing the product development timeframe, and providing a more realistic assessment of longer-term product performance.

Although these efforts have proven to be quite significant and very helpful in creating better products for the automotive finishes marketplace, the level of performance prediction often tends to be based on bulk chemistry and bulk properties. As product differentiation in the marketplace becomes more subtle, the ability to go beyond bulk characterization is believed to be key for the future. This chapter will provide one view of the future and the analytical tools and sampling techniques that will allow investigators to routinely conduct locus specific investigations.[1-7] The focus herein will be on chemical surface/near-surface, interface, and depth profiling methodologies.[8-21]

Case studies that are shown are primarily taken from the research and development work associated with DuPont's Performance Coatings automotive finishes (OEM/Refinish) businesses. Examples highlighting the utility of the Stage-1 analytical toolbox[2] will also come from plant/field problem solving (or defect root cause analysis) and material characterization studies. Manufacturing (OEM) and weathering-related automotive coating defect studies have previously been reviewed in varying detail,[24-30] including general and specialized measurement technologies. The currently reported work is considered to be additive to the accomplishments of the earlier investigators, particularly from the context of analytical toolbox essentials.

EXPERIMENTAL (STAGE-1 ANALYTICAL TOOLBOX)

Microscopy

OPTICAL MICROSCOPY (OM): The OM analyses can be readily conducted by viewing from top-down (i.e., surface, microtomed section, or isolated particle) or cross-section (normal or oblique cut). The OM images were obtained using a Leica model DM RXA™ microscope, type 020-516.014 III97, equipped with an MTI 3CCD color video camera model DC 330™. A ring light or gooseneck lamp was used for surface illumination of samples. Backlighting was also used for thin cross-sections to help differentiate layers or defects.[21]

ENVIRONMENTAL SCANNING ELECTRON MICROSCOPY (ESEM): The ESEM effectively retains most of the performance advantages of a conventional SEM, but removes the high vacuum constraint from the sample environment. Samples can be potted (epoxy or acrylic). Oily, dirty, and nonconductive samples may be examined in their natural state without modification. The ESEM offers high resolution secondary electron imaging in a gaseous environment of practically any composition and at pressures as high as ~50 Torr. A Philips (now FEI) XL 30 ESEM Spectrometer has been used in all reported studies.

ENERGY DISPERSIVE [X-RAY] SPECTROSCOPY (EDS): EDS microanalysis is frequently very effective in detailed study of microscopic samples, including defect particles, domains, or interface/interphase regions.[22,23] It has an advantage of being sensitive to low concentrations where minimum detection limits (MDLs) are below 0.1% in the best cases and typically less than 1%. Also, the dynamic range runs from the MDL to 100%, with a relative precision of 1 to 5% throughout the range.

Spectroscopy

IR ATR-MODE AND TRANSMISSION-MODE ANALYSES: Infrared (IR) analysis of surface (top-down) chemistry, revealed surface chemistry (via slab microtomy) or isolated material chemistry (extraction, needle harvest) was done either directly using an attenuated total reflectance (ATR) technique or by IR-microscopy in transmission-mode. Depth profiling of the general chemistry through a coating layer or multiple layers was done by ATR analysis of slab microtomed sections cut in sequence. Additional sample preparation was not required. An important advancement in recent years has been the design and application of internal reflection elements (IRE) such as ZnSe, which has a thin layer (wafer) of diamond on the surface. Here, the IRE effectively acts as if it is ZnSe, but with a much more durable surface. Under sufficient contact pressure the material opposite the diamond will deform to provide uniform surface contact, optimizing the quality of the resultant ATR spectrum. The ATR-mode analysis was done using a Nicolet Nexus 470™ FTIR ESP spectrophotometer equipped with a Smart Dura Sampl*IR*™ module. Transmission-mode analysis was done using a Nicolet model 20SXC™ FTIR spectrophotometer equipped with a microscope and XY-translating stage. ATR spectral corrections were made using Nicolet's OMNIC.

UV-Vis Analysis

Slab microtomed sections can be readily solvent extracted to obtain materials such as UV-screeners (benzotriazole or triazine type UVAs), contaminants, or degradation species. Spectroscopic grade methylene chloride (CH_2Cl_2) is the solvent of choice since it is largely UV-Vis transmissive and very effective for swelling and extraction of all automotive coating system layers. The typical size of the slab microtomed section extracted is ~7–10 μm thick and ~2 x 2 cm^2. A 5 mL aliquot of solvent was used to do the extraction over a period of ~1 day. These solutions were run without further dilution or concentration in a standard quartz 1 x 1 x 4 cm^3 cell with cap. This experiment is illustrated in *Figure* 1. The UV-Vis analyses were performed using an HP model 8452A diode array spectrophotometer equipped with the UV-Visible ChemStation Rev. A.06.03 (48).

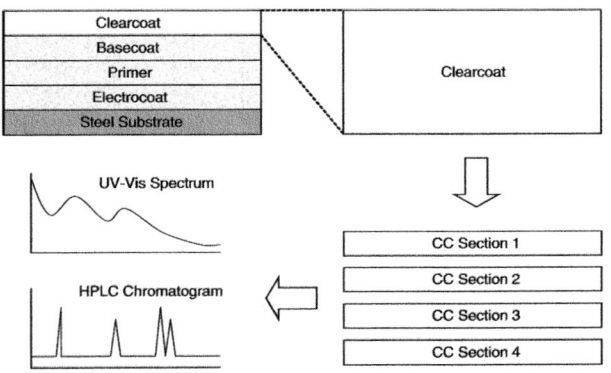

Figure 1—Diagrams of the UVA depth profiling experiments are shown. The CC from a multi-layered automotive coating system is sectioned co-planar to the surface at steps in the ~7–10 micron thickness. Each section is weighed and extracted with 5 mL CH_2Cl_2 solvent in a closed vial. The UVA extract solutions are analyzed by HPLC or UV-Vis techniques. Observed UVA concentrations are normalized to the weight of each section.

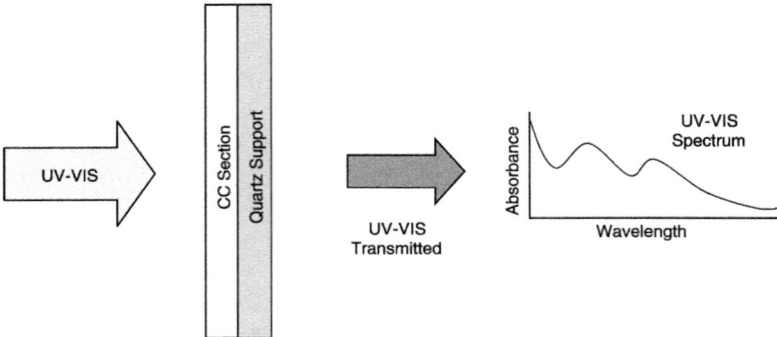

Figure 2—Direct UV-Vis analysis of UVA (benzotriazole) content in thin slab microtomed CC sections. A top-down sequence of these microtomed sections will determine a relative UVA content depth profile. Note that each section is typically cut at the same thickness, thus additional normalization is not required.

Direct UV-Vis analysis of slab microtomed CC sections in transmission-mode is illustrated in *Figure* 2. The sections are found to adhere well to a quartz (UV-transparent) support using a small amount of nonvolatile (UV-transparent) solvent such as n-octanol or n-decanol. Handling of these larger slab sections is relatively easy, relative to handling of the smaller standard microtome sections. This approach is generally limited to the CC

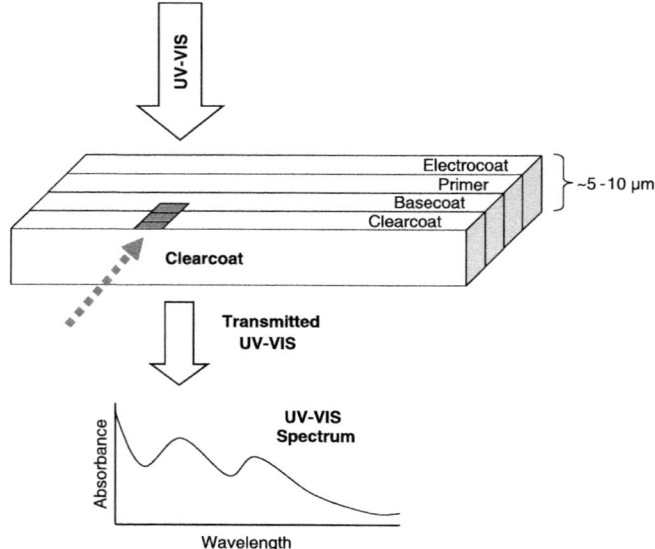

Figure 3—UVA depth profiling approach developed by Ford Scientific (Gerlock and colleagues). A thin cross-section of a coating system is obtained with a standard microtome, anchored on a quartz support. UVA analysis is done by a UV-Vis microscope equipped with an X-Y translating stage.

layer since the remaining layers contain a high loading of particles (i.e., colorant pigment, filler, mica, aluminum flake), which can cause extensive UV-Vis light absorption and/or scattering. A UV-Vis spectrophotometer with standard sample compartment configuration was used, as previously mentioned.

Depth profiling of additives, specifically benzotriazole- or triazine-type UV-absorbers, was also successfully done using UV-Vis spectroscopy. This approach was pioneered by Gerlock and colleagues at the Ford Scientific Laboratories (Ford Motor Company)[8] and is illustrated in *Figure* 3. A standard microtome is used to obtain relatively thin cross-sections (~5-10 µm thickness range) of the automotive coating systems. As previously noted, a nonvolatile (UV-transparent) liquid is used to anchor the small sections to a quartz substrate which is attached to an XY-translating UV-Vis microscope stage.

Sample Preparation Techniques

SURFACE/NEAR-SURFACE SAMPLING: A diamond cutter is frequently used to cut larger panels or parts down to a size that can be located on an OM or ESEM microscope translating XY stage. Typical sample sizes range from ~1 cm² to ~2.5 cm². OM stage top-down sample illumination is by a ring light or two adjustable gooseneck lamps, one on either side. Putty or double-sided tape is often used to anchor the samples onto a larger glass microscope slide, which in turn can be locked onto the XY translating stage.

CROSS-SECTION (DEPTH PROFILE) IMAGING: Normal (90°) cut cross-sections are used commonly for obtaining dimensional information and creating access to defect particles or regions of interest. The samples are typically cut so that they can be potted on edge, sanded, and polished to prepare a cross-section that can be easily imaged with microscopy techniques. Image analysis is routinely used to obtain dimensional information such as thickness of coating system layers or size, number, and location of defects. In the case of larger defects the original cut is made within them requiring a minimal number of polishing steps. The cross-sections are used in both optical microscopy and/or environmental scanning electron microscopy analyses. ESEM imaging also permits access to areas of interest for energy dispersive spectroscopy analysis for determining elemental composition of layers, particles, interphase regions, or other areas (i.e, maps) of interest.

BACK-POLISHING TO CROSS-SECTION DEFECTS: Defects that are both small (micron scale) and located within a coating system layer or at an interface (between layers) are often difficult to cross-section. A procedure that has a relatively high probability of success includes cross-sectioning just outside of the defect particle or area, followed by a number of very fine polishing steps. This approach is described as a type of back-polishing (*Figure* 4). After each of the polishing steps one can then use an optical microscope to determine proximity to or penetration into the defect. Defects that are in the 1–5 micron size range often require numerous fine polishing steps. This careful, but time consuming, approach tends to avoid accidental polishing through a smaller defect. It is common to employ as many as 20–30 of these polishing steps to target near to the middle of such defects.

378

Figure 4—Cross-sectioning and back-polishing sequence to access the buried defect particle. (A) Defect buried in sample volume element, with no direct access. (B) Sectioning to get just outside buried defect. (C) Back-polishing has just touched edge of defect. (D) Back-polishing has just entered defect. (E) Back-polishing attempts to maximize defect access (showing access to near center). An optical microscope is used to iteratively monitor stage of sampling (B–E).

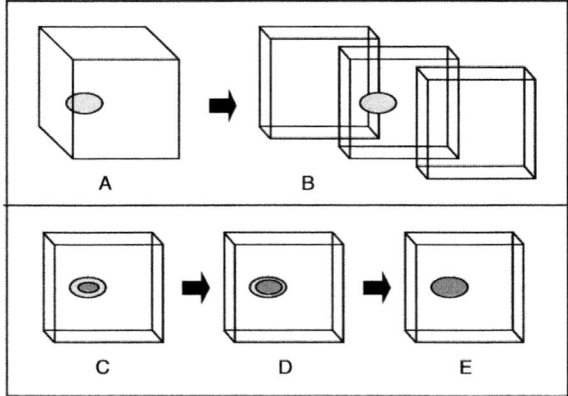

SLAB MICROTOMY SAMPLING FOR DEPTH PROFILING: A Leica Polycut model SM-2500E Slab Microtome was used in sectioning the automotive coating systems co-planar to the surface. The microtoming was effective in reliably cutting coating layers as thin as 5 μm and as thick as 15–20 μm. Once cut, sections can be conveniently stored between sheets of weighing paper. Cutting sections thinner than 5 μm (e.g., 3–4 μm range) often resulted in their shredding, thus destroying their integrity. Also, sections this thin tend not to support their own weight without tearing and, thus, cannot be handled easily with flat-bladed tweezers. However, if section integrity is not important, such as in the case with solvent extraction studies, then harvesting in steps as thin as ~2 μm proved attainable for some OEM automotive CC layers. *Figure 5* shows the Leica Slab Microtome configured with epoxy sample support and cutting block.

SOLVENT EXTRACTION OR WASHING FOR COMPONENT ID: Slab microtomed sections could be effectively extracted with solvents such as methylene chloride (CH_2Cl_2) or

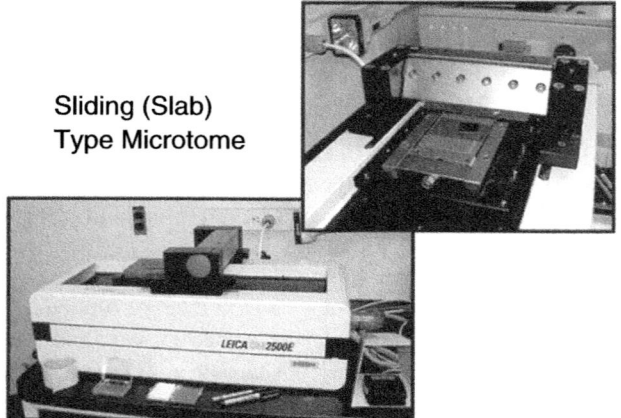

Sliding (Slab)
Type Microtome

Figure 5—Leica model SM 2500E slab microtome used in the microtomy studies. Sections were typically cut 7–10 microns thick for most automotive coating system CC studies.

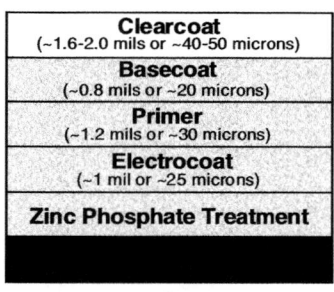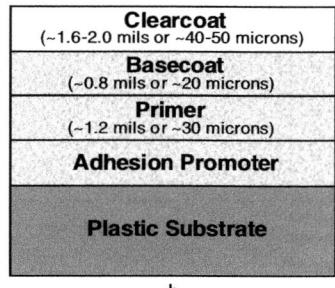

Clearcoat (~1.6-2.0 mils or ~40-50 microns)	Clearcoat (~1.6-2.0 mils or ~40-50 microns)
Basecoat (~0.8 mils or ~20 microns)	Basecoat (~0.8 mils or ~20 microns)
Primer (~1.2 mils or ~30 microns)	Primer (~1.2 mils or ~30 microns)
Electrocoat (~1 mil or ~25 microns)	Adhesion Promoter
Zinc Phosphate Treatment	Plastic Substrate
a	b

Figure 6—Normal cross-sections of most common automotive coating systems, including (a) metal (usually steel) and (b) plastic (usually TPO). Typical thicknesses are shown for the various coating layers.

tetrahydrofuran (THF). These solvents tend to swell the thin sections to a significant degree, allowing high yield extraction of various additives, contaminants, and degradation products. UV-screener (UVA) depth profiling is now routinely done using slab microtomy, solvent extraction, and UV-Vis or HPLC analysis.

Materials

The samples used throughout these studies are either multi-layered automotive coating systems (control or defect containing) or materials isolated from such systems (component aggregates, solvent wash residues, or contaminant materials). In many cases the defect-containing automotive coating systems were obtained from plant or field sources. Case studies include samples from both OEM and refinish product development as well as plant/field problem-solving efforts.

Figure 6 shows typical commercial automotive coating systems in cross-section, including those over metal (usually steel or aluminum) or plastic (usually TPO) substrates. Typical coating system layer configuration and thicknesses are given.

RESULTS AND DISCUSSION

The case studies presented below focus on application of IR, UV-Vis, OM, and/or ESEM/EDS measurement technologies, as well as the associated sampling techniques applied in problem solving or material characterization.

Case Study 1—Tracking Mass Loss

Tracking mass loss in automotive coating system layers, specifically the uppermost (usually CC) layer, is done easily by monitoring the layer thickness changes. This can be particularly useful when the study is done as a function of time and exposure conditions. OM analysis of coating system cross-sections (*Figure* 7) has been done easily using magnifications ranging from 10X to 535X.

Figure 7 illustrates the effect of longer-term exposure to outdoor weather conditions. Typically, CCs that are fully fortified with normal commercial levels of HALS and UVA additives tend to show little mass loss during the first five to seven years of vehicle own-

380

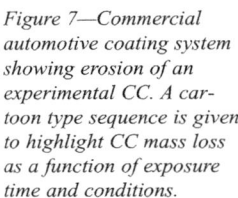

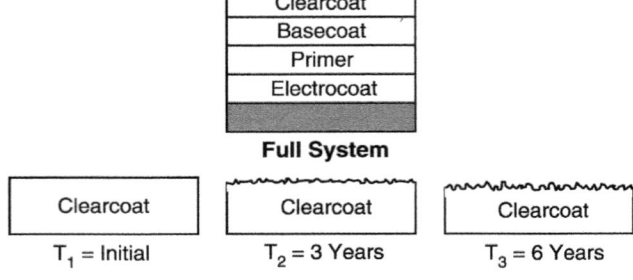

Figure 7—Commercial automotive coating system showing erosion of an experimental CC. A cartoon type sequence is given to highlight CC mass loss as a function of exposure time and conditions.

ership. However, CCs that are not appropriately fortified can result in significant mass loss. Acrylic-melamine-styrene type CCs often show mass loss in stages—initially a slight shrinkage due to loss of alkoxy groups from the melamine crosslinker (e.g., methyl/n-butyl/i-butyl melamine) and later due to erosion (network breakdown) at/near the surface. OM analysis of coating system cross-sections is ideally suited to track mass loss in CCs, or even other layers demonstrating change in thickness.

A common, albeit less direct, way to track appearance changes caused by erosion is through gloss loss type measurements. As erosion progresses the surface roughness tends to increase, resulting in gloss loss. Over the last decade (or so) commercial CC suppliers have identified UV-fortification packages that are very effective in slowing down the at/near surface degradation processes. Currently, hybrid crosslinking systems (e.g., employing both melamine and silane-based crosslinkers) have been successfully developed to work with today's available UV-fortification packages to provide long-term durability.

Figure 8—(a) UVA CC depth profile for virgin (unexposed) automotive coating system. (b) UVA CC depth profile for longer-term multi-year automotive coating system exposure. A slab microtome was used to obtain 7 micron thick sections. Each section was first weighed and extracted by 5 mL aliquot of CH_2Cl_2. A UV-Vis spectrophotometer was used to determine UVA content in each solvent solution, weight normalized.

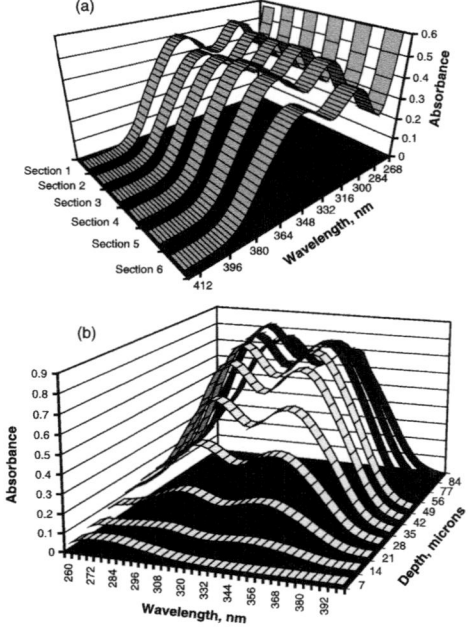

Case Study 2—UVA Depth Profiling

UVA depth profiling of automotive coating system CCs was done successfully using a variety of approaches:

(1) Sectioning co-planar to the surface with a slab microtome, weighing each section, CH_2Cl_2 solvent extraction, and UVA content tracking by HPLC chromatographic or UV-Vis spectrophotometric analysis resulted in easy to obtain and highly reproducible UVA depth profiles. This approach was frequently applied in these efforts and is illustrated in *Figure* 1. A (primarily CC) UVA depth profile from a virgin (unexposed) and longer-term automotive coating system exposure is shown in *Figure* 8a and b. Herein a sigmoid-like depth profile is observed for some longer-term exposures, showing markedly less UVA content at/near to the CC surface and significantly higher content at/near to the CC/BC interface.

(2) Direct (relative) UVA level tracking can be done by adsorbing a thin (~5–10 micron thick) microtomed section onto a quartz slide support followed by UV-Vis analysis (*Figure* 2). A depth profile can be readily obtained by analysis of a slab microtomed sequence (top to bottom) through a CC. Adhering the sections onto a quartz support with UV-Vis transmissive species such as n-octanol or n-decanol was found effective; however, there were considerably more (and variable) scattering results requiring spectral baseline corrections. A standard cuvette holder was configured to hold the CC sections anchored on quartz slides. This approach, since it is not extracting/isolating the UVA from the surrounding CC network, results in CC UVA depth profiles that include other UV-Vis absorbing species present.

(3) An early, but effective, approach to obtain CC UVA depth profiles was developed by investigators (Gerlock[12] and colleagues). Thin cross-sections of a coating system were obtained using a standard microtome, anchored on a quartz support, and transmission-mode UV-Vis analyses were done using a UV-Vis microscope equipped with a X-Y translating stage (*Figure* 3). As in the previous approach with direct UVA tracking, the profiles also include other UV-Vis absorbing species present.

Determination of UVA species loss and/or migration kinetics is done by careful weighing of each slab microtome CC section and thorough UVA extraction using a CC swelling and UVA solubilizing solvent such as CH_2Cl_2. Typically, a time-based study includes a virgin (unexposed) coating system and identical samples that have been exposed (aged) over selected periods of time. Each sample (virgin or aged) is then sectioned using slab microtomy. The concentration of UVA introduced in the virgin CC is known and is used to determine subsequent UVA reductions through loss (surface stripping, volatility), degradation, and/or migration. The weight of each section is used to normalize the UVA signals observed. UVA migration into underlying coating system layers usually can be done by slab microtomy. Occasionally certain coating systems subjected to longer-term aging prove difficult to microtome due to their increasingly fragile nature.

Case Study 3—Particles, Domains, and Interfaces/Interphases

The combination of ESEM and EDS methods allows for efficient analysis of very small (down to sub-micron) defect particles, localized domains, and interface/interphase regions. The electron beam (via ESEM imaging) impacts on a sample at/near to the surface resulting in emission of X-rays from this region whose energies and signal strength provide the relative abundance of each element. In summary, this category of microanalysis gives surface specific elemental composition, as well as micro-morphology detail.[22]

Note that this type of analysis is practically nondestructive in most cases and additional sample preparation steps (e.g., enhancing surface conductivity or forced degassing) are generally not required. However, certain materials such as binder domains or organic colorant materials show localized ablation type damage (i.e., appears as burn-in).

Applications have included documenting elevation of fluorine at the surface of a graffiti-resistant CC, mapping micro-domain damage of a melamine/silane crosslinked CC due to the Jacksonville (FL) acid rain environment, and tracking type and extent of residue materials depositing on panels subjected to outdoor exposure throughout Dade County (FL). The latter application was reported in a presentation at the 2007 ASTM Symposium on Weathering and Durability (January 23, 2007; Ft. Lauderdale, FL).

Case Study 4—Component Mixing and Migration

A melamine/styrene (ratio) depth profile for an automotive CC layer was determined to monitor interlayer mixing and migration. *Figure* 9 shows a melamine/styrene (ratio) depth profile for an experimental CC over a black BC. The automotive coating system had the top two layers, CC and BC, applied "wet-on-wet" and then fully cured. In this case the melamine crosslinker was present only in the BC. As the CC layer is sprayed over a tacky (solvent flashed) BC layer there is some mixing and migration of the melamine into the topmost CC. Styrene was uniformly distributed in the CC and gave (effectively) an internal reference to ratio the melamine against.

Slab microtomy was used as the sampling technique. Each microtomed section was 8 microns thick. IR (ATR-mode with ZeSe IRE) analysis was done from the topside of each section. The IR spectra are thus representative of the top ~3–5 microns of each section. There were six CC sections (~48 microns) cut prior to encountering the BC. This was consistent with the amount of CC applied. Note that transmission-mode experiments for CC sections can also be done successfully; however, in our investigations the upper thickness limit was in the ~7–8 μm range, depending on the innate absorption of the DuPont CCs.

Various CC technologies were explored to determine the extent of mixing and migration. The combination of slab microtomy and IR (ATR-mode) analysis was a relatively

Figure 9—Melamine/styrene ratio depth profile for experimental CC over black BC. Each slab microtomed section was 8 micron thick. IR (ATR-mode) analysis was done from the topside of all sections. Microtomy was done co-planar to the panel surface.

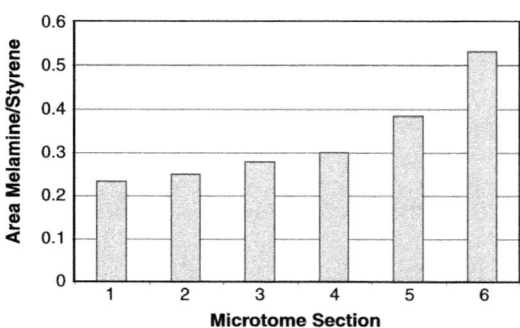

quick and efficient way to obtain the melamine depth profile. If necessary, then even thinner sections of CC can be prepared. The practical limit (in our experience) was ~5 micron thick CC sections. Attempts to cut in the 3–4 micron thick range resulted in some CC shredding. All of the sectioning was done at room temperature with no cooling system.

In this case the melamine/styrene area ratios are used to develop the depth profile histogram (*Figure* 9) and, thus, used as an indicator of relative melamine concentration changes as a function of CC section. It should be recognized that this approach does not directly quantify the melamine content in each section. Also, ATR-mode experiments in these studies were not used to generate a depth profile per se; rather, slab microtome sectioning was used to enable depth locus sampling. IR transmission-mode experiments would work in a similar way.

Although not documented in the current report, "wet-on-dry" CC over BC application is an effective way to determine interpenetration of local free BC melamine into the CC layer. However, the majority of the BC melamine would be incorporated into the BC binder network. The "wet-on-wet" applications can incur considerable material mixing and subsequent migration, particularly at/near to the CC/BC interface. Here the melamine incorporation is not fully realized until the thermal cure is completed.

Work in progress involves monitoring any further changes with respect to the melamine crosslinker as a function of time and exposure conditions. This will provide input on the longer-term stability of the overall coating system. The melamine depth profiling will give locus specific changes and insight on improving coating system design.

Case Study 5—Refinish Repair Studies

A Refinish repair is used as an example showing the similarity and difference between cross-section view layering obtained from OM and ESEM imaging. *Figure* 10a gives the OM cross-section indicating eight distinct layers. *Figure* 10b gives the ESEM cross-section of the same repair area indicating nine distinct layers. All OM layers, except for BC-1, match up with corresponding ESEM layers. The OM BC-1 layer is actually made up of two distinct ESEM layers. *Figure* 11a shows an EDS spectrum giving the elemental composition of the second layer from the top. *Figure* 11b shows an EDS spectrum giving the elemental composition of the third layer from the top.

The third layer has a relatively high chlorine content, suggesting that it may be an adhesion promoter or sealant layer. This was not expected for such a Refinish coating system. One of the middle layers contained fluorine, which was also not expected and could be problematic with regard to longer-term adhesion. In summary, this study was effective at detailing the overall system layering and composition. Inappropriate layers or materials could be readily identified.

This study demonstrates an important point where an investigator may need to exercise caution in assuming that the OM imaging is always sufficient to detail the complete coating system layering. Herein, ESEM could readily identify the number of layers based on electron beam interaction with the polished surface of the cross-section. Also, the associated EDS analysis could clearly differentiate the two layers based on their elemental composition. Note that both OM and ESEM imaging allows one to determine apparent coating system layers thicknesses.

384

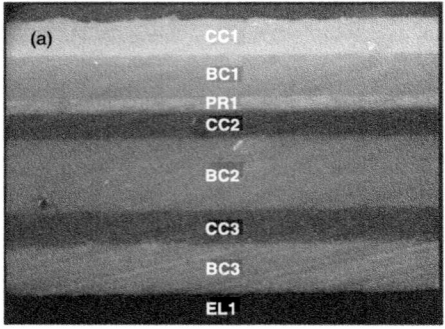

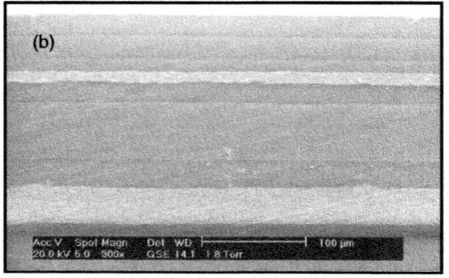

Figure 10—(a) Optical microscopy (OM) cross-section (normal 90° cut) of Refinish field repair showing the apparent layers based on visible range lighting. The layer thicknesses are as follows: CC-1 = 30 μms; BC-1 = 38 μms; PR-1 = 13 μms; CC-2 = 23 μms; BC-2 = 65 μms; CC-3 = 30 μms; BC-3 = 43 μms; EL-1 = 10 μms. (b) ESEM image of same cross-section. Note BC-1 from OM image is now showing two distinct layers in ESEM image.

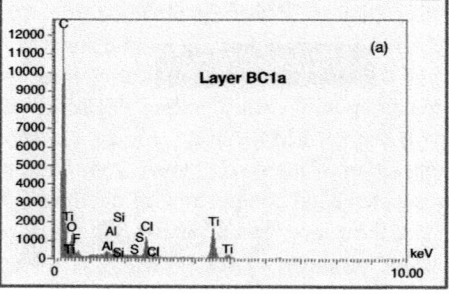

Figure 11—EDS spectra from two layers that compose the OM BC-1 layer. (a) BC-1a is second layer from top. (b) BC-1b is third layer from top. ESEM/EDS analysis can readily differentiate two layers based on interaction with the electron beam and elemental composition.

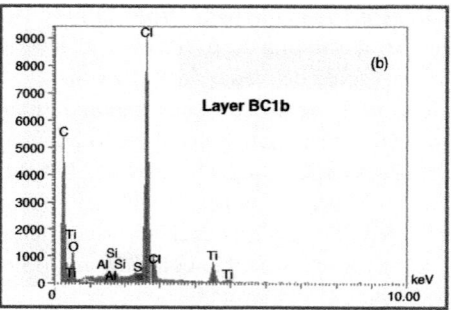

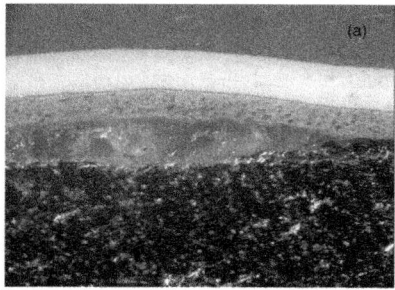 

Figure 12— Coach bus-coating system showing blister formation. The top (white) layer is a monocoat. The second (gray) layer is a primer. The third (dark gray) layer is an anti-corrosion primer. The substrate is an aluminum alloy. Image (a) cross-section shows evidence for corrosion under the primer layer. Image (b) cross-section shows a normal (no defect) paint system.

Case Study 6—Blisters and Corrosion

Blister formation due to substrate corrosion can be determined from the study of the defect cross-section images and by systematic monitoring elemental composition of all apparent layers or regions. *Figures* 12a and b give OM images showing the onset of oxidation of a painted aluminum substrate in the field and same system prior to corrosion, respectively. ESEM/EDS analysis of the region just above the substrate provides an intimate look at the type of corrosion product, aluminum oxide in this case. *Figure* 13 gives OM panoramic images showing oxidation of a steel substrate below a multi-layered coating system. The surface views clearly show blister formation, while cross-section views show the locus and extent of the substrate corrosion.

Even the relatively early stages of blister formation, which manifest as surface roughness (morphology) changes, can be monitored with gloss or profilometry measurements.

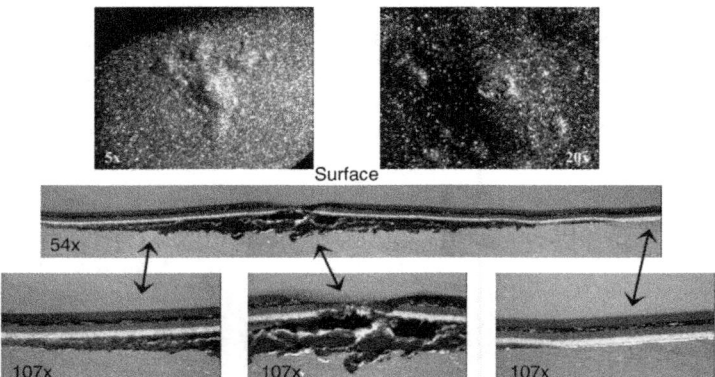

Figure 13—A standard automotive OEM coating system exhibiting a corrosion-type failure. Optical microscopy surface (topdown) views show examples of localized blisters. Cross-section views provide a panoramic overview of a single blister showing rust formation and system disruption.

However, obtaining a cross-section will clearly document the underlying substrate corrosion. This helps determine if the anti-corrosion layer components are appropriate for longer-term coating system durability or if a localized breakdown (e.g., micro-cracks, missing or thin EC layer, trapped water in PR or EC layers, or substrate surface contamination) had occurred.

Case Study 7—Bubble Type Defects

Bubble type defects can be imaged in cross-section using OM or ESEM methods. *Figure* 14a provides an OM cross-section of a bubble-containing anti-corrosion primer over an aluminum substrate. This bubble was introduced during primer application and was quickly trapped while undergoing cure. Bubbles of this type can be significantly smaller and appear as if suspended within a given layer. The common indication of these defects is a slight rise of the layer surface over top of the bubble. *Figure* 14b provides an OM cross-section of a multi-layered automotive coating system over a steel substrate containing a bubble in the BC (colorant) layer. This imaging clearly shows the distribution of aluminum flake within the BC, as well as the shape and size of the bubble defect. The inside surface of the bubble contains a thin film of gray-white aluminum oxide. This was readily determined by using the cross-section in an ESEM/EDS analysis. The probable cause of this defect is degassing due to poor or inappropriate aluminum flake passivation. This sample contained many such defects suggesting that there was a gas producing reaction between the aluminum flake and binder components.

The presence of bubbles can be problematic for a coating system in several ways. The initial diagnostic would be surface irregularities, but the leveling (rheology) of upper layers may initially conceal smaller bubbles. A cross-section of a defect area will clearly show the size and locus of any bubbles. Over time a bubble may allow water easier access to a metal substrate. This could result in subsequent and often early onset of corrosion. Crack formation and propagation may result from bubbles at/near an interface. Thermal

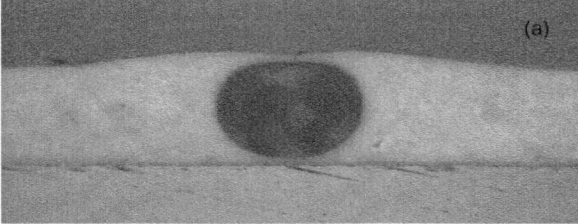
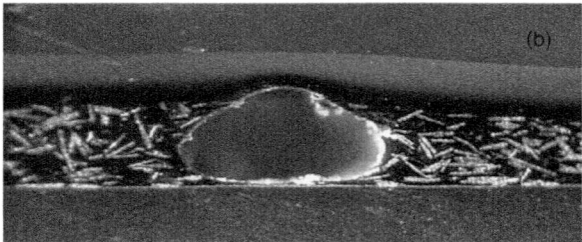

Figure 14—(a) OM cross-section image of anti-corrosion primer layer over metal (aluminum) substrate containing bubble type defect. The bubble extends from the substrate to the surface of the primer; (b) OM cross-section image of multi-layered automotive coating system over metal (steel) substrate containing bubble type defect. The bubble extends from the bottom to the top of the aluminum flake BC layer. A thin film of gray-white aluminum oxide is shown along on the inside walls of the bubble.

cycling (e.g., daily or seasonal) may accelerate such a breakdown. Degassing can be caused by poor application of an anti-corrosion layer (e.g., zinc metal shards in a zinc phosphate treatment layer). Rapid degassing can cause micro-channels to form, resulting in water/contaminant access to the metal substrate.

Case Study 8—Adhesion Failure Studies

Adhesion failures are observed to fall into several categories, including adhesive, cohesive, or mixed (adhesive and cohesive). A case detailing adhesive failure of a paint system over plastic substrate is shown in *Figure* 15a. The cross-section view clearly reveals the delamination of the coating system from the plastic substrate. After (easily induced) delamination of the coating system from substrate, a solvent wash using (spectroscopic grade) petroleum ether was done, the solvent was evaporated leaving an oily residue, and an IR analysis was done to identify the material. *Figure* 16a shows the IR spectrum, which was consistent with (primarily) poly dimethyl siloxane (PDMS). A case detailing adhesive failure of the upper three layers (CC/BC/PR) from the EC layer is shown in *Figure* 15b. The cross-section view clearly reveals the locus of failure. After (easily induced) delamination of the top layers from the EC layer, both IR and EDS analyses were done of the material at the surface of the EC. *Figure* 16b shows the EDS spectrum revealing mostly caustic NaOH. *Figure* 16c shows the IR spectrum, which is also consistent with NaOH. The source of the thin layer of caustic was determined to be due to an incompletely water rinsed caustic wash solution. A case detailing cohesive failure of a PR layer within a full coating system is shown in *Figure* 15c. The cross-section view demonstrates the breakdown of the PR layer. Often a small region is pulled apart (forced delamination) and IR and/or EDS analysis is done to confirm the cohesive nature of the breakdown.

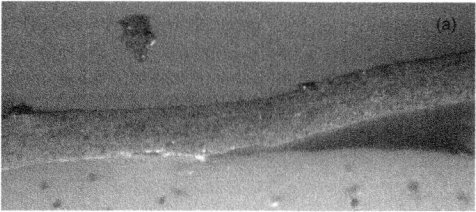

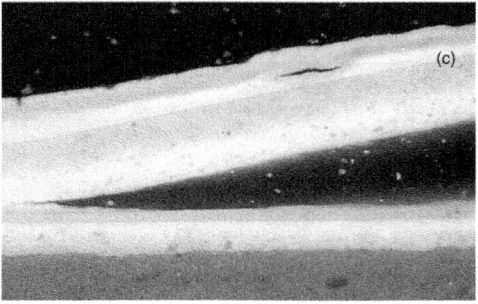

Figure 15—(a) OM cross-section image shows adhesive type adhesion (debonding) failure between plastic substrate and a single layer soft touch coating system; (b) OM cross-section image shows adhesive type adhesive (debonding) failure between coating system layers. The adhesion failure is between the EC and PR layers; (c) OM cross-section image shows a cohesive type failure in the PR layer. The breakdown is within the PR layer and not at either the BC/PR or PR/EC interfaces.

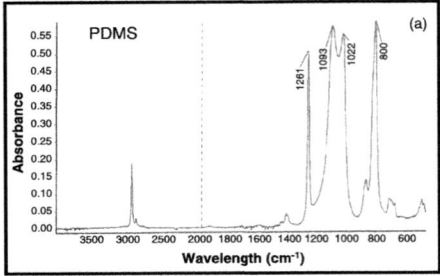

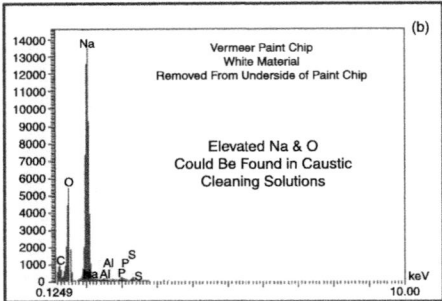

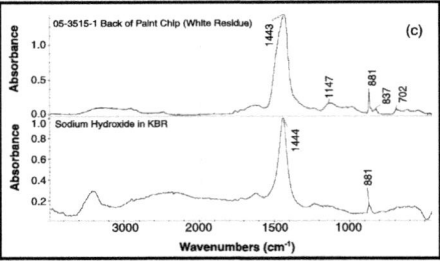

Figure 16—(a) IR spectrum obtained from petroleum ether (PE) solvent wash extract residue of plastic substrate revealed following debonding of the coating system. The spectrum is very similar to that of poly dimethyl siloxane (PDMS) oil; (b) EDS spectrum of interface revealed after debonding of paint chip containing CC/BC layers. Primary elements observed include Sodium (Na) and Oxygen (O); (c) IR spectrum of the same revealed interface and spectral match (NaOH) from MRL IR database.

Case Study 9—Photooxidation Index Studies

Ford Scientific[12,18,20] had developed a simple, yet effective, way to rate and rank automotive CC layer performance with respect to (primarily) photooxidation and hydrolysis. *Figure* 17 shows a photooxidation index (POI) study comparing the same CC with various levels of UV fortification. An unfortified CC (no UVA, no HALS) is compared with three levels of fortification, including UVA-only, HALS-only, and fully fortified (UVA and HALS). Essentially, the total peak envelopes of the hydroxy, amine, and carboxylic acid IR spectral components (peak areas), [–OH, –NH, –COOH], are in ratio over the methylenic peak envelop, [–CH], and monitored as a function of exposure time and conditions. This approach has been useful in helping optimize the UV fortification package to be applied in a given commercial CC. This type of analysis was amenable to depth profiling. *Figure* 18 shows a network cartoon with various stages of degradation (i.e., photooxidation and hydrolysis). Slab microtomy allowed sectioning coplanar to the CC surface, providing a series of slices throughout the CC. POI calculations are readily obtained from each IR spectrum in sequence. These results provide a detailed look at photooxidation and hydrolysis at a given CC depth. This permitted investigators a way to monitor the effectiveness of a specific UVA or HALS, a way to determine optimum UVA and HALS concentrations, as well as a way to compare co-polymerizable and co-condensable with standard (free add) UVA and HALS species.

Depth Profiling—The Bigger Picture

This chapter highlights a variety of depth profiling applications in the study of automotive finishes. *Figure* 19 provides a chart of common automotive coating system prob-

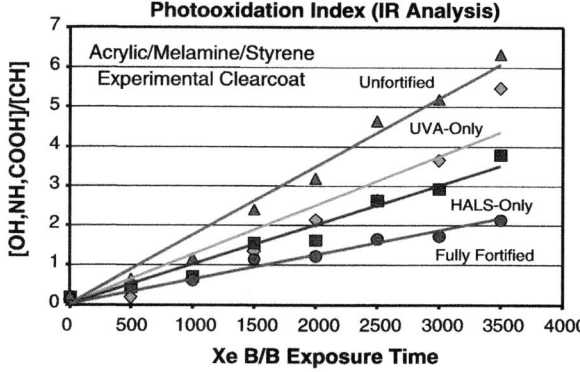

Figure 17—Photooxidation index (POI) study comparing a fully fortified (containing both HALS and UVA) experimental acrylic-melamine-styrene automotive coating system CC vs. the same with HALS-only fortification, UVA-only fortification, and unfortified (no HALS or UVA). The fortification levels are consistent with commercial systems.

lems or failures, underscoring those where such depth profiling techniques have proven effective in day-to-day problem resolution. Several categories were also given wherein depth profiling was key in root cause identification or defect material characterization.

Preparation of cross-sections for OM and/or ESEM/EDS analyses allows an efficient way to determine system details such as layer thickness, non-uniformity in layer thickness, strike-in (extensive mixing) between layers, number of coating system layers, and defect (size, shape, number, and elemental composition) analysis.

The information obtained (to date) by chemical depth profiling falls into various categories. Included among them have been the following: (a) general chemical composition; (b) pigment, microgel, mica, or metal flake type and concentration; (c) UVA, HALS, or catalyst type and concentration; (d) chemical degradation; (e) solvent trapping; (f) crosslinker distribution; (g) wet-on-wet binder/component mixing; and (h) component migration.

In the case of tracking UV-absorber depth profiles, a variety of issues have been involved. The key issues include the following: (a) performance of CC (or BC) in longer-term field applications; (b) photo-stability (primarily) and (in general) resistance to chemical degradation; (c) thermal permanence; (d) migration; (e) solubility; and (f) segregation.

A broad range of measurement technologies has been utilized thus far.

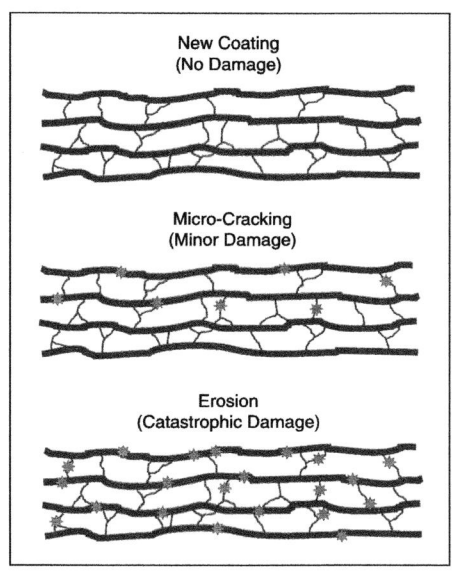

Figure 18—Cartoon illustration of automotive CC network that is undergoing degradation, leading to material erosion. The star symbol indicates scission of chemical bonds.

> Cratering	> Wrong Color
> ***Adhesion Failure***	> ***Bleaching***
> Dirt/Grit Particles	> Orange Peel
> Gel Particles	> ***Scratches***
> Pigment Aggregates	> ***A/B Etching***
> ***Blistering***	> Segregation
> ***Chipping***	> Poor Hiding
> ***Cracking***	> Spitting
> ***Mottling***	> Fibers
> ***Layer Thickness***	> Wrinkling
> Pinholing	> Bubbles
> Slow Drying	> ***Exudates***
> Telegraphing	> ***Migration***
> Flake Orientation	> ***Staining***
> ***Oil/Wax Residue***	> ***Corrosion***

Figure 19—Summary of common automotive coating system problems or failures. Those items that are in bold/underscored are periodically associated with aging, durability, and SLP issues.

They include IR and Raman (for tracking general chemistry); UV-Vis and HPLC (for tracking UV-screener type/content); AA and ICP (for tracking catalyst or crosslinker levels); DSC (for tracking T_g values or crosslinker type); potassium ionization of desorbed species (K^+IDS) mass spectrometry (for tracking species migration and degradation products); ToF-SIMS and ESCA (for tracking degradation products); optical microscopy (for tracking 3D morphology, layer thickness, and defect location); ESEM (for tracking

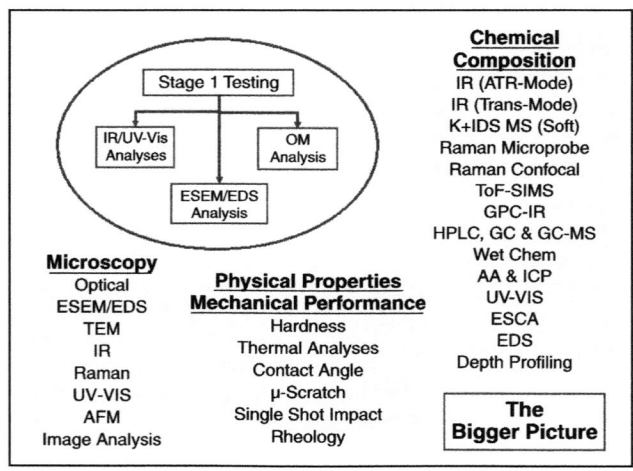

Figure 20—An overview for an automotive coating system and associated defect analyses that go beyond the Stage-1 testing. These category listings are not meant to be all-inclusive, but to give a sense of the range of measurement technologies employed.

3D morphology, layer thickness, and defect location); and EDS (for tracking elemental composition at a given locus or in area mapping). *Figure* 20 offers an overview of measurement technologies available that go well beyond the Stage-1 analytical toolbox.

SUMMARY

Locus specific analysis of multi-layered automotive coating systems has permitted investigators access to various types of depth dependent information. Studies included in this manuscript focused on the following reasons for chemical profiling as a function of depth: surface/near-surface composition elucidation; component distribution, gradient or migration monitoring; domain mapping in 2D or 3D; interface or interphase characterization; network crosslinking density profile determination; locus and mode of adhesion failure analysis; contaminant identification and mapping; and polymer, network, or additive degradation tracking. The Stage-1 analytical toolbox includes a simple, yet effective, group of sampling techniques and measurement technologies. These capabilities are now business critical in reducing the paint/coating product development timeframe, permitting more realistic service life predictions, and troubleshooting in day-to-day plant/field problem resolution.

Many of the cases presented show how the Stage-1 analytical toolbox can give investigators the ability to better understand the impact of various material locus factors on coating system function and longevity. The combination of the IR, UV-Vis, OM, and ESEM/EDS measurement technologies has proven effective and critical to success in the automotive finishes markets.

ACKNOWLEDGMENTS

The research reported here involves a number of collaborations and detailed discussions involving DuPont R&D laboratories (Marshall R&D Laboratory, Experimental Station, Troy R&D Laboratory, AJAX R&D Laboratory), DuPont suppliers (Ciba, Cytec), and a DuPont OEM customer (Ford). Bob Matheson (formerly DuPont Performance Coatings) has been a long-time advocate and mentor for these research efforts to build a better "Toolbox." John Gerlock, David Bauer, Roscoe Carter, Mark Nichols, and Cindy Peters of Ford have been very important in underscoring the materials characterization needs and in developing measurement technologies for determining durability of automotive finishes. Nancy Cliff and Mouhcine Kanouni of Ciba were key in many discussions, as well as partners in many depth profiling efforts. Gottfried Haacke (Research Fellow, now retired) of Cytec deserves much of the credit in our effectively using slab microtomy in many of our depth profiling applications. Kate Stika, Barbara Wood, Dennis Walls, Greg Blackman, Kathy Lloyd, and Dennis Swartzfager of DuPont's Experimental Station were significant contributors during the last decade in these (depth profiling) efforts. Ken Leavell, Jim Halpin, Jenny Campbell, Joanne Paci, John Mclaughlin, Lenny Abbott, Dom Barsotti, Bill Simonsick, and Lance Litty of DuPont's Marshall R&D Laboratory are credited in assisting on various depth profiling projects.

References

(1) Haacke, G., *Book of Abstracts*, 215th ACS National Meeting, Dallas, TX, March 29–April 2 (1998).

(2) Adamsons, K., "3D Characterization of Multi-Layered Automotive Coating Systems: Stage-1 Analytical Toolbox for Surfaces, Interfaces and Depth Profiles," *PMSE Preprints*, 95, p. 198 (2006).

(3) Adamsons, K., "Chemical Depth Profiling of Multi-Layer Automotive Coating Systems," *Prog. Org. Coat.*, 45 (2-3), p. 69–81 (2002).

(4) Adamsons, K., "Chemical Depth Profiling of Automotive Coating Systems Using Slab Microtome Sectioning with IR/UV-Vis Spectroscopy and Optical Microscopy," *J. Coat. Technol.*, 74, No. 924, 47-54 (2002).

(5) Adamsons, K., "Chemical Depth Profiling of Automotive Coating Systems Using IR, UV-Vis and HPLC Methods," In *Service Life Prediction: Methodology and Metrologies*, ACS Symposium Series, 805, pp. 185-211 (2002).

(6) Adamsons, K., Litty, L., Lloyd, K., Stika, K., Swartzfager, D., Walls, D., and Wood, B., "Depth Profiling of Automotive Coating Systems on the Micrometer Scale," *Service Life Prediction of Organic Coatings: A Systems Approach*, ACS Symposium Series 722, pp. 257-287 (1999).

(7) Adamsons, K., Lloyd, K., Stika, K., Swartzfager, D., Walls, D., and Wood, B., "Characterization of Multi-Layered Automotive Paint Systems Including Depth Profiling and Interface Analysis, Symposium on Interfacial Aspects of Multicomponent Polymer Materials," *Symposium Proceedings*, pp. 279-300 (1997).

(8) Haacke, G., Brinen, J.S., and Larkin, P.J., "Depth Profiling of Acrylic/Melamine Formaldehyde Coatings," *J. Coat. Technol.*, 67, No. 843, 29-34 (1995).

(9) Urban, M.W., "Multi-Dimensional Surface and Interfacial Analysis of Polymers and Coatings: ATR, Step-Scan Photoacoustic, FT-IR/FT-Raman Imaging," *Polym. Mater. Sci. Eng.*, 78, pp. 18-19 (1998).

(10) Ochial, S., "Analysis of Depth Profile of Coating Films by Fourier Transform IR Spectroscopy," *Toso Kogaku*, 20 (5), pp. 192-195 (1985).

(11) Cliff, N., Kanouni, M., Adamsons, K., and Yaneff, P., *Proc. International Waterborne, High-Solids, and Powder Coatings Symp.*, pp. 29-46. (2003).

(12) Gerlock, J.L., Smith, C.A., Cooper, V.A., Dusbiber, T.G., and Weber, W.H., "On the Use of Fourier Transform Infrared Spectroscopy and Ultraviolet Spectroscopy to Assess the Weathering Performance of Isolated Clearcoats from Different Chemical Families," *Polym. Degrad. Stab.*, 62 (2), pp. 225-234 (1998).

(13) Gerlock, J.L., Prater, T.J., Kaberline, S.L., Dupuie, J.L., Blais, E.J., and Rardon, D.E., "18O Time-of-Flight Secondary Ion Mass Spectrometry Technique to Map the Relative Photo-oxidation Resistance of Automotive Paint Systems," *Polym. Degrad. Stab.*, 65 (1), pp. 37-45 (1999).

(14) DeVries, J.E., Haack, L.P., Prater, T.J., DeBolt, M., Gerlock, J.L., Holubka, J.W., and Dickie, R.A., "Characterization of Polymer/Substrate Interfacial Chemistry by Spatially Resolved Surface Analytical Methodologies," *Polym. Mater. Sci. Eng.*, 67, pp. 69-70 (1992).

(15) Pennington, B.D., Ryntz, R.A., and Urban, M.W., "The Effect of Thermoplastic FT-IR Spectroscopic Studies," *Book of Abstracts*, 214th ACS National Meeting, Las Vegas, NV, Sept. 7-11, PMSE-337 (1997).

(16) Carter III, R.O. and Bauer, D.R., "Infrared Methods for Coating Analysis," *Polym. Mater. Sci. Eng.*, 57, pp. 875-879 (1987).

(17) McEwen, D.J. and Cheever, G.D., "Infrared Microscopic Analysis of Multiple Layers of Automotive Paints," *J. Coat. Technol.*, 65, No. 819, 35-41 (1993).

(18) Gerlock, J.L., Smith, C.A., Carter III, R.O., Dearth, M.A., Korniski, T.J., and Dusbiber, T.G., "Paint Weathering Tests: Transition from Art to Science," *Surf. Coat. Aust.*, 34 (7), pp. 14-16 (1997).

(19) Carter III, R.O., "Ultraviolet Photoacoustic Spectroscopy Evaluation of the Distribution of Ultraviolet Absorber of Paint Additives as a Result of Processing," *Opt. Eng.*, 36 (2), pp. 326-331 (1997).

(20) Smith, C.A., Gerlock, J.L., Kucherov, A.V., Misovski, T., Seubert, C.M., Carter R.O. III, and Nichols, M.E., "Evaluation of Accelerated Weathering Tests for Automotive Clearcoats Using Transmission Fourier Transform Infrared Spectroscopy," *Proc. 80th Annual Meeting Technical Program of the FSCT*, pp. 1-33 (2002).

(21) Benko, J.J., "Investigating Coating Defects with Optical Microscopy and FT-IR Microspectrometry," *Microscope*, 47 (3), pp. 141-146 (1999).

(22) Vaughan D. (Ed.), *An introduction to Energy-Dispersive X-ray Microanalysis*, Kevex Instruments, Inc., 355 Shoreway Rd., San Carlos, CA, 94070-1308, 1989.

(23) Keene, L.T., G.P. Halada, G.P., and C.R. Clayton, C.R., "Failure of Navy Coating Systems 1:

Chemical Depth Profiling of Artificially and Naturally Weathered High-Solids Aliphatic Poly(ester-urethane) Military Coating Systems," *Prog. Org. Coat.*, 52 (3), pp. 173-186 (2005).

(24) Schoff, C.K., "Automotive Coatings Defects Part 1: In-Plant and Field Defects," *JCT CoatingsTech*, 1, No. 2, pp. 34-39 (2004).

(25) Schoff, C.K., "Automotive Coatings Defects Part 2: Weathering Processes and Their Effects on Coating Properties," *JCT CoatingsTech*, 1, No. 3, pp. 22-26 (2004).

(26) Schoff, C.K., "Painting Problems," *Mater. Eng.*, 21 (Coatings of Polymers and Plastics), Marcel Dekker, Inc., pp. 203-241 (2003).

(27) Schoff, C.K., "Surface Defects: Diagnosis and Cure," *J. Coat. Technol.*, 71, No. 888, pp. 56-73 (1999).

(28) Reinartz, R., Reader, J., des Courtis, F., and Gauvin, N., "Elimination of Surface Defects in Waterborne Automotive Refinish Paints," *European Coat.*, 82 (4), pp. 5-10 (2006).

(29) Vessot, S., Andrieu, J., Laurent, P., Galy, J., and Gerard, J.F., "Curing Study and Optimization of a Polyurethane-Based Model Paint Coated on Sheet Molding Compound, Part II: Drying Defects Related to Curing Conditions," *Drying Technology*, 18 (1&2), pp. 219-236 (2000).

(30) Brenda, M., Doring, R., and Schernau, U., "Investigation of Organic Coatings and Coating Defects with the Help of Time-of-Flight Secondary Ion Mass Spectrometry (ToF-SIMS)," *Prog. Org. Coat.*, 35 (1-4), pp. 183-189 (1999).

High Throughput Measurements, Combinatorial Methods, Informatics

Chapter 26

Development and Deployment of a High Throughput Exterior Durability Program for Architectural Paint Coatings

Edward A. Schmitt

Rohm and Haas Company, Architectural and Functional Coatings
727 Norristown Rd., Spring House, PA 19477

This chapter describes a new high throughput system for evaluating the exterior durability of architectural coatings undergoing natural weathering. The motivation for the design and deployment of this system is the ability to better understand the range and distribution of real world performances by testing more exposure scenarios than would be possible by traditional means. The system integrates inventory management, sample handling and tracking with automated data acquisition, analysis and archival capabilities including the use of environmental sensors, machine vision, and informatics. One of the unique characteristics of this system is that not only does it facilitate the evaluation of more samples per unit time per unit resource compared to traditional methods, but produces more objective data per sample analysis.

INTRODUCTION

There are many challenges to developing architectural coatings for exterior applications. The array of surfaces to be coated in architectural applications can be particularly challenging, ranging from wood to cementitious to metallic. For any one of these, there are a diverse number of subtypes and unique challenges that exist. For example, many species of wood with widely varying properties, which ultimately influence service life, have been used in construction over the years.[1] Furthermore, any of these types of wood may be a part of pristine new construction or remodeling efforts where the integrity of the surface may be compromised through aging or a buildup of previous coatings. A superior coating must first adhere to the substrate being coated and then withstand the sun's radiation, airborne contaminants, wet and/or dry conditions, thermal cycling, mechanical stress, and biological attack. Climates vary considerably as a function of geography, as well as from year to year at a single location. Moreover, a single geographic location may contain several microclimates dependent upon direction or orientation of a coated surface or even proximity to specific structures or landforms. There are numerous descriptions of these conditions in the literature with a recent review given by Martin[2] and new details regarding the complications of angular dependence given by Hardcastle.[3]

397

In order to assess the performance of an architectural coating and its suitability to a particular application, one often exposes the coating of interest to either artificial or natural weathering conditions while monitoring physical or chemical changes in the coating. The nature of these changes is dependent upon the mode of failure and failure mechanism. Examples of failure modes include blistering, cracking, mold growth, chalking, tannin staining, gloss loss, and tint loss. The mechanisms leading to these failures are inevitably complex, and in many cases there are competing pathways that are influenced by chemistry, formulation, the particular application, and a changing environment.

In the case of artificial or laboratory exposures, one of the goals has been to create apparatus and protocols to, in many cases, accelerate the onset of failure while producing repeatable experimental conditions. The achievement of this goal has been met with varying degrees of success.[4,5] Ultimately, one would like to be able to accurately predict from these experiments the real world performance of a particular coating.

For natural weathering, climatic variation over time and location makes it virtually impossible to repeat an experiment, and, hence, the absolute performance results. On a practical level, however, the process of developing new and hopefully better architectural coatings often involves simultaneously testing several coatings, typically the result of an experimental design to probe the effects of specific variables like chemistry or formulation. In this case, the relative performance or rank order of the coating's performance with respect to the experimental design is of interest. The information from this type of experiment is easily augmented by the inclusion of specific extensively studied coatings that are used as pass or fail controls for failure modes of interest.

An operative definition of an exposure scenario is a coating over a single substrate type that is exposed at a single location and direction. Over time, a sample exposed in this way will be subjected to a changing environment that could be described in terms of variables such as temperature, moisture, radiation, airborne contamination, and many others. If the multidimensional space defined by these variables is referred to as exposure space then the evolution of these variables over time for a coating being tested may be referred to as an exposure trajectory.

Because of the previously described diversity of application and environmental variables, it is often important to study a coating of interest under a number of exposure scenarios. The number of individual exposure scenarios and, likewise, tests, grows geometrically as more variables are probed. The number of individual exposure tests would be given by the following equation.

$$N_{exposure\ tests} = (N_{coatings}) \times (N_{locations}) \times (N_{directions}) \times (N_{substrates})$$

For a modest hypothetical study involving 32 coatings on eight different substrates exposed at four locations, in four different directions, there are 4096 individual exposure tests being conducted. Each coating in this study is in fact subjected to 128 different exposure scenarios defined by 128 unique combinations of location, direction, and substrate. A distribution of absolute performances as a function of time for a single coating or relative performances for all coatings is thusly defined over the exposure space sampled by the 128 individual trajectories. Re-exposing the samples under the same exposure scenarios gives rise to new trajectories, helping to better define the range and distribution of performances. In principle, these distributions are analogous to the distributions one might find in real world applications. In practice, however, the choice of exposure scenarios is often weighted more towards those extreme conditions that are known to elicit the presentation of particular failure modes.

One's ability to predict the range or distribution of "real world" performances of an architectural coating, based upon the results of a single exposure trajectory whether natural or artificial, is limited to those cases where there is a dominant well-understood failure mechanism. Similarly, the prediction of a range or distribution of "real world" performances under different exposure scenarios where there are competing degradation pathways and multiple modes of failure is likely to require the sampling of multiple trajectories over a larger exposure space.

A coating's performance is often assessed quantitatively over time using either instrumental methods or by a rating system in conjunction with human observation. In the first case, there are the usual sources of noise in the measurement that can be minimized by a number of best practices like regular calibration. In the latter case, there are additional sources of variability associated with any subjective rating system that relies upon human observation. For example, there is variability in one person's ability to rate performance due to factors like lighting conditions and fatigue. Another source of variability results from the way different people perceive and consequently rate a coating's performance. To some extent this can be minimized through extensive training, round robin testing of replicate samples, and retraining. However, there will always be an element of subjectivity involved. Unfortunately, many failure modes observed for architectural coatings can only be rated visually.

In order to improve ones ability to understand, and hopefully one day model and predict the range and distributions of an architectural coating's "real world" performance, a number of capabilities are needed. A short list of some of these needs follows.

(1) The ability to sample large representative exposure spaces over multiple trajectories

(2) The ability to quantitatively characterize those trajectories

(3) The ability to assess performance objectively and reproducibly

(4) The ability to assess performance frequently

(5) The development of models that explain and predict the evolution of a coating's performance from an exposure trajectory

In short, items (1) through (4) call for more experiments with higher quality assessments that are made more often. At a time when costs are constantly being driven out of the development process, this can only be accomplished by improvements in efficiency. This chapter describes a high throughput system aimed at meeting these first four needs. With additional work, over time, the data generated should facilitate the development of new, better models for predicting the range and distribution of an architectural coatings performance.

DESCRIPTION OF THE SYSTEM

Rohm and Haas Company has developed a high throughput system, eXposure Vision™, for the analysis of a coating's exterior durability.[6] The product of any successful high throughput system is actionable information that can be used to make sound decisions during a new material's discovery and development cycle. For a large exterior exposure program, achieving high levels of sustainable throughput requires a highly integrated comprehensive systems oriented approach. The eXposure Vision system integrates inventory management, sample handling and tracking with automated data acquisition, and analysis and archival capabilities including the use of environmental sensors and machine vision. Results are accessed through the system's informatics capabilities, providing for the rendering and visualization of the data generated along with contextually relevant chemistry and formulation data.

Sample Management

In its present configuration, the system is optimized to handle coated substrates which have nominal dimensions of 36" x 6" x 0.5". Each substrate is divided into six distinct test areas. Information about the substrate type, coatings being tested, and exposure scenario are associated with a unique identity for each substrate and stored in a primary SQL Server database. An RFID (radio frequency identification) tag is attached to the back of each substrate, providing a machine-readable identity and linkage to the information in the primary database. Samples are transported using specially designed carriers that hold 12 substrates each and golf carts modified to transport four carriers simultaneously. There is a one-to-one correspondence between this configuration and the physical layout of the racks where the samples are exposed.

Environmental Sensors

The basis for the environmental sensor subsystem is a HOBO™ weather station produced by the Onset Corporation that measures ambient temperature, humidity, rain fall, wind direction, and speed. HOBO silicon pyranometers, leaf wetness sensors, and temperature probes mounted in black copper discs are mounted in vertical south facing, 45°

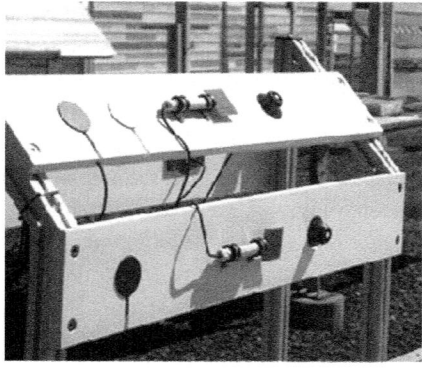

Figure 1—Weather station and directionally dependent sensors.

south facing, and vertical north facing orientations and directions. This weather station and directional dependent sensors are shown in *Figure* 1. The pyranometers have a spectral response in the 300 to 1100 nm range. Readings from each of these sensors are logged hourly and transferred to the primary database twice a day.

Automated Board Scanner

The ABS (Automated Board Scanner), depicted in *Figure* 2, acquires images of the samples and stores them in a separate image database. A carrier containing 12 substrates (72 samples) is manually placed in the automated board scanner, and the "Go" button is clicked on the software. The ABS selects one substrate at a time from the carrier, identifies it via the integrated RFID reader, and moves it into a controlled lighting environment where each sample area is imaged by one of six Pixelink model PL-A782 6.6 megapixel color firewire cameras. The images are automatically stored in the database and the substrate returned to the carrier. The ABS then proceeds to the next substrate. The entire process proceeds under computer control. The total time required to image the 72 samples is approximately 10 minutes.

Machine Vision

The MV_Engine is a computer program that analyzes images in the image database and stores the results in the primary system database. It also reduces the resolution of the images to approximately one-fourth of the resolution of the original image and stores these de-resolved images in the primary system database. The MV_Engine is designed to operate in a distributed computing environment. Multiple instances of the program can run on multiple CPUs of a single computer and/or on multiple computers within the system's high-speed 1 gigabite network. A scheduling program, which is aware of these instances, dynamically assigns the analysis of images to each instance.

Figure 2—Automated Board Scanner, board carrier, and cameras.

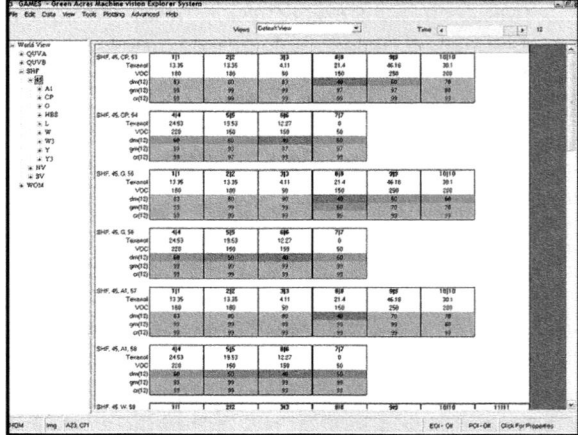

Figure 3—GAMES rendering of chemistry and performance as a panel layout where performance metrics are colored to indicate performance. If the user moves the time slider forwards and backwards in time, the display is instantly re-rendered.

There are two discrete paths for analyzing the performance of coatings from images. The first involves the analysis of the overall characteristics (background) of the coatings appearance. For example, environmental contamination on a white coating can be tracked by monitoring the whiteness of the sample relative to its original state over time. Tint retention, staining, and yellowing can be tracked analogously. The second analysis path involves identification and classification of discrete features or defects in the image. This is accomplished by first separating the defects from the background using an adaptive thresholding algorithm. Once identified, a number of descriptive metrics are calculated for each defect. Select metrics are then used to classify each defect. Examples of

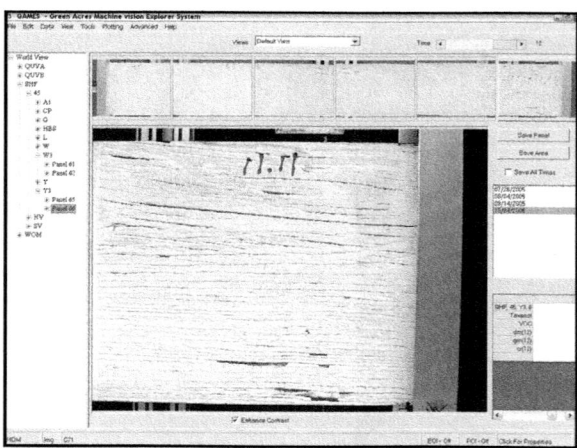

Figure 4—Example of the depiction of images in GAMES.

defect classes include cracks, flakes, and spotty mildew. The number, size, and spatial distribution of specific defects are used to assign an objective rating from 0 (worst) to 100 (best) for a particular failure mode. Additional information, related to the statistical distribution of defect characteristics like size, may be calculated if desired.

Informatics

All of the data is accessed using a desktop client, GAMES (Green Acres Machine Vision Explorer Suite). A user may use GAMES's database searching capabilities to find series of interest in the primary system database. Once selected, the data for that series is downloaded by GAMES, organized, and rendered to the screen depending upon a person's particular needs. All data is organized according to an exposure hierarchy (location, direction, substrate type, panel number, and panel area), and this hierarchy is then rendered to the user interface as a familiar navigation tree. The user navigates through the data by selecting branches and nodes of the tree as desired. GAMES renders the data as either a panel layout, spreadsheet, graphically, or as images. GAMES provides access to chemistry, formulation, and other properties of the coatings, allowing the user to select the properties of interest and add them to the particular rendering. An example of this is given in *Figure* 3. By drilling down to the panel level in the navigation tree, the images for that panel are rendered as shown in *Figure* 4. The user may choose to look at a specific test area, and may go forward and backwards in time to watch the evolution of defects over time.

DEPLOYMENT SUMMARY

Presently, over 5,000 samples are undergoing testing in the system, with over 100,000 images collected thus far. It is anticipated that this sample load and throughput level will triple over the next two years as new samples enter the system for testing. As a matter of practice samples being tested facing south at 45° are imaged approximately every 30 days, with north and south vertical exposures imaged at approximately 60-day intervals. Many of these images are currently being used to calibrate and optimize the MV_Engine. The details of this process will be the subject of a future publication.

While still in the early stages of deployment, there have been a few unforeseen benefits of the system, mostly related to the increased frequency of measurement. For example, on a number of occasions, coating failure has been detected months earlier than it would have been if semi-annual or annual measurements were being made. For some highly specialized studies, weekly measurements have been made with the observation of relatively dramatic coating changes within a week's time period. Moreover, there is at least anecdotal evidence at this point suggesting that these changes may be related to specific climatic events.

ACKNOWLEDGMENTS

The author wishes to thank the entire Exposure Group for many useful discussions. A great debt of gratitude is also owed to Mike Linsen, who was an instrumental contributor in the design and development of this technology. Last but not least, the contributions made by Ryan Schmitt and Emily Schmitt are immeasurable.

References

(1) Williams, R.S., Jourdain, C., Daisey, G.I., Springate, R.W., "Wood Properties Affecting Finish Service Life," *J. Coat. Technol.*, 72, No. 902, 35 (2000).

(2) Martin, J.W., *Service Life Prediction: Methodology and Metrologies,* Martin, J.W. and Bauer, D.R. (Eds.), Chapt. 1, American Chemical Society, Washington, D.C., 2002.

(3) Hardcastle, H.K. III, *Service Life Prediction: Methodology and Metrologies*, Martin, J.W. and Bauer, D.R. (Eds.), Chapt. 3, American Chemical Society, Washington, D.C., 2002.

(4) Martin, J.W., Saunders, S.C., Floyd, F.L., and Wineburg, J.P., *Methodologies for Predicting the Service Lives of Coating Systems*, Federation of Societies for Coatings Technology, Blue Bell, PA, 1996.

(5) Meeker, W.Q. and Escobar, L.A., *Statistical Methods for Reliability Data*, John Wiley & Sons, New York, 1998.

(6) eXposure Vision is a registered trademark of the Rohm and Haas Company with patents pending in the U.S., EU, Mexico, and China.

Chapter 27

Comparing Transport Properties of Coatings Using High Throughput Methods

B. Hinderliter, V. Bonitz, K. Allahar, G. Bierwagen, and S. Croll

North Dakota State University, Fargo, ND 58105

A high throughput experimental procedure and analysis method based on electrochemical impedance spectroscopy (EIS) was developed to evaluate barrier coatings. This procedure is designed to generate parameters for bulk coating simulations, as a combinatorial method to rank coatings, and a means of understanding percolation within coating materials. The ability to examine large numbers of samples allows comparison of the variation in coating quality. The procedure begins with a single frequency measurement of the impedances as water is added to the cell. The EIS response is based on water intrusion into the coating. The second stage is a standard EIS spectrum taken after the coating has been exposed to water for an extended period of time and the coating has achieved saturation. Analysis of the time evolution of the single frequency data is accomplished with a computer code written to regress the single frequency impedance to estimate such parameters as saturation volume fraction of water, diffusion coefficient (and any anomalous behavior related to electrolyte transport), and relative dielectric coefficient. The potentiostatic frequency spectrum measures the bulk property of resistivity and pore resistance.

INTRODUCTION

The quality of a coating is often based on its barrier properties. The topcoat lifetime ends when water and oxygen access the substrate in sufficient quantity and with sufficient continuity to allow corrosion at a metal substrate or delamination or other breakdown mechanism of various nonmetallic substrate material. The advances in robotics and combinatorial methods have opened opportunities to develop methods to quickly quantify responses of large numbers of coating samples. Experimental procedures that can be done quickly with many samples concurrently need to be developed to quantify coating quality, allowing at very least a rapid ranking. Several properties that are considered representative of the initial quality of a barrier coating are diffusion coefficient, saturation water volume fraction, and electrical properties such as relative dielectric constant and resistivity. In this chapter, the focus will be restricted to those properties assessed through single frequency measurements, since electrochemical impedance spectroscopy (EIS) is a standard technique for ranking coatings (for example, Scull,[1] Thomas,[2] Thu,[3] and DeRosa[4]).

For Fickian diffusion, the water flux, J, is proportional to the diffusion coefficient, D, and the gradient in the water concentration, C, as given in equation (1).

$$J = -D\nabla C \qquad (1)$$

When the flux of water is the limiting ingredient to corrosion, the corrosion rate is proportional to the flux and the water that reaches the substrate is immediately consumed (zero concentration boundary condition), causing corrosion or other substrate degradation. Thus, as a particular limiting case, corrosion rate is proportional to the diffusion coefficient.

The saturation concentration, when all water reacts with the substrate ($C_{substrate} \sim 0$), is also proportional to the flux of water and thus corrosion rate [equation (2)].

$$J_{Steady-State} = -D\frac{C_{saturation} - C_{substrate}}{L} \qquad (2)$$

The diffusion coefficient and saturated water volume fraction can be measured by single frequency EIS. Additional information on the initial quality of a coating can be ascertained from the potentiostatic spectra; in particular, the low frequency defines the accessibility of ion transport to the substrate. Thus, in the simplest case described above, the saturated water volume fraction, diffusion coefficient, and electrical properties can be used to rank the initial quality of coatings.

Single frequency measurements are taken at high frequency and these result in low impedance. Thus, small cells are possible, whereas low frequency measurements are not possible for high impedance coatings with small area cells due to low current detection limits. Low impedance measurements are necessary in order to measure with the multiplexer, which due to internal impedances reduce the lower current level by roughly an order of magnitude. The high frequency method thus allows the use of small sample sizes and normal coating thicknesses, which is not possible for low frequency samples.

METHOD—EXPERIMENTAL

The combinatorial device developed to measure up to 12 coated samples concurrently is shown in *Figure* 1. The coated metal panels are the working electrode and contact is made by the springs on the lower plate. The gasket seals the upper portion of the well to the coated metal panels. The Pt-counter electrodes and small reference electrodes are set in from on top and are not shown.

An extended life fluoro-polyurethane topcoat (DEFT 99-GY-1 ELT) and a polyurethane unicoat (DEFT 03-GY-374 Navy TT-P-2756 PUR) coating have been tested. The metal panels (Al 2024) were 20 mm x 40 mm. The coated metal panels were spincoated to minimize thickness variations over the panel. The coatings were cured according to the manufacturer's specifications. Dilute Harrison's solution was used as the electrolyte for EIS measurements.

The test procedure consisted of single frequency (1 KHz) measurements of the impedance in 1-min increments over a period of 24 h, beginning as electrolyte is added to the cell. The EIS response was interpreted based on water intrusion into the coating. After

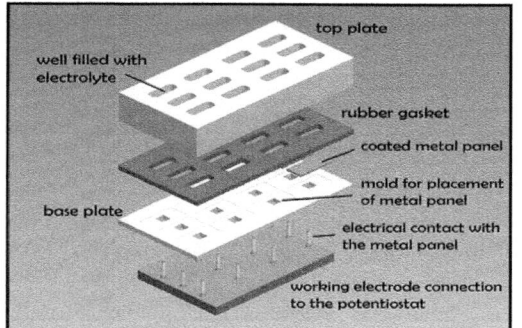

Figure 1—Electrochemical set up used for high throughput EIS measurements.

the 24 h to saturate, or approach saturation, a potentiostatic spectrum was taken. Future tests may include accelerated weathering of the panels followed by a repeat of the above procedure to identify degradation in these three measured parameters.

METHOD—NUMERICAL

The diffusion coefficients were calculated from the single frequency capacitance measurements taken over time at room temperature. Water ingress into a planar system was solved for a fixed surface concentration and there was no flux at the substrate interface using a series solution,[5] as shown in equation (3). Coating systems are well represented by the one-dimensional diffusional equation due to the nearly infinite lateral extent relative to the thickness.

$$\frac{M(t)}{M_{saturation}} = 1 - \frac{8}{\pi^2} \sum_{n=0}^{\infty} \frac{1}{(2n+1)^2} \exp\left(\frac{-(2n+1)^2 D\pi^2}{4L^2} t\right) \qquad (3)$$

M is the mass of diffusant (water), L is the thickness of the coating, D is the diffusion coefficient, and t is time. The subscripted variable indicates that property under one of the following conditions: *saturation* refers to steady-state water concentration in the coating under immersion, and *dry* refers to steady-state water concentration (bound water only) at the given temperature and in a desiccated environment. The additional term suggested by van Westing[6] for swelling, SC_c, (notation from original source) is included in our program setup but not used in this study. This leads to the following equation:

$$\frac{M(t)}{M_{saturation}} = 1 - \frac{8}{\pi^2} \sum_{n=0}^{\infty} \frac{1}{(2n+1)^2} \exp\left(\frac{-(2n+1)^2 D\pi^2}{4L^2} t\right) + SC_c t \qquad (4)$$

The premise of the model is that the polymer contains void inclusions that are initially filled with air or exist empty. During the diffusional process the voids become filled with water and are assumed not to change size appreciably.[7-8] Usually referred to as the Brasher-Kingsbury approximation, the assumptions necessary are that the water is distributed as homogeneous and spherical inclusions. The volume fraction (φ) of void replaced by water is given in equation (5).

$$\varphi(t) = \frac{\ln(C(t)/C_{dry})}{\ln(\varepsilon_R)} \tag{5}$$

C is the capacitance and ε_R is the relative dielectric constant. Equation (5) is normalized to the volume fraction of water at saturation, resulting in the first portion of equation (6). It is assumed that the volume fraction of water is equal to the mass fraction of water. Solving equation (5) for the capacitance as a function of time results in equation (6), and is also interpreted as the logarithmic average of the dry polymer capacitance and the water-saturated polymer capacitance.

$$\frac{\log\left(C(t)/C_{dry}\right)}{\log\left(C_{saturation}/C_{dry}\right)} = \frac{\varphi(t)}{\varphi_{saturation}} \cong \frac{M(t)}{M_{saturation}} \tag{6}$$

The single frequency capacitance with time is fit to the combined equations (3) and (7), which generates predictions for the saturated and dry capacitance of the polymer, the diffusion coefficient, the "swelling coefficient" of van Westing et al., and by applying equation (3), the water volume fraction.

$$\log\left(C(t)\right) = \frac{M(t)}{M_{saturation}}\log\left(C_{saturation}\right) + \left(1 - \frac{M(t)}{M_{saturation}}\right)\log\left(C_{dry}\right) \tag{7}$$

RESULTS

Thus far, numerous thicknesses of an extended life fluoro-polyurethane topcoat (DEFT 99-GY-1) denoted as ELT, and a polyurethane unicoat (DEFT 03-GY-374 Navy TT-P-2756) denoted as PUR, coating have been tested. The small size of the test panels (20 mm x 40 mm) and the possibility of testing multiple samples simultaneously allow for a systematic investigation of EIS parameters on the dependence of film thickness.

It appears that coatings with thicknesses below 50 μm have reasonably constant diffusion coefficients, while those of greater thickness have a diffusion coefficient that is increasing with thickness (see *Figure* 2). This result is counterintuitive since one might expect that the thicker the coating, the lower its permeability. One possible explanation for this trend may be the presence of more prominent solvent escape paths formed during the coating's drying period that might flaw thicker coatings. Other possible causes will be discussed later.

The water volume fractions, shown in *Figure* 3, for various thicknesses are nearly constant. The water volume fraction is slightly less for the ELT compared to the PUR. The slight increase in water volume fraction for the thick samples is consistent with the

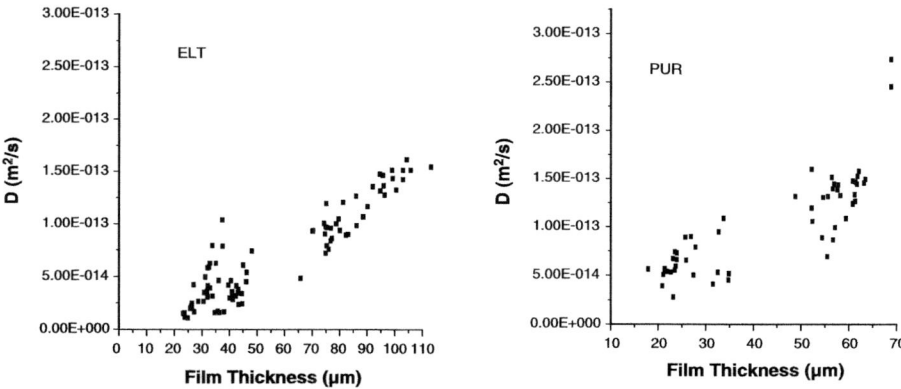

Figure 2—Diffusion coefficient based on EIS response for various coating thicknesses.

proposition that solvent escape paths are larger for thicker coatings. The scatter in the water volume fractions may allow the identification of samples containing flaws, and when large numbers of replicates are measured, the propensity of a coating system to produce defects can be analyzed statistically. This would allow the estimate of number of flaws expected per unit area for a given coating thickness, which is another useful but often overlooked metric of coating quality.

The dielectric constant of the coatings also changes with thickness (*Figure* 4), showing higher values for thinner coating samples. All samples have been coated via spin coating, and it might be plausible that the centrifugal forces led to a distribution profile of the pigments that might be different for different thicknesses. This may explain the higher relative dielectric at lower thickness since the dielectric constant is a material property that encompasses contributions from pigments with very high relative dielectric constant, binders with very low relative dielectric constant, and water inclusions. Since the trend is also observed for the relatively dry coatings at the beginning of electrolyte

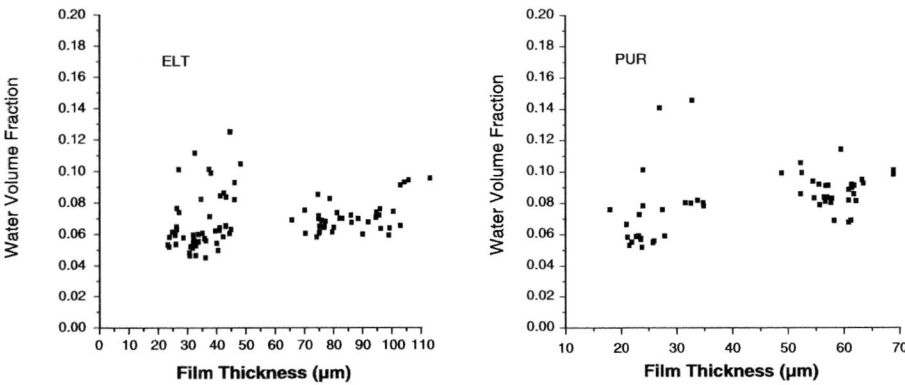

Figure 3—Calculated water volume fraction based on Brasher-Kingsbury formulation.

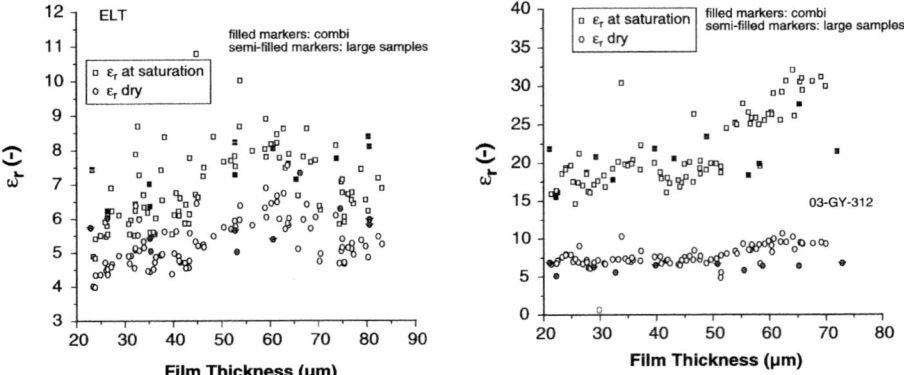

Figure 4—Relative dielectric coefficient based on the measured capacitance extrapolated back to dry conditions and fit to the measured saturated capacitance.

immersion, there is a possibility that this trend is not necessarily a result of the coating's pore structure. For *Figure* 4, ELT relative dielectric constant and thickness above 65 µm was actually generated in two batches. One batch consisted of coating thicknesses from 65 to 85 µm, which showed very low dielectric, 3–3.5 for the dry condition, and 4–5 for the saturated condition. One possible explanation is that this batch had been insufficiently dispersed, and was depleted in pigments. This is being investigated.

A further issue that might be relevant to all the properties is that our EIS method is based on the assumptions of the Brasher-Kingsbury formulation, which necessitates a homogeneous distribution of water in the coating, as well as discrete water inclusions. However, if our coating systems exhibit pore structures that differ from the assumed model (at least for certain film thicknesses), the result might be a trend such as observed in the present data sets. Possible issues include pore formation (perhaps due to solvent escape or pigment clumping), which overestimates water volume fraction.[9] Other possible explanations include Type II diffusion, rough surfaces impacting diffusion and EIS response, and bound water. Bound water is the water that is significantly impacted by hydrogen bonding with the polymer matrix and thus is less able to reorient under the influence of the alternating electric field resulting in a lower relative dielectric for this water. Diffusion in hydrophilic polymers increases with increase in water concentration[10] and, conversely, the diffusion coefficient would decrease with water concentration in hydrophobic polymers. Another mechanism for changing the water diffusion in polymers is the lowering of the glass transition temperature—plasticization—which as a feedback mechanism allows more water into the polymer. This is often identified as Type II diffusion. As the coating swells and increases in thickness, the capacitance would decrease for the same water concentration based on the simple planar capacitor. The water volume fraction must therefore increase at a greater than linear rate in order to result in a net increase in capacitance for this explanation to be plausible.

SUMMARY

The combinatorial EIS method is an effective means to quantify various coatings' parameters. The ability to test large numbers of samples is leading to insights into water distribution as a function of coating thickness. Replicates allow quantification of the variability of the coating between panels and help to quantify coverage over surfaces. This allows the comparison of coatings based on their quality of coverage as noted for the lower thickness ELT coatings.

This series of tests was designed to validate the method and combinatorial instrument as a means to rank coatings. Though the method continues to give insights and satisfies the criterion of a relatively high number of samples measured concurrently necessary for combinatorial measurements, the results thus far have also elicited questions. Further investigations continue regarding whether the dependence of the parameters investigated on thickness is real or an aspect of the method. Preliminary gravimetric measurements support the thickness variation in diffusion coefficient.

The objective of the combinatorial method is to compare coatings, and from the results, the ELT has lower diffusion rates and lower water volume fraction at saturation with less variance. The measurement of two coatings predicts the ELT to exceed the PUR in barrier performance, at least initially, due to its lower diffusion coefficient and water volume fraction at saturation. The secondary aspect of the investigation leads to the apparent dependence of ranking parameters on the thickness of the coating.

ACKNOWLEDGMENTS

The Air Force Office of Scientific Research supported this work under Grant No. FA9550-04-1-0368.

References

(1) Scully, J., "Electrochemical Impedance of Organic-Coated Steel: Correlation of Impedance Parameters with Long-Term Coating Deterioration," *J. Electrochem. Soc.*, 136 (4), 979-989 (1989).
(2) Thomas, N.L., "The Barrier Properties of Paint Coatings," *Prog. Org. Coat.*, 19, 101-121 (1991).
(3) Thu, Q.L., Bierwagen, G.P., and Touzain, S., "EIS and ENM Measurements for Three Different Organic Coatings on Aluminum," *Prog. Org. Coat.*, 42, 179-187 (2001).
(4) DeRosa, L., Monetta, T., Mitton, D.B., and Belluci, F., "Monitoring Degradation of Single and Multilayer Organic Coatings," *J. Electrochem. Soc.*, 145 (11), 3830-3838 (1998).
(5) Crank J., *The Mathematics of Diffusion*, 2nd Ed., Oxford Science Publications, 1975.
(6) van Westing E.P.M., Ferrari, G.M., and DeWit, J.H.W., "The Determination of Coating Performance with Impedance Measurements – II. Water Uptake in Coatings," *Corrosion Science*, 36, 957-977 (1994).
(7) Hartshorn, L., Megson, N.J.L., and Rushton, E., "The Structure and Electrical Proeprties of Protective Films," *J. Soc. Chem. Ind.*, 56, 266-270 (1937).
(8) Brasher, D.M. and Kingsbury, A.H., "Electrical Measurements in the Study of Immersed Paint Coatings on Metal," *J. Appl. Chem.*, 4, 62-71 (1954).
(9) Stafford, O.A., Hinderliter, B.R., and Croll, S.G. "Electrochemical Impedance Spectroscopy Response by Finite Element Methods," *Electrochimica Acta*, 51 (17), 3558-3565 (2006).
(10) Sangaj, N.S. and Malshe, V.C., "Permeability of Polymers in Protective Organic Coatings," *Prog. Org. Coat.*, 50, 28-39 (2004).

Chapter 28

A High Throughput System for Accelerated Weathering and Automated Characterization of Polymeric Materials

Joannie Chin, Eric Byrd, Brian Dickens, Robert Clemenzi,
Rusty Hettenhouser, Art Ellison, Jason Garver, Debbie Stanley,
Vincent Colomb, and Jonathan Martin

Polymeric Materials Group, Materials and Construction Research Div., Building and
Fire Research Laboratory, National Institute of Standards and Technology,
100 Bureau Dr., Gaithersburg, MD 20899

In response to industry requests for a service life prediction methodology capable of generating timely, accurate, and precise service life estimates for polymeric materials, a high throughput system for controlled, uniform accelerated weathering has been implemented and an automated material characterization laboratory is being developed at NIST. The weathering device, known as the NIST SPHERE (Simulated Photodegradation via High Energy Radiant Exposure), has the capability of irradiating >500 specimens with uniform, ultra-high intensity ultraviolet (UV) radiation while simultaneously and independently subjecting them to a wide range of precisely controlled temperature and relative humidity environments. Chemical and physical changes in the exposed specimens will be analyzed in an automated analytical laboratory equipped with multiple instruments that is currently under development. An informatics system is being designed and developed to control and monitor the operation of the SPHERE and the automated analytical laboratory, as well as to store and analyze the collected environmental and analytical data collected from the UV weathering device and the analytical instruments. This chapter describes the integrating sphere-based weathering system and the anticipated capabilities of the automated analytical laboratory and the informatics system.

INTRODUCTION

In the materials industry today, there is a growing need to efficiently and rapidly introduce new products into the marketplace. This need has spurred the development of high throughput research in product development and testing, which has built upon methods originally pioneered by the pharmaceutical industry. In the last five years, an upswing has occurred in high throughput and combinatorial methods applied to polymeric materials.[1-6] Many of these methods involve fabrication of specimen libraries and screening for desired properties.

413

In response to industry requests for a service life prediction methodology capable of generating timely, accurate, and precise service life estimates for polymeric materials, new technology for high throughput accelerated weathering has been developed at NIST. This advanced high throughput system is comprised of an integrating sphere-based UV weathering device, an automated analytical laboratory, and an informatics system.

The integrating sphere-based weathering device, known as the NIST SPHERE (Simulated Photodegradation via High Energy Radiant Exposure), is based on integrating sphere technology that ensures that all of the exposed specimens are uniformly irradiated. The NIST SPHERE has the capability of irradiating >500 specimens with uniform, ultra-high intensity ultraviolet (UV) radiation while simultaneously and independently subjecting them to a wide range of precisely controlled temperature and relative humidity environments.

Chemical and physical changes in the specimens exposed on the SPHERE are currently analyzed manually using gloss measurements, infrared spectroscopy, and UV-visible spectroscopy. The time required for the manual analysis of 500+ specimens is immense and will occupy a great deal of researcher and technician time. In order to circumvent this bottleneck, a high throughput automated analytical laboratory (AAL) equipped with multiple analytical instruments has been designed and is being built at NIST. It will measure chemical and physical changes in the specimens following accelerated weathering on the SPHERE.

In order to handle the extremely large amounts of data and numerous spectra that will eventually be generated in the AAL, an informatics system is being developed that will control and monitor the operation of the SPHERE and the AAL, and also store and analyze the collected data from both the SPHERE and the analytical instruments. A user interface will allow researchers to retrieve, process, and analyze the data archived in the informatics system.

This chapter describes the capabilities of the SPHERE, the AAL, and the informatics system, as well as the anticipated impacts of the NIST high throughput system for accelerated weathering and automated characterization of polymeric materials.

INTEGRATING SPHERE-BASED WEATHERING DEVICE

The NIST SPHERE is an accelerated weathering device based on a 2 m diameter integrating sphere equipped with a high intensity UV light source and environmental chambers, as shown in *Figure* 1. The integrating sphere is constructed from modular panels, allowing individual panels to be removed as needed for modification or repair. The exterior shell of the sphere is aluminum, and the interior surface is lined with poly(tetrafluoroethylene) (PTFE), which has the highest Lambertian reflectance of any known material in the spectral range between 290 nm and 400 nm. The sphere currently contains thirty-two 11.2 cm diameter ports, and a 61 cm diameter top port to accommodate the UV light source.

UV LAMP SYSTEM: The lamp system that serves as the source of UV radiation is a microwave-powered lamp system with an output in the region between 290 nm and 400 nm. Six lamp modules are incorporated into a custom-engineered light shield and are symmetrically arranged around the top port of the sphere. The total output flux of the six lamps operating at 100% power is approximately 8400 W in the spectral range between 290 nm and 400 nm. The electrodeless bulbs used in the lamp system consist of sealed quartz tubes that are filled with mercury vapor, argon, and trace amounts of metal halides, and are powered by microwave excitation. Each bulb is located inside an elliptical reflector assembly coated with a proprietary dichroic coating that removes 80–90% of the infrared and visible emissions. A longpass filter is also installed in the optical path between the UV lamp system and the integrating sphere to remove radiation

Figure 1—NIST SPHERE, showing UV light source and environmental chambers.

below 290 nm, the value generally accepted as the terrestrial cut-off wavelength.[7]

Collimation and conveyance of the highly uniform radiation emitted from the sphere port to the specimen chambers is accomplished with minimal loss of uniformity and intensity using compound parabolic concentrators (CPCs). The narrow end of the CPC is located at the exit port on the surface of the SPHERE, while specimen holders are secured on the wide end.

Using both a NIST-calibrated UV-visible spectrometer and a commercial hand-held radiometer, the total integrated output intensity at the end of the CPC has been measured to be (479.0 ± 0.2) W/m^2 in the range between 300 nm and 400 nm. As a basis for comparison, the integrated direct normal spectral irradiance of the sun between 300 nm and 400 is roughly 22 W/m^2, according to ASTM.[8] Spectral UV intensity measurements were taken at 98 points systematically distributed over the output plane of several CPCs with all six lamps running at 100% power. Results showed that the average uniformity over all 98 points within the exposure area is 94.0% $\pm$ 3.4%. *Figure* 2a shows that the UV-visible spectra measured at the center points of 14 different CPCs are virtually non-distinguishable. The integrated intensities at the center points of these ports fall within an extremely narrow range, as seen in *Figure* 2b. The data from this series of measurements led to the conclusions that both the intra-port uniformity and inter-port irradiance uniformity of the integrating sphere output are extremely high.

ENVIRONMENTAL CHAMBERS: Characterization of UV radiation effects on materials requires that specimens be irradiated over a range of exposure conditions. This has been accomplished by equipping each port with a specimen holder enclosed in an environmental chamber in which temperature, relative humidity, and UV-visible irradiance can be precisely and independently controlled. A schematic of a typical specimen holder is shown in *Figure* 3. Similarly-constructed filter holders are often paired with specimen holders, shown in *Figure* 3b, so that both the intensity and the spectral properties of the incident UV radiation can be manipulated as described in the previous paragraph. Both specimen and filter holders can be custom-engineered to a particular material or experiment. To reduce human error and improve the quality of the data, the samples will be tracked by barcodes during the manufacturing process, SPHERE exposure, and analytical measurements.

Within the environmental chamber, temperature can be controlled between 25°C and 75°C with a precision of ± 0.1°C. Relative humidity is controlled by the controlled mixing of saturated and dry air streams and can be varied between 0% RH and 95% RH with a precision of ± 2% RH. The spectral irradiance to which each specimen is exposed can be modified by inserting bandpass filters and/or neutral density filters in front of the specimen holders, to allow the effect of wavelength and intensity, respectively, to be studied. With 32 ports and the capability of exposing 17 or more specimens in each port, the responses from all of the treatment levels in a designed experiment show a multiplicity of environmental conditions that can be evaluated simultaneously.

MICROPROCESSOR CONTROL SYSTEM: Temperature and relative humidity are monitored and controlled via a custom software package implemented using a microprocessor contained in a microcontroller assembly. At the present time, each microcontroller has an address and is on a RS485 communications network; efforts are underway to upgrade to a TCP/IP network system. Temperature, relative humidity, and ultraviolet radiation irradiance measurements are recorded every 6 min and archived in the informatics system. The total time that the UV-visible radiation is incident on the specimens is also recorded.

SPHERE SAFETY: The performance of the SPHERE is monitored continuously by measuring the output of a multitude of sensors connected to a microprocessor. Pressure,

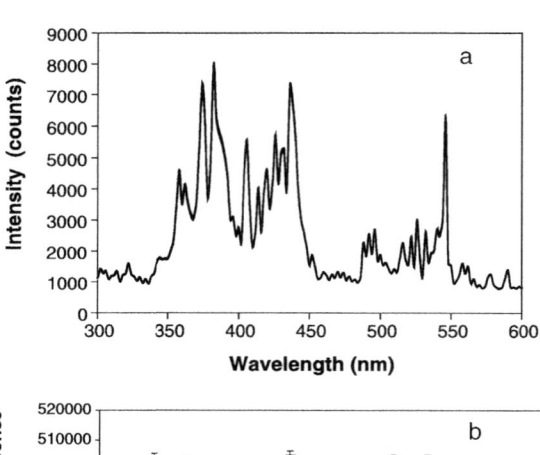

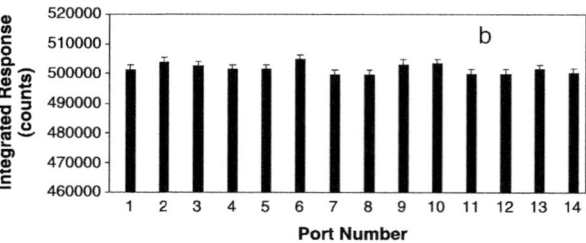

Figure 2—(a) UV spectral intensity distribution measured at the center points of 14 CPCs, showing all curves superimposed, and (b) total integrated intensity at the center points of 14 CPCs. Error bars represent ± one standard deviation. No error bars are shown on the spectral intensity distribution diagram since these spectral exhibited essentially no variation.

air flow, and temperature at multiple points in the SPHERE system are measured and stored. Fail-safe conditions such as high temperatures, component failure, or zero air flow will trigger an alarm, cause the SPHERE system to shut down, and display an alert message.

AUTOMATED ANALYTICAL LABORATORY

Since its fabrication, the NIST SPHERE, in conjunction with a reliability-based service life prediction methodology developed at NIST, has been used in weathering studies of polymeric materials.[9-12] At the present time, specimens exposed on the SPHERE must be periodically removed and mounted on a series of analytical instruments for characterization and analysis. The large number of specimens exposed on the SPHERE at any given time and having each analytical instrument located in a different laboratory makes the analyses extremely time consuming. The goal of the automated analytical laboratory (AAL) is to automate and accelerate the processing of these analytical measurements. It is estimated that the AAL in its fully functioning configuration can analyze the same number of specimens in 1 h that a single operator can manually analyze in 8 h. In the late 1990s, a prototype instrument for automated analysis was developed at NIST, on which the current design is based.[13]

A schematic of the AAL under development, which is physically located next to the SPHERE, is shown in *Figure* 4. Prior to analysis, specimen and filter holders are manually removed from the SPHERE and placed into a specimen handler. The specimen handler is part of the AAL but resides in the SPHERE laboratory. The function of the specimen handler is to identify the specimen and filter holders via their bar codes and track them during the measurement process. At any given time, specimen and filter holders can be awaiting analysis, actively being analyzed, partially analyzed and waiting for an analytical instrument to become available, or fully analyzed and waiting to be returned to SPHERE for further exposure.

The majority of the AAL consists of an array of independently operating instrument assemblies mounted on the modular laboratory tables. Each instrument assembly is com-

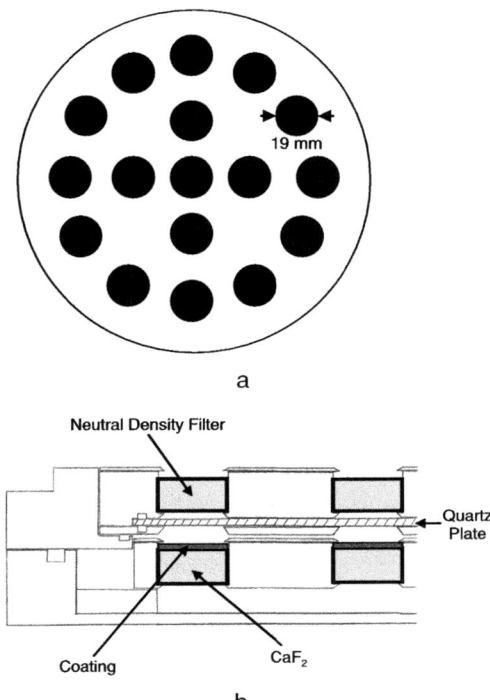

Figure 3—(a) Specimen and filter holder configuration. (b) Cutaway view of assembled exposure cell, showing relative position of specimen and filter holders.

418

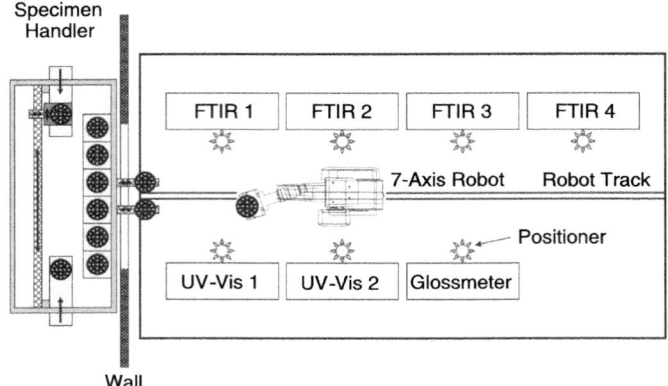

Figure 4—Schematic of automated analytical laboratory, showing speciment handler, robot, and analytical instruments.

prised of an analytical instrument, a specimen positioner, and control electronics. Instruments currently installed in the AAL are four Fourier transform infrared (FTIR) spectrometers with custom optical light guides, two photodiode array UV-visible spectrometers with fiber-optic coupled collimating lenses, and a custom integrating sphere gloss meter. Measurement efficiency is achieved through redundancies in the analytical devices; that is, by populating the table with more analytical devices requiring the longest measurement time and with only one instrument for measurements that are rapid. Measurements on the FTIR take twice as long as the UV-visible measurements; therefore, more FTIR instruments are required than UV-visible spectrometers. Future instruments to be incorporated are FTIR spectrometers with attenuated total reflectance (ATR) attachments, total reflectance (integrating sphere) UV-visible/near-infrared (NIR) spectrometers, Raman spectrometers, and optical microscopes.

The robot is used to move specimens to and from the specimen handler as well as between instruments. It is mounted on a 4 m track embedded into the modular laboratory tables that hold the analytical instruments, and has a reach of 86.3 cm, payload capacity of 3 kg, repeatability of ± 0.005 mm, and can position specimens anywhere on the table. When an instrument assembly is ready to analyze a specimen, the robot obtains the specimen holder from the specimen handler, moves toward that instrument assembly, and places the specimen holder in the instrument's specimen positioner. When an analysis is completed on a particular instrument, the robot will retrieve the specimen holder and either proceed to the next instrument, or return it to the specimen handler, from which the specimen holder can be retrieved and returned to the SPHERE for further exposure.

Before measurement data is added to the informatics system, their veracity is checked based on the history of the specimen and expected trends. Measurements that are suspected to be erroneous or that reveal that the specimen has failed are flagged and reported to a log file. To re-confirm specimen identity and minimize errors, the robot scans the specimen barcode at each stage of the analysis.

INFORMATICS SYSTEM

A UV weathering system of this complexity produces more raw data than can be manually processed. To address this impending issue, a custom laboratory informatics system is being designed to greatly facilitate data collection and analysis. The objectives of this informatics system are to:

- Aid in the design of new experiments.
- Control and record data from SPHERE environmental chambers (temperature, humidity, and UV exposure).
- Inform researchers when a sample holder needs to be removed from its environmental chamber and be characterized by the analytical instruments.
- Control the AAL robot and record data collected by various analytical instruments.
- Perform basic quality control as the data is collected (to flag erroneous data, a failed sample, or a faulty instrument).
- Provide tools for data processing, analysis, and presentation.

The informatics system will ultimately consist of various stand-alone applications and web-based interfaces, which will allow the data to be accessed without regard to the type of computer that is being used. Firebird (an open source, relational database with ANSI SQL-92 features that runs on Linux, Windows, and Unix platforms) is used to store the data on a server that is accessed by all the various modules.

The experimental matrix design program is the entry point into the informatics system. It allows researchers to design a SPHERE experiment and specify the number of replicates, temperature and relative humidity, and bandpass filter and neutral density filter ranges. Specimens are logged into the database by running them through a spectral uniformity check, and a database record for that project is generated. Specimens are assigned to specimen holders and filters are assigned to filter holders. Specimen wheel/filter holder pairs are then assigned to open, non-used ports on the SPHERE as the ports become available. The SPHERE controller program takes the temperature and relative humidity set points from the experimental conditions specified in the experimental matrix design program.

The raw analytical and environmental measurements are archived in the informatics system but must be processed to provide quantities that are more meaningful to service life prediction. Various spectral processing programs have been written to allow researchers to analyze the data. The key programs are used to assess spectral quality, calculate UV dosage, assess spectral changes, determine the relationship between damage and dosage for SPHERE and outdoor exposures, and to assess relationships such as additivity, reciprocity, and time to failure. Both raw data and spectral processing programs are available via a web browser to researchers on the local network.

Various safeguards (based on groups and passwords) are in place to limit access to only those who are qualified to perform a specific task or to those who are allowed to have access to the data. Once data is collected, it can be modified only by pre-specified personnel. In order to maintain data integrity, there will be an audit trail identifying who made the change and exactly what was changed.

SUMMARY

A novel, high throughput system for the accelerated UV weathering of polymeric materials has been developed and is being fabricated at NIST. This system consists of an integrating sphere-based high intensity UV source, an automated analytical laboratory for specimen characterization, and an informatics system for storing, processing, analyzing, and retrieving data on the exposure environment and the specimen. It is anticipated that a system such as the one described in this chapter will facilitate and greatly expedite the research and development process for new polymer products and formulations, particularly in regards to the characterization of weathering and degradation behavior.

References

(1) Karim, A., Sehgal, A., Amis, E.J., and Meredith, J.C., in *Experimental Design for Combinatorial and High Throughput Materials Development,* Cawse, J.N. (Ed.), Wiley-Interscience, 2003.

(2) Meredith, J., *J. Mater. Sci.*, 38, 4427 (2003).

(3) Schmatloch, S. and Schubert, U.S., *Macromol. Rapid Commun.*, 25, 69 (2004).

(4) Potyrailo, R.A., Olson, D.R., Medford, G., and Brennan, M.J., *Analytical Chemistry*, 74 (21), 5676 (2002).

(5) Potyrailo, R.A. and Pickett, J.E., *Angewandte Chemie*, 41 (22), 4230 (2002).

(6) Chisholm, B.J. and Webster, D.C., "The Development of Coatings Using Combinatorial/High Throughout Methods: A Review of the Current Status," *J. Coat. Technol. Res.*, 4 (1), 1 (2007).

(7) Baker, R.E., *Photochem. Photobio.*, 7 (1968).

(8) ASTM E 891-92, "Standard Tables for Terrestrial Direct Normal Spectral Irradiance for Air Mass 1.5," American Society for Testing and Materials, 1992.

(9) Chin, J., Byrd, E., Martin, J., and Nguyen, T., "Validation of the Reciprocity Law for Coating Photodegradation," *J. Coat. Technol. Res.*, 2 (7), 499 (2005).

(10) Chin, J., Nguyen, T., Byrd, E., and Martin, J., in *Natural and Artificial Aging of Polymers*, Reichert, T. (Ed.), Gesellschaft fur Umweltsimulation, Gothenberg, SE, 2005.

(11) Granier, A., Nguyen, T., Shapiro, A., and Martin, J.W., *Proc. Adhesion Society*, 228 (2007).

(12) Nguyen, T., Granier, A., Steffens, C., Lee, H., Shapiro, A., and Martin, J.W., *Proc. FutureCoat! Conference*, Federation of the Societies for Coatings Technology, Blue Bell, PA, October 2007.

(13) White, C.C., Embree, E., Byrd, W.E., and Patel, A.R., *J. Res. Nat. Institute of Standards and Technology*, 109 (5), 465 (2004).

Contributions of Pigments, Additives, and Fillers

Chapter 29

Investigating Pigment Photoreactivity for Coatings Applications: Methods Development

Stephanie S. Watson, I-Hsiang Tseng, Amanda Forster, Joannie Chin, and Li-Piin Sung

Polymeric Materials Group, Materials and Construction Research Division, National Institute of Standards and Technology, Gaithersburg, MD

Titanium dioxide (TiO₂) is used in building and construction applications as a pigment or filler for polymeric products to improve their appearance and mechanical properties. TiO₂ pigments exhibit a wide range of photoreactivity including reactivities that would be detrimental to the service life of polymeric materials. The lack of a scientifically-based metrology for measuring photoreactivity has hindered innovation and acceptance of new pigments, especially nanostructured pigments. The goal of this research is to develop scientifically-based metrologies for the measurement of photoreactivity, including candidate techniques such as photoconductivity, spectrophotometric assays, electron paramagnetic resonance (EPR) spectroscopy, and ultraviolet (UV) weathering exposures of pigment-filled polymer systems. Preliminary data from spectrophotometric assays, EPR, and UV exposure of a filled polymer system are presented.

INTRODUCTION

Large quantities of titanium dioxide (TiO_2) are used in coatings, sealants, plastics, and paper products for opacification, hiding, and UV absorption. TiO_2 is the most effective pigment for hiding or opacity due to its high refractive index and excellent ultraviolet (UV) absorption properties.[1] However, TiO_2 is also a photoreactive material in that absorption of UV radiation promotes electrons from the valence band into the conduction band, leaving behind a positively charged species, or hole, in the valence band, as shown schematically in *Figure* 1. These electron-hole pairs are extremely reactive and are capable of participating directly in oxidation-reduction (redox) reactions with organic materials and/or undergoing interfacial charge transfer with surface or adsorbed species to form reactive radical species.[2] Within the last decade, the ability of TiO_2 to decompose organic compounds has been exploited in applications such as air cleaning, water purification, and self-cleaning/self-disinfecting surfaces.[3,4] Photostable, nanoparticle TiO_2, on the other hand, appears to improve the long-term performance and initial mechanical and durability properties of coatings without affecting appearance.[5,6]

423

424

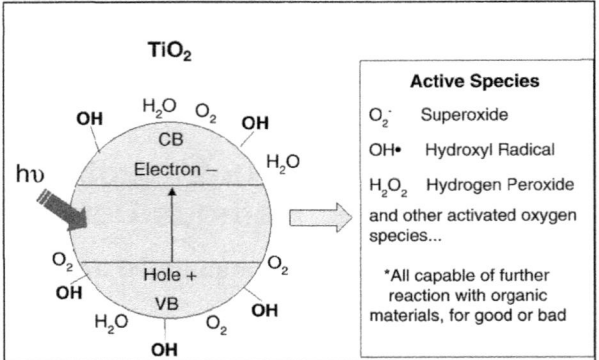

Figure 1—Schematic diagram of the photoreactive process in TiO₂, illustrating bandgap irradiation, promotion of an electron from the valence band (VB) to the conduction band (CB), and the generation of reactive oxygen species.

Commercial TiO_2 materials exhibit a wide range of photoreactivities, depending on crystal phase, manufacturing method, and post-processing steps employed.[7] Presently, no standardized, quantitative measurement techniques exist for assessing the photoreactivity of TiO_2. Instead, qualitative or product-based tests are used, which often do not provide fundamental information about the underlying mechanisms and are not standardized across all industries.[8-10] Current test protocols can range from gas or liquid phase catalytic probe reactions, bacterial inactivation methods, or UV exposure studies of pigment-filled polymer films. Observing the performance of a pigment or catalyst in the final product, while a necessary part of any product development cycle, does not provide insight into the mechanistic processes involved. Moreover, competing effects of pigment flocculation and pigment-binder compatibility are not always accounted for in the performance of the final product. Additionally, as the use of nanoscale metal oxide fillers for nanocomposites increases, the question arises of whether conventional methods of measuring photoreactivity designed for pigmentary particles (≥ 250 nm) are valid for nanoparticles (≤ 100 nm). There are also many new products on the market that tout the benefit of nanoparticle inclusion to achieve new or better end-product properties.[5,6] However, no standards are available for their comparison to conventional products.

The objective of this research is to obtain a comprehensive understanding of the fundamental properties and mechanisms controlling TiO_2 photoreactivity and the effect that photoreactivity has on the service life of polymeric materials. The general goals associated with this objective are:

- Develop novel metrologies for the measurement of photoreactivity, including non-contact methods and methods for use with nanostructured materials.
- Develop analytical techniques for characterizing bulk and surface properties of nanostructured TiO_2 materials.
- Establish correlation(s) between semiconductor photoreactivity, material properties, and heterogeneous photochemistry.

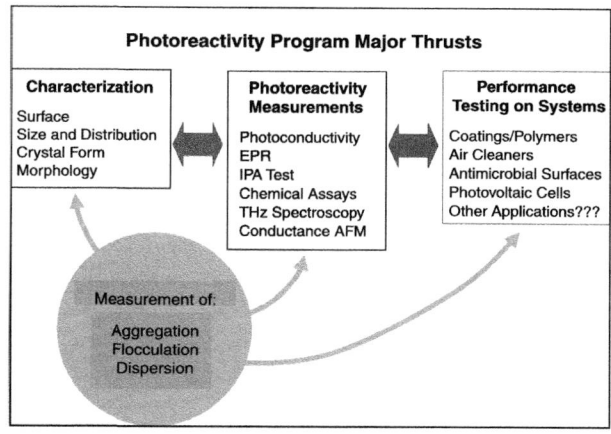

Figure 2—Major thrusts of the NIST Photoreactivity Program.

The major thrusts of this program are outlined in *Figure* 2. It was determined that the measurement of the pigment aggregation, flocculation, and dispersion for all of the investigated metrologies must be included to more accurately describe the observed photoreactivity.

This chapter reviews a number of methodologies and metrologies that are being explored for quantifying photoreactivity in TiO_2 and other semiconductor metal oxides, including photoconductivity, electron paramagnetic resonance (EPR) spectroscopy, chemical assays, and UV exposure studies on pigment-filled polymer films. Data from the chemical assays, EPR methods, and UV-weathering exposures are presented.

METHODS FOR THE MEASUREMENT OF TiO_2 PHOTOREACTIVITY

TiO_2 photoreactivity can be roughly divided into three distinct stages (*Figure* 3): (1) charge carrier generation, (2) interfacial charge transfer to surface species, and (3) redox reactions between active species and the organic material of interest. To obtain a complete understanding of photoreactivity, the fundamental processes that take place in each

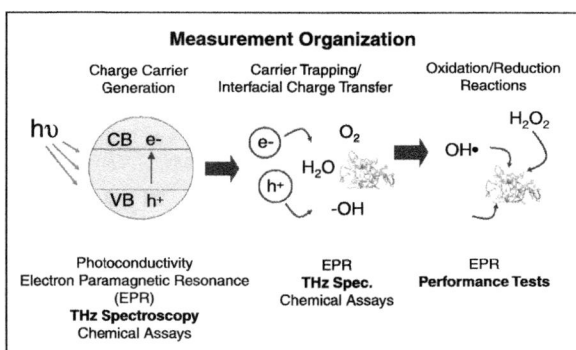

Figure 3—A schematic of the three stages of TiO_2 photoreactivity.

stage must be studied using analytical techniques that are capable of focusing on the specific reactions involved. The techniques that are discussed in this section are currently being investigated at NIST for their utility and potential in quantitatively assessing photoreactivity in TiO_2. Experimental results will be presented in a later section.

Photoconductivity

Photoconductivity measurements can be used to provide information about the population and lifetime of charge carriers in TiO_2 during band gap irradiation. TiO_2 is an n-type semiconductor; therefore, conductivity is attributed to electronic transport. For the appropriate wavelength of radiation, the valence band electrons of TiO_2 are promoted to the conduction band, increasing the population of conduction band electrons or free electrons. For an external direct current (DC) voltage applied to a specimen during irradiation, free electrons in the conduction band accelerate and migrate to the positive electrode and generate a measurable current, I, that can be used to calculate the electrical conductivity σ of the specimen via[11,12]:

$$\sigma = I \, l/VA \tag{1}$$

where l is the length of the specimen, A is the cross-sectional area of the specimen, and V is the applied voltage. In general, the conductivity σ of an insulator or semiconductor is related to charge carrier density and mobility through the expression:

$$\sigma = e \, (n\mu_n + p\mu_p) \tag{2}$$

where e is the electronic charge, n and p are the density of free electrons and free holes, respectively, and μ_n and μ_p are the drift mobilities of the electrons and holes, respectively. Changes in conductivity due to irradiation can be related to an increase in the number of free electrons and holes.

The magnitude of photoconductivity is a function of the competition between charge carrier generation, recombination of the charge carriers, and trapping of the charge carriers by lattice or surface defects. Measurements of photoconductivity have been used to study electronic transport properties and photocatalytic reaction kinetics in semiconductors and insulators.[13-16] Since the generation of electrons and holes is the first step in any

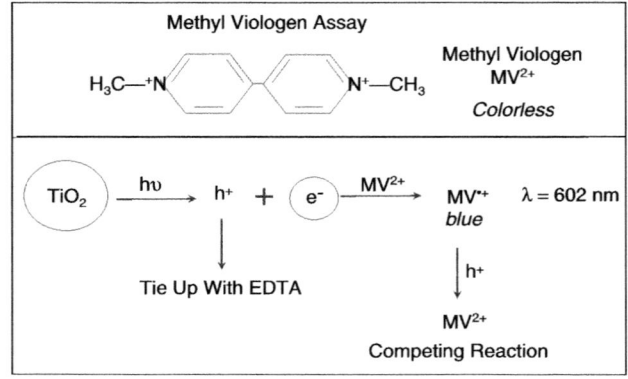

Figure 4—Schematic representation of the reactions observed in the methyl viologen assay.

photocatalytic process, measurement of photoconductivity provides fundamental information on the density, mobility, and the lifetimes of these species.

Low or no measurable photoconductivity in a TiO_2 specimen is indicative of a low photoreactive pigment, also referred to as a "high durability" or "low chalking" pigment. If charge carriers are not generated in significant quantities, become trapped in defect states, or recombine rapidly, they cannot react directly with the organic matrix nor can they undergo subsequent interfacial charge transfer reactions with surface species to generate additional redox agents. On the other hand, high photoconductivity does not necessarily imply high photoreactivity, but only provides an indication of the *potential* that a pigment has for initiating photoreaction. Photoconductivity measurements do not provide information on interfacial charge transfer reactions between the charge carriers and surface species or on charge carrier-redox agent reactions. Results of the photoconductivity studies are described elsewhere[17] and will not be discussed here.

Chemical Assays

Spectrophotometric assays quantitatively measure the concentration of holes and electrons generated during band gap irradiation or of products arising from the reactions of holes and electrons with a surrounding matrix, such as hydrogen peroxide. A number of chemical assays for assessing the photoreactivity of TiO_2 exist and are documented in the technical and patent literature.[20-22] One type of chemical assay involves the use of specific compounds that react directly and stoichiometrically with either holes or electrons.[18-21] Another type of assay uses probe compounds that react stoichiometrically with species such as hydrogen peroxide, which results from the reaction of holes and/or electrons with TiO_2 surface species.[22-24] Various spectroscopic measurements are then used to measure the decrease in concentration of the probe compound or the formation of products resulting from the reaction of the probe compound and other active species.

For this research, two assays were chosen to investigate the two types of methods described above. The methyl viologen (MV) assay is based on the reaction of methyl viologen, an electron acceptor, with conduction band electrons. The chemical structure of methyl viologen di-cation and the reaction schematic for the assay is shown in *Figure* 4. The methyl viologen di-cation, a colorless compound, is easily reduced to a relatively stable cation radical form, which is blue in color with strong absorption at 602 nm.[18] Upon reacting with a free electron generated via UV irradiation of TiO_2, a methyl viologen cation is reduced to a methyl viologen cation radical. The methyl viologen cation radical also reacts with holes to regenerate the initial methyl viologen di-cation in a competing reaction; thus, ethylenediaminetetraacetic acid (EDTA) is used to inactivate the photogenerated holes.[19] The MV assay will be compared to the isopropanol-to-acetone conversion method (IPA) that is commonly used in the pigment industry.[20-21] The photooxidation of TiO_2 is determined by the amount of acetone generated in isopropanol-pigment slurries exposed to UV. In industry, the acetone concentration is used to rank the photoreactivity of pigments.

The amount of hydrogen peroxide produced from reactions of TiO_2 electrons and holes with particle surface species was measured using a leuco-crystal violet/horseradish peroxidase (LCV) assay, as shown schematically in *Figure* 5. The concentration of hydrogen peroxide produced during UV irradiation of aqueous pigment slurries has been

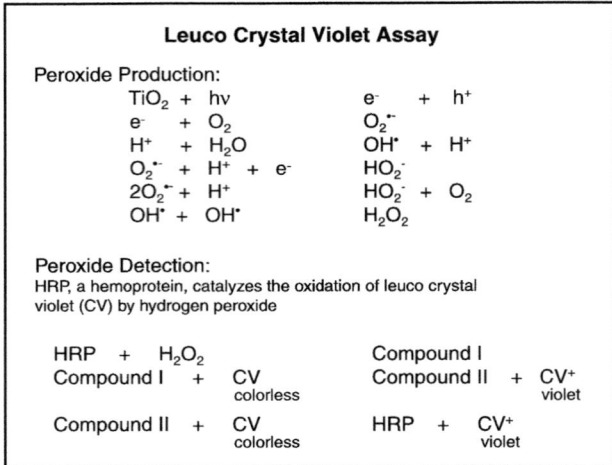

Figure 5—Schematic representation of the reactions observed in the leuco crystal violet assay for peroxide determination.

used to quantify the photoreactivity of the pigment.[19,21] Horseradish peroxidase catalyzes the oxidation of leuco crystal violet dye with the produced hydrogen peroxide and its effectiveness is dependent upon buffer pH and composition.[23,24] The intensity of the maximum absorption peak of the oxidized leuco crystal violet dye at 596 nm is used to quantify the hydrogen peroxide concentration.

Electron Paramagnetic Resonance (EPR) Spectroscopy

EPR spectroscopy, also known as electron spin resonance (ESR) spectroscopy, is capable of detecting short-lived transient paramagnetic species (molecules with one or more unpaired electrons) such as free radicals and several transition metal ions. In an EPR experiment, a sample is exposed to a static magnetic field and bombarded with monochromatic microwave radiation generated by a klystron tube or Gunn Diode. The absorption of microwave radiation by a paramagnetic species occurs at the resonance frequency, and an EPR spectrum is obtained by measuring the microwave absorption as a function of magnetic field strength.[25,26] To extend the lifetimes of the short-lived species, experiments are often conducted in liquid helium (He) (4.2 K) or liquid nitrogen (N_2) (77 K) atmosphere.

In the study of TiO_2 photoreactivity, EPR is used to study each of the three stages of the photoreactive process. Researchers have documented that trapping of photogenerated holes and electrons in the TiO_2 lattice and on surface species can be detected using EPR spectroscopy.[27-31] Spin trapping with stable free radicals such as nitroxide compounds has been used to indirectly detect OH and HO_2 radicals as well as the formation of the Ti^{+3} species.[30,31] Direct observation of free radical or ionic species on UV irradiated TiO_2 has also been reported by a number of researchers.[32-34] Finally, the reaction of various photogenerated species within organic materials, such as coatings, has also been studied with EPR spectroscopy.[35-38]

UV Exposure of Pigment-Filled Polymer Films

The practice of evaluating the performance of a paint formulation using outdoor weathering methods has been utilized as early as the 1930s.[39] Accelerated weathering methods were developed shortly thereafter and continue today to be optimized in order to obtain a prediction of paint durability, especially those formulations containing pigments.[40-44] Our group's research has focused primarily on pure resin systems for about the past eight years.[45] The research is now moving toward studying pigment-filled polymer systems. For most applications, these components are added to increase the opacity and improve the appearance of a coating system. However, the addition of pigments has been found to affect the properties of the coating, especially its durability.[46-48] Basically, pigments can interact with their polymer binder in three different (but not mutually exclusive) ways:

(a) Protect the resin from direct photochemical degradation. Some polymers are particularly susceptible to direct UV degradation, which has sufficient energy to break chemical bonds within the resin. In this case, the absorption of UV by TiO_2 mitigates direct photochemical attack. The absorption of UV by the pigment continues and will lead to photocatalytic attack as well, but is an initially less favored process.

(b) Degrade the resin by photocatalytic degradation. Radicals generated by the pigment oxidize the polymeric binder and is the process pigment manufacturers attempt to minimize via surface treatments of the pigment.

(c) Promote physical interactions with the resin. Degradation of a coating can lead to chalking, which is the physical separation/exposure of the pigment. Another factor, which is particularly important, especially in binders susceptible to photochemical attack, is the relative degree of dispersion of the pigment particles. Well-dispersed pigments absorb more UV and give a smoother surface, resulting in enhanced gloss retention.[41]

The goal of this component of the photoreactivity program is to examine the effects of pigment surface treatment and pigment concentration on the degradation of a resin system. A less durable, but well-characterized amine-cured epoxy system was chosen for the investigation to minimize the time needed to detect degradation. The epoxy and pigment were analyzed separately to gain an understanding of their contributions to chemical changes observed for the entire pigmented epoxy system. A number of studies on the weathering of polymers have focused on monitoring the mechanical properties and the surface morphology of the polymers using SEM.[46-48] From our evaluation of the literature, no study has reported changes in the chemistry of a resin system as a function of the type of pigment, only the durability (chalking or gloss retention) of the resin. A surface-sensitive analytical technique, X-ray photoelectron spectroscopy (XPS), was utilized to monitor the degradation reactions that occur at the surface (~10 nm) of the resin. Attenuated total reflectance Fourier transform infrared spectroscopy (ATR-FTIR), which monitors the near surface (ca. 500 nm), was also used to help assign the elemental chemical shifts observed in the XPS spectra. Chemical analyses of the surface of the polymer systems[49-54] and the pigment,[55-57] separately, have been made using ATR-FTIR and XPS. This study is one of the first to look at chemistry of polymer photodegradation and pigment data in conjunction.

EXPERIMENTAL PROCEDURE

Chemical Assays

METHYL VIOLOGEN ASSAY: A 1.5 g/L suspension of TiO_2 with 2.5×10^{-3} mol/L in methyl viologen dichloride and 3.0×10^{-2} mol/L in disodium ethylenediaminetetraacetic acid (EDTA), buffered to pH = 6.0 with a 0.2 mol/L phosphate buffer, was prepared. In other methyl viologen assay experiments, a potassium hydrogen phthalate (KHP) buffer was used to elucidate results with no phosphates present. The suspension was purged with argon (Ar) in a sealed sparging reactor for 20 min, and then irradiated with UV radiation with a 100 W mercury (Hg) lamp for 2 h. During the UV irradiation period, the TiO_2 suspension was mixed with a magnetic stir bar. In an Ar-purged glove bag, 3 mL of the irradiated suspension was transferred into a cuvette for UV-visible (UV-VIS) spectroscopy using a syringe equipped with a 0.2-μm filter. The absorbance of the methyl viologen cation radical at 602 nm was measured on a UV-VIS spectrometer. The standard uncertainty associated with UV-VIS spectroscopy absorbance measurements is typically ± 2%.

HYDROGEN PEROXIDE ASSAY WITH LEUCO-CRYSTAL VIOLET: 250 mg of TiO_2 pigment was mixed with 25 mL of 1 mole/L acetate buffer (pH = 4.4) and irradiated with UV radiation with a 100 W Hg lamp for 2 h. During the UV irradiation period, the TiO_2 suspension was mixed with a magnetic stir bar. The irradiated pigment suspension was then filtered and 8.5 mL of the filtrate was mixed with 0.5 mg/mL horseradish peroxidase and 1 mL leuco-crystal violet (LCV) solution (100 mg LCV dye in 200 mL of 0.5% by volume hydrochloric acid). A solution of 0.3% by volume solution of hydrogen peroxide was used as a standard for calibration. The absorbance of the leuco-crystal violet solution at 596 nm was immediately measured using UV-VIS spectroscopy.

ISOPROPYL ALCOHOL TEST: TiO_2 suspensions in isopropanol were prepared from TiO_2 dried at 120°C for 2 h and purged with compressed air (500 mL/min) for 1 min prior to UV irradiation. Pigment loadings varied and were based on the pigment surface treatment: 1 g for alumina (Al_2O_3)-coated pigments and 4 g for silica (SiO_2)-coated pigments. A jacketed beaker containing the TiO_2 suspension was illuminated from the top with UV light from a 100 W Hg lamp. Suspensions were stirred with a magnetic stirbar to maintain the suspension of the pigment. After 2 h of UV exposure, the suspension was filtered and the acetone content of the filtrate was determined by gas chromatography with flame ionization detection. A blank measurement was also performed using isopropanol alone to establish the initial amount of acetone present in the isopropanol reagent.

LIGHT SCATTERING MEASUREMENTS: In this study, dynamic light scattering (DLS) was used to estimate the particle/cluster size of TiO_2 in the various assays. Light scattering measurements were carried out using a variable angle light scattering instrument. The instrument was calibrated with NIST traceable polymer sphere standards ranging in diameter from 90 nm to 304 nm. The laser wavelength used in this study was 532 nm. Scattering angles of 15° and 30° were used for the measurements. Suspensions consisted of TiO_2 and the spectrophotometric assay components as described in the above sections. Suspensions were mixed for 3 min by shaking followed by sonication for 30 min. 2 mL aliquots of the assay suspension were transferred into a 15-mm diameter DLS vial. The assay suspension was measured at 15° for 50 min followed by measurements at 30° for 90 min. The assay suspensions remained stable after 50 min, but some pigment settled after 90 min. Thus, all data was collected in a 50 min window. Some of the assay suspensions were diluted from 90% to 10% by volume of their original TiO_2 concentration with their corresponding buffer solutions.

Dynamic fluctuations of the scattered light from the particles in the assay matrix were measured. The time dependence of the scattered light at a particular scattering angle (θ) was analyzed by an autocorrelation function of the scattered light.[58] For monodisperse particles in solution, the correlation function is characterized by an exponential decay. However, for a polydisperse sample, the correlation function no longer has a simple exponential function. In this study, non-linear least squares regression analysis was used to fit the experimental autocorrelation functions to determine the hydrodynamic radius of particles in the TiO_2 suspensions.

EPR Spectroscopy

SPIN TRAP MATERIALS: 5, 5-dimethyl-1-pyrroline-N-oxide (DMPO), 3-carboxyproxyl (CP), 3-aminoproxyl (AP), and 3-carbamoylproxyl (CM) used in this study were purchased from Aldrich without further purification. Stock solutions (500 µmole/L) of each spin trap were prepared in deionized water and stored at 7°C. Stock preparations (0.2 g/L) of TiO_2 suspensions in deionized water or in solvent were also prepared for spin trap studies. For each spin trap experiment, a new mixture of spin trap stock solution and TiO_2 suspension was prepared to monitor the photo-produced radicals under UV irradiation at ambient temperature. 50 µL aliquots of the spin trap/TiO_2 suspension mixture were placed into EPR capillary tubes for UV irradiation in the EPR instrument cavity. Different TiO_2 spin trap molar ratios were also under investigation in this study.

For solid state EPR studies, TiO_2 with different sizes, crystalline forms, and modified surfaces were investigated. The same volume of powder for each TiO_2 sample was placed into an EPR tube and dried under vacuum for at least 8 h to reduce the moisture. A comparison experiment of the TiO_2 suspensions (0.2 g/L) in deionized water was performed as well. 1 µL aliquots of the TiO_2 suspension were used in a quartz capillary tube contained within a conventional EPR tube for in-situ UV irradiation.

EPR MEASUREMENTS: All EPR spectra for the spin trap studies were recorded at ambient conditions with a Bruker Elexsys E500 EPR spectrometer. Most measurements were carried out using a resonant frequency of 9.38 GHz and a microwave power of 0.6 mW to 10 mW. The time constant, conversion time, sweep time, and signal receiver gain were adjusted to obtain optimum signal resolution. For some weak-signal experiments, higher microwave power, receiver gain, and signal-to-noise ratio were applied. Samples of spin trap/TiO_2 suspension mixtures were placed in 50 µL capillary tubes that were positioned in a conventional EPR tube and directly irradiated with a Xe arc lamp at 500 W. EPR spectra were recorded at regular time intervals during the UV irradiation period. Several control experiments were carried out to insure that the observed signals did not arise from photolysis or oxidation products of the spin traps themselves, as well as to monitor the

EPR signal stability in the dark and under illumination. In addition, the stability of spin trap solutions without TiO_2 over time (eight weeks at 7°C) was monitored.

All solid-state EPR spectra were recorded at a resonant frequency of 9.3 GHz, microwave power of 10 mW, signal receiver gain of 30 dB to 70 dB, modulation amplitude of 10 G, and modulation frequency of 100 kHz. The time constant, conversion time, sweep time, and scan times were adjusted to obtain optimum signal resolution. Quartz EPR tubes were employed to avoid the background signal during measurement. A liquid helium (He) cryostat was used in the EPR instrument to achieve lower sample temperatures (77 K) in the cavity. The samples were directly irradiated in the cavity with the Xe arc lamp at 500W for in-situ measurements.

UV Exposure of Pigment-Filled Polymer Films

MATERIALS: Commercially available TiO_2 particles were selected for the TiO_2-epoxy films. In general, the reactivity of the TiO_2 particles is directly related to surface treatment. Two pigment volume concentrations (PVC), 2.5% and 10%, were used and an unfilled control sample was also prepared. The epoxy matrix was formed by using a 2.4:1 ratio of commercial diglycidyl ether of bisphenol A (epoxide equivalent mass: 185-192; Resolution Performance Products) to polyetheramine curing agent (Huntsman). Films were prepared using a Dispermat (BYK Gardner) mixer by first dispersing the pigment into the resin and then mixing a curing agent into the suspension. The mixtures were degassed for 2 h and films were drawn down with a 1.016 mm bar on release paper. Films were cured at room temperature for 48 h, followed by a 2 h postcure at 130°C. Circular free-film samples having a 19-mm diameter were cut from each of the films. The resulting film thicknesses ranged from 70 μm to 450 μm.

UV EXPOSURE EXPERIMENTS: Epoxy samples described above were exposed to ultraviolet radiation using the Simulated Photodegradation by High Energy Radiant Exposure (SPHERE) at ambient laboratory conditions of temperature and relative humidity (nominally 25°C and 35% RH). Complete details regarding this instrument and its configuration are described elsewhere.[59,60] The SPHERE average irradiance between 300 nm and 400 nm over the exposure period was 102 W/m^2. Total incident dose was calculated using the average irradiance, exposure time, and the sample size. To accommodate the destructive measurements used in this study, the experiment was designed so that a new set of specimens was analyzed after each time interval, instead of measuring the same degraded specimens.

FILM CHARACTERIZATION MEASUREMENTS:

XPS—All XPS measurements were performed with a Kratos Axis Ultra photoelectron spectrometer. Experiments were conducted at room temperature with a base pressure in the 1.3×10^{-6} Pa range. The monochromatic Al Kα x-ray source was operated at 140 W (14 kV, 10 mA). The energy scale was calibrated with reference to the Cu $2p_{3/2}$ and Ag $3d_{5/2}$ peaks at binding energies (BE) of 932.7 eV and 368.3 eV. A coaxial charge neutralization system provided charge compensation. Initial and degraded polymer films were attached to the XPS sample holder using double-stick tape. The analysis area for the high-resolution spectra was 2 mm x 1 mm. The O 1s, Ti 2p, N 1s, C 1s, and Si 2p spectra were acquired at a pass energy (PE) of 20 eV and a maximum acquisition time of 8 min per element. Peak BEs were determined by referencing to the adventitious C 1s photoelectron peak at 285.0 eV. Quantitative XPS analysis was performed with the Kratos VISION software (version 2.1.2). The atomic concentrations were calculated from the photoelectron peak areas by subtracting a linear-type background. The O 1s, N 1s, and C 1s regions were deconvoluted using mixed 70% Gaussian/30% Lorentzian components.

ATR-FTIR: ATR-FTIR was performed on all the epoxy resin samples using a Magna-IR 560 (Thermo Nicolet) with a mercury cadmium telluride (MCT) detector and an ATR accessory with a diamond crystal (Durascope). Each specimen was sampled at three different locations with 128 spectra collected at each location at 4 cm^{-1} resolution. All spectra were imported into a custom software package for analysis.

RESULTS AND DISCUSSION

All components of this research program focused primarily on the use of commercial TiO_2 products. A range of pigment surface treatments, crystallinity, and particle sizes were examined as these are some of the many factors that influence pigment photoreactivity. *Table* 1 lists the TiO_2 pigments used in this research program and their properties. For comparison, an in-house prepared TiO_2 specimen (sample K) was also studied. In some components of this research program, a subset of the pigments listed in *Table* 1 was used.

Chemical Assays

The chemical assay study showed that the photoreactivity response from TiO_2 analyzed in the spectrophotometric assays varied with experimental conditions. Light scattering measurements were performed to examine the effects of assay experimental conditions on the pigment cluster size to test the hypothesis that cluster size is one variable

Table 1—TiO$_2$ Pigments[a]

Label	Crystallinity	Particle Size (nm)	pH of Suspension	Zeta Potential (pH)	Surface Treatment
A	anatase (70%) rutile (30%)	20	6.6	6.0	none
B	anatase	7-8	4.4		none
C	anatase	35-40			Al_2O_3/SiO_2 (25%)
D	rutile	10-30		6.5-8.0	Al_2O_3 (9–14%) SiO_2 (1–7 %)
E	rutile	250	6.0	7.5	Al_2O_3 (6%)
F	rutile	250	6.7	7.5	Al_2O_3/SiO_2 (12%)
G	anatase	300	6.3		none
H	rutile	>400	6.0		none
I	rutile	30-50	6.3	6.5-8.0	Al_2O_3 (10–15%) ZrO_2 (2–5%)
J	anatase	7	6.2		none
K	anatase	50	6.7		none

(a) Values were provided by manufacturer, except for pH of suspension, which was measured in deionized water.

that can affect the photoreactivity response for the various assays. For example, the surface treatment on pigments affects the zeta potential of the base pigment, which in turn affects the electrostatic interactions of the pigments in solution and, therefore, the pigment cluster size. Thus, larger clusters shield inner nanoparticles from UV irradiation and produce a lower photoreactivity value, while smaller cluster sizes result in a higher pigment surface area for UV interaction.

PHOTOREACTIVITY RANKINGS:

Standard Experimental Conditions—For the methyl viologen (MV) assay, the colorimetric assay is based on the reaction of methyl viologen, an electron acceptor, with conduction band electrons. UV-VIS spectroscopy analysis of the TiO_2 suspensions assayed with methyl viologen showed an intense absorbance at 602 nm for a number of TiO_2 samples. Lower absorbance values were observed for some of the surface-treated TiO_2 pigments (*Figure* 6). These MV results are indicative of the greater net production of free electrons by catalytic TiO_2 relative to coated TiO_2. Based on the MV results, the uncoated, nanosized, anatase TiO_2 pigments are more photoreactive than the rutile TiO_2 pigments.

The amount of hydrogen peroxide produced from reactions of TiO_2 electrons and holes with particle surface species was measured using a leuco crystal violet/horseradish peroxidase (LCV) assay. Via the intensity of the absorbance peak at 596 nm, a different photoreactivity ranking relative to the MV assay was observed (*Figure* 7). A 3% by volume hydrogen peroxide solution is also shown as a standard sample in the figure. Pigments without surface treatments showed a greater peroxide production in the LCV assay. However, among the uncoated TiO_2 pigments, the larger particle size anatase pigment showed the greatest peroxide concentration as compared to the nanosized anatase pigment in the MV assay.

Both MV and LCV assay results were also compared to a conventional industrial method used to measure photoreactivity, the isopropyl alcohol conversion to acetone (IPA) test (*Figure* 8). The IPA test is another measure of conduction band electrons from the TiO_2 particles and so should be closest to the MV assay results. Again, the photoreactivity ranking differed, but anatase pigments were still the most photoreactive, as

Figure 6—Photoreactivity comparison of absorbance values obtained using the methyl viologen assay with TiO_2 pigments. Standard uncertainty in the absorbance values is ± 2%.

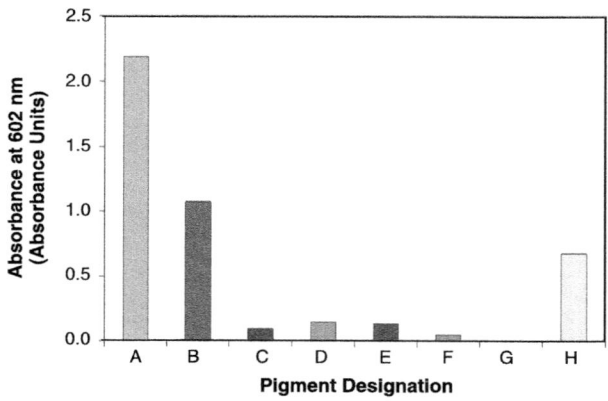

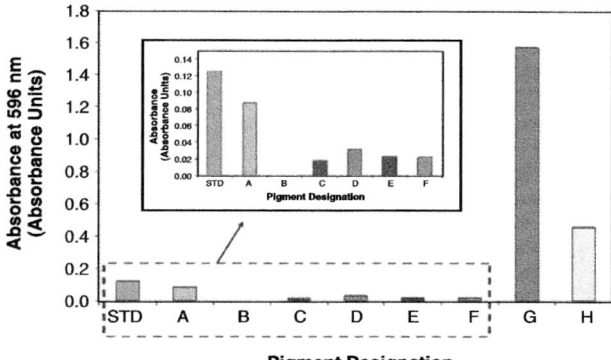

Figure 7—Photoreactivity comparison based on peroxide concentration using absorbance from the leuco crystal violet assay with TiO_2 pigments. The standard sample (STD) is a 0.3% by volume hydrogen peroxide solution. The inset figure is a magnified version of the original graph to view the lower concentrations (specimens A–F). Standard uncertainty in the absorbance values is ± 2 %.

observed in the MV assay. This is a possible indication that the product mechanism for each assay provides a different measure of photoreactivity. However, differences observed between each assay could be due to the different assay matrix (buffer composition, pH, and solvent), causing the nanoparticles to flocculate and form different-sized particle clusters. In turn, larger clusters may shield inner nanoparticles from UV irradiation and produce a lower apparent photoreactivity.

Modified Experimental Conditions—Each assay examined in this study used a different solvent, buffer pH, and pigment loading. To more accurately compare the photore-

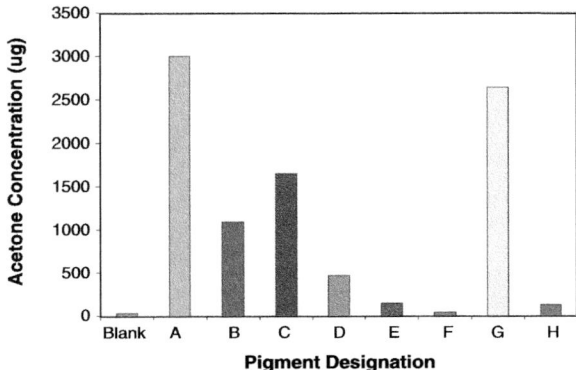

Figure 8—Isopropyl alcohol conversion results for the series of TiO_2 specimens. The blank sample is a measure of acetone present initially in the isopropanol. Standard uncertainty in the acetone values is ± 2%.

436

activity results from the assays, the experimental conditions of the MV assay and the LCV assay were changed so that all assays were similar. Particularly, the same pH and the same pigment loadings were used. Buffer composition was also examined in the case of the MV assay. A phosphate buffer is the standard buffer used in the MV assay. However, phosphate addition has been reported to negatively influence the reactivity of titanium dioxide.[61,62] Phosphates are known to modify alumina, a common surface treatment on pigments and nanoparticle catalysts.[63,64] In this study, potassium hydrogen phthalate (KHP) was used as the non-phosphate buffer. At the time of this writing, the investigation was limited to TiO_2 A, E, and F. The results for the entire set of pigments will be reported in a future publication.

Experiments with the MV assay revealed that experimental conditions influenced the absorbance intensity for the pigments investigated. *Figure* 9 shows the results of the MV assay with phosphate buffer and a change in buffer pH from the standard pH 6.0 to pH 4.4. The results indicate that pH 6.0 is the optimum condition, producing the highest intensity, especially for TiO_2 A. *Figure* 10 shows the results of buffer composition for the MV assay, switching phosphate buffer to KHP buffer. KHP buffer experiments were run at pH 6.0, the optimum pH revealed in *Figure* 9. KHP buffer results for the MV assay reveal that KHP buffer produces a higher intensity for all pigments, especially for TiO_2 E, an alumina surface treated pigment. This result indicates that the phosphate groups may be exchanging with the hydrous alumina surface treatment on the pigment and forming aluminum phosphate under standard phosphate conditions in the MV assay, thus reducing the photoreactivity response.

Figure 11 shows the effects of pigment loading on the MV assay response. For all pigments investigated, an increase in the pigment loading increased the MV assay response. The response for TiO_2 F was very low for both pigment concentrations. The solutions with greater pigment loading for TiO_2 A resulted in a deeply colored solution, which was off of the absorbance scale for the UV-VIS instrument. Dilutions with deionized water were necessary to keep the absorbance readings on scale.

Results from the LCV assay revealed that assay matrix experimental conditions strongly influence the absorbance intensity at 596 nm for the pigments investigated. It has been reported that the effectiveness of horseradish peroxidase is dependent upon buffer pH and composition.[23] In addition, dissolution of the leuco crystal violet dye is

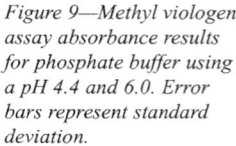

Figure 9—Methyl viologen assay absorbance results for phosphate buffer using a pH 4.4 and 6.0. Error bars represent standard deviation.

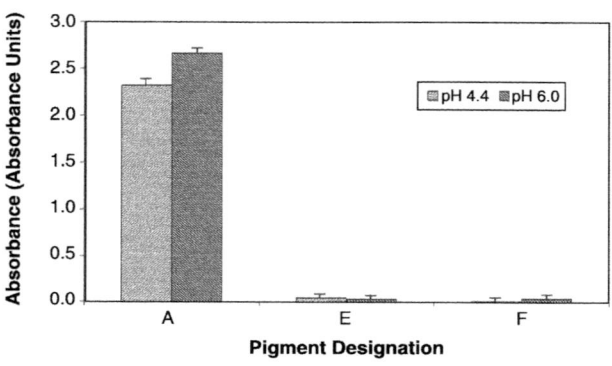

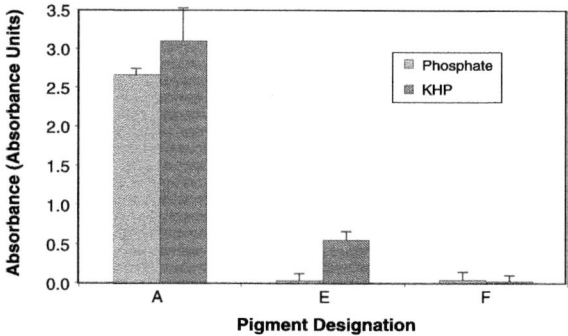

Figure 10—Methyl viologen assay absorbance results using phosphate buffer and KHP buffer at pH 6.0. Error bars represent standard deviation.

also very sensitive to pH.[24] *Figure* 12 shows the results of the experiments with the LCV assay using acetone buffer. The pH 5.8 was the highest pH possible with acetone buffer and so an exact match for pH 6.0 was not possible in this system. Regardless, the assay results showed that increasing the pH from 4.4 to 5.8 increased the absorbance intensity for the standard assay conditions of a 2 h UV-irradiation exposure time at 10 mg/mL pigment loading. However, close inspection of the assay solutions at pH 5.8 revealed a high degree of cloudiness without any increase in the purple color produced by the reduced leuco crystal violet species. The cloudiness caused an artificial increase in absorbance. The increase in cloudiness is a result of the lack of dissolution of the leuco crystal violet at the higher pH.

In general, decreasing the pigment loading caused a large drop in the LCV assay response. There was no absorbance intensity observed for TiO_2 E and F after the standard 2 h UV irradiation exposure time. In fact, the lower pigment loading solutions containing surface treated pigments, TiO_2 E and F, has to be run for 12 h overnight to obtain a measurable absorbance intensity. Even after the overnight run, the response for the surface treated pigments did not reach the response obtained for the 10 mg/mL pigment loading at the standard 2 h UV irradiation exposure time. The response for TiO_2 A at 1 mg/mL was different from that obtained under similar conditions for the surface treated pigments. For TiO_2 A, the absorbance obtained after the standard assay condition of 2 h UV irradiation exposure time was slightly greater than that obtained for an overnight

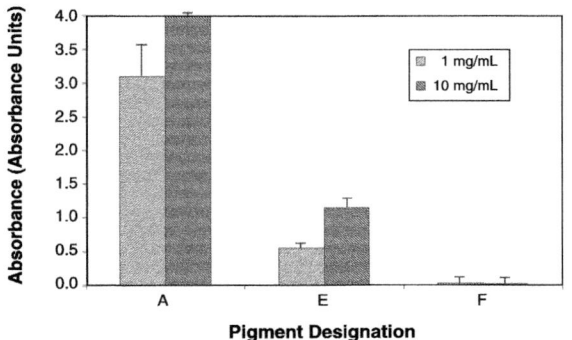

Figure 11—Methyl viologen assay absorbance results with KHP buffer at pH 6.0 and using two pigment loadings, 1 mg/mL and 10 mg/mL. Error bars represent standard deviation.

438

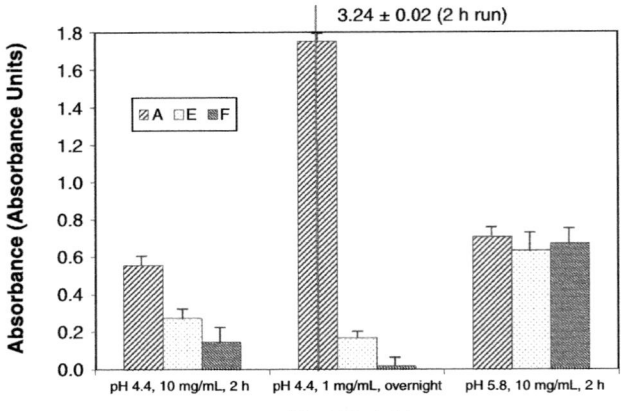

Figure 12—Leuco crystal violet assay absorbance results with acetone buffer using pH of 4.4 and 5.8 with 10 mg/mL run for 2 h and at pH 4.4 with 1 mg/mL as noted. Pigments are shown as: A (////), E (dots), and F (\\\). The line in the pH 4.4, 1 mg/mL pigment loading experiments presents pigment A run for the standard 2 h UV irradiation exposure time. Error bars represent standard deviation.

run. Furthermore, both responses obtained for TiO_2 A at the different UV irradiation exposure times at the lower pigment loading were significantly greater than that obtained for the 10 mg/mL pigment loading.

LIGHT SCATTERING MEASUREMENTS: Dynamic light scattering (DLS) measurements were performed to measure particle cluster size of TiO_2 A, E, and F in the various assays. The results for the entire set of pigments will be reported in a future publication. Here, a particle cluster represents a floc, which is a loose structure formed by primary particles, aggregates, and agglomerates from the pigments.[65] Aggregates are considered to be primary particles joined at their faces with a specific surface area that is significantly less than the sum of the area of the constituents. Aggregates are very difficult to separate.

Table 2—Measured Hydrodynamic Diameter Values for MV Assay in Phosphate Buffer Versus pH[a]

	Measured Hydrodynamic Diameter (nm)	
	pH	
Pigment	4.4	6.0
A	206	360
E	1270	676
F	1204	mm[b]

(a) Standard uncertainty in the particle diameter values is ± 5%.
(b) Denotes unstable suspensions in terms of intensity resulting from large clusters.

Table 3—Measured Hydrodynamic Diameter Values for MV Assay at pH 6.0 for Phosphate and KHP Buffers and Versus Pigment Loading for KHP Buffer[a]

	Measured Hydrodynamic Diameter (nm)			
	Buffer at 1 mg/mL Pigment Loading		Pigment Loading in KHP	
Pigment	Phosphate	KHP	1 mg/mL	10 mg/mL
A	360	962	962	1118
E	676	1140	1140	1130
F	mm[b]	1210	1210	1030

(a) Standard uncertainty in the particle diameter values is ± 5%.
(b) Denotes unstable suspensions in terms of intensity resulting from large particle clusters

Agglomerates are a collection of primary particles or aggregates joined at their edges or corners, with a specific surface area not markedly different from the sum of the areas of the constituents. Agglomerates can be separated into their primary particles.

For the MV assays, the DLS data revealed that experimental conditions of the assay influenced the particle cluster size of the pigments. The shapes of all DLS curves are most similar to the shape of polydisperse suspensions for the MV assay using phosphate buffer at pH 6.0 and 4.4. *Table 2* summarizes the measured hydrodynamic diameter or particle cluster results obtained for the MV assay. The measured pigment cluster sizes do not agree with the pigment particle size based on crystallite size reported by the manufacturer. These results indicate that nano- and pigmentary-sized particles have a strong tendency to form clusters in the MV assay matrix. The largest change was observed for TiO_2 E, whose cluster size nearly doubled from pH 6.0 to 4.4. The opposite trend was observed for the cluster size of TiO_2 A and F, which decreased with decreasing pH. However, the large measured cluster size for TiO_2 F at pH 6.0 was a result of very large clusters that produced an unstable suspension and, therefore, unstable intensity during the DLS measurements. Buffer composition also affects the cluster size for MV assays as the shapes of the DLS curves differ in the sharpness of the turnover for each pigment between the two buffers. A new feature at longer lag times in the DLS curves for KHP buffer results was observed and may represent a secondary, larger-sized, pigment cluster set. Interpretation of the data is continuing.

Pigment loading has the greatest effect for TiO_2 A in the MV assay. The DLS curve for TiO_2 A at higher pigment loading showed an increase in amplitude, an indication of increased organization of the clusters, but the shapes of the curves for the surface-treated pigments do not change with increasing pigment loading. *Table 3* summarizes the measured hydrodynamic diameter or particle cluster results obtained for the MV assay after changing buffer composition and pigment loading. In general, all pigments more than double in cluster size with a change in buffer from phosphate to KHP. No large clusters were observed for the KHP buffer for any pigments as was observed for pigment F in the phosphate buffer. For the KHP buffer all pigments had similar cluster size, regardless of surface treatment and initial particle size. Decreasing pigment concentration in KHP buffer did not affect the surface coated pigments.

Table 4—Measured Hydrodynamic Diameter Values for LCV Assay in Acetone Buffer at pH 4.4 Versus pH and Pigment Loading[a]

	Measured Hydrodynamic Diameter (nm)			
	Buffer at 10 mg/mL Pigment Loading		Pigment Loading at pH 4.4	
Pigment	4.4	5.8	1 mg/mL	10 mg/mL
A	1206	934	386	1206
E	862	1158	mm[b]	862
F	676	1558	mm[b]	676

(a) Standard uncertainty in the particle diameter values is ± 5%.
(b) Denotes unstable suspensions in terms of intensity resulting from large particle clusters

For the LCV assays, the DLS data revealed that the experimental conditions of the assay also influence the cluster size of the pigments. Upon increasing the pH to 5.8 at 10 mg/mL pigment loading for the LCV assay, the DLS curves for the surface-treated pigments become more polydisperse. New peaks at longer lag times were also observed for surface treated TiO_2 E and F for pH 5.8 at 10 mg/mL pigment loading, an indication of a possible secondary, larger-sized, cluster set. However, the DLS curve for TiO_2 A remains close to monodisperse despite an increase to pH 5.8 at 10 mg/mL pigment loading, an indication of increased organization of the clusters. Lowering the pigment loading in the LCV assay produces curves that are similarly shaped for all pigments, regardless of pigment surface treatment. *Table* 4 summarizes the measured hydrodynamic diameter or particle cluster results obtained for the LCV assay after a change in buffer pH and pigment loading. In general, TiO_2 E and F show similar trends in cluster size. The cluster size for TiO_2 E and F increased with increasing pH. The cluster size of TiO_2 A slightly decreases with increasing pH. Decreasing the pigment loading markedly increased the cluster size of the surface treated pigments, TiO_2 E and F. The large cluster size calculated for TiO_2 E and F at 1 mg/mL was a result of very large clusters that produced an unstable suspension and therefore unstable intensity during the DLS measurements. The lower pigment loading had the opposite effect on TiO_2 A such that the cluster size decreased by a factor of 3.

EPR Spectroscopy

SPIN TRAP METHOD: EPR is used to detect free radicals formed in a TiO_2 system to determine its photoreactivity. For short-lived radical species, a spin trap technique is often applied by mixing an appropriate spin trap or radical trap in the sample solution. Nitroxyl radicals (or nitroxides) are a typical spin trap and have been reported to react with hydroxyl radicals (which are known to initiate oxidation in most polymers[1-3]), resulting in a stable radical adduct and a loss in the initial nitroxide EPR signal.[66-68] However, the nitroxide compounds can also react with other radicals present in solution, such as superoxide anion radicals, and cause a similar decrease in the nitroxide EPR signal. Therefore, the decay of the initial nitroxide concentration indicates a non-selective reaction of all radicals from the TiO$_2$ suspension and spin traps, but provides a general indication of photoreactivity of the TiO$_2$ system. A more selective spin trap, 5, 5-dimethyl-1-pyrroline-N-oxide (DMPO), is widely used as a spin trap reagent for individual radical species. DMPO is a diamagnetic compound (no EPR signal) but reacts with other radicals to form paramagnetic spin-adducts, which produces an EPR signal.[69,70] When DMPO spin-adducts form, a distinctive EPR spectrum corresponding to a specific radical is produced.

The feasibility of the spin trap technique was investigated to determine the species in UV-illuminated TiO$_2$ suspensions. In addition, the effect of electrostatic interactions of the nitroxide spin trap with the TiO$_2$ particle in suspension was studied. The sensitivity of nitroxide radicals to the target molecule is related to the chemical structure of the spin trap itself as well as the environment of solution.[66,71] Therefore, three nitroxide spin traps of varying electrostatic charge (*Figure* 13), 3-aminoproxyl (AP), 3-carboxyproxyl (CP), and 3-carbamoylproxyl (CM), were used to monitor the generation of radicals during irradiation. All of the nitroxide traps chosen for this study have different functional groups on the nitroxide ring to produce a different surface charge in solution. In aqueous solutions, AP (pKa ≈ 9) is positively charged, CP (pKa ≈ 4) is negatively charged, and CM is neutral. DMPO was also investigated to determine the presence of specific radicals in the TiO$_2$ suspensions after UV irradiation. The chemical structure of DMPO (*Figure* 13) is different from that of the other nitroxide traps studied, but this was not taken into consideration for the study at this time.

EPR spin trap measurements in this study were considered in-situ measurements as the various TiO$_2$ suspensions were directly illuminated while in the EPR cavity, and the resulting EPR signal was monitored as a function of UV irradiation time. Blank tests were run to determine that the UV source did not significantly affect the EPR signal, that the EPR tubes and deionized water used in the study did not result in an EPR signal. The stability of the nitroxide solutions prior to their addition to the TiO$_2$ suspension was also investigated. Five hundred μmole/L solutions of nitroxyl spin trap for AP, CP, and CM showed no EPR signal decay for 30 min for both dark conditions and after UV illumination. Fresh DMPO solutions (with no TiO$_2$ present) showed no EPR signal initially or after UV irradiation. Hence, the stability of all spin trap solutions under UV irradiation was considered acceptable for the measurement period. Any decay or increase in the EPR intensity of the spin traps was considered a result from the TiO$_2$ reactions.

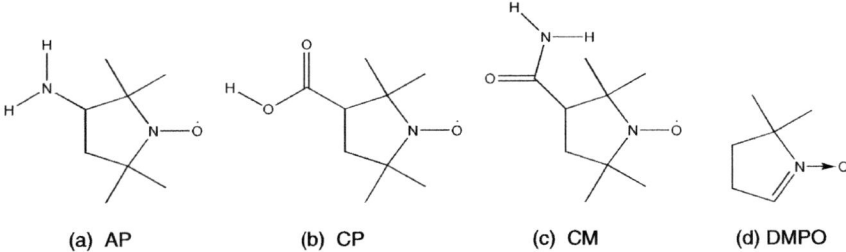

(a) AP   (b) CP   (c) CM   (d) DMPO

Figure 13—Chemical structures of spin traps used in the EPR studies: (a) AP= 3- aminoproxyl, (b) CP= 3-carboxylproxyl, (c) CM= 3-carbamoylproxyl, (d) DMPO= 5, 5-dimethyl-1-pyrroline-N-oxide.

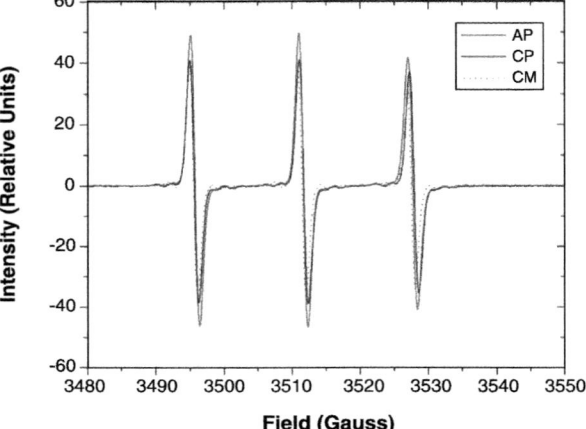

Figure 14—EPR spectra of nitroxide spin traps before UV irradiation.

EPR spectra for fresh solutions of nitroxide spin traps, AP, CP, and CM, without TiO_2, are shown in *Figure* 14. The spectra show the three characteristic nitrogen hyperfine lines,[66] and for similar spin trap concentration the shape and the position of the spectrum are similar for all spin traps despite the different functional groups. Quantification of the intensity of the EPR spectrum, which is a 1st derivative of the EPR absorption signal, was done by double integration of a specific EPR line. The doubly-integrated intensity of an unsaturated EPR spectrum is believed to be to proportional to the concentration of the radicals in the solution.[72] Hence, in this study, reported EPR intensity for most spin traps is the 2nd-low-field line (middle peak) after double integration.

EPR intensity decay plots were generated for the nitroxide spin traps to compare the hydroxyl generation and, hence, the photoreactivity for the TiO_2 specimens. A large decrease in the EPR intensity as a function of irradiation time indicates a highly photoreactive TiO_2 speciman. Reproducibility of the spin trap method was tested as shown in *Figure* 15. The concentrations of spin trap and TiO_2 in each system were kept constant

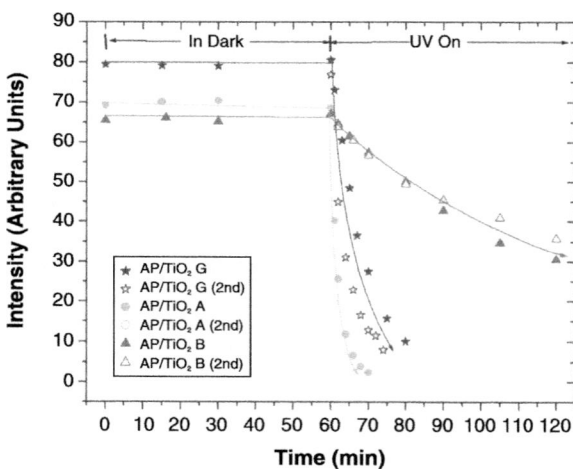

Figure 15—EPR intensity decay plots for TiO_2 specimens in the AP spin trap. Filled symbols represent the first trial. Open symbols represent the second trial. Standard uncertainty in the intensity values is ± 2%.

for the two different trials and the EPR spectrum of each TiO_2/spin trap system recorded under dark conditions did not change in intensity for 1 h. The EPR intensity of the spin traps decreased significantly with UV irradiation and all three TiO_2 specimens presented reproducible decay trends. Moreover, the various TiO_2 pigments showed different effects on spin trap AP consumption during UV irradiation. The EPR intensity decay plots for the series of TiO_2 systems for all three nitroxide traps are shown in *Figures* 16-18. The majority of the TiO_2 specimens show a decrease in the concentration of the nitroxide trap over UV irradiation time, regardless of the type of spin trap. However, there was a different ordering of photoreactivity for the series of TiO_2 specimens depending on the spin trap used, as shown in *Table* 5. In all cases, TiO_2 A showed the greatest photoreactivity. The lack of spread in the curves for the CP and CM spin traps resulted in many of the TiO_2 specimens with the same ranking as their EPR intensity decay curves overlapped. These differences in photoreactivity ranking may be due to the flocculation of some TiO_2 specimens at the particular pH of the spin trap. This hypothesis is currently being investigated using light scattering techniques.

A difference in the slopes for the EPR intensity decay curves generated for each TiO_2 specimen was observed as a function of the specific nitroxide spin trap used. This difference is an indication that the electrostatic attraction of the spin trap to the TiO_2 particle can affect the resulting photoreactivity value. The zeta potential of most pure TiO_2 specimens is pH $\approx$ 4. Therefore, for suspensions in which the pH > 4 the TiO_2 surface is negatively charged, for pH = 4 the TiO_2 surface has no charge, and for pH < 4 the TiO_2 surface is positively charged. However, there are some TiO_2 specimens with surface treatments of aluminum oxide, silicon oxide, or zirconium oxide, which can raise the initial TiO_2 zeta potential from pH = 4 to between pH = 6.5 to 8.0. As a result, for the AP spin trap studies all TiO_2 specimens have a positively charged surface as the pH of the AP/TiO_2 systems ranged from 7.7 to 8.1; but for the CP spin trap studies, where the pH of the CP/TiO_2 systems ranged from 5.1 to 5.9, there is a mix of negatively to positively charged TiO_2 surfaces depending on the TiO_2 specimen surface treatment. The pH of the CM/TiO_2 systems ranged from 3.7 to 3.8, but the charge on the TiO_2 surfaces in these systems was considered negligible as there is no charge on the CM spin trap.

Figure 19 illustrates the effects of electrostatic attraction of the spin trap to the TiO_2 particle before UV irradiation on the EPR intensity. It is not surprising that there is some

Table 5—Photoreactivity Ranking for TiO₂ Specimens based on Spin Trap Measurements

	Spin Trap		
	AP	CP	CM
Most Photoreactive	A	A	A
	G	E	G
	J	G	J
	H	I	B
	K	B	H
	B	K	F
	E	F	E
Least Photoreactive	F	J	I
	I	H	K

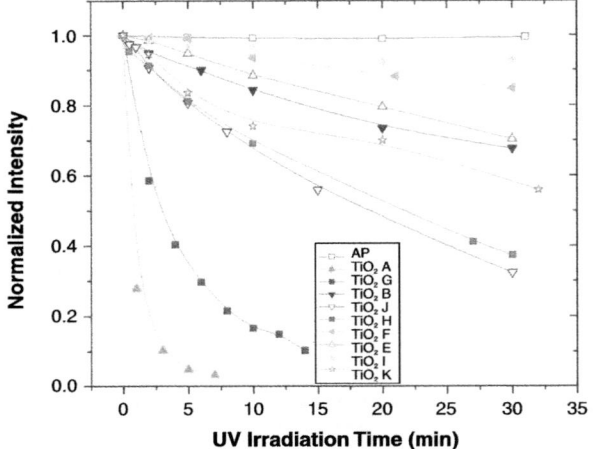

Figure 16—EPR intensity decay plots for TiO₂ specimens in the AP spin trap. Standard uncertainty in the intensity values is ± 2%.

variation in the EPR intensity depending on the TiO_2 specimen and its surface treatment, but a general trend is apparent. The stronger electrostatic attraction for the AP/TiO_2 system results in a greater decrease in the initial concentration of the AP spin trap after TiO_2 addition. The change in concentration observed for the CP/TiO_2 systems is markedly lower, an indication of a weak electrostatic interaction. If the spin trap is not physically close to the surface of the TiO_2 particle, the free radicals generated from the TiO_2 particle cannot be efficiently trapped or titrated. More work is underway to further understand this electrostatic attraction phenomenon and possible particle flocculation using light

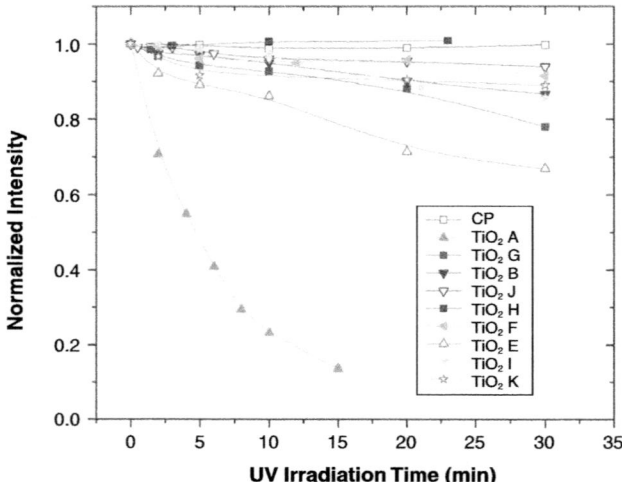

Figure 17—EPR intensity decay plots for TiO₂ specimens in the CP spin trap. Standard uncertainty in the intensity values is ± 2%.

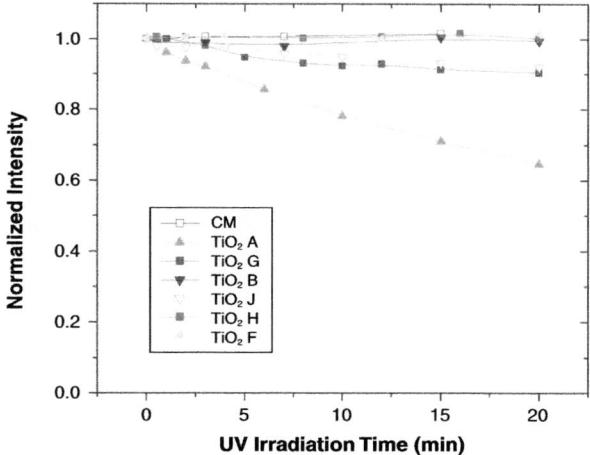

Figure 18—EPR intensity decay plots for TiO₂ specimens in the CM spin trap. Standard uncertainty in the intensity values is ± 2%.

scattering techniques. However, hereinafter the results from the AP spin trap system will generally be used to rank the general photoreactivity for the TiO₂ series.

The DMPO spin trap was also investigated to identify the specific free radicals generated during UV irradiation in aqueous solution and other solvents. In this work, only the formation of hydroxyl radicals is discussed. *Figure 20* shows the EPR spectra for DMPO spin-adducts after reaction with specific radicals. No EPR spectrum is observed with DMPO initially, but after radical reactions with DMPO, distinct EPR spectra are

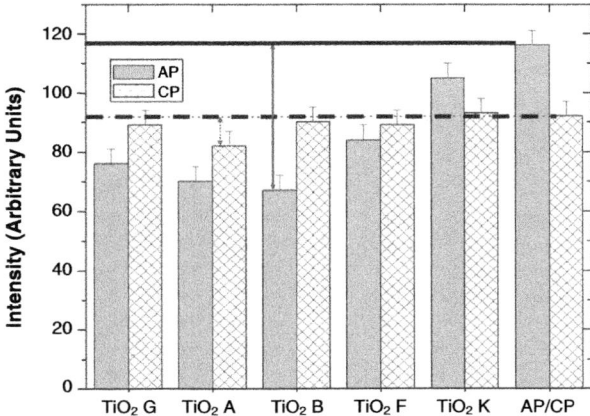

Figure 19—Nitroxide spin trap adsorption when added to suspensions of TiO₂ specimen. AP represented by solid bar and CP represented by bar with diamond pattern. The solid horizontal line represents the initial AP concentration and the dashed horizontal line represents the initial CP concentration before TiO₂ addition. Standard uncertainty in the intensity values is ± 2%.

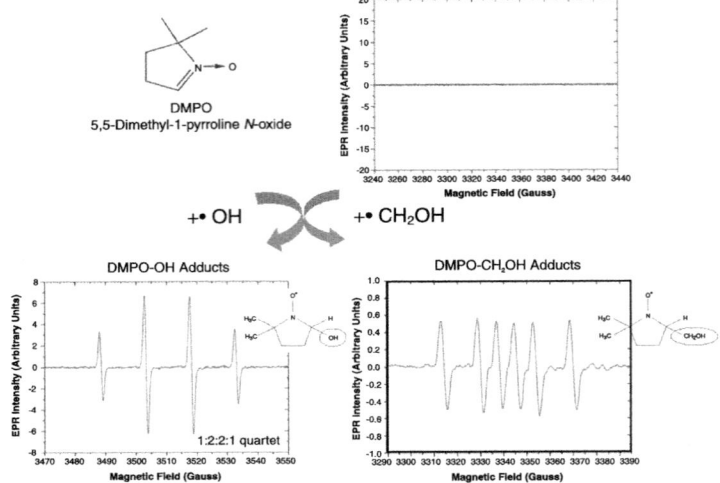

Figure 20—Schematic for DMPO spin-adduct formation after specific radical reactions.

seen for the different DMPO spin-adducts. Out of the entire TiO_2 series, only TiO_2 G, A, B, and K were found to produce the characteristic EPR spectrum for the DMPO hydroxyl spin-adduct. It was also observed that EPR intensity of the DMPO hydroxyl spin-adduct decreased with increasing UV irradiation time as shown in *Figure* 21. This decay, which was also reported by Grela et al.,[70] was affected by both initial DMPO and TiO_2 concentration and believed to be a reaction of the DMPO hydroxyl spin-adduct with excess hydroxyl radicals present in solution. An optimum concentration of DMPO (0.25 millimole/L) was found to provide the greatest concentration and least decay for the DMPO hydroxyl spin-adduct. Only TiO_2 G and A produced the highest concentra-

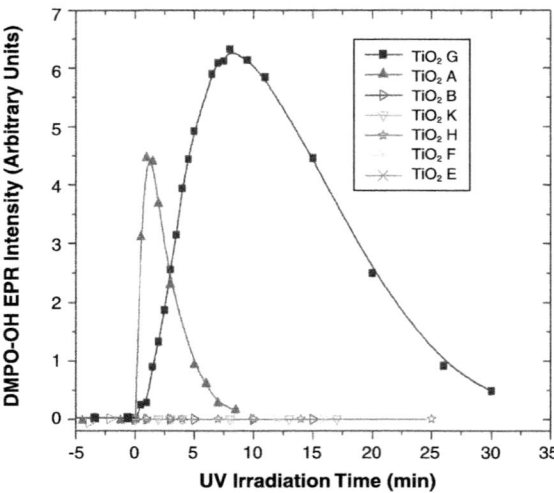

Figure 21—Plots of EPR intensity versus UV irradiation time for DMPO/TiO$_2$ specimens. Standard uncertainty in the intensity values is ± 2%.

tions of the DMPO hydroxyl spin-adduct, a further indication that these TiO_2 specimens are highly photoreactive. The DMPO studies showed that there is only a small subset of the TiO_2 series that generate detectable amounts of hydroxyl radicals in the aqueous suspensions. Interpretation of this finding is underway. Additionally, experiments to examine the effect of solvents on the radical species generated for the TiO_2 series have begun.

SOLID STATE METHOD: The EPR technique was also used to directly measure free radical generation in TiO_2 materials using solid state measurements at lower temperature (77 K). Signal intensity in an EPR spectrum provides evidence of unpaired electrons in the sample. The position of the EPR signal or g-factor can indicate the type of unpaired electron; that is, if it is organic or inorganic in nature. The g-factor, g, is defined as

$$g = h\nu / \mu_B B_0 \qquad (3)$$

where h is Planck's constant, ν is the frequency, μ_B is the Bohr magnetron, and B_0 is external magnetic field. The hyperfine interactions observed by the hyperfine structure within an EPR spectrum can reveal the molecular structure and environment near the unpaired electron. The linewidth and/or lineshape of the EPR spectrum can also provide information about the molecular motion in a sample with the unpaired electrons.

With this information in mind, the EPR spectra of the TiO_2 series were examined at 77 K before and after UV irradiation. The g-factors for the EPR instrument used in this study were calibrated using a standard reference compound, 2, 2-diphenyl-1-picrylhydrazyl (DPPH), which has a known g-factor value of 2.00037.[29] *Figure 22* illustrates the change in the EPR signal for TiO_2 A. An increase in the EPR intensity at g = 2.014 and g = 1.979 is observed with increasing UV irradiation time. After the UV source is turned off, the EPR intensity of the peak at g = 2.014 decreased more rapidly than the EPR peak at g = 1.979. The EPR peak at g = 2.014 has been reported to be a component of the hole-centers produced near the TiO_2 surface.[73,74] The EPR signal located at g-factor = 1.979 has been assigned as a species related to the electron trapping sites, which is usually rep-

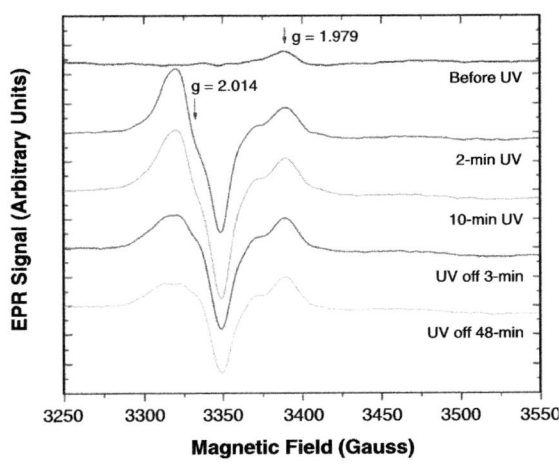

Figure 22—EPR spectra for TiO_2 A as a function of UV irradiation time.

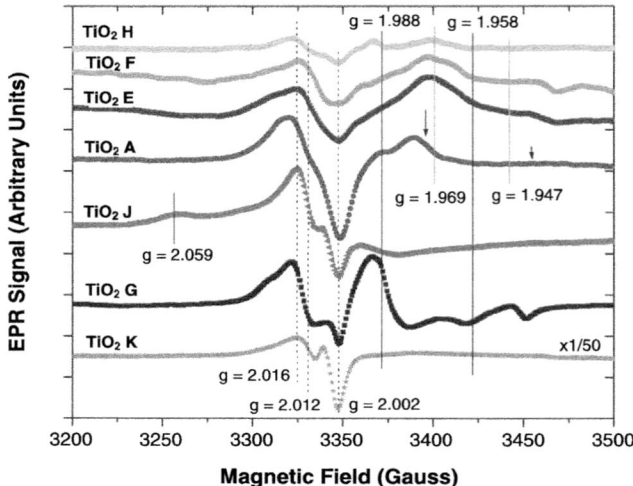

Figure 23—EPR spectra for a series of TiO₂ specimens after UV irradiation. The EPR spectrum for TiO₂ K is reduced by 1/50 to achieve the same scale as other TiO₂ specimens.

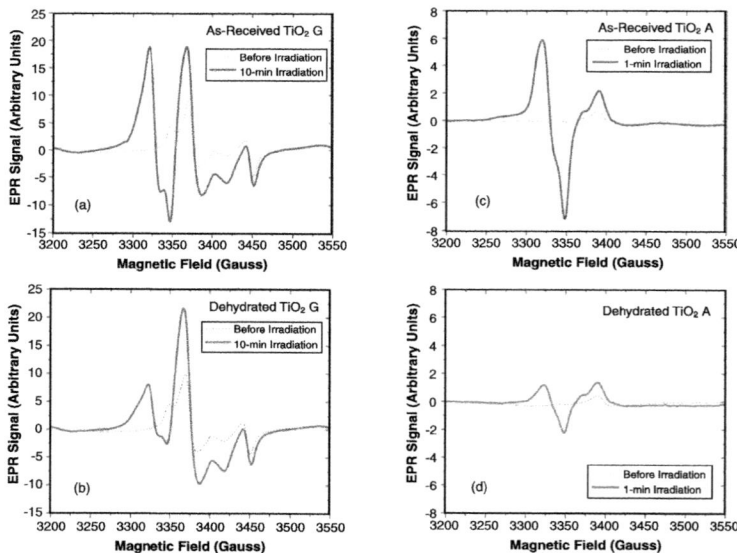

Figure 24—EPR spectra for TiO₂ specimens as a function of UV irradiation time and extent of TiO₂ hydration. Dashed-line spectra represent the TiO₂ specimens before UV irradiation and solid-line spectra represent the TiO₂ specimens after UV irradiation. Spectra (a) and (b) represent TiO₂ G in the as-received state and after dehydration (under vacuum at 85°C for 10 d), respectively. Spectra (c) and (d) represent TiO₂ A as-received and after dehydration, respectively.

resented as a Ti^{+3} group on TiO_2 particles.[75,76] *Figure* 22 suggests that the electron trapping sites are more robust and longer lasting than the hole-centers for TiO_2 A. *Figure* 23 shows the EPR spectra for the series of TiO_2 specimens after UV irradiation. Each TiO_2 specimen has a different EPR spectrum; however, all have EPR peaks in the positions where hole-centers and electron trapping sites are located. Most of the anatase specimens have more complex EPR peaks in the hole-center region. The lineshape and linewidth of the EPR spectra should reveal information about the solid-state physics of the photoreactivity for the TiO_2 samples and how it differs for the various manufacturing processes. For example, the width of the hole-center EPR peak is sharper for some TiO_2 systems, particularly TiO_2 J, than others. This could indicate possible neighboring nuclei and slower relaxation times for the electrons in TiO_2 J. The electron-trapping sites for TiO_2 G are markedly different than those observed on the other TiO_2 systems. The peak is located at a higher g-factor, $g = 1.988$ versus $g = 1.969$, indicating a different spin state and coordination environment near the unpaired electron. The width of the peak is also broader, an indication of rapidly relaxing electrons. More detailed interpretation of these EPR spectra are in progress.

Lastly, the extent of hydration of the TiO_2 system was investigated using the solid-state EPR technique. As has been reported,[1-3] the generation of hydroxyl radicals from pigment fillers is believed to lead to the degradation of a surrounding polymer matrix. One condition under which hydroxyl radicals form is in the presence of adsorbed water. Therefore, the effect of adsorbed water on the radical generation was monitored for the TiO_2 systems. TiO_2 powders were dehydrated by evacuation under vacuum at 85°C for 10 days. The dehydrated TiO_2 powder was sealed in an EPR tube and immediately analyzed in the EPR instrument and compared with the EPR spectrum obtained for the as-received TiO_2 specimens (*Figure* 24). Dehydration of the TiO_2 systems appears to reduce the amount of hole-centers produced as shown by the reduction of the EPR peak at lowest magnetic field. For TiO_2 G, the EPR intensity for the hole-center region decreased by 60% compared to the as-received sample. For TiO_2 A, there was an 80% decrease in the hole-center region compared to the as-received sample. However, the EPR intensity for the region of the electron-trapping sites was not significantly affected in either sample. This type of experiment on TiO_2 can provide insight into the solid-state physics of photoreactivity and how the TiO_2 manufacturing process affects hole-center and electron-trapping site production. This is the direction in which our research is heading.

UV Exposure of Pigment-Filled Polymer Films

The aim of this study is to examine the effects of surface treatment and concentration of pigment on the degradation of a resin system and determine the extent of protection that pigment addition contributes to polymer durability. A less durable, but well-characterized amine-cured epoxy was chosen for the investigation to minimize the time necessary to detect degradation. The epoxy and pigment were analyzed separately to obtain an understanding of their individual contributions to chemical changes observed for the pigmented system. Three commercially available TiO_2 particle types were selected, having properties shown in *Table* 6. In general, the reactivity of the TiO_2 particles is directly related to surface treatment as determined by the chemical assay and EPR measurements.

Table 6—TiO$_2$ Properties Used in Epoxy Polymer Films

Designation	Reported Particle Diameter (nm)	Surface Treatment	Photoreactivity
TiO$_2$ F	≈250	Al$_2$O$_3$ and SiO$_2$ (12%)	Low
TiO$_2$ E	≈250	Al$_2$O$_3$ (6%)	Med
TiO$_2$ A	≈20	None	High

XPS was used to monitor the changes in the surface chemistry of the pigmented poly-mer as a function of UV irradiation time. The hypothesis was that a highly photoreactive pigment would degrade the surface of the polymer film at a greater rate than a more durable or less photoreactive pigment. The dispersion of the pigment in the epoxy film was not considered for the results in this work, but is known to significantly affect the degradation of the polymer system and will be discussed in a future publication. Using XPS, the atomic concentrations of carbon (C), oxygen (O), and nitrogen (N) can be fol-lowed. Additionally, the chemical components in the pigment filler, i.e., titanium (Ti), sil-icon (Si), and aluminum (Al), can also be examined upon severe degradation or chalk-ing. XPS results showed that oxidation in the epoxy films increased with increasing UV irradiance time as indicated by an increase in O 1s atomic concentration from 17%–18% initially to 28% –36%. These results indicate that the pigment did not provide any notable protection from photodegradation as these values were not significantly different from the unpigmented epoxy film. The N 1s atomic concentration also increased by about 40%, which reflects a change in the crosslinking in the polymer. These changes coincide with the decreases in the C 1s concentration in all epoxy films. The decrease in C 1s atomic concentration ranged from 14%–17% from the initial value and a slightly lower decrease; 12% was observed for the unpigmented epoxy film.

Figure 25 shows the high-resolution C 1s spectra for the epoxy films with TiO$_2$ F and TiO$_2$ A, extremes in pigment photoreactivity. Changes in the C 1s peak shape are observed with increasing UV irradiation, and the same peaks were observed for the

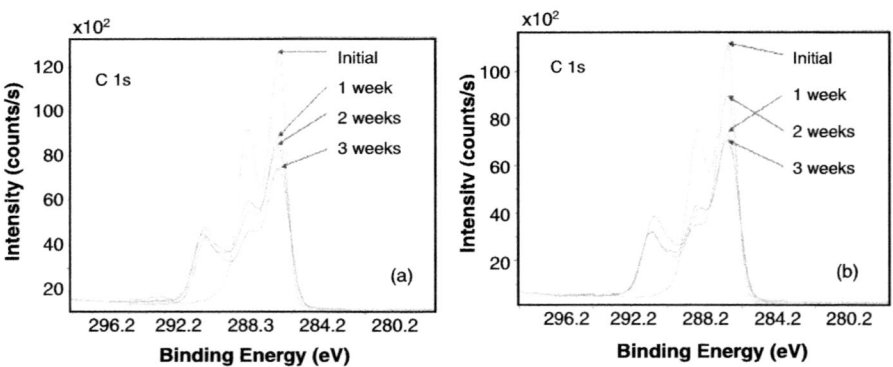

Figure 25—XPS spectra of the C 1s region for pigmented epoxy films with increasing UV exposure. Spectrum (a): epoxy film with TiO$_2$ A. Spectrum (b): epoxy film with TiO$_2$ F. The green spectrum rep-resents the initial pigmented epoxy film before UV exposure.

epoxy films regardless of the pigments and pigment concentrations used. *Figure 25* shows a slightly greater initial reduction in the C 1s region for the epoxy containing the more photoreactive pigment, TiO_2 A, but these changes level off for both pigments after the second week of UV exposure. A new C band appears at higher binding energy with increasing photodegradation. This shift of 3.9 eV relative to the adventitious carbon peak at 285.0 eV is indicative of carbonyl species containing C–O–C=O (4.0 eV shift) or HO–C=O (4.3 eV shift) or carbonyl species associated with nitrogen, i.e., O=C–N–C=O (3.6 eV shift) or N–C=O–N (3.8 eV shift).[77] Clark and Munro[78,79] also noted similar changes in the carbon region during photooxidation for a bisphenol A polysulphone (the epoxy component). An increase of ca. 1.0 eV in the full-wide at half maximum (FWHM) in the O 1s region with UV irradiance was observed for all epoxy films. This is an indication of new oxygen species forming during photodegradation. Curve-fitting of all XPS high resolution element regions will be completed for further interpretation of the degradation results. This will help to identify the chemical species in the film formed and/or lost during the photodegradation process.

Traces of pigment were not observed on the sample surface until after five weeks of UV exposure (*Figure 26*). It should be noted that a trace amount of silicon (0.5%) observed in the XPS results was due to the interaction of the epoxy with the silicone release paper. For the epoxy film with TiO_2 F (silica treated), a marked increase in the Si concentration (2.0%–2.4%) was observed in the third week of UV exposure (*Figure 26b*). This high Si concentration compared to the maximum Ti concentration (0.2%) is not unusual as XPS is more surface sensitive, but it indicates that more pigment is exposed

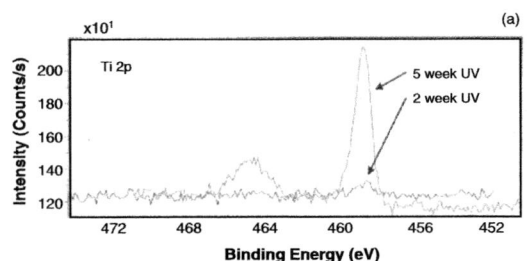

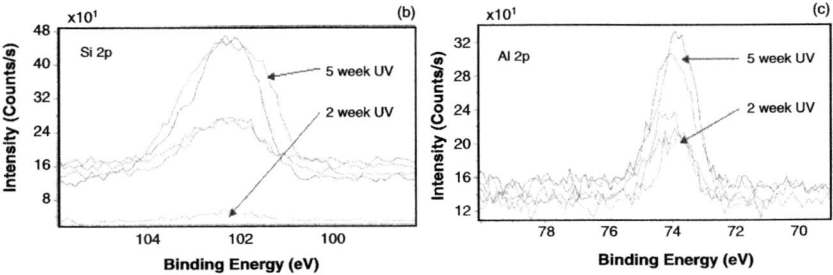

Figure 26—XPS spectra of the different chemical regions for pigmented epoxy films with increasing UV exposure. Spectrum (a): Ti 2p region for epoxy film with TiO_2 A. Spectrum (b): Si 2p region epoxy film with TiO_2 F. Spectrum (c): Al 2p region for epoxy film with TiO_2 E.

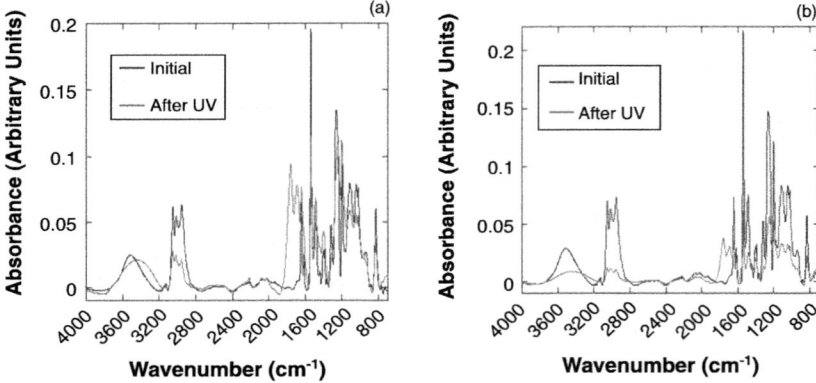

Figure 27—ATR-FTIR spectra for epoxy films before (black) and after UV irradiation (red): (a) epoxy film with no pigment added, (b) epoxy film with TiO₂ A.

in this epoxy film. Epoxy films with alumina-treated TiO_2 E also show evidence of the pigment on the surface of the epoxy film after a slightly shorter period of UV irradiation (*Figure* 26c). Hints of the titanium oxide present on epoxy films with TiO_2 A (no surface treatment) are also evident after two weeks of UV exposure with larger contributions observed after five weeks of UV exposure. All the results are evidence of pigment chalking that occurs with increasing photodegradation of the polymer matrix. In general, XPS results have shown that there appears to be a similar and rapid increase in the degradation of the epoxy system regardless of incorporated pigment, an indication that this epoxy system may be too photo-unstable to accurately assess the effect of pigment photoreactivity.

ATR-FTIR was used to confirm the assignment of XPS C 1s, O 1s, and N 1s peaks and to monitor the chemical degradation of the epoxy films. While ATR-FTIR is considered a surface technique, it is not as surface sensitive as XPS and so some variance between the two results are expected. An example of ATR-FTIR spectra is shown in *Figure* 27. ATR-FTIR results before UV exposure show the presence of a ketone carbonyl band at 1720 cm^{-1}, of a C=N band at 1654 cm^{-1}, and of an aromatic ester band at 1510 cm^{-1}. With increasing UV irradiance time, the formation of a peak at 1762 cm^{-1} and increases in the 1654 cm^{-1} band were observed. Additionally, there was a marked decrease in the 1510 cm^{-1} and 2952 cm^{-1} bands, an indication of chain scission in the backbone of the epoxy system. The ketone C=O stretch at 1762 cm^{-1} and the chain scission of the backbone at C–H at 2952 cm^{-1} for all epoxy films is plotted as a function of UV dosage in *Figures* 28 and 29, respectively. No marked differences in the rates of chemical change were observed between the various TiO_2 particles as may be expected given the range in the particle reactivity as a function of surface treatment. However, the rate of oxidation of the epoxy system was about 1.7 times greater than the rate for the decrease in the mass loss. It appears that the unfilled epoxy system showed the greatest extent of oxidation while the epoxy filled with TiO_2 F (least UV reactive) showed the least oxidation (half that of the unfilled epoxy). These results are similar to the XPS results observed after longer UV exposure times. Chain scission was greatest in the

Figure 28—Extent of UV degradation in the epoxy system as monitored by ATR-FTIR. An increase in the oxidation products is shown by the C=O stretch at 1762 cm⁻¹ for epoxy films with TiO₂ F (diamond), with TiO₂ E (square), and with TiO₂ A (triangle). Error bars represent the estimated uncertainty of K=2 (95%).

epoxy-TiO_2 A system (1.5× greater than for the unfilled epoxy). These results are consistent with those reported in the literature.[44,80] Quantification of the ATR-FTIR results will be done using a peak ratio method in a future publication. This study showed that the photostability of the polymer matrix is an important variable that needs to be taken into account. The epoxy system used in this study degraded too quickly to be able to distinguish the photoreactivity effects of the various TiO_2 particles. A study using an acrylic urethane polymer, a more photostable system, and similar pigments has been completed and the results will be reported in a future publication.

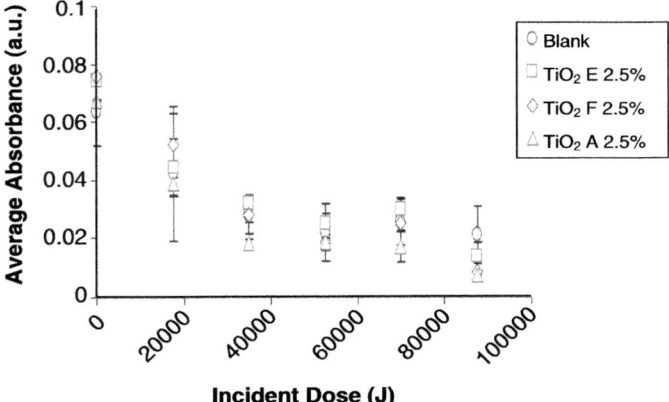

Figure 29—Extent of UV degradation in the epoxy system as monitored by ATR-FTIR. A decrease in mass loss by the C–H stretch at 2952 cm⁻¹ for epoxy films with particle A (diamond), with TiO₂ F (square), and with TiO₂ A (triangle). Error bars represent the estimated uncertainty of K=2 (95%).

SUMMARY AND CONCLUSIONS

Photoreactivity measurements of TiO_2 using spectrophotometric assays, EPR spectroscopy, and performance tests of UV exposed pigmented epoxy systems were presented. Spectophotometric assay methods appear to be useful for measuring and ranking TiO_2 pigment photoreactivity. The reaction mechanism of the assay must be taken into account to accurately rank the pigments for photoreactivity. However, the spectrophotometric assays are highly susceptible to a large number of variables that have historically neither been controlled nor measured. The assay matrix appears to affect TiO_2 particle cluster size, which in turn may influence photoreactivity values. In particular, the pH of an assay system can affect particle cluster formation. In addition to cluster size of pigments, other variables such as buffer composition and TiO_2 concentration were also shown to affect photoreactivity. From a practical standpoint, until measurement and control metrologies and protocols are imposed, the results from such analyses are suspect. The results from this study will be used to optimize the standard assay protocol to account for the interactions of pigments with the assay matrix.

EPR spectroscopy has been shown to be a powerful technique capable of monitoring small concentrations of free radicals by use of spin trapping methods or directly using low temperature, solid-state EPR measurements. For spin trap methods, electrostatic interactions of the spin trap molecule with the surface of the TiO_2 particle must be taken into consideration. However, in cases for which the electrostatic attraction and concentrations of both spin trap and TiO_2 specimen are optimized, the spin trap method can provide a sensitive measure for a wide range of TiO_2 samples with varying photoreactivity. Solid-state EPR measurements on TiO_2 samples have the potential to directly compare free radical generation and solid-state physics properties over a range of TiO_2 photoreactivity. By using the EPR signal intensity as evidence for unpaired electrons, the position (g-factor) of the EPR signal as an indication of the type of unpaired electron, the hyperfine structure of the EPR signal for information on the molecular structure and environment near the unpaired electron, and the linewidth/lineshape of the EPR signal for information about the molecular motion in a sample with the unpaired electrons, one can determine the type of hole-centers and electron-trapping sites found in each type of TiO_2 sample.

Finally, a performance test involving UV exposure of pigmented polymer systems was also used to examine the effects of TiO_2 photoreactivity on the durability of a polymer matrix. Surface-sensitive spectroscopic techniques were used to follow the degradation of the polymer matrix. Preliminary results show that in the case of photo-unstable epoxy polymer, the photoreactivity of the TiO_2 filler did not have any significant effect on the degradation of the polymer. The studied pigmented epoxy systems showed similar trends in oxidation and mass loss, regardless of TiO_2 photoreactivity or TiO_2 concentration. Future studies will use a more photostable polymer matrix where accelerated rates of degradation can be linked to particle reactivity and will include characterization of particle dispersion.

References

(1) Solomon, D. and Hawthorne, D., in *Chemistry of Pigments and Fillers*, Krieger Publishing Co., Malabar, 1991.
(2) Diebold, M., *Surf. Coatings Intern.*, 6, 250-256 (1995).
(3) Hoffman, M., Martin, S., Choi, W., and Bahnemann, D., *Chem. Rev.*, 95, 69-96 (1995).
(4) Blake, D., Maness, P.-C., Huang, Z., Wolfrum, E., Huang, J., and Jacoby, W. "Separation and Purification Methods," 28, 1-50 (1999).
(5) Pilotek, S. and Tabellion, F., *European Coat. J.*, 170-176 (2005).
(6) Vaia, R.A. and Maguire, J.F., *Chem. Mater.*, 19, 2736-2751 (2007).
(7) Sciafani, A., Palmisano, L., and Schiavello, M., *J. Phys. Chem.*, 94, 829-832 (1990).
(8) Brand, J., *Plastics Compounding*, 27-29 (1985).
(9) Diebold, M., Kwoka, R., Mehr, S., and Vargas, R., *Proc. 81st Annual Meeting of the FSCT*, 2003.
(10) Stieg, F.B., "Accelerating the Accelerated Weathering Test," *J. Paint Technol.*, 38, No. 492, 29-36 (1966).
(11) Joshhi, N., *Photoconductivity: Art, Science and Technology*, Marcel Dekker, Inc., New York, 1990.
(12) Bube, R., *Photoconductivity of Solids*, John Wiley and Sons, New York, 1960.
(13) Runyan, W. and Shaffner, T., *Semiconductor Measurements and Instrumentation*, 2nd Ed., McGraw Hill, New York, 1998.
(14) Hermann, J., Disdier, J., Mozzanega, M., and Pichat, P., *J. Catal.*, 60, 369-377 (1979).
(15) Hermann, J., Disdier, J., and Pichat, P., in *7th International Vacuum Congress and 3rd International Conference on Solid Surfaces*, Vienna, pp. 951-954, 1977.
(16) Eppler, A., Ballard, I., and Nelson, J., *Physica E*, 14, 197-202, 2002.
(17) Chin, J., Scierka, S., and Forster, A., in *Service Life Prediction: Challenging the Status Quo*, Martin, J.W., Ryntz, R.A., and Dickie, R.A. (Eds.), Federation of Societies for Coatings Technology, Blue Bell, PA, pp. 229-240, 2005.
(18) Maruszewski, K., Jasiorski, M., Hreniak, D., and Strek, W., *J. Molec. Struct.*, 597, 273- 277 (2001).
(19) Degani, Y. and Heller, A., AT&T Bell Labs, 1990.
(20) Bickely, R.I. and Jayanty, R.K.M., *Faraday Discussion*, 585, 194-204 (1974).
(21) Egerton, T.A. and King, C.J., *J. Oil Colour Chem. Assoc.*, 62, 386 (1979).
(22) Pappas, S. and Fischer, R., "Photo-Chemistry of Pigments: Studies on the Mechanism of Chalking," *J. Paint Technol.*, 46, No. 599, 65-72 (1974).
(23) Worthington, V., *Worthington Enzyme Manual: Enzymes and Related Biochemicals*, The Worthington Biochemical Corp., NJ, pp. 293-299, 1993.
(24) Mottola, H.A., Simpson, B.E., and Gorin, G., *Anal. Chem.*, 42, 410-411 (1970).
(25) Eaton, G., Eaton, S., and Salikhov, K., *Foundations of Modern EPR*, World Scientific Publishing Co., 1998.
(26) Che, M. and Giamello, E., in *Electron Paramagnetic Resonance: Principles and Applications to Catalysis*, Imelik, B. and Vedrini, J. (Eds.), Plenum Press, New York, pp. 131-197, 1994.
(27) Micic, O., Zhang, Y., Cromack, K., Trifunac, A., and Thurnauer, M., *J. Phys. Chem.*, 97, 7277-7283 (1993).
(28) Coronado, J., Maira, A., Conesa, J., Yeung, K., Auguliaro, V., and Soria, J., *Langmuir*, 17, 5368-5374 (2001).
(29) Howe, R., *Adv. Col. Interface Sci.*, 18, 1 (1982).
(30) Jeager, C. and Bard. A., *J. Phys. Chem.*, 83, 3146 (1979).
(31) Harbour, J., Tramp, J., and Hair, M., *Can. J. Chem.*, 63, 204-208 (1985).
(32) Gonzalez-Elipe, A., Munuera, G., and Soria, J., *J. Chem. Soc., Faraday Trans.*, 75, 749 (1979).
(33) Anpo, M., Shima, T., and Kubokawa, Y., *Chem. Lett.*, 1799-1802 (1985).
(34) Howe, R. and Gratzel, M., *J. Phys. Chem.*, 91, 3906-3909 (1987).
(35) Christaans, M., Wienk, M., Hal, P., Kroon, J., and Janssen, R., *Synthetic Metals*, 101, 265-266 (1999).
(36) Cao, H., He, Y., Zhang, R., Yuan, J., Sandreczki, T., Jean, Y., and Nielsen, B., *J. Polym. Sci.: Part B, Polym. Phys.*, 37, 1289-1305 (1998).
(37) Schlick, S., Kruczala, K., Motyakin, M., and Gerlock, J., *Polym. Degrad. Stab.*, 73, 471-475 (2001).
(38) Pientka, M., Wisch, J., Boger, S., Parisi, J., Dyakonov, V., Rogach, A., Talapin, D., and Weller, H., *Thin Solid Films*, 451-452, 48-53 (2004).
(39) Jacobsen, A.E., *Ind. Eng. Chem.*, 41, 523-526 (1949).
(40) Steig, F.B. Jr., "Accelerating the Accelerated Weathering Test," *J. Paint Technol.*, 38, No. 492, 29-36 (1966).

456

(41) Kaempf, G., Papenroth, W., and Holm, R., "Degradation Processes in TiO_2–Pigmented Paint Films on Exposure to Weathering," *J. Paint Technol.*, 46, No. 598, 56-63 (1974).
(42) Gaumet, S., Siampiringue, N., Lemaire, J., and Pacaud, B., *Surf. Coatings Intern.*, 8, 367-372 (1997).
(43) Wernstahl, K.M. and Carlsson, B., "Durability Assessment of Automotive Coatings—Design and Evaluation of Accelerated Tests," *J. Coat. Technol.*, 69, No. 865, 69-75 (1997).
(44) Allen, N.S., Edge, M., Ortega, A., Sandoval, G., Liauw, C.M., Verran, J., Stratton, J., and McIntyre, R.B., *Polym. Degrad. Stab.*, 85, 927-946 (2004).
(45) Martin, J.W., in *ACS Symposium Series 722*, Bauer, D.R. and Martin, J.W. (Eds.), ACS Publishing, pp. 1-20, 1999.
(46) Balfour, J.G., *JOCCA-Surf. Coatings Intern.*, 12, 478 (1990).
(47) Gesenhues, U., *Double Liaison: Physique, Chimie et Economie des Peintures et Adhesifs*, 479-480, X (1996).
(48) Simpson, L.A., *Australian OCCA Proc. and News*, 6 (1983).
(49) Dilks, A., in *Degradation and Stability of Polymers*, Vol. 1, Jellinek, H.H.G. (Ed.), Elsevier, Amsterdam, 601-628, 1983.
(50) Clark, D.T., *Pure Appl. Chem.*, 57, 941 (1985).
(51) Hawkridge, A.M., Gardella, J.A. Jr., and Toselli, M., *Macromolecules*, 35, 6533 (2002).
(52) Motyakin, M.V. and Schlick, S., *Polym. Degrad. Stab.*, 76, 25 (2002).
(53) Haverkamp, R.G., Siew, D.C.W., and Barton, T.F., *Surf. Interface Anal.*, 33, 330 (2002).
(54) Walters, K.B., Schwark, D.W., and Hirt, D.E., *Langmuir*, 19, 5851, (2003).
(55) Egerton, T.A., Parfitt, G.D., Kang, Y., and Wightman, J.P., *Coll. Surf.*, 7, 311 (1983).
(56) Diebold, U. and Madey, T.E., *Surf. Sci. Spec.*, 4, 227 (1997).
(57) Erdem, B., Hunsicker, R.A., Simmsons, G.W., Sudol, E.D., Dimonie, V.L., and El-Asser, M.S., *Langmuir*, 17, 2664 (2001).
(58) Berne, B.J. and Pecora, R., *Dynamic Light Scattering with Applications to Chemistry, Biology, and Physics*, Dover Publications, Inc., New York, 2000.
(59) Chin, J.W., Byrd, W.E., Embree, E.J., and Martin, J.W., *Service Life Prediction: Methodology and Metrologies*, ACS Symposium Series 805, Martin, J.W. and Bauer, D.R. (Eds.), American Chemical Society, pp. 144-160, 2002.
(60) Martin, J.W., Chin, J.W., Byrd, W.E., Embree, E.J., and Kraft, K.M., *Polym. Degrad. Stab.*, 63, 297-304 (1999).
(61) Hidaka, H., Zhao, J., Satoh, Y., Nohara, K., Pelizzetti, E., and Serpone, N., *J. Molec. Catal.*, 88, 239-248 (1994).
(62) Ortiz-Islas, E., Lopez, T., Gomez, R., and Navarrete, J., *J. Sol-Gel Sci. Tech.*, 37, 165-168 (2006).
(63) Parida, K.M., Acharya, M., Samantaray, S.K., and Mishra, T., *J. Coll. Interf. Sci.*, 217, 388-394 (1999).
(64) Li, W. Li, S., Zhang, M., and Tao, K., *Colloids Surf. A: Physicochem. Eng. Aspects*, 272, 189-193 (2006).
(65) Ross, S. and Morrison, I.D., *Colloidal Systems and Interfaces*, John Wiley and Sons, NY, 1988.
(66) Schwarz, P.F., Turro, N.J., Bossmann, S.H., Braun, A.M., Abdel Wahab, A.A., and Durr, H., *J. Phys. Chem. B*, 101, 7127-7134 (1997).
(67) Itoh, O., Obara, H., Aoyama, M., Ohya, H., and Kamada, H., *Anal. Sci.*, 17, i1515-i1517 (2001).
(68) Kocherginsky, N. and Swartz, H.M., *Nitroxide Spin Labels: Reactions in Biology and Chemistry*, CRC Press, NY, 1996.
(69) Buettner, G.R. and Oberley, L.W., *Biochem.Biophys. Res. Commun.*, 83, 69-74 (1978).
(70) Grela, M.A., Coronel, M.E., and Colussi, A.J., *J. Phys. Chem.*, 100, 16940-16946 (1996).
(71) Saracino, G.A., Tedeschi, A., D'Errico, G., Improta, R., Franco, L., Ruzzi, M., Corvaia, C., and Barone, V., *J. Phys. Chem. A*, 106, 10700-10706 (2002).
(72) Bales, B.L. and Peric, M., *J. Phys. Chem B*, 101, 8707-8716 (1997).
(73) Nakaoka, Y. and Nosaka, Y., *J. Photochem. Photobio. A: Chem.*, 110, 229-305 (1997).
(74) Hirakawa, T., Kominami, H., Ohtani, B., and Nosaka, Y., *J. Phys. Chem.*, 105, 6993-6999 (2001).
(75) Howe, R.F. and Gratzel, M., *J. Phys. Chem.*, 89, 4495-4499 (1985).
(76) Gratzel, M. and Howe, R.F., *J. Phys. Chem.*, 94, 2566-2572 (1990).
(77) Beamson, G. and Briggs, D., *High Resolution XPS of Organic Polymers: The Scienta ESCA300 Database*, John Wiley and Sons, NY, 1992.
(78) Munro, H.S. and Clark, D.T., *Polym. Degrad. Stab.*, 11, 211 (1985).
(79) Munro, H.S. and Clark, D.T., *Polym. Degrad. Stab.*, 11, 225 (1985).
(80) Bellenger, V. and Verdu, J., *J. Appl. Polym. Sci.*, 30, 363 (1985).

Chapter 30

A Quantitative Model for Weathering-induced Mass Loss in Thermoplastic Paints

Kurt A. Wood* and Ségolène de Robien

Arkema, Inc., 900 First Ave., King of Prussia, PA 19406

We explore the mass loss behavior of model acrylic latex paint, using the framework of a simple conceptual model distinguishing different contributions to the photochemical mass loss rate. By independently monitoring the physical thickness and coating mass loss, the mass gain from oxygen uptake in clearcoats can be distinguished from an actual net mass loss. The degradation rate near the top surface was found to be greatly enhanced versus the rate in the bulk, both for clearcoats and pigmented coatings. Increasing levels of rutile TiO_2 cause a reduction in the rate of steady state mass loss, as would be expected from pigment screening effects; however, at the lowest pigment level studied (2% by volume), the mass loss rate is a factor of 3–4 higher than what would be expected considering the clearcoat mass loss rate in combination with the measured UV attenuation from the pigment as a function of coating thickness.

INTRODUCTION

The past decade has seen significant advances made in the ability to predict the service life of thermoset clearcoats. Much of this work has been done under the auspices of the J. Martin group at NIST (National Institute of Standards and Technology). Many of the early studies focused on melamine[1]- and urethane[2]-crosslinked acrylics, inspired by automotive clearcoats, while more recent intensive studies have used a model epoxy system.[3] A key concept of this approach has been to use the dosage (absorbed UV-visible energy) as a metric to put on a common scale the chemical damage done to the system under different exposure conditions.

In the NIST studies, infrared absorbance data, as measured by transmission FTIR spectroscopy, have typically been used as measures of the chemical damage done to the system. The transmission absorbance bands are a measure of the total chemical change occurring in the bulk of the sample, whereas, at least for the case of the epoxy system exposed outdoors, AFM microscopy shows that surface damage develops in a very heterogeneous fashion. Nevertheless, the use of bulk infrared (IR) data as a proxy damage metric seems to be successful in this case, in terms of the ability to generate "universal" curves linking chemical damage and dosage, regardless of the exposure conditions.

---

*Author to whom correspondence should be addressed.

For thermoplastic pigmented paints, it is evident that some kind of different damage metric besides IR transmission will be required. A number of other vibrational spectroscopy techniques are available which can be used with infrared opaque systems, and which could potentially give some degree of depth resolution to the damage. Here, we report on experiments to explore whether mass loss measurements might successfully serve as proxy measure of damage, for model clear and pigmented thermoplastic paint systems using an acrylic latex.

A focus on mass loss for thermoplastic paints is thought to be appropriate for several reasons: a measurable mass loss is often concurrent with the period when significant gloss loss occurs; physical changes resulting from mass loss should result directly in gloss effects, so that modeling gloss loss is possible in principle; and the photooxidative pathways of many acrylic resins are known to include chain scission and "unzipping" reactions which lead to mass loss through the volatilization of the reaction by-products.[4,5]

In a major 1981 review paper, Colling and Dunderdale[6] pointed out that, even if the pigment in a paint only has a protective (UV light blocking) function, the evolution of gloss loss and chalking will be substantially different depending on whether the coating binder degrades primarily at the air interface ("erosion"), or throughout the bulk as far as light is able to penetrate ("contraction"). From a chemical kinetics standpoint, this means that the bulk binder may undergo photodegradation at some particular rate, but there might also be an enhancement of the rate at or very near the top surface of the coating. Since pigments, and particularly the most important pigment in coatings, TiO_2, can also have some photocatalytic surface activity, this could be a third mechanism contributing to the photodegradation rate. Colling and Dunderdale presented considerable evidence for the contraction mechanism being the dominant one in paint systems where low photoactivity grades of TiO_2 were used.[7]

Based on this conceptual distinction between different photodegradation pathways, we have developed a quantitative model for the mass loss rate of a paint film, as a function of parameters like the pigment volume concentration (PVC) and coating thickness (see Appendix for details).

The quantitative model is based on the contraction model premise that pigment particles which absorb and scatter radiation will reduce the "bulk" mass loss rate by attenuating the penetration of UV light into the film. On a "macroscopic" scale (i.e., a scale large enough that the particulate nature of the pigment is not important), a Beers Law-type expression can be used to describe the light intensity as a function of depth, and for a given pigment dispersion, an effective "extinction coefficient," ε can be defined relating the depth intensity of the light penetration to the pigment volume concentration, c [equation (A10)]. Contributions to the mass loss rate from surface effects, and from pigment photoactivity, are added to the bulk mass loss rate to give the total mass loss rate.

To our knowledge, this is the first time a quantitative model of this generality for the mass loss rate has been attempted. An early model by Sperry and Mercurio[8] considered erosion effects only, though it had the advantage of generating an analytical expression for the evolution of the surface roughness, which could be directly linked to gloss.

In this chapter we present the results of some experiments to explore the potential and limitations of the quantitative mass loss model. Model paints were based on an acrylic latex binder. The photodegradation process was accelerated by using a UV fluorescent cabinet, with UV-B (313 nm) bulbs. To test the quantitative model, paints were prepared at a variety of PVC levels (2%, 8%, and 24%) and thicknesses. The lower PVC levels were chosen deliberately to allow for greater penetration of light into the bulk. This would be expected to increase the bulk mass loss contribution.

EXPERIMENTAL

The acrylic latex used was Rhoplex™ AC-3001 from the Rohm and Haas Company. This latex was chosen because it is an all-acrylic latex with a particle size and minimum film formation temperature (MFFT) (27°C) very similar to PVDF-acrylic hybrid materials which Arkema, Inc. has recently developed. The acrylic formulations used ethylene glycol butyl ether (EB) as the coalescent, following manufacturer starting point formulations, at a level of 10 wt% on resin solids. Pigmented formulations were prepared by adding to a clearcoat formulation the appropriate amount of a pigment dispersion, made using the following recipe. The dispersion was prepared using a high speed disperser, to a Hegman reading of 7+.

Deionized water...135 g
Disperbyk®-180 (BYK-Chemie) ....12.5 g
Ammonia (diluted at 7% in water)...0.2 g
TegoFoamex™ 810 (Degussa)..1.2 g
Triton™ CF-10 Surfactant (Dow Chemical) ....5 g
TiPure® R-960 (DuPont) ....500 g

The paints were drawn down on top of chromated aluminum panels using a variety of wire wound rods, to achieve paint films of different thicknesses. For the clearcoat formulations, a second set of coatings was also prepared over a baked 70% PVDF black paint to look at the effect of the intensity of the light inside the film. The film samples were air-dried overnight and then were annealed at 60°C for three days to remove as much residual coalescent and other volatiles as possible, prior to exposure testing.

The annealed samples were then exposed in a UV fluorescent cabinet (ASTM G 53-88) using UV-B (313 nm) bulbs and temperature conditions according to ASTM D 4587 Condition D (8 h light at 60°C plus 4 h condensation at 45°C). The coated panel size was 15 cm x 7.5 cm, of which an area of 9.5 cm x 6.3 cm, was exposed to UV light. Measurements of the coating thickness, panel mass, and 60° gloss were taken at suitable intervals. To minimize the experimental scatter in the mass loss data, the measurements were only taken near the end of the light (dry) cycle.

The mass loss data, presented in the following section, show two distinct types of behavior in different time regimes. After the first several hundred hours, all samples were observed to lose mass at a constant rate, over the remaining time of the test (1100–1200 h total exposure time). We designate this time period as the linear mass loss regime. In this regime, the mass loss rates can be determined with high precision from the slope of the mass versus time curve. The analyses presented below are for the behavior in the linear mass loss regime (approx. 300–1100 h cabinet exposure).

During the first few hundred hours of exposure, a higher degree of mass loss variability is observed. The analysis of the kinetics during this transient period will be presented in a subsequent paper.

EXPERIMENTAL RESULTS FOR ACRYLIC CLEARCOATS

Extent of Linear Mass Loss Regime and Substrate Effects

As mentioned above, under the UV-B test cabinet conditions employed, after an initial transient or "induction" period, both pigmented paints and clearcoats of the acrylic latex were observed to lose mass at a constant rate over the next 1000 or more hours of exposure. Representative mass loss curves are shown in *Figure* 1; the mass loss rate is equal to the slope of the line. The reproducibility of the mass loss rates at a given thickness, for panels prepared and exposed at different times, turns out to be very good (one example is shown). It should be noted that the initial mass of the exposed portion of a 38 micron clearcoat is about 275 mg, so that the total mass loss at 1100 h exposure, about 40 mg, is about 14% of the initial mass.

The mass loss curves for the same clearcoats over a black PVDF substrate are shown in *Figure* 2. They reveal a somewhat greater transient mass loss in the first 300 h or so of exposure, especially for the thicker films. At least some of the transient mass loss may be attributed to incomplete physical drying of the samples prior to beginning the exposure testing (including some material which may have been effectively trapped in the black PVDF primer layer). Some other evidence for this is discussed in the next section.

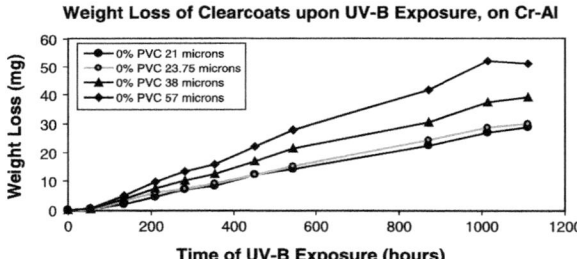

Figure 1—Weight loss of acrylic coatings upon UV-B exposure, on chromated aluminum.

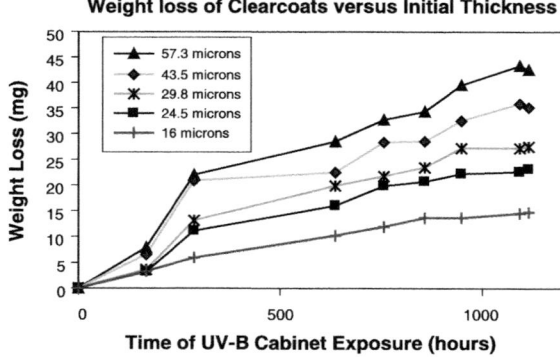

Figure 2—Weight loss of acrylic clearcoats on top of black PVDF primer.

After the transient period, a constant mass loss rate (constant slope) is again observed, through at least 1000 h of exposure. In the spirit of equation (A3) of the Appendix, *Figure* 3 compares the steady state mass loss rates for the two substrates, as a function of initial dry acrylic film thickness. The rates were determined using linear regression, over the period 300–1000 h UV-B cabinet exposure. It can be seen that the slopes of the lines, which are proportional to the bulk mass loss rate per unit volume, are about twice as large for the samples on chromated aluminum as for the samples on the black substrate.

We believe that this difference can most easily be attributed to differences in UV light intensity within the bulk of the clearcoat films. The clearcoat has a high rate of transmission for UV light at most wavelengths, so light penetrating into

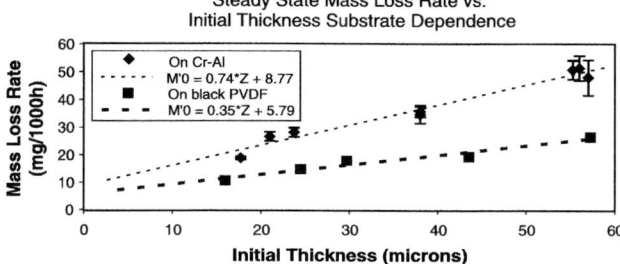

Steady State Mass Loss Rate vs. Initial Thickness Substrate Dependence

Figure 3—Comparison of mass loss rates for acrylic clearcoats. Error bars indicate experimental uncertainties in the mass loss rates (the least square slopes of the curves in Figure 1*), at a 95% confidence level.*

the coating from the top surface will normally travel substantially unattenuated through the clearcoat, to the substrate interface. If that light is reflected rather than absorbed, it will again travel through the clearcoat and exit the film; consequently, the internal light intensity should be roughly twice as large for the case of a reflecting substrate compared to an absorbing one. This would then be expected to cause a doubling of the photodegradation rate.

Based on the least squares fit calculations, the y-axis intercept in *Figure* 3 (corresponding to the "surface" mass loss rate contribution) is also somewhat larger for the case of the reflecting substrate; although, as is evident from the graph, the data are not precise enough to say whether there is a real difference. It should be emphasized that this "surface" contribution is not simply an additional one-time quantity of material (in milligrams) that is lost from the coating, but is an *ongoing* contribution to the mass loss rate during this period of the experiment, over and above what is contributed to the rate simply by having more material present (i.e., making the coating thicker). The size of the contribution is surprisingly large, being a quarter to half the size of the total rate.

The quantitative mass loss model predicts that, because of pigment shielding effects, the mass loss rate will be reduced relative to the bulk clearcoat mass loss rate. We would expect that in these calculations, the clearcoat rate on the absorbing substrate would be the "correct" one to use as a reference rate.

Mechanistic Insights

Some insight into the chemical mechanism of degradation in the UV-B fluorescent chamber was obtained by other physical and chemical measurements. As a complement to the mass loss measurements, the evolution of the thickness of the clearcoats on chromated aluminum was measured using a Permascope unit. The Permascope results are shown in *Figure* 4; as would be expected, the same general trends are observed as are seen in the mass loss measurements, though with less precision.

The two sets of results can be put on the same scale (i.e., mass loss or thickness loss), and compared quantitatively if the coating density as a function of time is known. As a first approximation, one might assume that the density is constant. The results of the comparison, assuming a constant acrylic density of 1.20 (an estimate derived from the

462

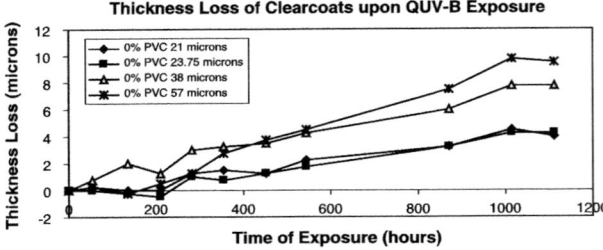

Figure 4—*Thickness evolution of clearcoats as measured with a Permascope unit.*

latex wet density and solids), are shown in *Figure* 5. A systematic discrepancy between the mass loss and thickness loss results can be seen, suggesting that the assumption of constant acrylic density is not correct. In fact, the direction of the discrepancy—the observed mass loss rate not being as high as would be expected based on the thickness loss rate—indicates that the density of the clearcoat is increasing somewhat over time. An increase in the density of 3.3% (1.20 to 1.244) over the course of 1100 h exposure would account fully for the observed differences.

An increase in the density due to UV-B exposure in air is not unreasonable, since one would expect the gradual incorporation of oxygen into the binder as various photooxidation products (e.g., hydroperoxides, ketones, carboxylic acids) are formed. Direct evidence confirming this was obtained spectroscopically. Attenuated total reflection infrared spectra of the unexposed and exposed regions of the sample panels show clear evidence in the 3500 and 1700 cm^{-1} regions for a significant increase in the concentration of oxygenated species in the near-surface regions. The surface oxygen content for this sample was also measured directly for one sample using XPS. For this sample after 1100 h UV-B exposure, the atomic O:C ratio went from 27% to 38%. The incorporation of this much oxygen throughout the bulk could easily increase the bulk density by several percent, which would fully account for the different rates inferred for mass loss between the Permascope and the direct measurements.

RESULTS FOR PIGMENTED COATINGS

According to the contraction model for gloss loss and chalking, pigment particles which absorb and scatter radiation can reduce the "bulk" mass loss rate by attenuating the

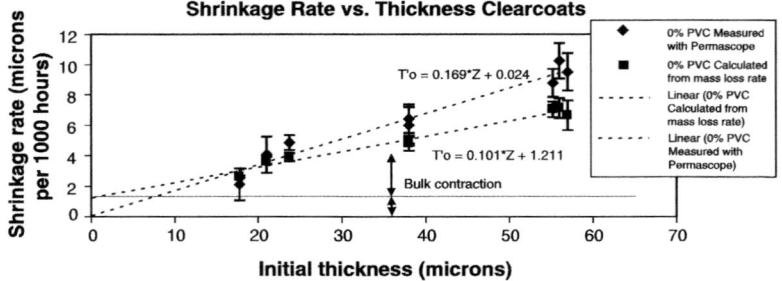

Figure 5—*Shrinkage rate (i.e., rate of loss of physical film thickness) from Permascope measurements and calculated from mass loss data.*

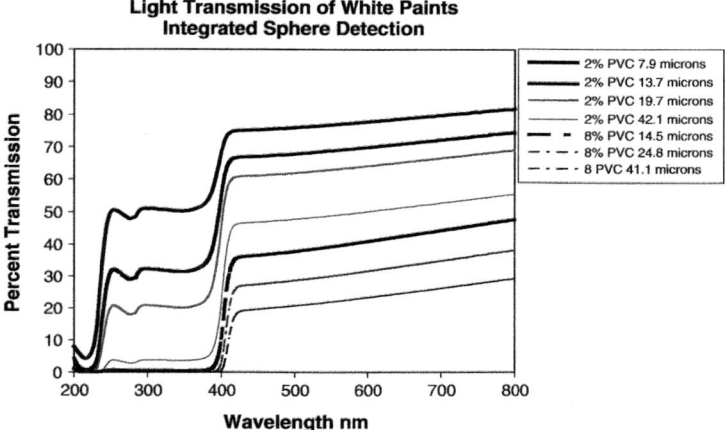

Figure 6—UV-Vis transmission of light through pigmented free films as obtained using an integrating sphere detector by a Lambda-850 UV/Vis spectrophotometer (curves follow the order of the legend).

penetration of UV light into the bulk. On a "macroscopic" scale (i.e., a scale large enough that the particulate nature of the pigment is not important), a Beers-Lambert-type expression would be expected to describe the light intensity as a function of depth, and for a given pigment dispersion, an effective "extinction coefficient," ε, can be defined relating the depth intensity of the light penetration to the pigment volume concentration, c, [equation (A10)]. To test these relationships, paints were prepared at a variety of PVC levels (2%, 8%, and 24%) and thicknesses. The lower PVC levels were chosen deliberately in order to allow for greater penetration of light into the bulk. This would be expected to increase the bulk mass loss contribution. In addition, at very low PVC, there is a minimal contribution from particle packing and CPVC effects,[9] which are not treated in the basic model.

Using free films of the paints, the UV-visible transmission of light through the films can be measured directly using a UV-visible spectrometer. Using a Lambda-850 UV/Vis spectrometer (Perkin Elmer, Inc.), we have measured the UV-visible transmission spectrum of several films. An integrating sphere attachment was used to collect the transmitted light at all angles. The results are shown in *Figure* 6. The angle-integrated percent light transmission is substantially greater than the transmission rate measured using the typical UV/Vis setup for transparent samples and dyes, where only the collimated transmitted light is collected.

The data in *Figure* 6 show that for the 2% PVC sample with a thickness of 13.7 microns, the percent transmission through the film in the UV region (300-400 nm) is about 32% of the incident intensity. A large reduction in the transmission rate can be noted below about 400 nm, the approximate value of the band gap for rutile TiO_2. Defining a Beers-Lambert extinction coefficient, ε_T, for transmission through the expression

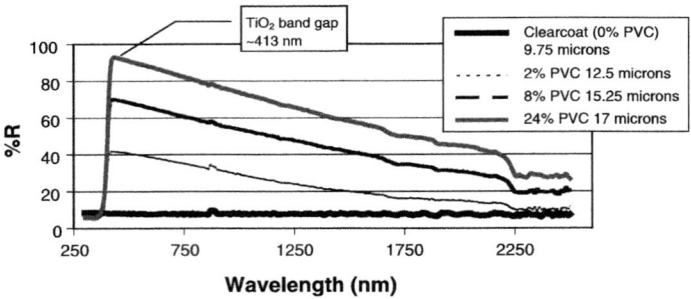

Figure 7—UV-Vis-Near IR reflectance of light through clearcoat and pigmented-free films as measured with a Perkin Elmer UV-Vis-IR reflectometer.

where c is the pigment volume concentration and z is the thickness, and noting that about 7% of the incident intensity is reflected from the incident surface and does not actually enter the film (*Figure* 7), the extinction coefficient, ε_T, in the UV-B spectral region can be calculated for each of the films of *Figure* 6. For each PVC level, very consistent values for ε_T were calculated independent of the thickness, validating the use of a Beers-Lambert model. Average values of ε_T were 3.93 ± 0.07 μm^{-1} at 2% PVC, and 4.58 ± 0.06 μm^{-1} at 8% PVC (with the error bars denoting 95% confidence limits for the averages, based on the standard deviation of the individual values for the averages). It is of interest to note that the extinction coefficient appears to be about 15% higher at 8% PVC compared to 2% PVC. This might be due to multiple particle scattering effects, or could likewise be a consequence of subtle differences in the pigment dispersion efficiency between the two formulations.

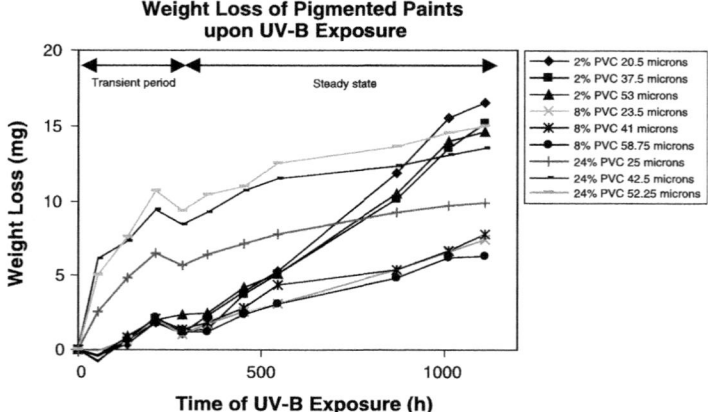

Figure 8—Mass loss of acrylic coatings upon UV-B exposure on chromated aluminum.

An estimate of the relative contributions of scattering and absorption as a function of wavelength can be gained by considering the reflectance spectra of the films, as shown in *Figure 7*, as well as a comparison of the light transmission for collimated (unscattered) light versus light transmitted at any angle. In the visible spectral region, scattering dominates. An effective absorption/scattering cross-section can be calculated by multiplying the collimated transmission coefficient, ε_c, times the pigment particle volume. In the visible region, with ε_c estimated at about 7 μm^{-1}, the visible scattering cross-section is almost exactly equal to the physical cross-sectional area of a 250 nm particle diameter. A value this high also implies that the quality of the pigment dispersion in the dry paint film is reasonably good (a result also to be expected given the good gloss and hiding power of the paint).

In the UV spectral region, absorption dominates over scattering—as would be expected given that the TiO_2 bandgap sits at about 400 nm. The average distance that normal UV-B light will travel into a film prior to absorption or scattering, given by $(1/\varepsilon_c c)$, is on the order of 13 μm at 2% PVC, 3 μm at 8% PVC, and 1.0 μm at 24% PVC. At 24% PVC, this value is starting to approach the diameter of the TiO_2 particles themselves, where the continuum extinction coefficient model breaks down—a good reminder that, at typical PVC levels for many commercial industrial coatings, a more realistic model of degradation will be needed which treats pigment particle effects properly.[10] The distance estimates are consistent with the physical expectation that above 10% PVC or so, most of the photochemical/weathering activity will be localized in the top few microns of the paint film.

Figure 8 shows the mass loss curves for the pigmented films on chromated aluminum. The films at 2% and 8% PVC can be seen to have a mass loss profile similar to that of the clearcoat; i.e., there is a sort of induction period where the initial mass loss rate is lower than the rate during the steady state period. The 24% PVC films, on the other hand, show a more substantial initial loss of mass, before the steady state conditions begin. For the 24% PVC films, some additional experiments, not reported here, strongly suggest that the additional mass loss is due to incomplete volatilization of residual coalescent and other low molecular weight materials during the pre-exposure annealing of the film.

The steady state mass loss rates (300–1000 h UV-B) for the pigmented paint films, as a function of coating thickness, are shown in *Figure 9*. Above a threshold thickness, (vis-

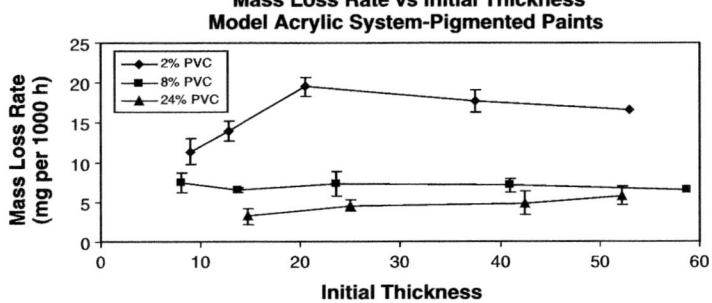

Figure 9—Steady state mass loss rates for pigmented acrylic films.

ible in the case of the 2% PVC films), the rates are independent of thickness, consistent with the expectation that for opaque coatings, significant photochemistry can only occur in the topmost layer.

The coating gloss loss over the same period is shown in *Figure* 10. It may be seen that most of the gloss is lost during the 1100 h exposure period, with the most rapid loss of gloss occurring during the initial transient period. Some differences in behavior may also be seen between the 24% PVC coatings (in a typical range for many commercial paints) and the lower PVC coatings. In particular, the 24% PVC coatings with a substantial transient mass loss, also have a much "flatter" gloss loss profile, while the low PVC coatings, like the clearcoat, show more rapid initial gloss loss, but with reduced mass loss during the initial transient period.

According to the mass loss model in the Appendix, considering a PVC-independent photocatalytic contribution and a bulk contribution whose size depends on pigment shielding effects, the PVC dependence of the mass loss rate can be expressed as:

$$M'_c = \frac{(1-c)}{c} \Sigma \frac{\Phi}{\varepsilon_{const}} + (1-c)\Phi_{surf} + \Phi_{photocat} \tag{A8}$$

For low enough PVC, where the surface erosion contribution is approximately constant (i.e., $(1-c) \approx 1$), this equation turns into a linear equation in the variable $(1-c)/c$, the "inverse PVC."

Figure 11 shows that a linear relation is indeed obtained for the steady state mass loss rates over the PVC range of interest. The magnitude of the intercept, which captures the mass loss rate contributions from surface erosion and photocatalytic activity, is somewhat lower than the "surface" erosion value measured for clearcoats. This value seems to be very reasonable, since the TiO_2 grade used here is considered to be highly weatherable, and the TiO_2 particles near the surface should have some screening effect.

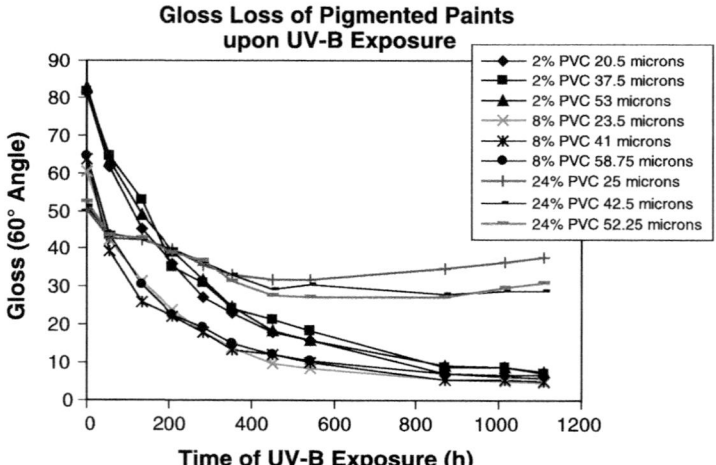

Figure 10—Gloss loss of acrylic coatings upon UV-B exposure.

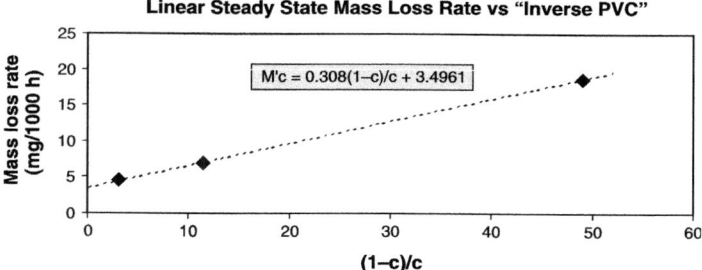

Figure 11—Linear relation for steady state mass loss rate versus "inverse PVC."

The value of the slope, 0.31 mg/1000 h exposure, corresponds [using equation (A10)] to a value $\varepsilon_{eff} \approx 1.14 \, \mu^{-1}$. This value is much more surprising, as it is more than a factor of three lower than the measured value for the actual UV intensity transmitted through the coating, $\varepsilon_T \approx 4 \, \mu^{-1}$. The lower value for ε_{eff} implies that the pigment-mediated steady state bulk mass loss rate contribution is substantially higher than would be expected from light attenuation effects alone (i.e., it is as if the UV light is penetrating deeper into the bulk of the coating than would be predicted from the measured UV transmission values). Possible reasons for this discrepancy are discussed in the next section.

DISCUSSION

The data presented above show that a simple conceptual model distinguishing different contributions to the photochemical mass loss rate can yield valuable insights into the physics and chemistry of the degradation process. In particular, the data for the acrylic latex model system being considered here show that:

• Initial transient effects can be very important and may reflect different kinetics from later, "steady state" photodegradation processes

• The degradation rate at the top surface is greatly enhanced versus the rate in the bulk, both for clearcoats and pigmented coatings

• Oxygen uptake and coating mass loss rates can be independently monitored by carefully looking at both the coating mass and the physical thickness

A number of limitations of the simple conceptual model are also apparent from the data and the physical considerations discussed in the previous sections. For "normal" coating pigment loadings, nearly all the relevant photochemistry is happening in the topmost 1–2 microns of the coating, in regions that are either in the immediate vicinity of the topmost pigment particles, or which may be considered pigment-free regions lying between the topmost pigment particles and the top surface. So, a truly quantitative model will need to treat more realistically various effects related to the particulate nature of pigments, and to their distribution in these regions.

However, by far the biggest puzzle in the above data is surely the surprisingly low value of ε_{eff} , i.e., the surprisingly high value of the slope in *Figure* 11. Because the abscissa of the graph varies essentially as $1/c$, a high value for the slope can be seen to correspond to a high value for the steady state mass loss rate at the lowest (2%) PVC value. As mentioned above, this mass loss rate is a factor of 3–4 higher than what would be expected considering the clearcoat mass loss rate, plus the measured UV attenuation from the pigment as a function of coating thickness.

Although it is highly speculative at this point, one might imagine several possible physical mechanisms that could help to explain the discrepancy:

• Based on the transmission curve of *Figure* 6, photons near the TiO_2 bandgap (around 400 nm) penetrate much more deeply into the coating bulk than UV photons. If there is a significant contribution to the degradation photochemistry from this wavelength, it could account for much of the difference. The acrylic resin itself is not photosensitive at this wavelength; however, since we are considering here the behavior after some exposure time, the mass loss contribution could be coming mainly from other species, e.g., photooxidation products produced during the initial transient period from the acrylic and from other formulation components.

• The assumption of film homogeneity may not be correct; in particular, even if the pigment distribution is approximately uniform as a function of depth, it is likely some kind of surface enrichment of mobile surface-active species happens during latex film formation.[11] Being near the surface, these species would not see significant pigment shielding (however, if this mechanism is dominant, one might expect to see some kind of increase in the rate with increasing film thickness, which is not observed).

• Though this would be a minor effect, the actual light intensity inside the film should actually be somewhat higher than the transmission spectra of free films suggest, since light reaching the back side of the film at an oblique enough angle (due to scattering) will be reflected back into the film through internal total reflection, and so not be measured by the detector.

• The physics of photon scattering from isolated pigment particles can in principle lead to localized high intensity "hot spots" in the vicinity of the particle, where the binder degradation rate would be enhanced.[12]

We are continuing to study this phenomenon further, beginning with more detailed studies of the "transient" period. The transient period is critical for several reasons; first, gloss loss is already occurring during this period; second, as we have noted above, chemical changes during this period may be important in determining the later time photochemistry of the coating.

Even with a quantitative model that is able to correctly account both for the rate of mass loss in the coating and its depth dependence, an additional modeling step is required to link this mass loss to changes in gloss in the coating.[13] Several groups have done Monte Carlo studies along this line,[14] however there has been a tendency in some of these studies to assume that only an erosion mechanism is operative, i.e., that mass loss is only occurring at the surface. A more adequate modeling approach would consider also the effects of contraction processes, which would lead to a different near-surface pigment structure, as well as differences in the degree of surface roughness.

In any case, it appears that a better understanding of both the chemistry and the physics of the degradation process will be needed to truly make headway on predicting the photochemical gloss evolution of thermoplastic paints.

ACKNOWLEDGMENTS

We are grateful to Drs. Matthew Gebhard (Rohm and Haas Company), James Pickett (GE), Jon Graystone (PRA), and Stuart Croll (NDSU) for helpful discussions.

APPENDIX: QUANTITATIVE MODEL OF MASS LOSS

This model for the coating mass loss rate does not depend on the mass loss rate being constant versus exposure time; however, it is often experimentally easier to measure the instantaneous mass loss rate precisely when that rate is constant over a period of time, as is the case for the UV-B exposures we report on in this paper. It does disregard some of the features of pigmented coatings related to pigment particle packing effects (notably excluded are volume effects such as critical pigment volume concentration corrections); hence, it will have the best chance of quantitative success for films with low PVC levels.

We assume that the film/coating is uniform in the x-y plane, of thickness $\mathbf{Z}$, with the z (film thickness) direction defined so that $z = 0$ is the top (exposed) surface, and $z = \mathbf{Z}$ is the bottom surface of the film. The mass loss rate M' is obtained by integrating the mass loss rate/unit volume over the exposed volume of the film. The mass loss rate/unit volume is assumed to be proportional to the light intensity I at each location. Again, assuming that the light intensity is uniform in the x-y dimension and only varies with depth, the volume integration can be converted in to an integration in the z-dimension, i.e.,

$$\mathbf{M'} = \Sigma \int_0^Z dz \int d\lambda \; \varphi(\lambda) I(\lambda, z) \tag{A1}$$

where

Σ = exposed surface area of the coating; λ = wavelength of light; $\varphi(\lambda)$ = the "quantum efficiency"–i.e., mass loss rate/unit volume/unit light intensity at λ.

Case of a Clearcoat

If there is effectively no light attenuation through the clearcoat, i.e.,

$$I(\lambda, z) \approx I(\lambda) \equiv I_{z=0}(\lambda)$$

with $I_{z=0}(\lambda)$ being the incident light intensity at the top surface, then

$$\mathbf{M'}_0 = \Sigma \; Z \int d\lambda \; \varphi(\lambda) I_{z=0}(\lambda) \equiv \Sigma Z \Phi \tag{A2}$$

i.e., $\Phi \equiv \int d\lambda \; \varphi(\lambda) I_{z=0}(\lambda)$ and for a collection of clear coatings of different thicknesses, Z, the slope of the curve $\mathbf{M'}_0$ versus Z is equal to $\Sigma \Phi$.

If there is an enhancement of the rate of clearcoat mass loss, near the top surface (an "erosion" component), then it is necessary to add a term to the mass loss rate to describe this. For thick enough coatings, the erosion contribution will be a constant value, M'_{surf}, independent of the coating thickness, Z, but since the surface mass loss rate must be zero in the limit of zero coating thickness, and since the surface enhancement mechanism may be operative over some skin depth, ζ_{surf}, then in general,

$$M'_{surf}(Z) = M'_{surf}\left[1-\exp(-\zeta_{surf}^{-1}Z)\right]$$

in analogy with equation (A13) below. The more general case is then given by equation (A3):

$$M'_0(Z) = M'_{surf}\left[1-\exp(-\zeta_{surf}^{-1}Z)\right]+\Sigma\Phi Z \tag{A3}$$

This is shown schematically by the top, dotted gray line in *Figure* A1; the black middle line shows the case where M'_{surf} is negligible. The bottom, light gray line depicts a case where M'_{surf} is small, but where there is also a modest attenuation of the bulk rate, deep inside the coating, due, for instance, to a slight absorption of the light, or a reduced level of some reactant (e.g., oxygen) deep inside the film.

It should be pointed out that both the gray line cases, if sampled in the region denoted by the dotted-line box, will give approximately linear plots with a non-zero intercept. To distinguish the two cases, a broader range of thicknesses would be required.

Pigmented Case

When the binder contains a pigment at volume fraction c $(0 < c < 1)$, then due to light scattering and absorption from the pigment, the light intensity inside the coating is attenuated. At "low" pigment volume concentrations where the effects of near-neighbor interactions of the absorption and scattering are not important, and for depths which are large relative to the pigment particle size, a Beers-Lambert expression can be used to describe the attenuation as a function of depth:

$$I(\lambda,z) = I_{z=0}(\lambda)\exp(-\varepsilon(\lambda)cz) \tag{A4}$$

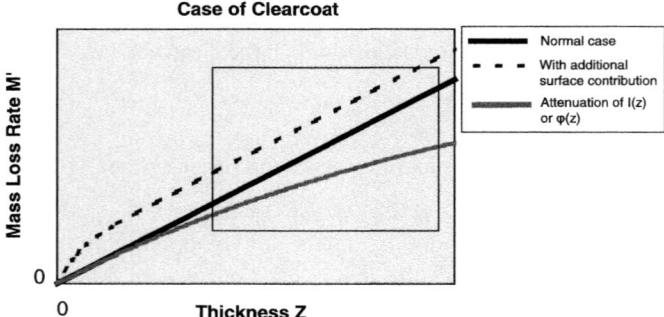

Figure A1—Mass loss rate as a function of film thickness for a clearcoat.

with $\varepsilon(\lambda)$ being the extinction coefficient per unit pigment volume fraction, with units of inverse length. This quantity can in principle be measured using a UV-Vis transmission device, as long as representative optical conditions are employed (specifically, the transmitted light should be collected at all angles using something like an integrated sphere; also, the angular distribution of the incident light should match the exposure test conditions since light penetration will on the average be deeper for normal incident light, versus light impinging on the film at an oblique angle). However, achieving these optical conditions may not be easy in practice.

Then from equation (A1) the mass loss rate at pigment volume fraction c, taking into account also that the volume fraction of the binder in the coating is now $(1-c)$, and also assuming for the moment that the pigment only plays a light screening role and does not contribute to the mass loss in any way, is:

$$\mathbf{M'}_c = (1-c)\Sigma \int_0^Z dz \int d\lambda \; \varphi(\lambda)I_{z=0}(\lambda)\exp(-\varepsilon(\lambda)cz)$$

$$\mathbf{M'}_c = (1-c)\Sigma \int d\lambda \varphi(\lambda) \, I_{z=0}(\lambda)\int_0^Z dz \; \exp(-\varepsilon(\lambda)cz) \tag{A5}$$

A. High Film Opacity Case: Considering the case where the pigment concentration is high enough, and/or the coating is thick enough, that there is full film opacity, the exponential term goes to zero for larger z, and the integration over thickness simplifies to:

$$\int_0^Z dz \; \exp(-\varepsilon(\lambda)cz) = \int_0^\infty dz\exp(-\varepsilon(\lambda)cz) = \frac{1}{\varepsilon(\lambda)\,c}$$

In this case the mass loss rate is independent of thickness Z:

$$\mathbf{M'}_c = \frac{(1-c)}{c}\Sigma \int d\lambda \frac{\varphi(\lambda)I_{z=0}(\lambda)}{\varepsilon(\lambda)} \qquad \textit{high opacity case} \tag{A6}$$

This can be simplified still further by noting that there should normally be a rather narrow spectral region where there is a significant contribution to the mass loss, i.e., the product $\varphi(\lambda)I(\lambda)$ will only be large for a small range of λ, (e.g., in the UV-B for an acrylic). Then if $1/\varepsilon(\lambda)$ is approximately constant over this region, we note:

$$\mathbf{M'}_c = \frac{(1-c)}{c}\Sigma \int d\lambda \frac{\varphi(\lambda)I_{z=0}(\lambda)}{\varepsilon_{const}} = \frac{(1-c)}{c}\Sigma \frac{\Phi}{\varepsilon_{const}} \qquad \textit{high opacity case} \tag{A7}$$

Adding in contributions from any surface-specific (erosion) component, and from any photocatalytic contribution from the pigment surface, the most general expression for the mass loss rate is then:

$$\mathbf{M'}_c = \frac{(1-c)}{c}\Sigma \frac{\Phi}{\varepsilon_{const}} + (1-c)\Phi_{surf} + \Phi_{photocat} \tag{A8}$$

We note that the photocatalytic contribution for opaque coatings is assumed to be independent of the pigment concentration, since the reaction rate is photon limited rather than

General Case, Small Z Region, Epsilon Effective

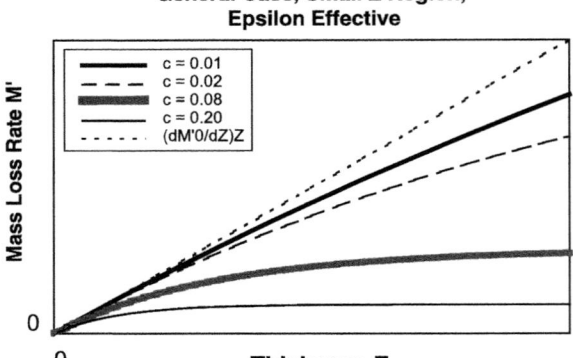

Figure A2—Mass loss rate as a function of film thickness and PVC.

pigment surface limited (the photon absorption rate is independent of the pigment concentration; essentially, every UV photon is absorbed by a pigment particle, and has some small probability of initiating photochemistry).

And in the spirit of equation (A4) we note that an effective penetration distance for degradation can be defined as:

$$\varsigma \equiv \frac{1}{c\varepsilon_{const}} \qquad (A9)$$

or more generally and empirically, for the high opacity case, an effective extinction coefficient, ε_{eff}, and penetration distance can be defined by combining data for the pigmented coating and the clearcoat:

$$\varepsilon_{eff} \equiv \frac{(dM_0'/dZ)}{dM_c'/d\left((1-c)/c\right)} \qquad (A10)$$

and for any particular concentration c,

$$\varsigma_{eff} \equiv \frac{1}{c\varepsilon_{eff}} = \frac{M_c' - (1-c)\Phi_{surf} - \Phi_{photocat}}{(dM_0'/dZ)} \qquad (A11)$$

The quantity ε_{eff} should be independent of the concentration; if it is, this would provide a verification that the assumptions of the model hold, and that the relevant photochemistry can be accounted for using a weighted average extinction coefficient ε_{eff}.

B. INCOMPLETE FILM OPACITY CASE: More generally, including at low concentration and/or lower film thickness (incomplete film opacity), equation (A5) can be integrated to give:

$$\mathbf{M'}_c = (1-c)\Sigma \int d\lambda \varphi(\lambda) I_{z=0}(\lambda) \left(\frac{-1}{\varepsilon(\lambda)c} \right) \left[\exp(-\varepsilon(\lambda)cZ) - 1 \right] \tag{A10}$$

And in cases where we can make the approximation $\varepsilon(\lambda) \simeq \varepsilon_{\text{eff}}$, this simplifies to:

$$M'_c = \Sigma \frac{(1-c)}{c} \frac{\int d\lambda \varphi(\lambda) I_0(\lambda)}{\varepsilon_{\text{eff}}} \left[1 - \exp(-\varepsilon_{\text{eff}} cZ) \right] \tag{A12}$$

$$M'_c(Z,c) = \frac{(1-c)}{c} \frac{\Sigma\Phi}{\varepsilon_{\text{eff}}} \left[1 - \exp(-\varepsilon_{\text{eff}} cZ) \right] \tag{A13}$$

Figure A2 shows the general shape of this function, for various values of c.

References

(1) Martin, J.W., Nguyen, T., Byrd, E., Dickens B., and Embree, N., "Relating Laboratory and Outdoor Exposures of Acrylic Melamine Coatings I. Cumulative Damage Model and Laboratory Exposure Apparatus," *Polym. Degrad. Stab.*, 75, 193-210 (2002).

(2) Nguyen, T., Martin, J.W., Byrd, E., and Embree E.J., "Effects of Spectral UV on Degradation of Acrylic-Urethane Coatings," *Proc. FSCT 80th Annual Meeting Technical Program*, New Orleans, LA, October 30-November 1, 2002.

(3) Rezig, A., Nguyen, T., Martin, D., Sung, L., Gu, X., Jamin, J., and Martin, J.W., "Relationship Between Chemical Degradation and Thickness Loss of an Amine-Cured Epoxy Coating Exposed to Different UV Environments," *J. Coat. Technol. Res.*, 3 (3), 173 (2006).

(4) Allen, N.S., Regan, C.J., Dunk, W.A.E., McIntyre, R., and Johnson, B., "Aspects of the Thermal, Photodegradation and Photostabilisation of Waterborne Fluorinated-Acrylic Coating Systems," *Polym. Degrad. Stab.*, 58, 149-157 (1997).

(5) Chiantore, O., Trossarelli, L., and Lazzari, M., "Photooxidative Degradation of Acrylic and Methacrylic Polymers," *Polymer*, 41, 1657-1668 (2000); Chiantore, O. and Lazzari, M., "Photo-oxidative Stability of Paraloid Acrylic Protective Polymers," *Polymer*, 42, 17-27 (2001).

(6) Colling, J.H. and Dunderdale, J., "The Durability of Paint Films Containing Titanium Dioxide-Contraction, Erosion, and Clear Layers Theories," *Prog. Org. Coat.*, 9, 47-84 (1981).

(7) In a recent study, we have also found considerable evidence for a contraction mechanism in PVDF-based thermoplastic coatings; see Faucheu, J., Wood, K.A., Sung, L.-P., and Martin, J. W. , "Relating Gloss Loss to Topographical Features of a PVDF Coating," *J. Coat. Technol. Res.*, 3 (1) 29-39 (2006).

(8) Sperry, P.R. and Mercurio, A., "Exterior Durability of TiO_2 Pigmented Acrylic Coatings," *Am. Chem. Soc. Div. Org. Coat. Plast. Chem.*, 30 (1), 400-417 (1970).

(9) Fitzwater, S. and Hook, J.W., "Dependent Scattering Theory: A New Approach to Predicting Scattering in Paints," *J. Coat. Technol. Res.*, 57 (721), 39-47 (1985).

(10) Del Rio, G. and Rudin, A., "Latex Particle Size and CPVC," *Prog. Org. Coat.*, 28, 259-270 (1996); Brown, R.F.G., Carr., C., and Taylor, M.E., "Effect of Pigment Volume Concentration and Latex Particle Size on Pigment Distribution," *Prog. Org. Coat.*, 30, 185-194 (1997); Bierwagen, G., Fishman, R., Storsved, T., and Johnson, J., "Recent Studies of Particle Packing in Organic Coatings," *Prog. Org. Coat.*, 35, 1-9 (1999); Al-Turaif, H. and Lepoutre, P., "Evolution of Surface Structure and Chemistry of Pigmented Coatings During Drying," *Prog. Org. Coat.*, 38, 43-52 (2000).

(11) Niu, B.-J. and Urban, M. W., "Recent Advances in Stratification and Film Formation of Latex Films: Attenuated Total Reflection and Step-Scan Photoacoustic FTIR Spectroscopic Studies," *J. Appl. Polym. Sci.*, 79 (7), 1321-1348 (1998).

(12) We are indebted to Dr. James Pickett of GE for this insight. He reports seeing microscopic evidence for this phenomenon on the lee ("shadow") side of TiO_2 particles in a polycarbonate matrix. Mie theory could be used to calculate the magnitude of this effect (cf. Ref. 9, also Born, M. and Wolf, E.,

Principles of Optics, 7th Ed., Cambridge University Press, New York, 1999). Since photodegradation rates are typically proportional to light intensity, and total integrated intensity is normally a conserved quantity in physics (because of the principle of conservation of energy), it is not clear that this mechanism should lead to a net increase in photochemically induced mass loss, but it would certainly localize the region where the photodegradation is occurring.

(13) Whitehouse, D.J., Bowen, D.K., Venkatesh, V.C., Lonardo, P., and Brown, C.A., "Gloss and Surface Topography," *Annals of the CIRP*, 43 (2), 541-549 (1994); Alexander-Katz, R. and Barrera, R.G., "Surface Correlation Effects on Gloss," *J. Polym Sci. Part B: Polym. Phys.*, 36, 1321-1334 (1998).

(14) Hunt, F.Y., Galler, M.A., and Martin, J.W., "Microstructure of Weathered Paint and Its Relation to Gloss Loss: Computer Simulation and Modeling," *J. Coat. Technol. Res.*, 70 , 45-54 (May 1998); Hinderliter, B. and Croll, S., "Monte Carlo Approach to Estimating the Photodegradation of Polymer Coatings," *J. Coat.Technol. Res.*, 2 (6), 483-491(2005); Hinderliter, B.R. and Croll, S., "Simulations of Nanoscale and Macroscopic Property Changes on Coatings with Weathering," *J. Coat. Technol. Res.*, 3 (3), 203-212 (2006).

Chapter 31

Effect of Pigment Dispersion on Durability of a TiO$_2$ Pigmented Epoxy Coating During Outdoor Exposure

Cyril Clerici,[1] Xiaohong Gu,[1] Li-Piin Sung,[1] Aaron M. Forster,[1] Derek L. Ho,[2] Paul Stutzman,[3] Tinh Nguyen,[1] and Jonathan W. Martin[1]

[1] Polymeric Materials Group, Building and Fire Research Laboratory, National Institute of Standards and Technology, Gaithersburg, MD 20899
[2] Electronic Materials Group, Materials Science and Engineering Laboratory, NIST
[3] Inorganic Materials Group, Building and Fire Research Laboratory, NIST

The effect of pigment dispersion on durability of a TiO$_2$ pigmented epoxy coating during outdoor exposure has been investigated. Well-dispersed and poorly dispersed coating samples were prepared through the addition or absence of a dispersant in the coating formulation. Ultra small angle neutron scattering (USANS) and scanning electron microscopy (SEM) showed that pigment aggregation occurs in the absence of dispersant. A thin, clear layer of epoxy was observed at the air/exposed surface interface in both the dispersed and non-dispersed samples. Chemical degradation and physical changes during UV exposure were measured by attenuated total reflectance Fourier transform infrared spectroscopy (ATR-FTIR), atomic force microscopy (AFM), and laser scanning confocal microscopy (LSCM). Results showed that the degree of pigment dispersion and the thickness of the clear layer contributed to weathering. Changes in surface topography and gloss loss during UV degradation were correlated with degree of pigment dispersion. Ripples and bumps on the top surface of the poorly dispersed coating greatly affected gloss. Bulk and surface mechanical properties were investigated using dynamic mechanical thermal analysis (DMTA) and instrumented indentation, respectively. Relative to the neat epoxy coatings, the addition of TiO$_2$ particles into the epoxy coatings increased elastic modulus but decreased the glass transition temperatures (T$_g$) of both of the pigmented coatings. Relationships between surface and bulk mechanical property changes and chemical degradation are discussed.

INTRODUCTION

Titanium dioxide (TiO$_2$) pigments are extensively used in polymeric products in the building and construction industry. Incorporation of pigments affects the appearance and mechanical properties of a coating. Pigments enhance gloss because of their high refractive index but decrease gloss through changes in a coating's surface topography.[1] In addition, a clear layer of binder has often been observed to form at the top surface of pigment-

475

ed coatings, as reported by Rutter.[2] Further, Colling and Dunderdale[3] proposed the clear layer theory which was based on its thickness and showed that gloss measurements were affected. The introduction of rigid particles in polymers is also known to affect bulk elastic moduli of materials, as well as the glass transition temperature (T_g).[4]

The degree of pigment dispersion affects appearance. Differences in degree of pigment dispersion come from two factors: (1) poor separation of the pigment particles during mixing, and (2) pigment flocculation after mixing. The degree of pigment dispersion can be improved through the introduction of additives, such as wetting and dispersing agents, into the formulations.[5,6]

During outdoor exposure, polymeric binders weather due to their susceptibility to UV radiation, temperature, and moisture. Metrologies exist for determining chemical, physical, and appearance changes with exposure. Fourier transform infrared spectroscopy in attenuated total reflectance mode (ATR-FTIR) is commonly used to measure chemical changes of coatings during UV degradation, including chain scission and formation of oxidized products.[7,8] Atomic force microscopy (AFM) and laser confocal scanning microscopy (LSCM) are also powerful techniques for quantifying topographic changes of polymeric coatings resulting from surface roughening, pitting, and cracking.[9-12] In combination, these two microscopic techniques cover the length scale range from microns to millimeters. During weathering, gloss is a relevant surface appearance property often correlated to coatings durability, which is not a function of just changes in the binder's topography with photodegradation, but also of pits formation due to loss of pigment.[13]

The objective of this study was to investigate the effect of pigment dispersion on surface morphology and appearance and mechanical properties of a TiO_2 filled amine-cured epoxy during outdoor exposure. Two levels of pigment dispersion were investigated— "well-dispersed" and poorly dispersed. Well-dispersed pigments were achieved through the addition of a dispersant; whereas poorly dispersed specimens were formulated without a dispersant. Pigment dispersion characterization and particle size analysis of pigmented coatings were performed using ultra small neutron scattering (USANS) and scanning electron microscopy (SEM), respectively. Changes in surface topography during outdoor exposure were followed using AFM and LSCM. The gloss loss, measured using a handheld glossmeter, was correlated to surface topographic changes and the different states of pigment dispersion. Chemical degradation was followed using ATR-FTIR. To determine the effect of pigment dispersion and UV degradation on mechanical properties, the characterization of bulk and surface mechanical properties was performed using DMTA and instrumented indentation, respectively.

EXPERIMENTAL[*]

Materials

The amine-cured TiO_2 pigmented epoxy system was a stoichiometric mixture of a pure diglycidyl ether of bisphenol A with an epoxy equivalent weight of 172 g/equiv

---

(DER 332, Dow Chemical) and 1,3-bis(aminomethyl)-cyclohexane (1,3 BAC, Aldrich). The titanium dioxide pigment was a commercial alumina- and silica-coated rutile having a mean particle size of 230 nm, and pigment volume concentration (PVC) was 15%. The solvent was n-butyl acetate and 1-methoxy-2-propanol acetate at a 4:1 ratio. The well-dispersed coating was formulated with 6% (by mass of pigment) of a commercial high molecular cationic wetting and dispersing agent added to the formulation (BYK-166, BYK-Chemie).

Sample Preparation

TiO_2 pigmented epoxy films were prepared using a Dispermat (BYK-Gardner) with 30 mm diameter impeller as follows:

(1) the epoxy resin was mixed with/without dispersant in the solvent at a mixing speed of 209.44 rad/s for 20 min;

(2) TiO_2 pigment was added and mixed at 366.52 rad/s for 60 min;

(3) the amine curing agent was added at the stoichiometric ratio with respect to epoxy resin, and mixed at 52.36 rad/s for 10 min; the mixture was degassed for 1 h and drawn down on release paper in a CO_2-free, dry air glove box;

(4) four replicate films for both the well-dispersed and poorly dispersed coatings were cured at room temperature for 24 h in the same glove box, followed by a 2 h postcure at 130°C in an air-circulation oven;

(5) film thicknesses were measured using a digital caliper and were found to be approximately 150 μm; and, finally,

(6) samples made from these films were randomly assigned to the weathering experiments.

Outdoor UV Exposure

Outdoor exposures were conducted on a National Institute of Standards and Technology (NIST) laboratory roof located in Gaithersburg, MD. Specimens were loaded in multiple-window exposure cells and placed in an outdoor environmental chamber at 5° from the horizontal plane facing south. The bottom of the chamber was made of black-anodized aluminum, and the top of the chamber was covered with "borofloat" glass; all four sides were enclosed with a moisture permeable fabric which prevented dust particles from entering the chamber. The exposure cell was equipped with a thermocouple and a relative humidity (RH) sensor. The exposure for this study was started in May 2006.

Pigment Dispersion Characterization

ULTRA SMALL ANGLE NEUTRON SCATTERING (USANS)—Pigment dispersion was characterized via USANS at the NIST Center for Neutron Research (NCNR) using a perfect crystal diffractometer (PCD). Details of this instrumentation are provided elsewhere.[14] The USANS data presented in this chapter are one-dimensional absolute scattered intensity profiles as a function of scattering wave vector q ($q = 4\pi \sin(\theta/2)/\lambda$, where θ is the scattering angle and λ is the wavelength). The scattering intensity, I(q), can be expressed as $I(q) = I_o P(q) S(q)$, where P(q) is the form factor, characteristic of the size and shape of the particle. S(q) is the structure factor, related to the correlation function describing the radial distribution function between particles.[15]

SCANNING ELECTRON MICROSCOPY (SEM)—Backscattering SEM images were collected on microtomed cross-sections with an accelerating voltage of 7 keV, at a magnification of 5000x. Microtomed surfaces were not coated with gold. The particle size distributions of the imaged specimens were estimated using ImageJ image analysis software.[16]

Surface Morphology Characterization

ATOMIC FORCE MICROSCOPY (AFM)—A Dimension 3100 AFM (Digital Instruments) was used to image the morphology and microstructure of TiO_2 pigmented epoxy coatings as a function of outdoor exposure. Images were obtained in tapping mode using commercial silicon probes having a resonance frequency of approximately 325 kHz, a force constant of 40 N/m, and a nominal tip radius of 5 nm–10 nm (values provided by Micromasch). The scan rate was 0.5–1.0 Hz. The AFM images are height and three-dimensional (3D) topographic images. Measurements were performed at almost the same location on the sample used, allowing us to track the sequence of morphological changes as a function of exposure time. The range of surface areas measured was from 5 μm x 5 μm to 50 μm x 50 μm.

LASER SCANNING CONFOCAL MICROSCOPY (LSCM)—A Zeiss model LSM510 reflection laser scanning confocal microscope was used to characterize the surface morphology over exposure time in a range of surface areas from 1.8 mm x 1.8 mm to 60 μm x 60 μm. The incident laser wavelength was 543 nm. LSCM images are two-dimensional (2D) intensity projections, which represent effectively the sum of all the light scattered back by different planar layers of the coating by moving the focal plane. The pixel intensity level represents the total amount of backscattered light. A z-step size of 0.5 μm was selected to obtain a series of overlapping optical slices using the objective of 5x, and 0.1 μm using the objective of 50x. The root-mean-square (RMS) surface roughness was determined from the 3D topographic profiles of coating surfaces. RMS roughness values were the average of four measurements on the same samples used for AFM measurements.

Chemical Changes

Chemical changes were followed by ATR-FTIR using a Nexus 670 FTIR (Thermo Nicolet) with a MCT detector and an ATR accessory with diamond crystal (Durascope). Pigmented and unpigmented coatings were sampled at four locations on each of four replicate samples. A background scan was collected before starting each new sample. All spectra were the average of 128 scans at 4 cm^{-1} resolution using dry air as the purge gas. Spectra analysis was performed using a custom software package.[17]

Mechanical and Material Properties

DYNAMIC MECHANICAL THERMAL ANALYSIS (DMTA)—The glass transition temperature (T_g) and storage modulus at 30°C (E′) were measured using a dynamic mechanical thermal analyzer (DMTA, TA Instruments RSA III). Measurements were performed from 30°C to 170°C at 5°C/min, and at a frequency of 1 Hz. Glass transition temperature was determined from the maximum of the tan δ peak. E′ and T_g values were the average of three measurements.

INSTRUMENTED INDENTATION TESTING (IIT)—IIT measurements were conducted using a commercial indenter (MTS Nanoinstruments NanoXP, Oak Ridge, TN). The exposed surface of the coating was indented to a depth of 2 μm at a constant strain rate of 0.05 s^{-1}. The indenter was a 90° diamond cone with a 10 μm diameter spherical tip. Tip area function was calibrated using a fused silica standard.[18] Contact stiffness was measured as a function of indentation depth by imposing a 2 nm, 45 Hz oscillation to the tip during loading. Coating modulus and hardness were calculated as a function of indentation depth from the measured contact stiffness and calibrated tip area function.[19] The modulus for each indent was averaged over a depth of 1 to 2 μm, which is the indentation depth range where the modulus was constant. Reported values of coating modulus are the average of 15 indentations, spaced over 5 × 2 array of indents (100-μm spacing) and a second 5 × 1 array (100-μm spacing) located several millimeters away from the first array.

Gloss Measurement

Specular gloss measurements were made using a handheld commercial glossmeter (Minolta, Multi-Gloss model 268). Gloss measurements at 20° and 60° angles of incidence were recorded. Reflectance areas for 20° and 60° gloss measurements were 9 mm × 9 mm and 9 mm × 18 mm, respectively. The reported gloss values are the average of 20 measurements.

RESULTS AND DISCUSSION

Pigment Dispersion Characterization and Surface Microstructure

Dispersion of the TiO$_2$ pigmented epoxy coatings prepared with and without dispersant was characterized using USANS, and the results are shown in *Figure* 1a. For the non-dispersed sample, a rather pronounced slope of ca. –3.7 is observed at q < 10^{-4} Å, corresponding to a length scale larger than 6 μm. In contrast, the USANS data suggest that the sample with dispersant exhibited much less aggregation. The USANS data from the sample prepared with dispersant also show a fairly good plateau regime in the Q

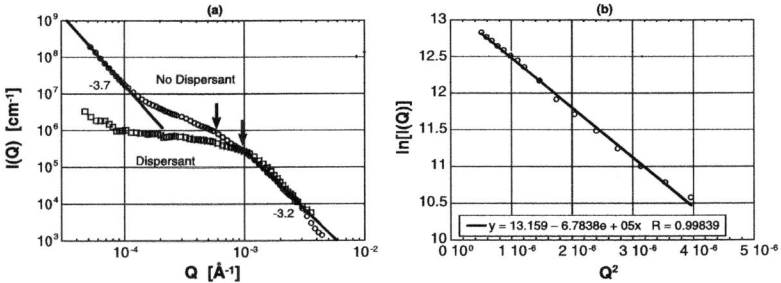

Figure 1—(a) USANS profiles of TiO$_2$ pigmented epoxy coatings prepared with and without dispersant. Error bars are smaller than the size of symbols. (b) Guinier plot and fitting to the USANS data for the well-dispersed sample.

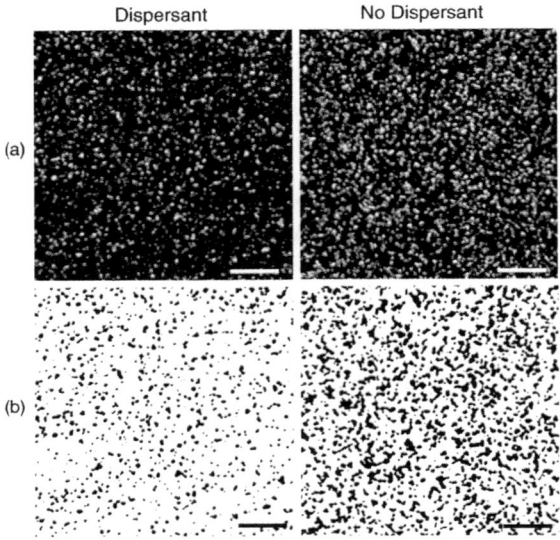

Dispersant   No Dispersant

(a)

(b)

Figure 2—(a) Cross-sectional SEM images in the center of
unexposed coatings prepared with and without dispersant.
Bright spots are TiO_2 particles. (b) Threshold images obtained
with ImageJ show pigment particles and aggregates. Dark
spots are TiO_2 particles (scale bar = 5 μm).

range of ca. 10^{-3} Å to 10^{-4} Å, corresponding to the so-called Guinier region,[20] indicating that the particle size distribution (polydispersity) is narrower than that of the non-dispersed samples. In addition, the averaged particle size with dispersant is smaller than that without dispersant shown by the point of change in the slope as indicated by the arrows in *Figure* 1a.

Fitting the data points of the sample formulated with dispersant in the Guinier region to the Guinier's equation,[20] $I(Q) = I(0)exp(-Q^2R_G^2/3)$ where R_G denotes the radius of gyration, as shown in *Figure* 1b, and yields an averaged R_G = 140 nm ± 10 nm (or an average diameter = 280 nm ±

20 nm) for the surface treated TiO_2 particles. For purposes of comparison, ultra small angle X-ray (USAXS) measurements were also performed on the sample with dispersant (profiles not shown) and the data fit to a polydisperse hard sphere model, giving an averaged diameter of 230 nm ± 10 nm with a polydispersity of 19% for only the TiO_2 particles due to the x-ray contrast. The average diameter obtained from USAXS is slightly smaller than that from USANS. However, these results are consistent with one another and show the great efficiency of both scattering techniques for pigment dispersion characterization.

Figure 2a shows cross-sectional SEM images in backscattering mode taken in the center of microtomed films prepared with and without dispersant. Under these conditions, pigment particles can be discerned by the bright spots in the SEM images. The image of the coating containing the dispersant shows near uniformly dispersed pigment particles, while pigment aggregates can be observed in the coating prepared without dispersant.

SEM images were computer image processed to obtain estimates of particle size and quality of dispersion. SEM images of both pigmented coatings were first smoothed through a median filter and thresholded at 145. Contrast images were obtained by assigning a white pixel value to all original pixel values lower than 145 (*Figure* 2b). Then, particle areas were determined with the particle analysis command of ImageJ, excluding particles touching the edge of the image. *Figures* 3a and 3b display the histograms of the particle size distribution for the coatings prepared with and without dispersant, expressed as the apparent diameter calculated from the measured area. In the absence of a dispersant, a broad particle size distribution was observed having a mean value of 454 nm ±

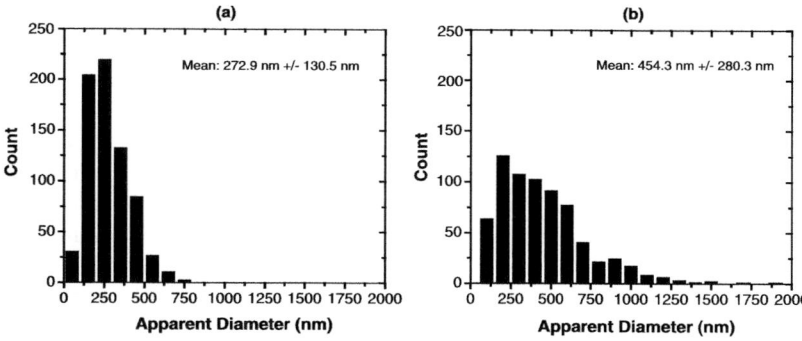

Figure 3—Pigment particle size distribution in the center of microtomed TiO_2 pigmented coatings prepared with dispersant (a) and without dispersant (b), determined from SEM images analysis. The bin size is 100.

280 nm. In *Figure* 3b, numerous aggregates having an apparent diameter greater than 750 nm, which are equivalent to aggregates formed with a minimum of three particles, were noticeable. Note that the largest pigment aggregate was observed to have a diameter greater than 2 μm (*Figure* 3b). On the other hand, the degree of aggregation was found to be significantly reduced in coatings prepared with dispersant. The addition of the dispersant led to a narrow particle size distribution with a mean particle diameter value of 273 nm ± 131 nm.

USANS data and SEM image analysis revealed significant differences in pigment dispersion and particle size distribution between coatings prepared with and without dispersant. Additionally, both pigmented systems showed a thick and well-defined clear layer where no TiO_2 particles were present near the exposed coating surface. From LSCM measurements, the thickness of the clear layer ranged from 1 to 3 μm for coatings prepared with dispersant, and from 3 to 6 μm for coatings prepared without dispersant.[21]

Figure 4 displays 10 μm x 10 μm AFM images of cross-sections near the surface of cryofractured pigmented coatings prepared with and without dispersant. The coating formulated with dispersant shows a clear layer thickness of approximately 2.5 μm, with the presence of

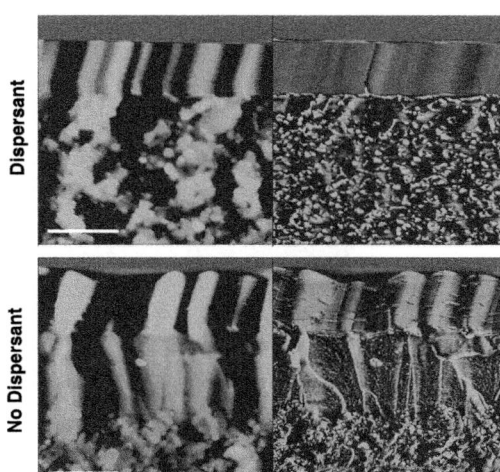

Figure 4—10 μm x 10 μm AFM height images (left) and phase images (right) near the surface of cryofractured pigmented coatings, revealing the clear layer on the top surface. Contrast variations from white to black are 300 nm for the height images and 25° for the phase images (scale bar = 3 μm).

a well-defined boundary. Below it, TiO_2 pigments appear to be randomly distributed, but some particle-free areas are noticeable at 3-μm deep, especially on the AFM phase image. Pigmented coatings prepared without dispersant present a thicker clear layer composed of two phases. The first top clear layer having a thickness of approximately 2.8 μm is very similar to that of the clear layer of coatings with dispersant, but is followed by a second layer below it containing few pigments, with a greater thickness of 3 μm. The presence of the dispersant did not cause the formation of the clear layer but its thickness was affected by the presence of the dispersant and the dispersion of pigments.

Surface Topography Characterization with AFM and LSCM

Figures 5 and 6 display tapping mode AFM and LSCM topographic images of the pigmented epoxy coatings prepared with and without dispersant at specified exposure times. AFM images provide topographic and microstructural changes with nanometer to micrometer resolution, while LSCM is a larger area profiling technique.

For unexposed pigmented epoxy coating samples, LSCM images were unremarkable in that they revealed little notable information. Surfaces of both epoxy systems appeared to be smooth, with a few defects resulting from the sample preparation. The intensity of LSCM images comes mostly from the backscattered light of TiO_2. AFM analysis of a 50 μm x 50 μm area revealed significant differences in the morphology of pigmented coatings prepared with or without dispersant, but no evidence of TiO_2 particles was observed,

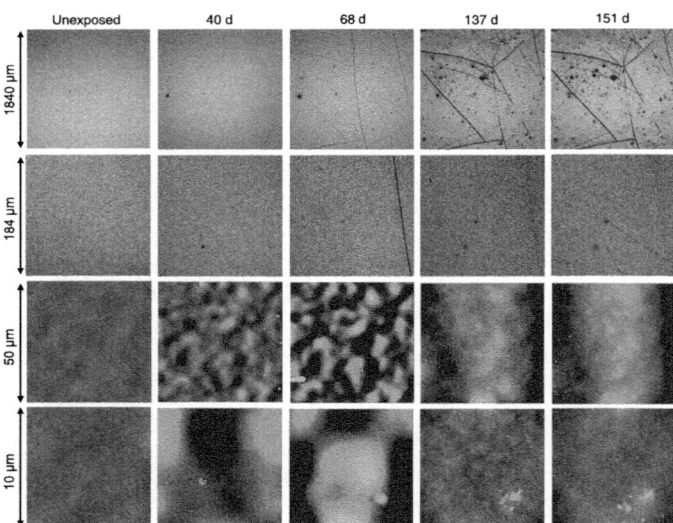

Figure 5—LSCM and AFM topographic images of the pigmented coating with dispersant during outdoor exposure in Gaithersburg, MD. Exposure times, from left to right, are 40 d, 68 d, 137 d, and 151 d. From top to bottom, lateral dimensions for LSCM are 1840 μm and 184 μm, and 50 μm and 10 μm for AFM. For each scale, approximately the same location was imaged. The height scales for 50 μm x 50 μm and 20 μm x 20 μm AFM images are 50 nm and 20 nm, respectively.

since the polymeric binder covers all pigment particles. It is important to note that both images have the same z-scale of 50 nm, allowing direct comparisons between the systems prepared with and without dispersant. For a well-dispersed pigmented film, the surface was relatively smooth and uniform, with an RMS roughness value of 1.02 nm ± 0.34 nm. On the other hand, the system containing no dispersant had a rough surface with large ripples. The roughness of this system was found to be much greater, with a value of 10.30 nm ± 1.89 nm. At higher magnification, this topographic difference becomes unnoticeable, as shown on 10 μm x 10 μm AFM images, indicating the magnitude of the large scale roughness is over the measurement scan size of 10 μm.

Macro-scale morphological changes due to photodegradation can be observed from the 1840 μm scanning size by LSCM. For the first 40 d of exposure, surfaces remained relatively smooth and featureless, which are similar to the unexposed specimens. After 68 d, cracks and few small protuberances, represented by dark spots, began to appear. At longer exposure times, these protuberances became numerous and bigger, and cracks became deeper and/or disappeared by erosion. During outdoor exposure, similar macro-scale physical changes can be observed on both pigmented systems, independent of the state of pigment dispersion. The RMS roughness based on 1840 μm x 1840 μm LSCM images varied little in the first 68 d of exposure, but increased quite rapidly thereafter, reaching a value of 2.23 μm ± 0.21 μm for pigmented epoxy coatings prepared with dispersant, and 2.29 μm ± 0.45 μm for the system prepared without dispersant after 150 d.

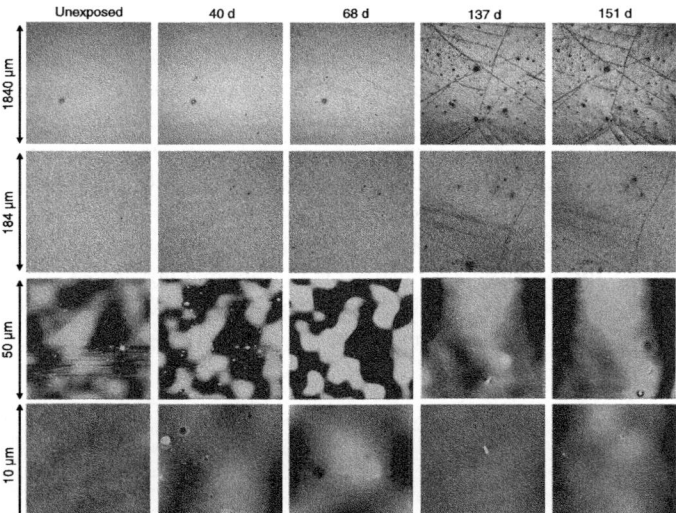

Figure 6—LSCM and AFM topographic images of the pigmented coating without dispersant during outdoor exposure in Gaithersburg, MD. Exposure times, from left to right, are 40 d, 68 d, 137 d, and 151 d. From top to bottom, lateral dimensions for LSCM are 1840 μm and 184 μm, and 50 μm and 10 μm for AFM. For each scale, approximately the same location was imaged. The height scale for 50 μm x 50 μm and 20 μm x 20 μm AFM images is 50 nm.

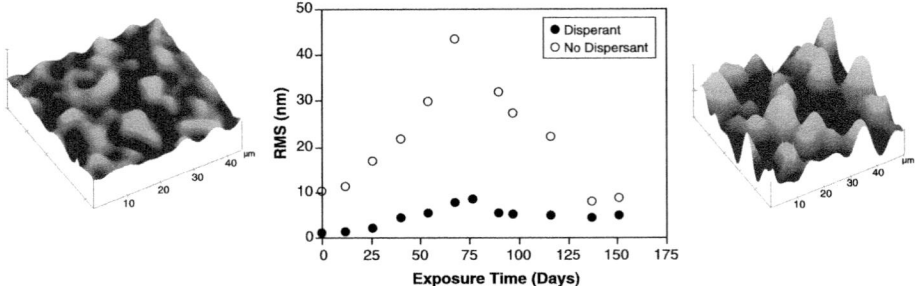

Figure 7—RMS roughness based on 50 μm x 50 μm AFM images for both pigmented systems as a function of outdoor exposure time, and 3D AFM topographic images at 68 d of pigmented coating prepared with dispersant (left) and without dispersant (right). Lateral dimension is 50 μm. Height scale is 100 nm.

Micro-scale morphological changes during photodegradation can be observed in the AFM 50 μm-length scan, while smaller scan sizes provide more detailed information on nanostructural changes. AFM topographic images of these materials during outdoor exposure revealed two stages. For the first 68 d, the surface of both materials shows the formation of circular protuberances and pits, both of which disappeared almost completely at 68 d of exposure. Such degradation features, which are numerous for the film prepared without dispersant, have previously been observed in unpigmented amine-cured epoxy samples exposed outdoors.[11] Both surfaces became rougher with the appearance of bumps or ripples, as seen on 3D topographic images of *Figure* 7. Surface defects for the pigmented coating prepared without dispersant were wide and high. Although numerous similar defects were observed in the well-dispersed coatings, these defects were generally smaller in every dimension. AFM roughness, based on 50 μm x 50 μm images, reached a value of 43.4 nm and 7.7 nm, respectively (*Figure* 7). At 137 d, both surfaces became relatively smooth again at the micro-scale level, with the loss of ripples revealing erosion and material loss from photodegradation.

Highly localized chalking was observed by AFM after 97 d of exposure. *Figure* 8 presents 5 μm x 5 μm AFM height images taken from one pigmented epoxy prepared with dispersant at different exposure times. TiO_2 particles were clearly visible protruding from the surface. These results show individual pigments having diameters of about 250

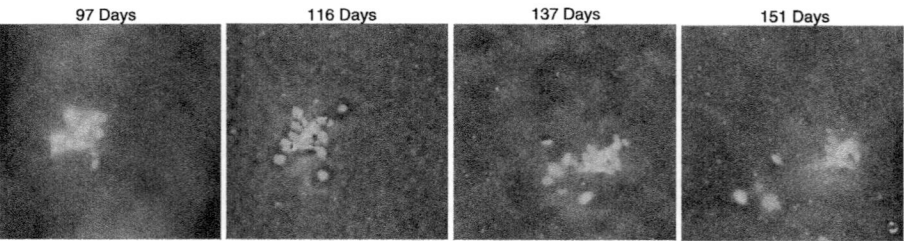

Figure 8—5 μm x 5 μm height AFM images of the pigmented coating prepared with dispersant showing localized chalking and rearrangement of pigment on the surface at different exposure times. Height scale is 20 nm.

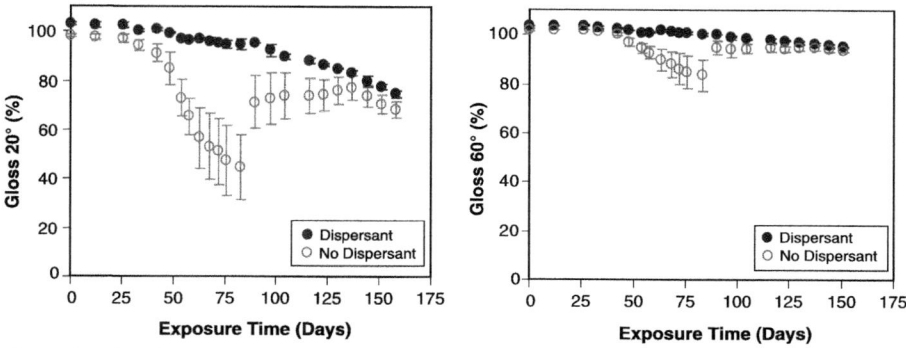

Figure 9—Gloss loss as function of exposure time for pigmented coatings prepared with and without dispersing agent at (a) 20° and (b) 60° of angles of incidence. Error bars represent one standard deviation.

nm. Some titania particles, that were initially held in place by residual polymer, became locally rearranged on the surface after further exposure. The removal of loose pigments left the top surface relatively smooth, unlike the behavior of systems formulated with photoreactive pigment where hole formation occurs.[13,22] However, this chalking process was highly localized and appears to be attributable to particles that initially resided within the clear binder layer. No comparison could be made with coatings without dispersant since no chalking was observed in these samples at this time.

Effect of Pigment Dispersion and Outdoor Exposure on Gloss

Gloss is an appearance property of materials related to the reflection of light by surfaces. In the coatings industry, gloss loss is often used as a measure of paint durability. *Figures* 9a and 9b show 20° and 60° gloss changes during outdoor exposure for the same samples used to characterize the surface topography changes. Gloss values observed for these pigmented films before exposure are a function of surface roughness and refractive index of the pigmented layer. For coatings with dispersant, the high gloss could be attributed to an increase of the refractive index of the pigmented layer and the relative surface smoothness. Poorly dispersed samples exhibited larger surface defects and, correspondingly, lower initial gloss values. During exposure, both coatings lost gloss, but at different rates. While gloss changes were not observed in the first 30 d of exposure, a significant drop in gloss for the system without dispersant is noticeable up to 80 d, followed by a rapid increase. For longer exposure times, gloss changes for both systems have similar trends.

The dramatic change in gloss for coatings prepared without dispersant is consistent with the two-stage topographic changes observed by AFM over the same exposure period. The rapid gloss loss occurs when the 50 μm-scale RMS roughness rapidly increases due to the formation of ripples, as seen in *Figure* 7. Then, the gloss increase correlates with the loss of these surface defects via erosion, leading to a decrease of surface roughness. However, this behavior is not observed in coatings with dispersant, even with the surface smoothening process. Gloss change for the dispersed system appears to be mainly dominated by an increase of the macro-scale surface roughness resulting from the formation of pits and cracks as observed by LSCM, and the contribution of the small structures observed by AFM. The correlation between gloss loss and macro- to micro-scale surface topography during outdoor exposure is currently being investigated and these results will be reported later.

Chemical Changes

Chemical changes in the exposed coatings were monitored by ATR-FTIR. Measurements were performed on free films at different exposure times. Spectra of coatings prepared with and without dispersant are shown in *Figure* 10. Note that the band at 1720 cm^{-1} is present in the spectra of the coating containing the dispersant prior to outdoor exposure. This band was absent for the pigmented system formulated without dispersant. This peak is attributed to the dispersant. To verify this assignment, a thin film of dispersant was applied by spin casting onto CaF$_2$ substrate, and the solvent was evaporated following the same curing procedure used for pigmented epoxy coatings. The dispersant specimen was measured by FTIR in transmission, showing a peak at 1720 cm^{-1} due to acetate C=O stretching (spectra not shown).

During outdoor exposure, intensities of existing bands at 1510 cm^{-1} and 2925 cm^{-1} due to benzene ring stretching and CH$_2$ stretching, respectively, decreased for all materials, indicating that chain scission and mass loss take place in the backbone of the epoxy matrix. In addition, the formation of oxidation products was indicated by the appearance of new bands at 1720 cm^{-1} and 1660 cm^{-1}, which were assigned to ketone C=O stretching and amide C=O stretching, respectively. Such chemical changes are characteristic of photodegradation reactions for this amine-cured epoxy.[23] In this study, the characterization of the photodegradation process was carried out by analyzing the bands at 1510 cm^{-1} and 2925 cm^{-1}, which are normalized by their initial intensities, and band intensity ratios 1510 cm^{-1}/1660 cm^{-1} and 1510 cm^{-1}/1720 cm^{-1}; these results are displayed in *Figure* 11. Each data point

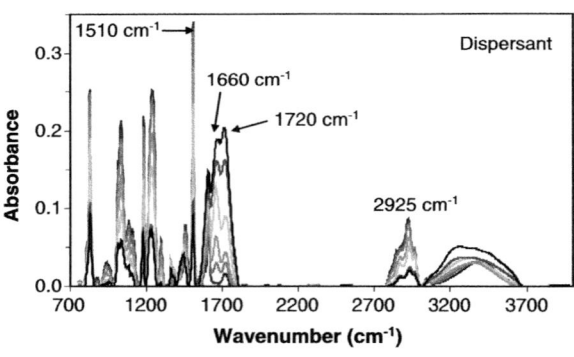

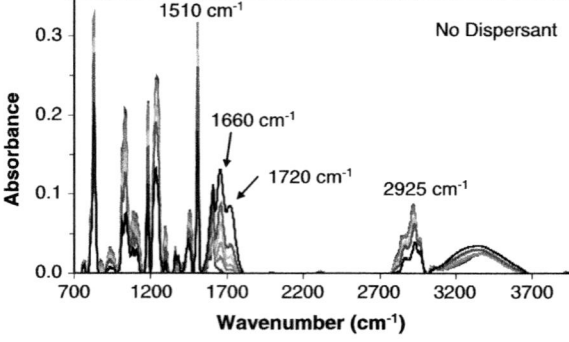

Figure 10—ATR-FTIR spectra taken after 0 d, 7 d, 12 d, 26 d, 44 d, and 54 d of outdoor exposure in Gaithersburg, MD, for pigmented coatings prepared with dispersant (upper graph) and without dispersant (lower graph).

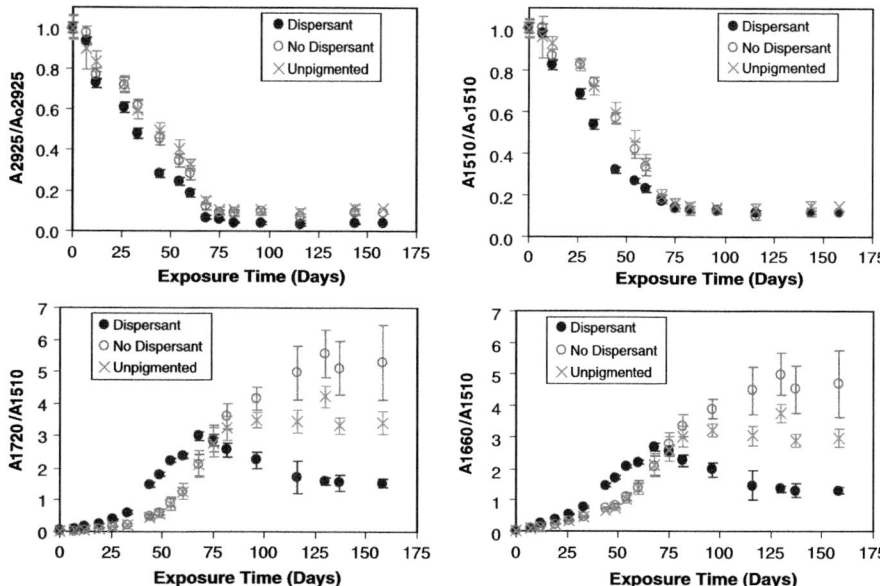

Figure 11—Upper graphs: relative change in FTIR-ATR absorbance peak at 2925 cm⁻¹ and 1510 cm⁻¹ as a function of outdoor exposure time for TiO₂ pigmented epoxy coatings prepared with and without dispersant, and the unpigmented epoxy. Bottom graphs: relative change in ratio of absorbance peaks 1720 cm⁻¹/1510 cm⁻¹ and 1660 cm⁻¹/1510 cm⁻¹. Error bars represent one standard deviation.

represents an average of 16 measurements, collected from four positions on each of four samples.

In the early stages of outdoor exposure, the percentage decrease of intensities of bands at 1510 cm⁻¹ and 2925 cm⁻¹ was greater for coatings prepared with dispersant than the ones without dispersant. In the absence of dispersant, the rate of photodegradation was observed to be similar to that of the unpigmented amine-cure epoxy. The relative loss of each peak after 75 d remained constant. The formation of oxidation products for coatings containing the dispersant appeared to be greater than for the other systems. Ratios of band intensities 1660 cm⁻¹/1510 cm⁻¹ and 1720 cm⁻¹/1510 cm⁻¹ were used to minimize the effect of contact area variability in the ATR-FTIR measurements. The results showed a maximum after approximately 75 d, and decreased thereafter for coatings with dispersant.

This sudden decrease in intensity ratios for the system formulated with dispersant occurred when the pigments started to appear on the surface of ATR-FTIR samples following surface erosion and removal of the binder from the surface of the film. The reduction of exposed polymer at the surface through chalking is most likely responsible for the decrease in intensity ratios.

For the pigmented coatings prepared without dispersant and for the unpigmented coatings, the evolution of these ratios is quite constant and consistent. Chemical changes as followed by ATR-FTIR reached a relatively steady state. However, photodegradation reactions continue with UV exposure, suggesting that the thicknesses of the degraded layer on the surface of coatings were greater than the penetration depth of IR beam in the

488

coatings. Higher values are observed for pigmented coatings prepared without dispersant relative to the unpigmented coatings, but are statistically similar. The greater experimental variability observed at later exposure times are probably due to a poor contact area between the surface of films and the ATR crystal. Since the ATR accessory used has a contact area of approximately 1 mm diameter, the surface roughness of the degraded films seen from LSCM results might have a negative effect on the measured absorption intensities.

Overall, a high degree of similarity between the unpigmented amine-cured epoxy and pigmented epoxy without dispersant may be explained by the thick, clear layer of binder on the top surface. The penetration depth of the IR beam in the film surface during ATR-FITR measurements is only up to a few microns. In such a case, FTIR results predominantly correspond to the clear layer in the system. Thus, chain scission and mass loss processes, as well as formation of oxidation products, are similar to those of the unpigmented coatings.

On the other hand, the faster degradation rate of pigmented coatings formulated with dispersant under UV exposure might be related to the high polarity of dispersant used to stabilize TiO_2 against flocculation in the binder. The dispersing agent could have increased the susceptibility of the system to UV and hydrolytic degradation. Additional investigations are being conducted to identify the role of the polar dispersant in the degradation of the epoxy.

Mechanical and Material Properties

BULK MECHANICAL PROPERTIES AND T_g—The effect of pigment dispersion on bulk mechanical properties during outdoor exposure was followed by DMTA. Before exposure, the glass transition temperature (T_g) of the unpigmented epoxy was found higher than that of the pigmented coatings formulated with and without dispersant. The T_g was 126.2°C ± 1.7°C for the clear epoxy coatings and 112.4°C ± 0.3°C and 113.3°C ± 2.4°C for pigmented coatings formulated with and without dispersant, respectively. In addition, the introduction of TiO_2 particles increased the elastic modulus, E′, from 3.94 MPa ± 0.13 MPa for the unpigmented epoxy, to 6.31 MPa ± 0.01 MPa for the pigmented epoxy with

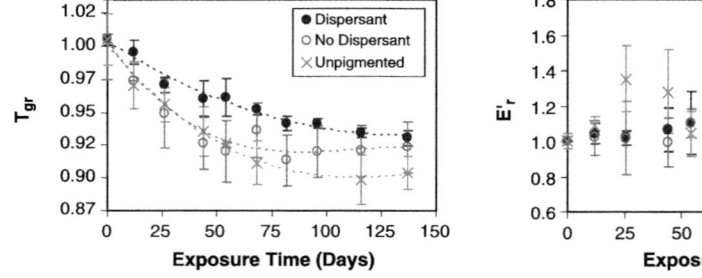

Figure 12—(a) Relative changes in glass transition temperatures of TiO_2 pigmented epoxy coatings prepared with and without dispersant, and the unpigmented epoxy as a function of outdoor exposure time. The lines serve as a guide. (b) Relative change in bulk elastic modulus versus exposure time measured by DMTA. Error bars represent one standard deviation.

dispersant. The elastic modulus for the pigmented epoxy prepared without dispersant is 6.55 MPa ± 0.47 MPa. However, the different degrees of dispersion had little effect on the glass transition temperatures and the elastic modulus of the pigmented coatings.

T_g and E' are known to be good indicators of crosslink density and degree of cure for crosslinked systems such as epoxy. These two properties are expected to increase with crosslink density. However, the presence of pigments in the epoxy coating leads to a heterogeneous system composed of three phases: the binder, the pigment, and the binder/pigment interphase. The increase in elastic modulus may be due to the reinforcing effect of pigment. The reduction of the glass transition temperatures may be related to a reduction of the crosslink density due to non-reacted epoxy and amine groups. The introduction of pigments in the formulation increases its viscosity, which could lead to a reduction in the mobility of reactive species.

Figures 12a and 12b show the relative changes in glass transition temperature (T_{gr}) and the relative change in elastic modulus (E'_r) for pigmented coatings and the unpigmented coating, respectively, as a function of exposure time. T_{gr} and E'_r are defined as the ratio of the glass transition temperature and the elastic modulus of photodegraded film to that of the unexposed film. Each data point represents the average of at least three measurements. Uncertainties are large but the trend is noticeable. All systems showed a decrease in T_g during UV exposure (*Figure* 12a). The pigmented coatings with dispersant experienced the lowest rate of decrease in T_g while the coatings without dispersant underwent a faster decrease in T_g during the first 50 d of exposure, similar to the unpigmented epoxy film. The decrease in glass transition temperature for all systems was approximately 11% for coatings without dispersant and 8% for coatings with dispersant over the total exposure.

The relative changes in elastic modulus with outdoor exposure time of unpigmented and pigmented epoxy prepared with and without dispersant presented in *Figure* 12b show that the bulk elastic modulus remains relatively unchanged, considering the great standard deviation associated with the measurements. In addition, the effect of pigment dispersion on the bulk mechanical properties measured by DMTA was not observed during the exposure period. Since the chemical changes in coatings during UV exposure are mainly based on a photochemical degradation occurring near the surface, bulk mechanical properties may not be sensitive to changes of a few microns deep in the 150-µm thick films.

Surface Mechanical Properties

The surface elastic modulus and its relative changes with exposure time for the two pigmented coatings and the unpigmented coating are shown in *Figures* 13a and 13b, respectively. The surface relative elastic modulus (E_r) is defined as the ratio of modulus of photodegraded film to that of the unexposed film. The reported values were the average of 15 indentations at different locations. As can be seen in *Figure* 13a for the unexposed films, pigment dispersion had little effect on the initial elastic modulus. However, the pigments present underneath the clear layer caused an increase in elastic modulus relative to the unpigmented epoxy coating.

Outdoor exposure led to an increase of the elastic modulus for all coatings up to 60 d. The unpigmented epoxy experienced the greatest increase in modulus, more than 50% at 54 d of exposure (*Figure* 13b). These results are consistent with those reported in the lit-

490

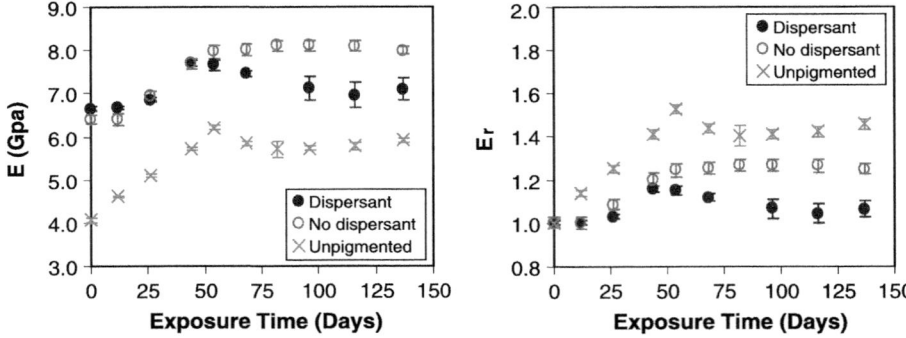

Figure 13—(a) Surface elastic modulus and (b) relative change in surface elastic modulus of TiO$_2$ pigmented epoxy coatings prepared with and without dispersant, and the unpigmented epoxy as a function of exposure time. Error bars represent one stanard deviation.

erature.[24] Thereafter, the surface modulus of the pigmented coatings prepared without dispersant remained constant while the pigmented coatings formulated with dispersant and the unpigmented coating experienced a slight decrease. The greater measurement variability for the system with dispersant may be related to the appearance of pigments at the surface, which led to an increase in the surface roughness. It is interesting to note that a similar trend was observed for ATR-FTIR results at the same exposure time. However, in the early stage of UV degradation, the increase of stiffness of the near-surface region was likely caused by further crosslinking. Thereafter, the competitive reaction was chain scission, which is thought to cause the small decrease of surface elastic modulus. In addition, the contribution of the subsurface pigments on the relative elastic modulus is also noticeable. Unpigmented coatings and pigmented coatings without dispersant had the same rate of photooxidation but the latter showed a lower increase in relative elastic modulus.

CONCLUSIONS

USANS and SEM measurements showed that the addition of the dispersant improved the degree of dispersion of TiO$_2$ pigment in epoxy coatings, resulting in a reduction of the pigment aggregation in the cured coating. In addition, the dispersant may have had an effect on the thickness of the clear layer at the surface, which played a significant role in the UV degradation process.

AFM and LSCM measured morphological changes in pigmented coatings surfaces during UV exposure, and showed that both pigment dispersion and thickness of the clear layer played a role in the resulting topography during UV exposure. In the early stages of exposure, the large pigment aggregates underneath the thick clear layer at the surface of coatings prepared without dispersant produced large ripples. The increase in RMS measured by AFM resulted in a significant decrease in gloss. On the other hand, gloss change for pigmented coatings formulated with dispersant appeared to be mainly dominated by the increase of surface roughness resulting from pits and cracks, as observed by LSCM.

Chemical degradation and mechanical property changes as a function of exposure time were measured by ATR-FTIR spectroscopy, instrumented indentation, and DMTA. The results showed that chemical changes in coatings during UV exposure occurred mainly near the surface and that the dispersant could have increased the susceptibility of the coatings surface to UV degradation. Instrumented indentation was used to follow changes in surface elastic modulus resulting from photodegradation. The increase of the surface elastic modulus observed with exposure time was likely due to further crosslinking. For the pigmented system formulated with dispersant, the appearance of pigments on the surface decreased the intensity of ATR-FTIR peaks associated with the epoxy and the surface elastic modulus, indicating that the surface might be dominated by exposed pigment in this case. Bulk mechanical property changes measured by DMTA showed that the incorporation of pigments increased the elastic modulus relative to the unpigmented epoxy. However, the effect of pigment dispersion on the bulk elastic modulus changes during UV exposure had no apparent effect.

ACKNOWLEDGMENT

The authors acknowledge the support of the National Institute and Technology, U.S. Department of Commerce, in providing the neutron research facilities used in this work, and Dr. Paul Butler of NIST for his help in conducting USANS experiments and data analysis. The authors would like to thank Germain Lespinasse for assistance with SEM measurements.

References

(1) Braun, J.H., "Gloss of Paint Films and the Mechanism of Pigment Involvement," *J. Coat. Technol.*, 63, No. 799, 43-51 (1991).

(2) Rutter, E.G., *J. Oil Colour Chem. Assoc.*, 28, 187 (1945).

(3) Colling, J.H. and Dunderdale, J., "The Durability of Paint Films Containing Titanium Dioxide— Contraction, Erosion, and Clear Layer Theories," *Prog. Org. Coat.*, 9, 47-84 (1981).

(4) Perera, D.Y., "Effect of Pigmentation on Organic Coating Characteristics," *Prog. Org. Coat.*, 50, 247-262 (2004).

(5) Hosseinpour, D., Guthrie, J.T., and Berg, J.C., "The Effect of Additives on the Quality of Dispersion and Physical Properties of an Automotive Coating Pigmented with TiO_2," *J. Adhes. Sci. Technol.*, 21, 141-151 (2007).

(6) Farrokhpay, S., Morris, G.E., Fornasiero, D., and Self, P., "Titania Pigment Particles Dispersion in Water-Based Paint Films," *J. Coat. Technol. Res.*, 3, No. 4, 275-283 (2006).

(7) Rabeck, J.F., *Polymer Photodegradation—Mechanisms and Experimental Method*, Chapman & Hall, New York, 1995.

(8) Luoma, G.A. and Rowland, R.D., "Environmental Degradation of Epoxy Resin Matrix," *J. Appl. Polym. Sci.*, 32, 5777-5790 (1986).

(9) VanLandingham, M.R., Nguyen, T., Byrd, W.E., and Martin, J.W., "On the Use of the Atomic Force Microscopy to Monitor Physical Degradation of Polymeric Coating Surfaces," *J. Coat. Technol.*, 73, No. 923, 43 (2001).

(10) Gu, X., Sung, L., Kidah, B., Oudina, M., Martin, D., Rezig, A., Nguyen, T., and Martin, J.W., "Relating Topographical Change to Gloss Loss of Polymer Coatings during UV Radiation," ACS Symposium Series, *Nanotechnology Applications in Coatings*, 2007.

(11) Gu, X., Nguyen, T., Oudina, M., Martin, D., Kidah, B., Jasmin, J., Rezig, A., Sung, L., Byrd, E., Martin, J.W., Ho, D.L., and Jean, Y.C., "Microstructure and Morphology of Amine-Cured Epoxy Coatings Before and After Outdoor Exposures—An AFM Study," *J. Coat. Technol.*, 2, No. 7, 547-556 (2005).

(12) Sung, L., Jasmin, J., Gu, X., Nguyen, T., and Martin, J.W., "Use of Laser Scanning Confocal Microscopy for Characterization Changes in Film Thickness and Local Surface Morphology of UV Exposed Polymer Coatings," *J. Coat. Technol. Res.*, 1, No. 4, 267-276 (2004).

(13) Faucheu, J., Wood, K.A., Sung, L., and Martin, J.W., "Relating Gloss Loss to Topographical Features of a PVDF Coating," *J. Coat. Technol. Res.*, 3, No. 1, 29-39 (2006).

(14) Barker, J.G., Glinka, C.J., Moyer, J.J., Kim, M.H., Drews, A.R., and Agamalian, M., "Design and Performance of a Thermal-neutron Double-crystal Diffractometer for USANS at NIST," *J. Appl. Crystallography*, 38, 1004-1011 (2005).

(15) Higgins, J.S. and Benoit, H.C., *Polymers and Neutron Scattering*, Oxford, New York, 1994.

(16) ImageJ is a free public domain image analysis program available from the U.S. National Institute of Health website (http://rsb.info.nih.gov/ij/).

(17) Dickens, B. in *Service Life Prediction: Methodology and Metrologies*, Martin, J.W. and Bauer , D.R. (Eds.), American Chemical Society, 2001.

(18) Oliver, W.C. and Pharr, G.M., *J. Mater. Res.*, 7, 156 (1992).

(19) Asif, S.A.S., Wahl, K.J., and Colton, R.J., *Review of Scientific Instruments*, 70, 2408 (1999).

(20) Guinier, A. and Fournet, G., *Small Angle Scattering of X-Rays*, Wiley, New York, 1955.

(21) Sung, L., Clerici, C., Hu, H., and Gu, X., "Relating Optical Properties to Pigment Dispersion of Weathered Pigmented Polymeric Coatings," *Proc. 4th International Symposium on Service Life Prediction of Coatings*, Key Largo, FL, 2006.

(22) Biggs, S., Lukey, C.A., Spinks, G.M., and Yau, S., "An Atomic Force Microscopy Study of Weathering of Polyester/melamine Paint Surfaces," *Prog. Org. Coat.*, 42, 49-58 (2001).

(23) Rezig, A., Nguyen, T., Martin, D., Sung, L., Gu, X., and Martin, J.W., "Relationship Between Chemical Degradation and Thickness Loss of an Amine-Cured Epoxy Coating Exposed to Different UV Environments," *J. Coat. Technol. Res.* 3, No. 3, 173-184 (2006).

(24) Forster, A.M., Waston, S.S., Forster, A.L., and Drzal, P.L., "Quantifying Surface Mechanical Properties of Filled Coatings Exposed to Accelerated Weathering Conditions," submitted to *J. Coat. Technol. Res.*

Chapter 32

Effect of the Environmental Stress and Polymer Microenvironment on Efficiency Trials and Fate of Stabilizers

J. Pospíšil, S. Nešpurek, J. Pilar

Institute of Macromolecular Chemistry, Academy of Sciences of the Czech Republic
162 06 Prague, Czech Republic

Oxidizing and acid components of the atmosphere or acidic polymer-borne impurities influence the material properties of stabilized polymers not only by a direct attack on the polymer matrix but also by interactions with stabilizers. The latter undergo sacrificial transformation/consumption, i.e., processes forming an unavoidable integral part of their stability mechanism. The extent of stabilizer consumption is affected by environmental stress. The performance and longevity of stabilizers under the harsh conditions of over-accelerated tests do not correspond to those taking place under natural environmental stress. This chapter outlines this finding, and presents examples of the depleting effect of acid impurities on basic stabilizers or effect of oxygenation arising from the photosensitizing activity of some colorants. Experimental data show the dynamics and heterogeneity in processes characteristic of aging of stabilized polymers that complicate a reasonable service life prediction.

INTRODUCTION AND BASIC CONSIDERATIONS

Determination of the efficiency and longevity of polymer stabilizers improves understanding of a complex of chemical and physical processes, some of them underrated from the point of view of polymer lifetime prediction. The material properties and durability of most polymers are subjected to changes resulting from degradation/aging during all phases of their service life. The mode, extent, and/or mechanism of degradation are strongly dependent on the intensity, duration, and combination of chemical and physical stresses (i.e., environmental factors). Moreover, they are influenced by the intrinsic sensitivity of the polymeric material to individual stresses, structural inhomogeneities in the polymer matrix arising from commercial synthesis, or resulting from degradation, adventitious active impurities or sensitizers formed during degradation/aging of the stabilized polymer containing various other additives (i.e., micro-environmental factors). Certainly, too many factors are involved. Their diversity is reflected in changes in chemical, physical, or mechanical properties of the material and ultimately accounts for the failure of the material.

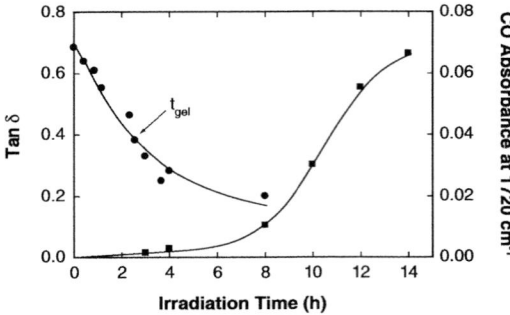

Figure 1—Time-dependent changes in EPDM films 100 μm thick upon irradiation at 35°C with artificial light source having λ > 300 nm. ●—Molecular changes: dynamic viscosity (tan δ 0.04 Hz); t_{gel} = gel point. ■—Chemical change: growth in carbonyl species at 1720 cm^{-1}.[1]

The changes develop gradually throughout the whole aging process. In a latent phase (an induction time, IT) the changes are rather difficult to monitor. The length of IT and the slope of the post-IT curve are indicative of the degradation progress and are influenced by the severity of the environmental attack together with a complex of micro-environmental factors. The changes in material properties are stepwise and time dependent, e.g., changes on the molecular level (formation of carbonyl species, changes of molecular weight due to crosslinking and chain scission, density, crystallinity, mechanical characteristics or physical properties (such as gloss, discoloration, and surface cracking).

The material lifetime of stabilized polymers is assessed according to failure criteria selected as crucial for the commercial application of the product. The aim of the assessment also requires an explanation of the gaps between knowledge of chemistry and physics of aging of stabilized polymers and the results of mechanical and physical tests. Links between the results of individual tests are often unclear; correlation of data is difficult or even impossible. It is sometimes questionable what data are the most suitable as criteria for lifetime prediction.

Information expected from the complex assessment of efficiency and longevity of stabilizers in polymers should include (1) data on common and principal causal atmospheric components during testing, dominance of the individual component, and their combination or periodical changes resulting in superposition of effects influencing the final result; (2) application of optimal assessment methods enabling a realistic correlation between measured parameters and material properties, considering heterogeneity of changes at the polymer molecular level and morphology, and elimination of over-accelerated testing making chemical and physical processes in polymers incomparable with natural conditions.

Conventional mechanical tests (tensile strength, elongation, loss modulus, tensile impact strength, Charpy and Izod impact strength), physical tests (time to embrittlement, color, haze, gloss, surface roughness and cracking, melt mass-flow rate, solution viscosity) and bulk spectral, chromatographic, or thermal analyses give overall data on the changes and their time dependence. Since the beginning of the investigation, we have been dealing with a heterogeneous system (heterogeneity in distribution of stabilizers, amorphous and crystalline regions in semi-crystalline polymers, phases in polymer blends) undergoing heterogeneous changes during aging/degradation [surface character of photooxidation, diffusion limited oxidation (DLO), depth dependence of radiation penetration, spatial gradients in stabilizer consumption, and formation of polymer degradation products]. Information about these changes, particularly those affecting the behavior of thick-walled articles, can be obtained by depth profiling using either the analyses of mechanically separated layers of the aged material or by surface/near surface analyses by various non-destructive methods, such as attenuated total reflection Fourier transform infrared spectroscopy (ATR FTIR), electron spin resonance imaging (ESRI), or photoacoustic spectroscopy.

Despite the fact that most of the changes proceed in concerted mechanisms, the onset and development of individual changes can be very different when compared on a time axis. Processes in photooxidized poly(ethylene-propylene-diene-monomer) (EPDM) are an example.[1] Gelation of the material due to gradual crosslinking starts at the very onset of oxidation and the material is fully crosslinked when only the very early stages of the development of carbonyl species are detected (*Figure* 1).

Commodity and engineering plastics, elastomers, elastomer-modified plastics, fibers, polymer blends, and coatings are commercial failures without additives in any common or special application. Additives, stabilizers among them, enable the processing of polymers, and shape and enhance their end-use performance.[2] Stabilizers also influence the development of polymeric materials making them value-added products helping to meet increasingly stringent and regionally differing environmental regulations. The stabilizer market is very cost-competitive, corresponding to the world growth of polymer production, and is standardized across the globe. Stabilizer application is under stringent legislation and environmental rules. The legal bans against particular stabilizers are globally sensitive. Application of a particular stabilizer or stabilizer combination must be declared in commercial polymers with appropriate details on the regulatory status of each stabilizer with respect to industrial hygiene and environmental impact. On the other hand, there is an understandable concern regarding cost-effective stabilization with stabilizers that have long service lives and that provide efficient protection against aging.

Modern stabilization systems make polymeric materials very resistant to environmental stresses and result in long-lasting products. To determine their durability, accelerated tests are necessary. To address philosophical and commercial concerns about stabilizer testing, reproducible and reliable results are required within acceptably short timeframes, thus allowing for the safe introduction of new high-performance products under application conditions. The data are usually essential for tailoring material properties and stabilizer formulations.

The testing parameters should provide a realistic estimation of the lifetime in an expected environment. The primary need includes data applicable for material lifetime prediction. In spite of the increased amount of information available, this is still a very difficult task. Various theoretical treatments of the problem are far from the reality because of gaps between knowledge of chemistry and physics of stabilized polymers, involvement of a complex of factors (common, causal) of atmospheric chemistry and environmental stress and their links to material engineering, including selection of proper analytical methods and instrumentation. Heterogeneity of processes in stressed materials, spatial distribution of changes, and formation of concentration gradients of products are not easily described by a simple kinetic treatment used for lifetime prediction.

Various accelerated tests have been developed to monitor degradation of particular polymeric materials. Some of them have been recommended as standards. Effects in thermal testing of stabilizers are rather well understood.[3] Assessment of photooxidation/weathering of stabilized polymers is still a mixture of empirical and applied science and, consequently, a rather difficult task.[4] The process is affected by "common" and "accidental" environmental factors stressing all components of the stabilized material in a seasonally and globally changing, but in long-term, non-reproducible manner.

Links between the behaviors of various components in the system are not easily predictable without an experiment. This is one of the problems limiting a better exploitation and generalization of results of mechanistic studies performed on simplified systems under fixed environmental conditions. Another problem is a commercial pressure to get fast information about the efficiency of stabilizers, mostly without any concern about the reliability of the result due to over-acceleration of the tests.

Mechanistic analyses concerned with stabilizer testing are mandatory for explaining why the application of a stabilizer at a specific environmental stress and particular composition of the matrix material was a success or a failure. This is the only way to understand adverse interactions between individual components of the stabilized system reflected in the final effect. The approach is also helpful in selecting an optimized additive system fitting requirements of the customer. Moreover, mechanistic analyses provide information about chemical changes in structure or reactivity of "minority" components (stabilizers) in the oxidizing stabilized polymer and on accumulation of their transformation products (structurally new compounds).

PROTECTION OF POLYMERS BY STABILIZERS AND STABILIZER FATE

There is a commercial concern to protect polymeric materials against degradation in all phases of their lifetime: production, processing, storage, and indoor and outdoor application. Current status of polymer stabilizers exploits all knowledge of mechanistic studies dealing with relations between the structure and efficiency. Processing additives, long-term heat and light stabilizers, and flame retardants of different structures are available.[2]

Polymer stabilizers are chemically more reactive than the polymer matrix itself. Their original structure is transformed during their service time as a consequence of interactions with polymer-borne free-radical intermediates (alkyls, alkoxyls, alkylperoxyls, acylperoxyls), polymer-derived hydroperoxides acting in combination with solar radiation. The role of polymer-borne O-centered radicals (POO·, arising in the chain oxidation of carbon-chain polymers) in sacrificial transformation of phenolic moieties in antioxidants and UV absorbers (UVAbs) or amino groups in aromatic amine antioxidants and antiozonants is a key process in their chain-breaking (CB) antioxidant mechanism and is well understood.[5-7] The process is considered as sacrificial stabilizer consumption and creates an integral part of the stabilizer active lifetime that cannot be avoided in any phase of the polymer application. The inherent chemistry of the stabilizer mechanism reflects harshness of the environmental stress and resistance of the polymer to degradation.

Similar to (photo)oxidative changes of the polymer on a molecular level, the sacrificial consumption of stabilizers has a heterogeneous character[6-9] with characteristic concentration gradients. This accounts for randomly distributed sites in the polymer bulk or near-surface layers with a high concentration of both oxidation products of the polymer and transformation products of the stabilizer.

In addition to sacrificial consumption, stabilizers are chemically attacked by atmospheric oxidizing (ozone, nitrogen oxides) or acid pollutants and by polymer-borne impurities (e.g., acid compounds from degrading polymers or additives), or residues of polymerization catalysts. An adverse effect of degradation-borne carbonyl groups on efficiency of UVAbs is known.[7]

Compounds having different structure, molecular weight, and properties formed from the originally added stabilizers accumulate gradually in the polymer matrix. Some of them discolor or stain the matrix, and may antagonize other additives and differ in environmental solubility or volatility.

Useful data on transformation chemistry of principal stabilizer classes have been summarized.[5-7] Determination of stabilizer consumption during the aging of polymers needs appropriate analytical methods and model compounds of the principal transformation products and typically is not a part of routine testing. The data are useful for assessing the influences of testing severity on stabilizer longevity. Analytically determined decrease of the added amount of the original additive does not always mean a loss of the stabilizing efficiency, as in the case where some of the transformation products act as active stabilizers (e.g., acid transformation products of thiosynergists) or as reservoirs of activity in subsequent stabilization steps.[5,6] Hindered amine stabilizers (HAS) are typical examples of the latter group.

The real depletion of the stabilizer activity accounts for irreversible transformation into products that are no longer able to protect the polymer. Structures of transformation products remaining in the matrix are mostly considered as polymer impurities, with some of them having properties of initiators or sensitizers of oxidation.[6,10] The decrease of the originally added amount of the stabilizer may be also due to physical losses as observed with UVAbs[11] and even with oligomeric HAS.[12]

The consumption of the active stabilizer form in later phases of the lifetime of the material results in reduction of the effective concentration of the stabilizing species below the level needed to protect the aged polymer (containing increased amount of chromophoric or catalytic impurities) against loss of service properties.

EFFECTS IN THERMAL TESTING OF STABILIZED POLYMERS

Thermal testing is common for the assessment of the processing stability of polymers and of long-term heat aging (LTHA). Test conditions for evaluation of performance of stabilizers during processing are kept very close to commercial conditions, i.e., without significant acceleration.[3] The performance and sacrificial chemistry of processing stabilizers for polyolefins (PO), i.e., hindered phenols, organic phosphates or lactone or various carboxylates, organotin stabilizers, and different co-stabilizers in poly(vinyl chloride) (PVC)[2,6] are comparable with those that are commercially processed.

Assessment of hindered phenolic antioxidants, thiosynergists, and HAS in protecting PO or elastomers during LTHA requires accelerated testing. Oven aging is typically used and provides samples for the complex determination of chemical, physical, or mechanical changes in the materials.[3] Oven testing is performed with solid samples reflecting polymer morphology and at increased temperatures that are reasonably close to those of the polymer end use. This allows a correlation of the stabilizer efficiency to real application conditions.

Two effects that influence the thermal testing of solid semi-crystalline PO should be mentioned. External stress during testing favors diffusion of oxygen into the polymer mass, thus enhancing oxidation. It also hinders molecular repair by recombination of free radicals formed by chain scission.[3] Accordingly, the efficiency of stabilizers drops with increasing stress due to increased sacrificial consumption.[13] An increase of oxygen pressure during LTHA forces oxygen to penetrate into deeper layer of the material. This testing approach increasing oxidation rate alters gradients in polymer oxidation products and stabilizer consumption in the near-surface and deeper layers. This reduces concentration level of antioxidants in these layers that are less protected than at atmospheric pressure. Moreover, the consumed stabilizers are only partially replenished by slow migration of fresh stabilizers from the polymer bulk. This is particularly reflected in testing of aromatic amines in elastomers and restricts correlation with natural LTHA.[14]

In addition to oven aging, testing methods such as measurements of oxygen uptake, differential scanning calorimetry (DSC), or chemiluminescence (CL) are used for testing thermal stability. DSC or CL detect the onset of the exothermic oxidation peak and its shift by antioxidants interfering with propagation and termination steps of the oxidation chain mechanism.[3] The DSC results are fast but measurements are performed at high temperatures, rather far from practical application. Isothermal DSC measurements were recommended to determine oxidation induction times (OIT), particularly in PO. This otherwise favorable method for monitoring the behavior of unstabilized polymers has a drawback: temperature of the oxidation onset depends on polymer sensitivity and is very high in oxidation-resistant semi-crystalline PO, particularly if they are effectively stabilized. Consequently, the tests have to be performed at temperatures above crystal melting points, too far from temperatures in practical use. Moreover, in the molten polymers the morphological effects of the tested material or distribution of stabilizers are not reflected. Testing above melting points of the material resembles more or less the behavior in a homogeneous solution and is unable to show specificity of the heterogeneous semi-crystalline solid system. In the solid semi-crystalline polymers, as tested by oven aging, the stabilizers are dissolved only in the amorphous phase. This affects patterns governing local concentration gradients of the tested antioxidants, particularly their participation in bimolecular coupling reactions with polymer-derived free radicals or autoreactions of antioxidant-derived reactive species.[6]

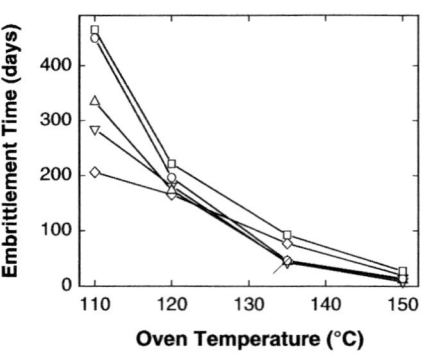

Figure 2—Effect of the oven aging temperature on efficiency of HAS in 1 mm thick PP plaques (base stabilization 0.15% of a 1:1 mixture of Irganox® 1010 with Irgafos® 168, 0.05% calcium stearate). ◊—HAS-free standard. Samples containing 0.1 % HAS: ▽—Cyasorb UV 3346, △—Chimassorb® 944, ○ —Tinuvin® 622 and ❑—Chimassorb 119.[17]

Temperatures used for DSC-OIT (180–220°C for PP[15] or 145°C for acrylonitrile-butadiene-styrene polymer (ABS)[16] accelerate disproportionally degradation processes with higher activation energies as compared with ambient conditions. The DSC-OIT was developed in times of dominance of phenolic CB antioxidants, tested conventionally by oven tests in PO above 100°C and was recommended as standard ASTM D 3895-94 for fast quality control of PO containing CB antioxidants. A linear regression analysis on a plot of the OIT versus antioxidant concentration is assumed. However, the

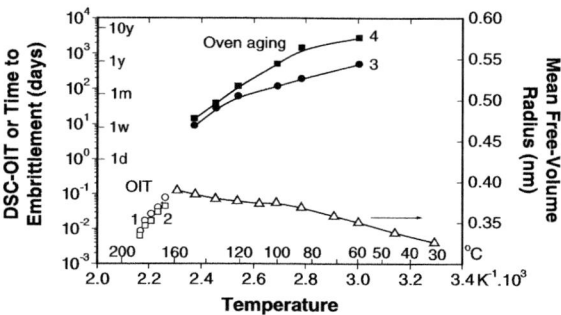

Figure 3—Comparison of DSC data measured in stabilized PP melts with oven aging embrittlement data of 120 μm thick PP plaques. 1, 3: Base stabilization with 0.1% of a 1:1 mixture of Irganox 1010 with Irgafos 168; 2, 4: Additional stabilization with 0.1% Chimassorb 944 and 0.1% Tinuvin 622.[18]

method is not ideal. Effects of material morphology, density or crystallinity are not considered. Various studies revealed problems encountering the application of DSC-OIT at high temperature testing.[3] It is limited to PO containing phenolic antioxidants having rather high ceiling temperature, and gives erroneous or unpredictable results in monitoring the contribution of secondary antioxidants. Moreover, even differentiation between various strong phenolic antioxidants (to determine relations between structure and efficiency) is questionable. The method fails completely in testing of modern HAS in their function as long-term heat stabilizers. The loss of HAS efficiency above ca 110°C was confirmed by oven testing (*Figure 2*).[17] Efficiency of HAS in polypropylene (PP) drops by increasing the testing temperature. Accordingly, it is insignificant at common DSC-OIT testing range. According to industrial experience,[15] the DSC-OIT method is useful for a continuous quality control in commercial PO production where samples containing the same polymer and stabilizer are compared. An application for selection of components for optimized stabilization package is questionable.

It is difficult to correlate the DSC-OIT data measured conventionally at 180–220°C with oven tests (performed at 100–150°C). Any correlation to common environmental temperatures is almost impossible due to nonlinearity in extrapolation of data if thermal intervals between measured and extrapolated temperatures are higher than 24–40°C. Arrhenius-like plots give overestimating lifetimes as exemplified by DSC data measured in stabilized PP melts compared with embrittlement data from oven aging of 120 μm PP plaques (*Figure 3*)[18] or as found by an attempt to extrapolate DSC and oven aging data measured on stabilized high density polyethylene (HDPE) insulating cables to application temperature.[19] Consequently the application of DSC-OIT for a correct lifetime prediction is questionable not only because of high temperatures incorrectly reflecting the performance of stabilizers (direct oxidation of phenolic antioxidants by oxygen, low performance if any of HAS over 110°C) but because of the temperature differentiation effect on material properties of the polymer, such as the effect of gradual annealing at temperatures between 100 and 150°C on density, flexural modulus, tensile strength, impact tensile strength, or elongation of PP.[17] Relaxation processes due to translational mobility in

500

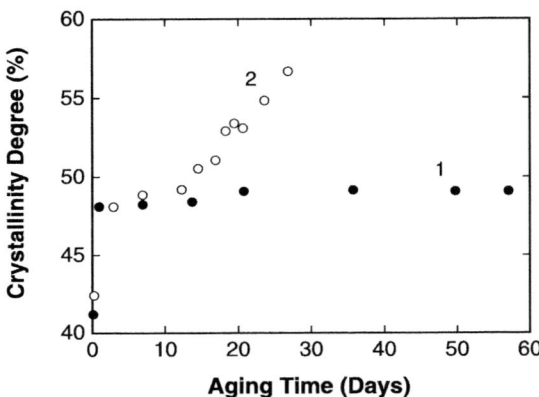

Figure 4—Change in degree of crystallinity of PP with oven aging time at 135°C affected by the heat stabilizer: 1— 0.05% Irganox 1010; 2—0.05% Chimassorb 944.[20]

the crystal phase or reorganization in the amorphous phase allow some deformation in the amorphous fraction and may influence physical delocalization of stabilizers, oxygen permeation, changes in free volume and polymer density, localized formation of polymer-borne oxidation species, and related stabilizer consumption.[3] The chemocrystallization process is stabilizer dependent (*Figure* 4)[20] and shows differences in the development of the degree of crystallinity in the PP homopolymer stabilized either by phenol Irganox 1010 or HAS Chimassorb 944 as a function of the oven aging time at 135°C.

Most of the interest in application of CL as a sensitive method for detection and determination of oxidation mechanisms of unstabilized carbon-chain polymers, PO in particular, has been academic. The method was mentioned in various papers. In principle, the CL method is based on the consequences of formation and recombination alkylperoxyls POO˙ in the oxidation chain process. This reflects sensitivity of the polymer matrix to oxidation. The process is most correctly understood for PO. It is theoretically suitable for detecting the influence of stabilizers scavenging POO˙ radicals (i.e., CB antioxidants: hindered phenols, 4-hydroxybenzoates, N,N'-dialkylhydroxylamines, and to some extent HAS in a part of their complex activity mechanism) on thermal oxidation. The principle of the assessment of the efficiency of CB stabilizers by DSC-OIT and CL-OIT is analogous. This also includes some of the previously mentioned disadvantages: a direct determination of stabilizers acting by other mechanisms than interference with POO˙ is difficult.

CL allows the determination of OIT at lower temperatures than DSC, i.e., closer to the temperature range when DSC is not sensitive enough to accurately detect the oxidation onset.[21] This makes CL-OIT a more suitable testing method for discriminating better between PO samples containing low concentration of antioxidants. CL-OIT may replace oven-aging tests in the case when monitoring of mechanical or physical properties is not the principal failure criterion or for fast and informative comparison between stabilizing systems in the same polymer substrate. The method can be used also for an informative determination of residual thermal stability of PO or for a rapid indirect screening of the relative performance of formulations of stabilizers in pre-aged coating.[22] However, the problem of the determination of efficiency of HAS in PO remains. The testing temperatures are still too high. In spite of excellent development in instrumentation,[23] the CL method is still far from industrial acceptance for general testing of stabilizer efficiency in PO. It is really difficult to predict potentials of tests in other synthetic conventional polymers because of mechanistic uncertainties in the explanation of the process. [The problem of a correct interpretation of measured data, necessity of simultaneous analyses (spectral in particular) excluding misinterpretation in correlation with changes on molecular level and changes of material properties are a serious barrier for a common application in testing of stabilizer performance and longevity in synthetic polymers other than PO.]

EFFECTS IN RADIATION TESTING OF STABILIZED POLYMERS

A substantial amount of data dealing with different aspects of photooxidation and weathering of any kind of stabilized organic materials is available, along with descriptions of testing devices. General information on the potential effects of solar radiation and those of artificial radiation sources is sufficient to explain the most common phenomena involved in polymer photooxidation.[24] There is new information about the response of light intensity on the development of physical properties of irradiated materials, effects of alternating dark/light period,[25] or effects of water spray on irradiated samples.[26]

In this part of the chapter we outline the effects of radiation on performance and durability of light stabilizers in exposed stabilized polymers and the effects of other components than radiation on the fate of stabilizers in the material under accelerated tests. This includes the results of model studies describing sacrificial chemistry of light stabilizers and their longevity, and the effects of superposition of radiation and temperature affecting fate of stabilizers. Effects of weather components other than radiation and oxygen (atmospheric pollutants) that shape the fate of antioxidants and light stabilizers in weathered materials are outlined in a separate section.

Common testing of weathering and photooxidation resistance involves monitoring of changes on a molecular level (formation of new functional groups, changes in molecular weight) and changes of physical and mechanical properties. More detailed investigation revealed changes in polymer morphology and heterogeneity of phototriggered processes.[27]

So far the best simulation of the terrestrial sunlight in accelerated tests of stabilized materials has been achieved with filtered xenon arcs in various weatherometers. Information on the fate of stabilizers and other additives was also obtained in experiments with fluorescent lamps emitting light in UV-B (UVB-313) or UV-A (UVA-340) regions or with testers using filtered medium- or high-pressure mercury lamps. Individual artificial sources may have a different response on the fate of light stabilizers due to differences in the emitted light energy.

Three classes of light stabilizers protecting polymers with different mechanism have been widely used[6,7]: UVAbs, photoantioxidants (HAS), and quenchers (Q). The most interest has been paid to UVAbs and HAS. Most UVAbs are phenolic derivatives of benzophenone, benzotriazole, and 1,3,5-triazine. Their phenolic character affects activity mechanism and tranformation fate. Excellent efficiency and longevity have been described by Excited State Intramolecular Proton Transfer (ESIPT) mechanism (*Figure 5*), based on strong intramolecular hydrogen bond (IMHB) (exemplified schematically together with involved keto-enol tautomers for benzophenone-type (**1**), benzotriazole-type (**2**) and triazine-type (**3**) UVAbs.[7]

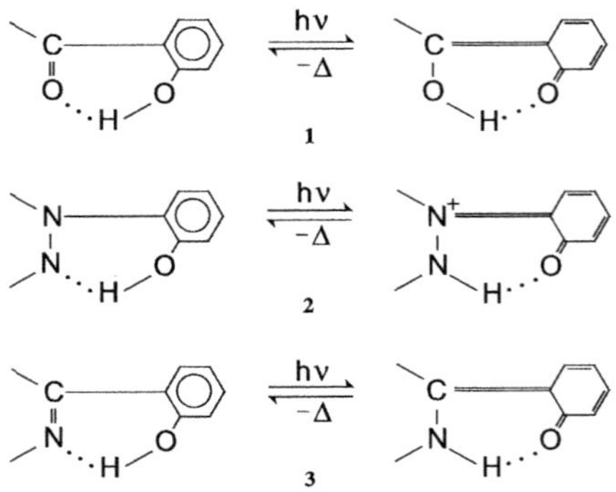

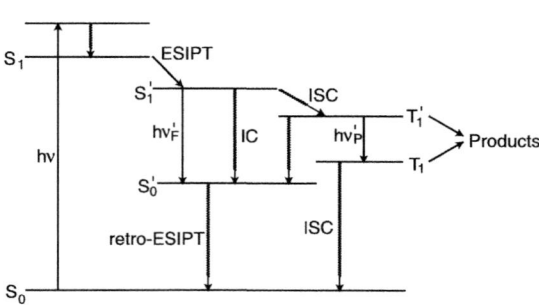

The ESIPT mechanism (*Figure* 5) involving keto/enol tautomerism of phenolic UVAbs describes the physical photoprocesses involved: absorption of a light quantum by UVAbs in its ground singlet state S_0, formation of phenolic first excited singlet state S_1, the ESIPT to zwitterionic (carbonyl) excited singlet state of the tautomer S'$_1$ that is transformed by non-radiative internal conversion (IC) accompanied with some fluorescence into ground state tautomer S'$_0$ returning into phenolic state S_0 (*Scheme* 1).

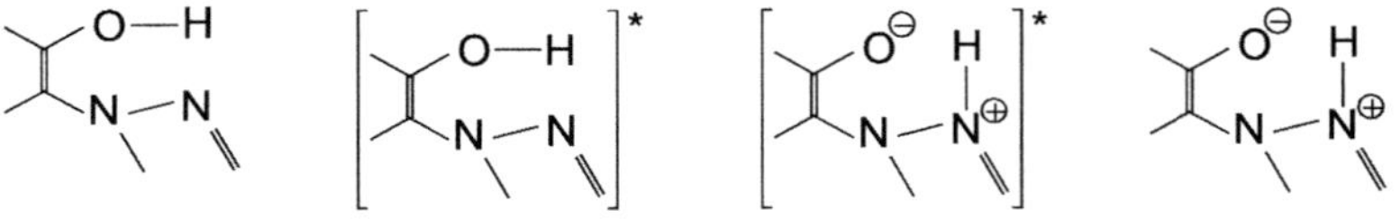

$$S_0 \qquad S_1 \qquad S'_1 \qquad S'_0$$

Scheme 1—Example of keto-enol tautomers in benzotriazole-based UV Abs within the ESIPT mechanism.

The analysis of the ESIPT mechanism explains at the same time the potentials of (photo)chemical depletion of the UVAbs efficiency and longevity after intersystem crossing (ISC) from the singlet state S'$_1$ and formation of long-living photoreactive triplet state T'$_1$ followed by phosphorescence accounting for T_1. A part of the UVAbs is converted by ISC from triplet states to the original ground state So. A part of the UVAbs is lost by depleting chemical transformations.

Some phototriggered processes interrupting the ESIPT mechanism and IMHB between hydrogen of the phenolic HO group and an oxygen-containing acceptor moiety (in benzophenone-trype UVAbs) or nitrogen-containing moiety (in benzotriazole- or triazine-besed UVAbs)

Figure 5—Scheme of the ESIPT[20] mechanism of phenolic UVAbs. S_o, S'_o and S_1, S'_1 = ground states and singlet states of the phenolic and tautomer keto forms, respectively. T_1, T'_1 = triples states. IC = internal conversion, ISC = intersystem crossing, hv'$_F$ = fluorescence, hv'$_P$ = phosphorescence.

bound intramolecularly are operative.[6,7] The phenolic moiety in UVAbs can act as a scavenger of POO˙ radicals by H-transfer from the HO group [equation (1)] in a mechanism typical for phenolic antioxidants,[6,28] with all the consequences: formation of a related phenoxyl and consecuvive molecular products having structures of peroxycyclohexadienones 4, benzoquinone 5 or 6 or quinone methide 7.

$$
\text{Subst.} \underset{OH}{—\bigcirc} \xrightarrow{\text{ROO˙}} \text{Subst.} \underset{O˙}{—\bigcirc} \longrightarrow \text{Products}
\tag{1}
$$

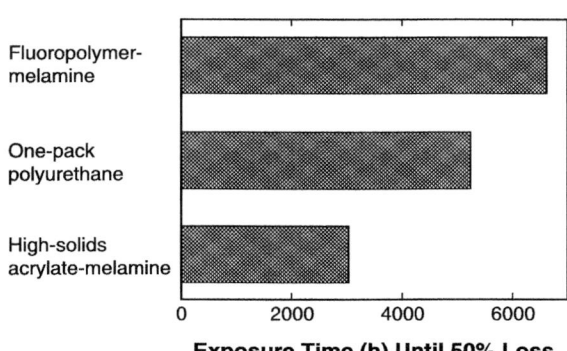

The necessary condition of the UVAbs performance—the presence of IMHB—is thus broken. This depleting free-radical assisted and phototriggered process is more pronounced at higher photooxidation temperatures, in oxidation sensitive matrices easily producing POO˙ radicals[29] and is affected by the structure of the UVAbs. Non-hindered phenolic HO groups in benzophenone-type UVAbs are more prone to oxidative depletion than in benzotriazole- or triazine-based UVAbs. Consequently, the weak "photoantioxidant" effect of phenolic UVAbs mentioned sometimes in the literature[30] means in reality an undesirable loss of the light stabilizing potential by decreasing the level of the IMHB-containing UVAbs species in the matrix, and, consequently, a lower screening effect according to Lambert-Beer law.

The free-radical assisted depleting process is more pronounced in easily oxidizing substrates. A time-dependent loss of 2-[2-hydroxy-3,5-bis (dimethylbenzyl) phenyl] benzotriazole at UVA-340 exposure in matrices differing in oxidative sensitivity[29] is an example (*Figure 6*).

A direct photolysis of the phenolic HO group[31] assists or

Figure 6—Effect of the sensitivity of binder films (~ 30 μm) to photooxidation (exposure with UVA-340 lamp at 40°C, dry conditions) on photo-triggered free radical assisted loss of UVAbs Tinuvin 234. Failure criterion: 50% loss of the UVAbs.[29]

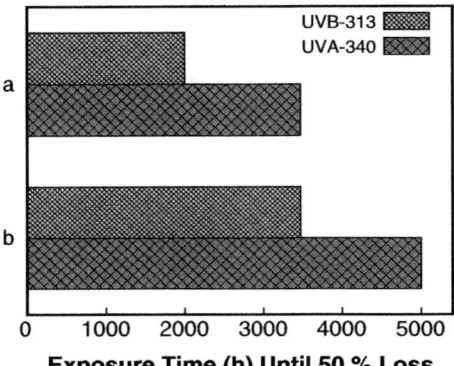

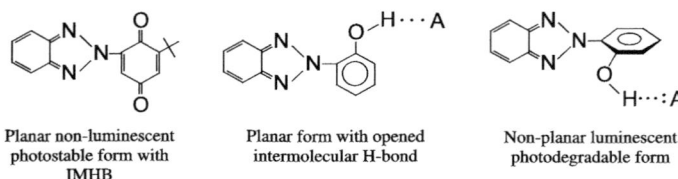

Figure 7—Photo-triggered loss of the efficiency of UVAbs Tinuvin 384 in 35 μm thick polyurethane films irradiated at 40°C with fluorescent lamps UVB-313 and UVA-340. Stabilization: a—1.5% UVAbs; b—1.5% UVAbs + 1.5% HAS Tinuvin 765. Failure criterion: 50% loss of gloss.[29]

enhances the UVAbs depletion. An example of the radiation-increased phototriggered loss of the benzotriazole-based effective UVAbs in polyurethane (PUR) films irradiated either by UVB-313 or UVA-340 lamps at 40°C indicates substantially faster UVAbs consumption at harsher irradiation conditions (*Figure 7*).[29] Application of the 313 nm source was depleting also for the UVAbs used in combination with HAS during photooxidation of PUR-acrylate coating.

A deeper photolytic destruction of the benzotriazole-based UVAbs accounting for fragmentation of the molecule has been described.[32] This process releases a phenolic fragment that can be easily oxidized into discoloring coupling products.

Another process affecting performance of UVAbs accounts for conversion of the intramolecular H-tunneling into an intermolecular one by interaction of the phenolic hydrogen with polymer-borne or a non-polymeric H-acceptor (A) present in the matrix. This includes interaction with polar construction units of the polymer or carbonyl moieties arising during polymer oxidation. The phenomenon was observed very early[33] and results in a gradual "consumption" of the intramolecular H-tunneling. This kind of interruption of IMHB and ESIPT mechanism in phenolic UVAbs has been explained[34] as the transformation of the efficient planar and non-luminescent photostable "closed" form with IMHB into a planar but open intermolecular H-tunnel with an acceptor molecule A, and finally into a non-planar luminescent photodegradable and even photosensitizing intermolecularly tunneled couple (*Scheme 2*).

Planar non-luminescent photostable form with IMHB

Planar form with opened intermolecular H-bond

Non-planar luminescent photodegradable form

Scheme 2—Examples of intramolecular and intermolecular H-tunnels in benzotriazole-based UVAbs.

The resistance against loss of the IMHB increases in the structural series of UVAbs benzophenone < benzotriazole < triazine and can be improved by a proper substitution of the phenolic moiety.[7]

The phenomenon of the IMHB conversion into the intermolecular form by tunneling to a "foreign" acceptor accounts for differences in performance and longevity of phenolic UVAbs in non-polar and polar polymers as well as for observed lower efficiency in recycled deeply oxidized PO containing carbonyl species in comparison with the virgin material in upgrading of recyclates.

Negative interferences between UVAbs with photoinitiators (PI) in photocuring of coatings or printing inks were discussed recently.[7] The problem can be solved by optimized selection of PI exploiting spectroscopically determined "curing windows" in spectral superposition of UVAbs and PI.[35] This problem affecting efficiency of UVAbs is not treated in this chapter.

Outdoor processes in stabilized polymers are controlled by the both radiation and temperature. Superposition of radiation and temperature effects in the assessment of stabilized polymers includes different aspects. Radiation is generally considered to have a major effect in the light period of the exposure temperature, certainly in dark periods. Thermooxidation may be a decisive factor not only in dark cycles, but also in irradiation of thick-walled samples accounting for oxidation effects on the non-irradiated surface where the thermal oxidation exploiting a sufficient near-surface availability of oxygen starts to be a decisive factor competing with photooxidation due to restricted radiation penetration to the non-irradiated side. Temperature during radiation exposure is an important factor due to significant thermal acceleration of processes with higher activation energies. However, the relationships in radiation/temperature effects are not fully explained. Because of the importance for service life prediction, the superposition attracts attention for a deeper understanding of temperature effects and their changes in testing sites at comparable radiation intensity, effects of backing accumulating environmental temperature, wet cycles reducing surface temperature, fixture type, or pigmentation.

Some interesting observations dealing with the phenomenon of thermal/radiation superposition have been published. Chemical changes of HDPE exposed in Bandung (Indonesia) and Tsukuda (Japan) clearly reflect the monthly changes of the environmental temperature[36] in Tsukuda when compared with a rather constant thermal stress in the Bandung exposure.

Differences between the measured environmental and polymer surface temperature due to different fixture types (black box, insulated box, open back exposure) indicate serious influences of the thermal stress at comparable irradiation.[37] The function of the spectral composition of the irradiation source on the surface temperature of differently colored specimens indicates a rather comparable thermal response of a scale of colors after the exposure with filtered Atlas xenon lamp and to outdoor Florida exposure (having a temperature difference between white and black color ca 20 and 23°C respectively) and a poor correlation with irradiation source UVA-340, due to lack of the contribution of the visible part of the light.[38]

Lifetimes of linear density polyethylene (LDPE) 200 μm films effectively stabilized with a benzophenone-based UVAbs and their combination with HAS in Florida weathering were lower if the samples were exposed on aluminum backing. The result reflects both the accumulation of heat and a potential mirror effect of the metal backing.

EFFECT OF ATMOSPHERIC POLLUTANTS AND MICRO-ENVIRONMENTAL IMPURITIES

Trace atmospheric pollutants arising from natural sources or anthropogenic activities attack together with common atmospheric components (oxygen, solar radiation, humidity) in phototriggered processes polymer matrices and change the environmental chemistry of stabilizers. The pollutants include oxidizing gases (ozone, nitrogen oxides), acid gaseous pollutants (oxides of sulfur and nitrogen in humid environment, gaseous hydrogen chloride), or acid deposits (rain, dew) and sensitizing pollutants (polynuclear aromatic hydrocarbons). The chemical and biological attack may be enhanced by abrasion of polymer surfaces by particular impurities (sand, dust).

Tropospheric ozone is a product of atmospheric photochemistry[7] and attacks preferentially unsaturated elastomers in concerted oxidation/ozonation processes.[39] The ozonation process has exclusively a surface character. Aromatic 1,4-phenylenediamines acting by an ozone scavenging mechanism are used to protect elastomers.[5,6] The antioxidant/antiozonant activity of diamines has a sacrificial character. The diamines are transformed into various, mostly dark colored, products absorbing at 420–580 nm, a part of them having structures of quinone imines **8** and **9**, including condensation products, e.g., **10** or **11**. Too strong acceleration in testing of stabilized rubber in oxygen bombs or ozone chambers results in too fast surface consumption of the diamines that cannot be replenished by slow migration of fresh diamines from the material bulk. The over-accelerated data indicate consequently lower antiozonant efficiency.

| 8 | 9 | 10 | 11 |

The attack of ozone on PO was considered more seriously only recently.[7,40] Results indicate participation of ozone in the initiation phase of PO oxidation and interaction with secondary and tertiary HAS and derived nitroxides after an electrophilic attack of ozone on the nitrogen of the HAS species resulting in an intermediary adduct accounting subsequently for (photo)ozonolysis of the HAS heterocycle and formation of acyclic degradation products, such as 2,6-dimethyl-2-hydroxy-6-nitroheptane.[40] This ozone scavenging by HAS depletes its photoantioxidant longevity in the surface/near surface layers of the HAS-stabilized material.

Atmospheric nitrogen oxides (NO$_x$) participate in the initiation phase of PO photooxidation[7] and are considered as responsible for gas fading (discoloration) of PO stabilized with phenolic antioxidants. Discoloration is the physical effect of a chemical depletion of the phenolic moiety, concerted with interruption of the CB activity. NO$_x$, nitrogen dioxide NO$_2$ in particular, transforms phenolic moieties in both antioxidants or UVAbs in photoassisted processes into NO$_2$-substituted cyclohexadienones **12-14**, photolyzing into substituted quinone methides (QM) e.g., stilbene qunone **15** or benzoquinones, such as **6**, all of them discoloring PO.[7] The actual CB efficiency of antioxidants or ESIPT mechanism in UVAbs is thus reduced.

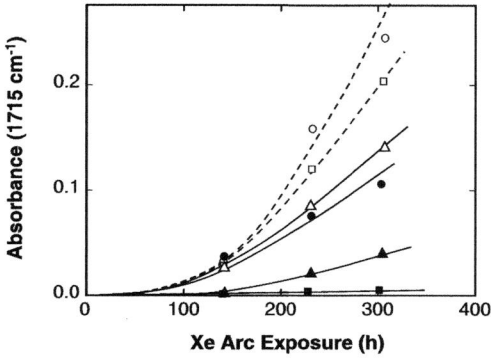

12   **13**   **14**   **15**

Atmospheric acid impurities[7] are a very dangerous deteriogen, particularly for pigmented automotive coatings. We consider here the influence of the impurities on the fate of stabilizers. It was evidenced in model experiments that a pretreatment of PP films stabilized with secondary HAS Tinuvin 770 with gaseous acid impurities reduces photostability of the matrix in weatherometer exposure at 38°C (*Figure 8*)[41] or after exposure in Q-Sun tester at 20°C.[42]

The prolonged exposure to acids increases the depleting effect on HAS. Weak acids like CO_2/H_2O or carboxylic acids do not deactivate HAS.

Atmospheric inorganic impurities are present in different amounts in various geographic sites and accordingly must be considered as a substantive factor responsible for differences observed in the service life of HAS-containing polymers.[40] The impurities may be reflected in the development of changes in oxidation-borne products or mechanical properties, different for unstabilized and HAS-containing materials exposed in different localities. For example, the course of photooxidation of unstabilized LDPE was essentially comparable after exposure either in Bandol (south France) or in Miami (Florida).[43] Differences in photooxidation were observed, however, between HAS-stabilized LDPE exposed in the two localities. The result can be explained by the contributions of "specific" environmental factors (atmospheric acid and oxidizing impurities) in the degradation of polymer and basic HAS (*Figure 9*).

Experimental results from Bandol field experiments using different model exposure modes of samples (opened or closed quartz or borosilicate glass vessels, direct weathering) performed with unstabilized LDPE, LDPE stabilized with a combination of UVAbs and Q, or LDPE containing oligomeric HAS Chimassorb 944 reveal[44] that proto-

Figure 8—Influence of the exposure of volatile acids on photooxidation of processing stabilizer free PP 30 μm thick films exposed at 30°C with Pyrex glass filtered xenon weatherometer. Stabilization: 0.15% Tinuvin 770. Films were exposed prior to irradiation at room temperature for 18 h to acid gases or vapors of aqueous acids. O— No HAS or acid treatment. HAS stabilized PP: ■— No inorganic acid or CO_2/H_2O, formic acid; ▲— SO_2/H_2O; ●—HCl; △—NO_2/H_2O; □—HBr.[41]

508

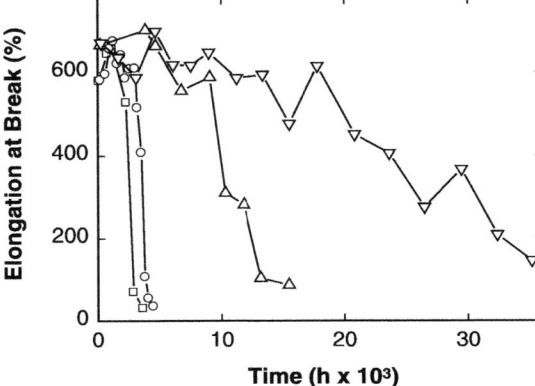

Figure 9—Effect of environmental parameters on degradation of LDPE 150 μm thick films during outdoor exposure in Bandol (south France, ❑, △) and in Miami (Florida, ○ , ▽). Unstabilized LDPE: ❑, ○ . Stabilization with 0.15% Chimassorb 944: △,▽. Failure criterion: elongation at break.[43]

cols on conditions of outdoor exposure and analyses of obtained results require more careful examination.

Efficiency and durability of stabilizers are seriously affected also by polymer-borne impurities, generated during thermal or photochemical degradation. Acids arising from thermolysis of PVC, acid transformation products of thiosynergists ("catalytic antioxidants"), or sulfur-containing PVC heat stabilizers, acid products from thermo-photolysis of brominated flame retardants (FR)[6,7,42,45] play a specific role together with acid fillers or pigments.

Activity of these compounds depletes basic stabilizers, HAS. Some experimental results are mentioned. According to reference 42, aromatic FR decabromodiphenylether used at concentration 4% initiated photooxidation of PP films in the Q-Sun tester (Figure 10). This effect cannot be eliminated by the addition of Chimassorb 944. The photolytic instability of the aromatic additive is considered to be the principal problem. The FR photolyses[42] under formation of atomic bromine, that converts the hydrocarbon substrate into carbon-centered polymeric alkyl radical P· (a photoinitiation process) prone to photooxidize, and hydrobromic acid, a strong deteriogen of HAS and of the related nitroxide. Aliphatic brominated FR, such as tris(tribromoneopentyl)phosphate is less dangerous because of its higher photostability and the photoantioxidant effect of HAS can be better exploited.[42] The negative effect of photolyzing brominated FR can be reduced by increased concentration of HAS in the system.[45] This indicates participation of a stoichiometric reaction between HBr

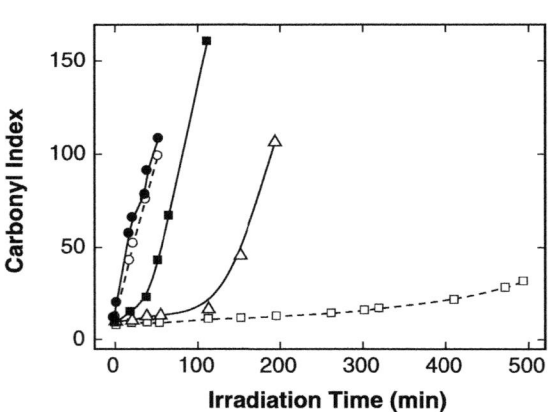

Figure 10—Photooxidation of 0.5 mm PP sheets containing decabromodiphenylether (I) or tris(tribromoneopemtyl)phosphate (II) as fire retardants. Exposure with Q-Sun xenon device at 65°C. Base stabilization: 0.1% of 1:1 mixture of Irganox 1010 with Irgafos 168. HAS: Chimassorb 944. △ neat PP, ●—4% I,■—4% I + 0.5 % HAS, ○ —4% II, ❑— 4% II + 0.5% HAS.[42]

developed during photolysis of the FR and HAS species. (Another improvement in the interaction between brominated FR and HAS based on chemical modification of HAS is mentioned later.)

The basicity of the secondary and tertiary HAS is considered to be the principal reason for their deactivation by acid polymer components. Values pK_a of the two classes of HAS are in the range 7.5 to 9.7. The related nitroxides have also high basicity.[5,6] This explains the negative effect of the environmental acidity on weatherometer exposed PP films stabilized with bis(2,2,6,6)tetramethyl-4-piperidinyl-*N*-oxyl[41] (*Figure* 11).

Formation of salts **16** (R=H, alkyl, X=anion of an inorganic or strong organic acid) from basic secondary or tertiary HAS hinders formation of nitroxides, the key intermediate in their stabilization mechanism. Salts **17** formed from nitroxides hinder their participation in HAS regenerative activity cycle.[5,6,40]

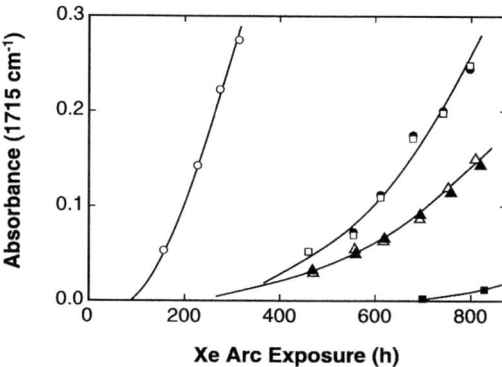

Figure 11—Influence of the exposure of volatile acids on photooxidation of processing stabilizer free PP 30 μm thick films exposed at 30°C with Pyrex glass filtered xenon weatherometer. Stabilization: 0.11% bis(2,2,6,6-tetramethyl-4-piperidinyl-N-oxyl)dodecane dioate. Films were exposed prior to irradiation at room temperature for 18 h to acid gases or vapors of aqueous acids. ○—Nitroxide-free PP, no acid treatment. Nitroxide stabilized PP: ■—No inorganic acid or exposure to CO_2/H_2O, formic acid; ▲—SO_2/H_2O; ●—HCl; △—NO_2/H_2O; ❑—HBr.[41]

$$\left[>N^+<{}^R_H X^-\right] \qquad \left[>N^+ = OX^-\right]$$

16   **17**

A part of the problem accounting for acid impurities was solved by development of "non-reactive" or "acid resistant" HAS having structures of *O*-alkylhydroxylamines (>NOR) and *N*-acylamines (>NCOCH_3), respectively[5-7] having lower pK_a values (4.2–4.4 and 2.0, respectively) and good applicability in plastics and coatings. A more efficient photostabilization of PP films fire retarded with decabromodiphenylether using the >NOR HAS Tinuvin 123 in comparison with tertiary HAS Tinuvin 765 is shown[45] in *Figure* 12. The problem of basicity of nitroxides derived from the both types of HAS remains, however.

ENVIRONMENTAL HYDROLYSIS OF STABILIZERS CONTAINING ESTER MOIETY

There is a potential danger of the effect of the environmental humidity or of a long-term contact of stabilized polymers with water (as with hot water pipes) on transformation/loss of propionate-type phenolic antioxidants and organic phosphites. Model exper-

510

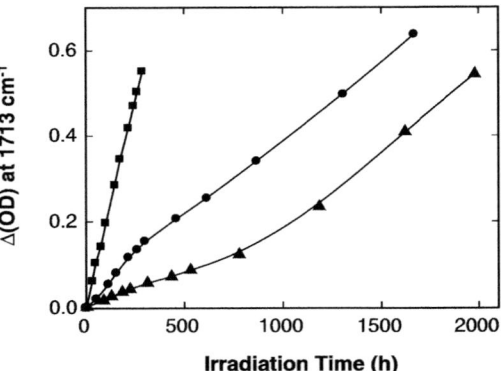

Figure 12—Influence of N-substitution (and related basicity) in the HAS molecule on photooxidation of 60 μm thick PP films fire retarded with 20% of decabromodiphenylether and 7% of Sb_2O_3. ■—No HAS; ●—1 % Tinuvin 765 + 5% TiO_2; ▲—Tinuvin 123 + 5% TiO_2.[45]

iments indicate a possibility of hydrolysis of phenols.[46] Free carboxylic acid **18** and alcohol $C_{18}H_{37}OH$ are formed from Irganox 1076 together with QM **19** and lactone **20**. Hydrolysis of polynuclear phenols such as Irganox 1010 is a gradual process resulting in **18–20** and phenolic products **21** of a partial hydrolysis. Pentaerythritol **22** arises in the ultimate step. Hydrolysis increases solubility of fragments in water. Losses by leaching into the environment reduce the level of the long-term protection of the material.

Fortunately, the hydrolysis of phenols embedded in the non-polar PO matrix is a restricted process. In spite of this, recent data[47] indicate an almost 6% loss of the originally added Irganox 1010 from medium density polyethylene films after four years of exposure to water. An increased (surface) hydrolysis can be expected in acid environment (acid rain).

Besides the formation of inactive salts, strong acid environment can also contribute to hydrolysis of ester-type HAS and formation of volatile fragments.[40]

Esters of trivalent phosphorus (phosphites and phosphonites), aliphatic and aryl-aliphatic in particular, hydrolyze in the presence of humidity.[6] Various acid intermediates are formed, phosphorous acid in the terminal step, corrode metallic parts of the processing equipment and deactivate HAS. The hydrolysis is catalyzed by acid environment. Sterically hindered aryl esters, a common component of PO processing package, are hydrolysis resistant. The hydrolysis itself is not typically accompanied by a loss of hydroperoxide decomposing antioxidant efficiency. However, volatile phenolic fragments and alcohols are released and may affect the environment.

PHOTOOXYGENATION OF STABILIZERS VIA EXCITED PIGMENTS

Pigments influence the fate of stabilizers by surface adsorption lowering their mobility and increasing heterogeneity in distribution of stabilizers in the matrix, by catalytic degradation of stabilizers, by increasing surface temperature of colored samples and enhancing thermal superposition over radiation effects, by generation of singlet molecular oxygen $^1O_2$, or by a direct interaction of the excited pigment with the phenolic moiety changing phototriggered and oxygen assisted processes in the integral stabilized system.[6,10,27] Formation of $^1O_2$ and its effect on chemistry of stabilizers containing phenolic moieties (antioxidants and UVAbs) was studied in detail.[10] Colorants (pigments and dyes) of the π,π^*-type supply excited electrons from the π orbital. Excited S_1 state of the colorant generates after ISC longer-living triplet state T_1 that participates in formation of $^1O_2$.

Reactive quenching of $^1O_2$ by phenolic moieties in antioxidants and UVAbs is postulated to have some free radical or transanular peroxidic intermediates and accounts in formation of a photooxygenation product having a general structure of hydroperoxycyclohexadienone [equation (2)]. The phenolic moiety can be also transformed by H-abstraction from the phenolic HO group by excited pigment via a phenoxyl into a QM [equations (3)] and in a series of concerted processes of the sensitized pigment with the polymeric substrate (PH) in the presence of oxygen into polymeric alkylperoxyl [equation (4)] and in alkylperoxycyclohexadienone in the final step [equation (5)]. General structures of the hydroperoxycyclohexadienone 23, QM 24, and alkylperoxycyclohexadienone 25 are shown ("Subst" in 23–25 means a residual part of the original antioxidant or UVAbs molecule). This indicates very complex involvement of colorants (and other effective chromophores) in the fate of stabilizers.

$$(2)$$

$$(3)$$

$$PH + S^* \longrightarrow {}^\cdot SH + P^\cdot \xrightarrow{\;O_2\;} POO^\cdot$$

$$(4)$$

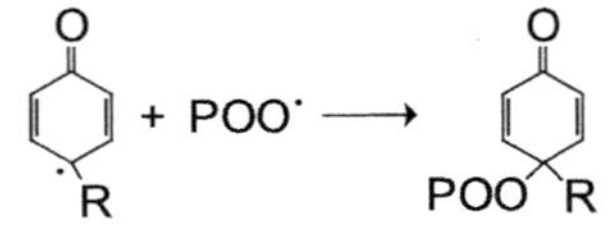

(5)

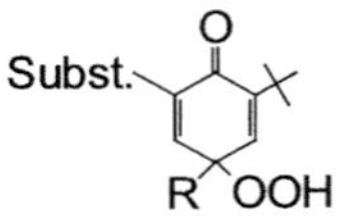

23 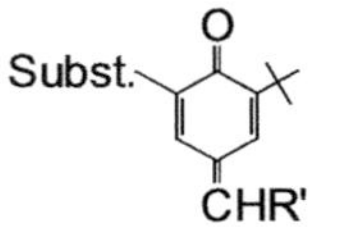   24 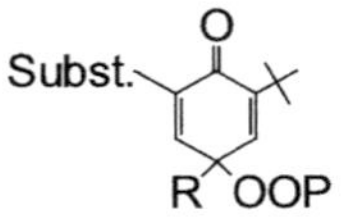   25

The principal structural types of transformation products **23–25** arising from sensitized photooxygenation have a specific influence on photooxidation of the polymer matrix as explained by model experiments. Thermolysis or photolysis of peroxidic cyclohexadienones XOO-CHD (**23** or **25**, X=H, R) accounts for homolytical formation of radical species. They have properties of initiators of oxidation of hydrocarbon substrates. Photolysis of 2,6-di-*tert*.butyl-4-*tert*.butylperoxy-2,5-cyclohexadiene was determined under the exposure with 333 and 436 nm light.[48] The homolysis of the peroxidic group is sensitized by the unsaturated carbonyl group system in conjugated cyclohexadiene molecule. Thermal and photochemical free radical homolysis of various XOO–CHD was evidenced by electron spin resonance (ESR).

The homolysis of XOO–CHD has two consequences[10]: (1) Formation of photostable, mostly low molecular weight compounds such as **15, 26–29** considered as impurities in the matrix; (2) Strong photoinitiation effect on photooxidation of hydrocarbon substrates: trimethylcyclohexane, squalane, atactic PP, or isotactic PP. In all experiments, alkylperoxycyclohexadienones of the general type **25** have a stronger photoinitiation effect at comparable concentration level than the hydroperoxyderivatives **23**.

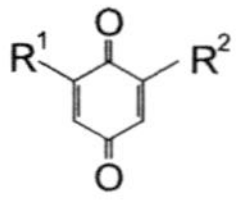

26 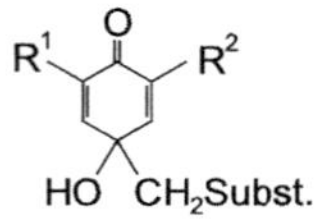   27 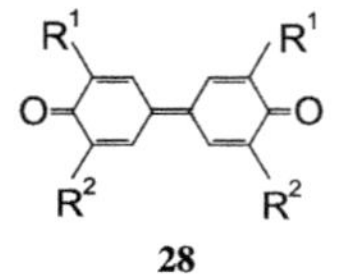   28

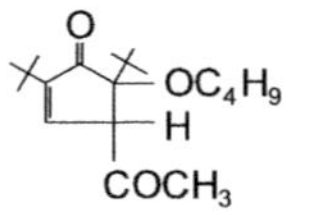

29 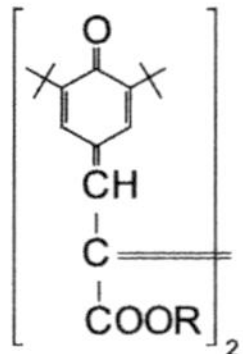   30

A very specific effect accounts for formation of various QM, e.g., **15**, general structure **24**, or diphenoquinone **28** arising either by a direct transformation of phenolic moieties by excited sensitizers or by photolysis of primary products of photooxygenation of phenolics. Comparable structures of QM are formed as common products of scavenging POO˙ radicals in the sacrificial CB antioxidant process.[49] QM, particularly those having high conjugated system of double bonds [e.g., stilbenequinone **15** or dimeric QM **30** (R=$C_{18}H_{37}$) derived from Irganox 1076] strongly discolor PO.[50] In any case, photooxygenation due to excited colorants has a depleting effect on efficiency of phenolic stabilizers.

HETEROGENEITY IN STABILIZER TRANSFORMATION/ CONSUMPTION

The heterogeneous character of processes proceeding in stabilized photooxidized/ weathered polymers has various causes: polymer morphology of semicrystalline materials and their geometry; spectral sensitivity of individual components of the material; localization of chromophores; defect structures or photoinitiating species; spatial distribution of oxidation-borne chemical polymer transformation products; physical concentration gradients in the matrix (internal stress); depth and intensity of radiation penetration into the material; distribution of stabilizers within the matrix or distribution of other additives (e.g., fillers or pigments) and their light scattering effect. Another factor contributing to heterogeneity of stabilization and degradation processes can be oxygen penetration into the polymer (DLO accounting for heterogeneous distribution of oxygen in the matrix, including oxygen deficiency localized in deeper layers of the matrix and influencing the relative participation of individual reactions of the general oxidation mechanism autoreactions of polymeric carbon centered free radicals in particular).

The individual causes of the heterogeneity do not remain static during photodegradation. Some develop with exposure time. For example, crystallinity in semicrystalline PO increases (chemorecrystallization) in surface layers after UV irradiation.[51] The changes in crystallinity degree of PP are influenced by the stabilizer as found by analyzing the differences between the effect of a hindered phenol Irganox 1010 and HAS Chimassorb 944.[20] Density of polyethylenes increases after exposure in weatherometer Atlas Ci 65 or to fluorescent lamp UVB-313. The changes are evidently accelerated by radiation intensity.[52] Concentration gradients of changes in surface chain scission of PP, crosslinking and related relation between chain scission/crosslinking,[53] growth in oxidation products,[54] weight average molecular weight in PP or the effect of stabilization by low-molecular weight and oligomeric HAS on the Mw dept profile in toughened grade PP[55] are examples.

Photochemical changes of the matrix affect spatial kinetics of oxidation as revealed in recent analyses. For example, high surface crosslinking in unsaturated diene-based polymer creates a barrier hindering oxygen permeation into deeper layers of the polymer during aging. On the other hand, the efficient surface stabilization limiting near-surface photodegradation by chemorecrystalization or crosslinking may allow more oxygen to penetrate into deeper layers of the materials.[56]

The formation of concentration gradients of products in degrading polymers occurs in concert with transformation/consumption of stabilizers. An extensive study was performed with HAS in PP and polystyrene (PS) plaques.[8,40] HAS have a characteristic but complicated transformation chemistry in active polymer protection. HAS-related nitroxides are formed in the primary activity step when deactivating polymeric hydroperoxides POOH and/or peroxyradicals POO·. The next step contains scavenging of macroalkyls

P'and cyclic regeneration of the nitroxide from O-alkylhydroxylamine $>NOP$ by reaction with alkylperoxyl POO˙ (*Scheme* 3).[5,6,40]

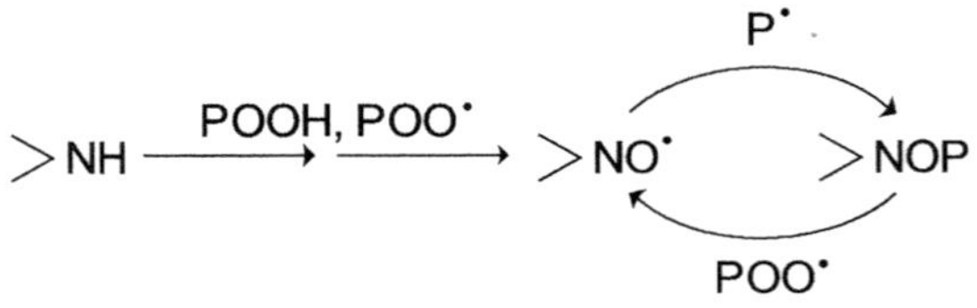

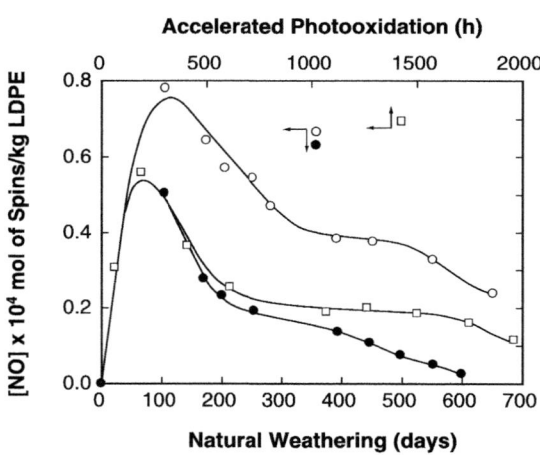

Scheme 3—Cyclic regenerative mechanism of HAS.

This mechanism is manifested in the stabilized matrix by an increase in nitroxide concentration as determined by ESR and a gradual formation of a nitroxide steady state concentration after reaching the maximum as found in HAS stabilized plastics and coatings.[8,12] The increased radiation harshness results in generation of higher concentration of nitroxides as found in acrylate isocyanate two-pack PUR exposed to a QUV device equipped with lamp UVB-313 or to a xenon Atlas Weather-Ometer[29] (*Figure* 13). The gradual decrease in nitroxide concentration was confirmed in epoxy-acid clearcoats,[9] acrylate-melamine clearcoat[57] or in experiments with photooxidized PP films doped with 2-hydroxy-2,2,6,6-tetramethylpiperidinyloxyl.[58]

Experiments with thick plaques of PP or PS stabilized with Tinuvin 770 or Tinuvin 123 irradiated in weatherometer Atlas Ci 3000+ clearly indicated formation of nitroxides on irradiated and non-irradiated sides of the sample (the amount of nitroxides was calculated for the normalized volume of the polymer sample cut from the exposed plaque and used for ESRI measurement).[8] Using the ESRI technique, the spatial distribution of nitroxides having U-shape profile with the highest concentration in surface areas was shown. The detailed analysis of the data revealed differences in the transformation fate of HAS Tinuvin 770 in PP and PS plaques exposed to weatherometer at 30°C (*Figure* 14). The reason is that the two polymers differ in morphology, with thermal sensitivity accounting for differences in nitroxide formation on irradiation and non-irradiation sides of the plaques, radiation transparency, and photosensitivity[8] (*Figure* 14). Comparative measurement with the same systems under thermal stress in an oven at 60°C confirmed unequiv-

Figure 13—Development in nitroxide formation in LDPE films stabilized with 1:1 combination of Chimassorb 944 with Tinuvin 622 exposed to natural weathering (○, ●) or accelerated photooxidation in mercury lamp equipped device SEPAP 12-24 (▢). Concentration of the combination of HAS: 0.2 (●), 0.6% (○, ▢).[12]

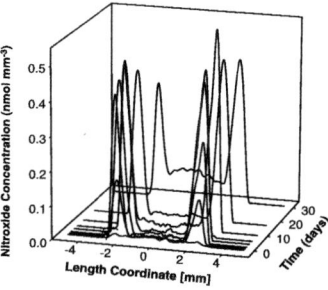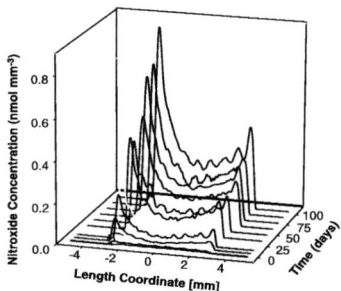

Figure 14—Dependence of nitroxide concentration profile measured by ESRI in PP (a) and PS (b) 6 mm thick plaques stabilized with 1% Tinuvin 770 as a sole additive on net exposure time with accelerated photooxidation with Atlas Weather-Ometer Ci 3000+.

ocally by determination of nitroxides the heat stabilizing effect of HAS in polymer matrix sensitive to thermal oxidation, i.e., PP in comparison with low sensitive PS.[8]

ESR measurements of nitroxide formation detect only one part of the stabilization cycle of HAS, including spatial distribution of nitroxides in the materials. The method indicates the steady state concentration of nitroxides in long-term exposed samples. At the same time, the method is able to detect if there is still some residual HAS-derived active species present in the material able to be converted into nitroxide. This analytical approach was confirmed in acrylate/isocyanate two-pack PUR clearcoat stabilized originally with Tinuvin 765 in combination with other stabilizers after five years' exposure in Florida.[29] The data on the stationary concentration of nitroxides itself do not provide information about the residual amount of HAS in the system. The important information can be obtained by ESR determination of the total amount of nitroxides converted from all HAS-active deposit forms present in the sample by p-nitroperbenzoic acid oxidation ("total HAS-based nittroxides") (*Figure 15*).[9,57]

The ESR method also confirmed the gradient in surface consumption of Tinuvin 123 in various clearcoats analyzed in mechanically separated layers. The heterogeneous HAS consumption indicates a more intensive

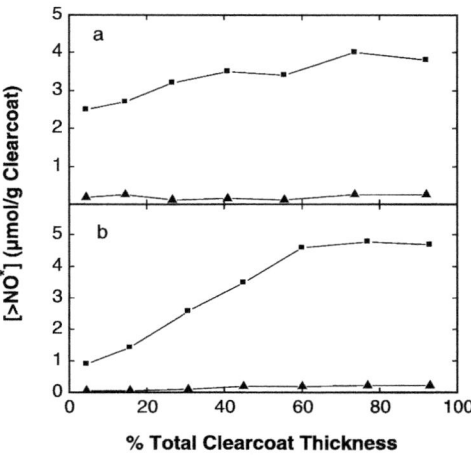

Figure 15—Steady state nitroxide concentration—▲ and "total HAS-based nitroxide concentration"—■ determined by ESR as a function of the depth of epoxy acid clearcoat over white basecoat paint system (a) or dark red paint system (b). Stabilization of the clearcoat: 0.3% Tinuvin 123 with 2.7% Tinuvin 328 for system (a), and with additional 3.0 Tinuvin 328 in the basecoat for system (b). Material analyzed after four years of Florida exposure.[9]

516

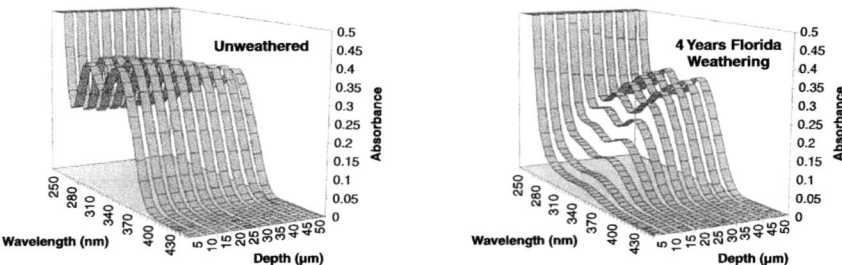

Figure 16—Concentration loss profile of UVAbs from acrylic/ melamine clearcoat stabilized with 1.9% of Tinuvin 234 over brown basecoat containing 1% of Tinuvin 328 determined by ultraviolet micro-spectroscopy at 5 μm increments through the clearcoat. (a) Unweathered sample; (b) Sample after four years' Florida exposure.[59]

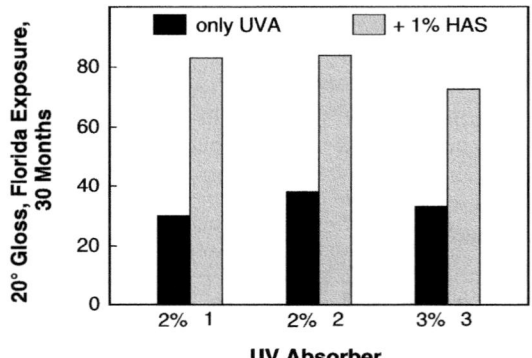

Figure 17—Effect of a combination of UVAbs with 1% HAS Tinuvin 765 on gloss retention of a HS two-pack polyure-thane clearcoat over waterborne silver metallic basecoat (cured for 30 min at 90°C) upon Florida exposure. UVAbs: 1—Tinuvin 384, 2—Tinuvin 400, 3—Sanduvor 3206.[29]

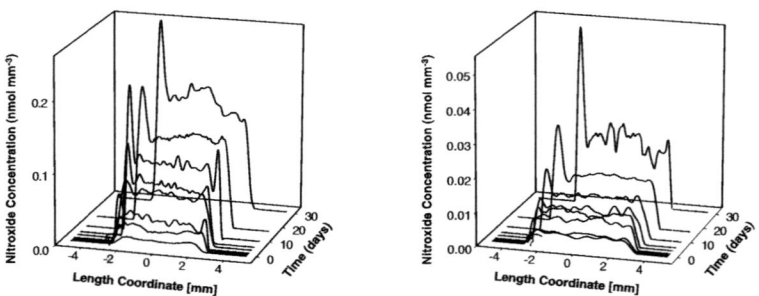

Figure 18—Dependence of nitroxide concentration profiles determined by ESRI in PP (a) and PS (b) 6 m plaques stabilized with 1% Tinuvin 770 and 0.5% Tinuvin 327 on net exposure time to accelerated photooxidation in Atlas Weather-Ometer Ci 3000+.

depletion of the HAS-active species in surface and near surface layers and reflects the integral composition of the stabilizer package. Compounds like **31–33**[5,6] are formed as stabilization inactive transformation products of HAS in terminal phases of its lifetime and accumulate in the polymer matrix.

Consumption gradients are characteristic also for UV absorbers used in organic substrates. UV-microspectroscopic analysis of nm increments of stabilized acrylic-melamine clearcoats exposed in Florida demonstrated the gradual loss of UVAbs (the data can reflect a combination of chemical and physical losses) (*Figure* 16).[59] Analogous UVAbs loss-gradients were found in acrylic-melamine clearcoats stabilized with a combination of phenolic benzotriazole-based UVAbs with Tinuvin 123.[60]

Generally, the changes indicating involvement of stabilizers in the photooxidized matrix and the transformation/consumption of light stabilizers can be demonstrated by U-shaped curves. Their shape and symmetry are influenced by sample thickness, sensitivity of the matrix to thermo-/photooxidation, effect of DLO, and radiation penetration.[56]

A combination of two stabilizers, UVAbs and HAS, is used commercially for a complex effective stabilization and exploits the difference in the mechanism of the two stabilizer classes.[6,7] Moreover, UVAbs protect HAS and derived nitroxides from photolysis. Conversely, the efficiency and longevity of UVAbs is increased by HAS by protecting the ESIPT mechanism from interruption by phototriggered and free-radical assisted processes (*Figure* 17).[29]

The combination of HAS with UVAbs has a specific influence on the U-shaped spatial distribution of nitroxides in PP and PS plaques exposed to Atlas Ci 3000+ Weather-Ometer. The combination of Tinuvin 770 and Tinuvin 327 was used (*Figure* 18).[8] The protection of the non-irradiated surface by the presence of UVAbs limiting light penetration through the sample is explicitly indicated, together with efficient integral stabilization.

CONCLUSIONS

- Chemical processes at molecular level are the principal factors responsible for the development of physical/mechanical changes and the durability of the stabilized polymer system.

- The rate and extent of the processes are affected by the environmental stress of common deteriogens, by rather unpredictable influences of casual pollutants/impurities and different polymer-borne impurities and photoactive components (additives) in the system.

- Stabilizers are the minority but the most reactive species in the polymer matrix. Consequently, their longevity and performance are more seriously affected by the harshness and complexity of the environmental stress than the matrix itself.

- Degradation/weathering of polymers and related consumption of stabilizers are not a static or homogeneous process. Heterogeneity of the matrix itself, heterogeneity due to diffusion limited oxidation, radiation penetration or temperature superposition affect material properties are sources of problems in simulation of aging of stabilized systems.

ACKNOWLEDGMENT

Financial support by grants FT-TA-2/048 from the Ministry of Industry and Trade of the Czech Republic, grant 106/06/0761 from the Grant Agency of the Czech Republic and grant 2B06097 from the Ministry of Education, Youth and Sports, and technical assistance of Ms. D. Dundrová are gratefully appreciated.

References

(1) Kumar, A., Commereuc, S., and Vernay, V., *Polym. Degrad. Stab.*, 85, 751-757 (2004).
(2) Zweifel, H. (Ed.), *Plastics Additives Handbook*, Hanser Publishers, Munich, 2001.
(3) Pospíšil, J., Horák, Z., Pilar, J., Billingham, N.C., Zweifel, H., and Nešpurek, S., *Polym. Degrad. Stab.*, 82, 145-152 (2003).
(4) Pospíšil, J. and Nešpurek, S., Keynote Lecture at the 4th Modification, Degradation and Stabilization of Polymers Conf., San Sebastian, 10–14 September 2006.
(5) Pospíšil, J., *Adv. Polym. Sci.*, 124, 87-189 (1995).
(6) Pospíšil, J. and Nešpurek, S., "Highlights in the Inherent Chemical Activity of Stabilizers," In: *Handbook of Polymer Degradation*, 2nd Ed., Hamid, H.S. (Ed.), Marcel Decker, New York, pp. 191-276, 2000.
(7) Pospíšil, J. and Nešpurek, S., *Prog. Polym. Sci.*, 25, 1261-1335 (2000).
(8) Marek, A., Kaprálková, L., Schmidt, P., Pfleger, J., Humlícek, J., Pospíšil, J., and Pilar, J., *Polym. Degrad. Stab.*, 91, 222-458 (2006).
(9) Gerlock, J.L., Kucherov, A.V., and Smith, C.A., *Polym. Degrad. Stab.*, 73, 201-210 (2001).
(10) Pospíšil, J., Nešpurek, S., Zweifel, H., Pilar, J., Rakušan, J., and Karásková, D., Lecture at the 4th Conf. on Modification Degradation and Stabilization of Polymers, San Sebastian, 10-14 September 2006.
(11) Pickett, J.E., "Permanence of UV Absorbers in Plastics and Coatings," In: Martin, J.W. and Bauer, D.R. (Eds.), *Service Life Prediction: Methodology and Metrologies*, ACS Symp. Ser. 805, 250-265 (2002).
(12) Scoponi, M., Simmino, C., and Kaci, M., *Polymer*, 41, 7969-7980 (2000).
(13) White, J.R. and Rappoport, N.Y., *Trends Polym. Sci.*, 2, 197-202 (1994).
(14) Spatz, G., *Polym. Test.*, 15, 381-395 (1996).
(15) Ciba-Geigy Publication No 28622/1/e, Basel 1992.
(16) Kovárová, J., Rosík, L., and Pospíšil, J., *Polym. Mater. Sci. Eng.*, 58, 215-219 (1988).
(17) Drake, W.O., Lecture 14th International Conf. on Advances in the Stabilization and Degradation of Polymers, Luzern, *Proceed.*, pp 57-73, June 25, 1992.
(18) Zweifel, H., *Stabilization of Polymeric Materials*, Springer Verlag, Berlin 1998.
(19) Bernstein, B.S. and Lee, P.N., Lecture 20th International Wire and Cable Symp., Cherry Hill, NJ, *Proceed.*, pp. 202-208, Nov. 1975.
(20) Schwarzenbach, K., Gilg, B., Knobloch, G., Pauquet, J.-R., Rota-Graciani, P., Schmitter, A., Zingg, J., and Kramer, E., "Antioxidants," In: *Plastics Additives Handbook*, Zweifel, H. (Ed.), Hanser Publishers, Munich, pp. 1-139, 2001.
(21) Billingham, N.C., Fearon, P., Whiteman, D.J., Marshall, N., and Bigger, S.P., Lecture 6th International Plastics Additives and Modifiers Conf., Addcon World 2000, October 24-26, 2000, Basel; *Proceed.*, paper 7.

(22) Dudler, V., Bole, T., and Rytz, G., *Polym. Degrad. Stab.*, 60, 351-365 (1998).
(23) Hamskog, M., Terzelius, B., and Gijsman, P., *Polym. Degrad. Stab.*, 82, 181-186 (2003).
(24) Rabek, J.F., *Photodegradation of Polymers*, Springer Verlag, Berlin 1996.
(25) Pickett, J.E., *Polym. Degrad. Stab.*, 85, 681-687 (2004).
(26) Real, L.P., Gardette, J.-L., and Rocha, A.P., *Polym. Degrad. Stab.*, 88, 357-362 (2005).
(27) Pospíšil, J., Pilar, J., Marek, A., Nešpurek, S., and Horák, Z., "Weathering of Polymers: Problems Encountering Weathering Assessment as a Tool for Lifetime Prediction," In: *Natural and Artificial Ageing of Polymers*, Reichert, T. (Ed.), Gesellschaft für Umweltsimulation Pfinztal, pp. 209-216, 2005.
(28) Pospíšil, J., *Adv. Polym. Sci.*, 36, 69-133 (1980).
(29) Valet, A., *Light Stabilizers for Paints*, C. R. Vincentz Verlag, Hannover 1997.
(30) Chakraborty, K.B. and Scott, G., *Eur. Polym. J.*, 15, 35-40 (1979).
(31) Pospíšil, J., "Photooxidation Reactions of Phenolic Antioxidants," In: *Developments in Polymer Photochemistry*, Vol. 2, Allen, N.S. (Ed.), Applied Science Publishers, London, pp. 53-133, 1981.
(32) Gerlock, J.L., Tang, W., Dearth, M.A., and Korniski, T.J., *Polym. Degrad. Stab.*, 48, 121-130 (1995).
(33) Sedlár, J. Petruj, J., and Pác, J., Lecture 11th Prague Microsymposium Mechanisms of Inhibition Processes in Polymers, September 4-7, 1972, Paper E4.
(34) Ghiggino, K.P., Scully, A.D., and Bigger, S.W., *ACS Symp. Ser.*, 381, 57-79 (1988).
(35) Valet, A. and Decker, C., *Mod. Paint. Coat.*, 91, 29-37 (2001).
(36) Satoto, R., Subowo, W.S., Yusiasih, R., Takane, Y., Watanabe, Y., and Hatabayara, T., *Polym. Degrad. Stab.*, 56, 275-279 (1997).
(37) Jacques, L.F.E., *Prog. Polym. Sci.*, 25, 1337-1362 (2000).
(38) Scott, K.P., ATLAS SunSpots 26 (53), 1-5 (1996).
(39) Pospíšil, J., "Aromatic Amine Antidegradants," In: *Developments in Polymer Stabilization*, Vol 7, Scott, G. (Ed.), Elsevier Applied Science Publishers, London, pp. 1-63, 1984.
(40) Pospíšil, J., Pilar, J., and Nešpurek, S., Lecture 17th International Plastics Additives and Modifiers Conf. Addcon World 2006, Cologne, October 17-18, 2006, Proc. 23/1-18.
(41) Carlsson, D.J., Zhang, C., and Wiles, D.M., *J. Appl. Polym. Sci.*, 33, 875-884 (1987).
(42) Antoš, K. and Sedlár, J., *Polym. Degrad. Stab.*, 90, 180-187 (2005).
(43) Gijsman, P. and Sampers, J., *Angew. Makromol. Chem.*, 262, 77-82 (1998).
(44) Sampers, J., *Polym. Degrad. Stab.*, 76, 455-465 (2002).
(45) Sinturel, Ch., Lemaire, J., and Gardette, J.-L., *Eur. Polym. J.*, 36, 1431-1443 (2000).
(46) Bertoldo, M. and Ciardelli, F., *Polymer*, 45, 8751-8759 (2004).
(47) Haider, N. and Karlsson, S., *J. Appl. Polym. Sci.*, 85, 974-988 (2002).
(48) Lerchová, J., Kotulak, L., Rotschová, J., Pilar, J., and Pospíšil, J., *J. Polym. Sci.*, Symposia, 57, 229-235 (1976).
(49) Pospíšil, J., Nešpurek, S., and Zweifel, H., *Polym. Degrad. Stab.*, 54, 7-14, 15-21 (1996).
(50) Pospíšil, J., Habicher, W.D., Pilar, J., Nešpurek, S., Kuthan, J., Piringer, G.O., and Zweifel, H., *Polym. Degrad. Stab.*, 77, 531-538 (2002).
(51) Rabello, M.S. and White, J.R., *J. Appl. Polym. Sci.*, 64, 2505-2517 (1997).
(52) Gulmine, J.V., Janissek, P.R., Heise, H.M., and Akcelrud, L., *Polym. Degrad. Stab.*, 79, 385-397 (2003).
(53) Shyichuk, A.I., White, J.R., Craig, I.H., and Syrotynska, I.D., *Polym. Degrad. Stab.*, 88, 415-419 (2005).
(54) Gonon, L., Vasseur, O.Y., and Gardette, J.-L., *Applied Spectroscopy*, 53, 157-163 (1999).
(55) Turton, T.J. and White, J.R., *Polym. Degrad. Stab.*, 74, 559-568 (2001).
(56) Pospíšil, J., Pilar, J., Billingham, N.C., Horák, Z., and Nešpurek, S., *Polym. Degrad. Stab.*, 91, 417-422 (2006).
(57) Kucherov, A.V., Gerlock, J.L., and Matheson, R.R., *Polym. Degrad. Stab.*, 69, 1-9 (2000).
(58) Pan, J.Q. and Yan, S., *Polym. Degrad. Stab.*, 32, 79-83 (1991).
(59) Smith, C.A., Gerlock, J.L., and Carter, R.O., *Polym. Degrad. Stab.*, 72, 89-97 (2001).
(60) Cliff, N., Kanonni, M., Peters, C., Yaneff, P.V., and Adamsons, K., *J. Coat. Technol. Res.*, 2, 371-387 (2005).

APPENDIX

Formulae of commercial stabilizers reported in the text:

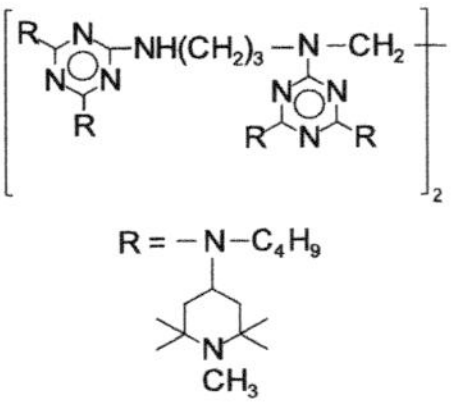

$R = -N-C_4H_9$

Chimassorb® 119

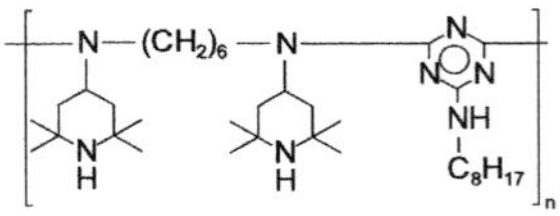

Chimassorb® 944

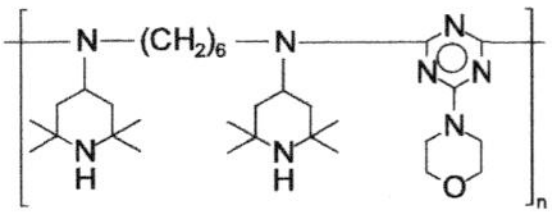

Cyasorb® UV 3346

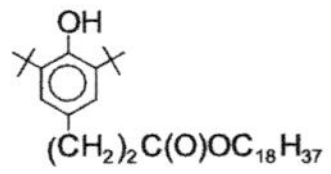

$(CH_2)_2C(O)OC_{18}H_{37}$

Irganox® 1076

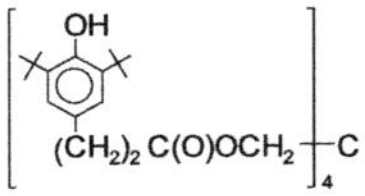

$[(CH_2)_2 C(O)OCH_2]_4 C$

Irganox® 1010

$P[O-]_3$

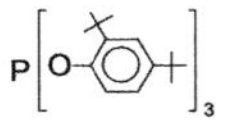

Irgafos® 168

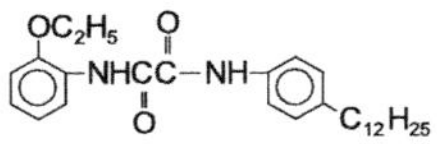

Sanduvor® 3206

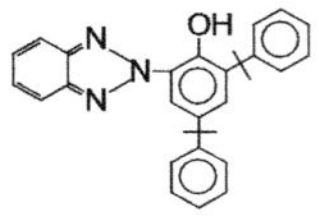

Tinuvin® 234

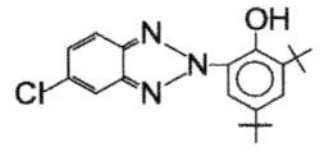

Tinuvin® 327

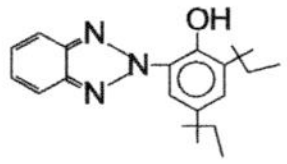

Tinuvin® 328

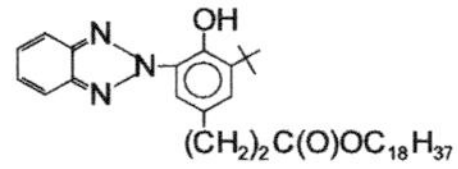

$(CH_2)_2C(O)OC_{18}H_{37}$

Tinuvin® 384

$OCH_2CH(OH)CH_2OC_{12}H_{25}/C_{13}H_{27}$

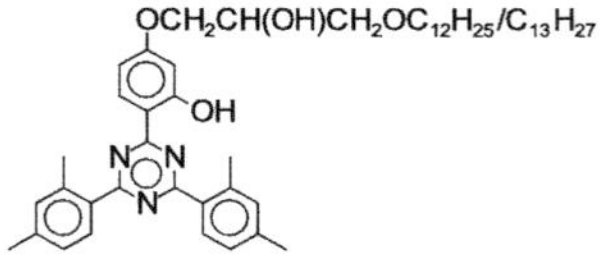

Tinuvin® 400

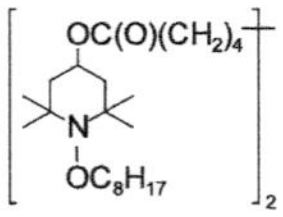

Tinuvin® 123

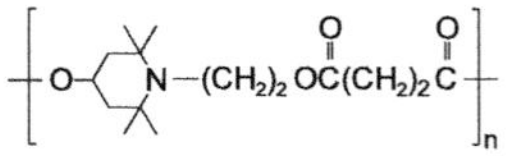

Tinuvin® 622

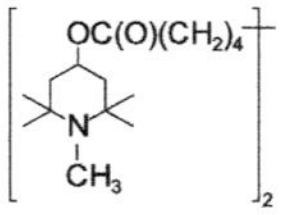

Tinuvin® 765

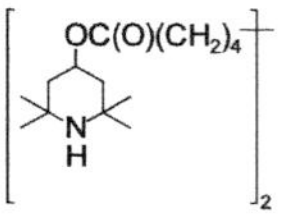

Tinuvin® 770

Wrap-Up: Implementation

Chapter 33

Implementing What We Have Learned

F. Louis Floyd

FLF Consulting, P.O. Box 31208, Independence, OH 44131

The 2006 Service Life Prediction (SLP) conference in Key Largo, FL, was the fourth in a series of international conferences addressing issues related to predicting the service life of polymeric materials. As has been our tradition, the Friday morning session was devoted to a review of the week's activities plus a look forward to consider how we might collectively start implementing what we have learned from these conferences. This chapter attempts to capture the essence of that session, and therefore departs from the strict scientific format of other papers in this book.

SUMMARY OF 2006 CONFERENCE

Attendees

The 81 attendees at this conference came from the USA, Canada, Sweden, Germany, France, UK, Australia, Japan, and the Czech Republic; and represented the following companies or organizations:

• GOVERNMENT LABS AND AGENCIES: NIST, USDA Forest Products Laboratory, Swedish National Testing & Research, Sandia National Labs, CISIRO Land & Water (Australia), Federal Institute for Materials Research & Testing (Germany), Fraunhofer Society (Germany), SP (Sweden)

• ACADEMIA: North Dakota State University, Drexel University, Iowa State University, Waseda University (Japan), Institute of Macromolecular Chemistry of Academy of Sciences (Prague, Czech Republic), Yale University, Laboratory de Photochimie (France), University of Newcastle Upon Tyne (UK)

• ASSOCIATION LAB: Paint Research Associates (UK)

• COATINGS MANUFACTURERS: Nippon Paint, Valspar, Sherwin-Williams, Carboline, DuPont, Rohm and Haas

• SUPPLIERS TO COATINGS MANUFACTURERS: BASF, Chevron-Phillips Chemical Co., DuPont, Cytec Surface Specialties, GE Silicones, Bayer MaterialScience, Sun Chemical, Noveon, Exatec LLC (Canada), Lyondell, Dow Chemical, Carisle SynTec, 3M, Arkema, Atlas Materials Testing Technology LLC, Q-Lab Corp

- END-USERS OF COATINGS: Philip Morris USA, Jeld-Wen Inc., BlueScope Steel (Australia), Ford Motor Co.
- OTHERS: Consultants, retirees, *Journal of Coatings Technology and Research*, *JCT CoatingsTech* (FSCT)

Sense of Community

As mentioned in the last *Service Life Prediction* book, the people attending these conferences have developed a strong sense of community, are collaborating with each other on larger projects, and are becoming more assertive regarding the implementation of what the community has learned. Several first-time attendees commented on the abundance of learning that had obviously already occurred.

This community recognizes that it is very unlikely that we will accomplish a single test for service life prediction. All our past progress has been a series of single steps forward, and there is no reason to expect this to be any different. *As a result, this community sees considerable merit in moving forward* now *with whatever pieces of the puzzle they can. A clear recognition of opportunity seems to be driving this attitude.*

Examples: at this conference it was reported that we finally seem to have gotten the light right—the community is eager to get this commercialized promptly. Meanwhile, it was suggested that we do not yet have the water right (see below). The community wants to learn what else we need to get exactly right before we can move ahead in other areas.

What Attendees Learned

As we have done at each of the earlier symposia, we opened our Friday morning discussion session with this question to individual attendees: *what have you personally learned from attending this conference? It does not have to be new to others, just to you. What are you taking back home from this conference?*

The responses were as follows:

+ SERVICE LIFE CONCEPTS
  - More to "service life" than weathering: hydrolysis, thermal stability, cyclic fatigue, corrosion, environmental fouling ("dirt"), acid rain, bio-fouling ("mildew"), substrate deterioration.
  - Corollary: "weathering" is far more than UV/water/temp resistance.

+ WEATHERING IS A *SYSTEM* PROPERTY, NOT A *MATERIAL* PROPERTY
  - This means that all ingredients, processing variables, substrate issues, and environmental variables must be considered if one is to arrive at true predictions of the natural result.
  - Additives can play a significant role in weathering. Consequences: (a) we need standard reference materials that can be employed in future testing; (b) we need to move on to the contribution that individual and combinations of ingredients have on the service life of a polymeric material in future conferences.
  - "System": there is a need to elaborate on the effect of the *distribution* of physical and chemical properties within the material, particularly as a function of dis-

tance from top surface and bottom interface. Most research treats only an average concentration effect, without regard to distribution within a film.

✦ GETTING IT RIGHT
- We do seem to be finally getting the light right, and it clearly is mattering.
- Thickness of degraded surface may be much thinner than previously thought, perhaps on the order of ~1 μ at most. So, the penetration depth needed for radiation is not all that great. This magnifies the importance of UV absorbers and stabilizers.
- Role of temperature in weathering is important! There are clear Arrhenius effects separate from irradiation effects.
- We may not be getting the water right—actual water inside the film may not be accurately predicted from various models and surrogates (like RH of atmosphere). *[Surrogate: representative property used to estimate another property—e.g., estimating water content of a film from measurement of surrounding water concentration such as relative humidity. Surrogates are frequently used because they are considerably easier to measure. Unfortunately, proper proofs that surrogates adequately predict desired property can be inadequate.]*
- We need to test/revise/improve existing models until we get each component right.

✦ LIMITS TO ACCELERATION AND ADDITIVITY
- There may be some real limits to our ability to accelerate the effects of weathering, and to detect consequences early on during the process. Some examples cited: a change in mechanism at higher temperature; diffusion-limited oxidation; numerous early "transient" events.
- Additivity—is our current spectral subset adequate?
- Is *rate* of dose important? (Reciprocity—do all properties obey similarly?)
- Confounding and interactions versus reciprocity.

✦ HYDROLYSIS
- The controlling factor for hydrolysis in films is the actual water content inside the film during weathering.
- The quantity of water in films is not adequately predicted from such surrogates as relative humidity of the surrounding environment. Separate experiments have shown that "hot" specimens are not in fact "wet" inside the film. Correspondingly, "wet" specimens are not necessarily "hot" as described by such things as black panel measurements. Therefore, better measures of actual moisture content (and temperature) inside films during weathering attempts are needed.

✦ ACID ETCH
- How difficult acid etch test is. Even in natural conditions, acid etching for a given year actually occurs on a relatively few days during the summer.
- Given the small window for acid etch, are there similar windows for other properties?

✦ Hydrophilic versus Hydrophobic

- In the debate over which is the best surface (hydrophilic or hydrophobic), hydrophilic is winning out because of water sheeting off the surface instead of beading. This sharply reduces deposition of fouling materials, and reduces chemistry driven by concentration during drying of beaded drops. Note that this is at odds with most previous thinking, particularly for tropical climates.

✦ Cyclic Processes and Annealing

- Polypropylene does exhibit some annealing behavior ("recovery").
- Cycles: daily, seasonal, annual. Problem: readings typically ignore one or more of these cycles. This means that the effective dose versus time for natural exposure can be quite different, depending on the observation interval and time of initial exposure. This may in turn limit attempts to predict service life performance based on lab tests. The solution may be continuous monitoring in both lab and field. (See the following section).

✦ Data Collection and Scale of Readings

- Readings: six months; while weather is minute-by-minute.
- Event-driven versus continuous degradation.
- Transport: oxygen, water, and reaction products all occur on a microscopic scale, not on a macro scale. The nanoscale is probably where we should focus for the future, both from observation and action viewpoints.
- Equilibrium stirred kinetics versus solid state behavior: Most of our chemistry training has been based on equilibrium behavior of stirred systems, while the items of interest for SLP are of the solid state, which is certainly not in equilibrium. More research is needed on this important difference, and its consequences for attempts to predict service life.
- How *sophisticated* monitoring has become (short interval).
- Need detailed spectral UV and temperature readings as a description of "weather." *[Attendee who submitted this appears unaware of numerous papers on both issues presented at previous conferences, which are covered in proceedings volumes referenced elsewhere.]*
- Chalking and gloss loss?

✦ The X-Axis of Weathering Degradation Plots

- Have not finished learning about x-axis.
- There does not appear to be a single x-axis, which limits the universality of any approach.
- Ultimately must translate x-axis back to time for commercial use.
- Natural weathering varies in dose so perhaps dosage, not time, is the correct metric.
- Y-axis may be more difficult than x-axis (but we do have the math).

✦ Remote Plasma Shows Promise

- Earlier work had used direct plasma, which caused all sorts of artifacts. However, remote source seems to be a marked improvement.

+ **COMPUTER IMAGING AND DATABASES**
  - Computer rendering provides an excellent way to visualize complex issues and communicate complex science. Vast databases used.

KEY SCIENTIFIC ISSUES THAT STILL NEED TO BE ADDRESSED

The attendees were asked to step back, reflect on both the proceedings and their own personal knowledge of service life prediction, and then list what they considered to be the key scientific issues that still need to be addressed. This is in no way intended to detract from the considerable excellent research that has been conducted to date.

+ The next steps need to be:
  - Testing for component (coating ingredient) contributions.
  - Expanding polymer classes beyond acrylic-melamine and epoxies to acrylic latex and aliphatic urethanes.

+ *SERVICE LIFE IS MORE THAN JUST WEATHERING*
  The effects of temperature, moisture, and UV irradiation are important, but only a start. The rest of the story needs to be included to achieve anything close to *service life* prediction. There are fouling effects, cyclic stress effects, annealing processes, etc. In addition, in some cases, we are not allowing enough time during cycles for diffusion-controlled processes to approach equilibrium.

+ How to assess real service conditions (presumably a form of continuous monitoring for multiple factors)?

+ **ROLE OF INFANT MORTALITY (EARLY) FAILURE ON SERVICE LIFE PREDICTION**
  (quality problem): substrate preparation, "robustness," early "transient" events. We have not really addressed the infant mortality effect on service life. All our efforts to date relate to the wearing-out process, and carefully screen out early failures as unrepresentative. (*See attached tutorial on Miner's Plot*).
  - Weathering actually produces heterogeneous effects on films, yet "average" assessments are made in an attempt to quantify the results. This may be inappropriate, analogous to infant mortality effects versus wearing out effects.
  - Tend to choose known robust test systems. This causes us to filter out robustness deficiencies by the way that we design our experiments.

+ All the science we have learned during our academic experiences has been either equilibrium or steady-state in nature. The real world is distinctly non-equilibrium and non-steady-state in nature. We need some help in dealing with this fact.
  - *Non-equilibrium* issues vs. *equilibrium* testing:
  - Instantaneous dose vs. cumulative dose
  - Discrete vs. continuous events
  - Fouling

+ The boundary between short-term and long-term mechanisms is presently unclear, and is potentially critical for understanding and achieving service life prediction.

+ How to limit issues in testing versus service life environment. Some significant attention needs to be given to compiling data that can be used to create new useful metrics (goal: simpler tests).
+ Use failure analysis techniques to differentiate between results of natural and artificial weathering. Use that information to highlight the problem differences between the two environments.

+ NIST should continue their work on origins of appearance.

+ Is it time to pause or pounce?
  • Definitely do not pause. We have made substantial progress, and do not want to lose momentum. However, we may want to change direction a bit.
  • **We should pounce on what we have learned to date and implement that in the form of improved (simplified) test protocols in the commercial arena.**
  • We should measure other polymeric materials (e.g., urethanes, thermoplastic acrylic latexes, vinyls) to determine their characteristic responses to the effects of temperature, UV irradiation, and moisture.
  • Service life is more than just weathering. For future research, we need to change direction by moving on to component (coatings ingredient) effects, fouling effects, cyclic fatigue effects, and annealing effects.
  • **Would like to see much more on prediction experiences from any field that has had SLP successes.**
  • Need to see more of the elephant!

WHAT CAN WE DO TO FOSTER IMPLEMENTATION OF SERVICE LIFE PREDICTION STRATEGIES?

One issue that is usually not covered in scientific symposia is that of implementing what has been learned. From the very beginning of the Service Life Prediction conferences, the organizers have strived to keep the implementation issue squarely in front of the community, in clear recognition that our ability to shorten R&D cycle times for coatings depends on rapidly implementing what has been learned to date regarding service life predictions. To this end, we asked the attendees to tell us what we can do to foster the implementation of what we've learned as a community.

+ FOSTER:
  • What steps can we take as a community to foster implementation of what we have learned to date?
  • What are the top few changes that would foster service life prediction the most?
  • What level in the food chain should handle each step?

+ IMPLEMENT:
  • "Universal" approach: single holistic test protocol that works for all systems at all times, and readily detects new modes of failure before they occur in the field.
  • System-specific approach: set of singular tests designed to detect sensitivity of a given system to given environmental conditions. Not capable of detecting new modes of failure, but viewed as "good enough."

- Realism would suggest substantial (but not exclusive) focus on system-specific steps forward—would produce commercial value now, to support longer range program.
- Any successful implementation must include acceptable risk management.

PERSPECTIVE OF THE ATTENDEES

✦ GET THE LIGHT, TEMPERATURE, AND MOISTURE CONTENT RIGHT IN CURRENT PRACTICES
  - Test equipment suppliers should immediately implement the "right light" in their equipment, and develop retrofits for existing equipment.
  - NIST should tell us what is required to get the temperature and moisture content right (actual moisture inside a film).
  - NIST should determine which issues are critical for other compositions—i.e., expand beyond melamine-acrylics and epoxies.

✦ THE COATINGS INDUSTRY AND THEIR END-USERS NEED TO DEFINE "FAILURE" FOR EACH SYSTEM
  - "Failure" is system-specific (also use-environment-specific).
  - For automotive coatings, scratch and mar resistance is currently more important than "weathering."
  - For exterior architectural coatings, biological fouling and cracking/peeling are currently the most important modes of failure.
  - As one mode of failure is improved markedly, other modes of failure will inevitably appear.
  - The definition process needs to be a continuously evolutionary one, driven by manufacturers and end-users of coatings.

✦ WEATHERING FACILITIES SHOULD IMPROVE QUALITY OF DATA COLLECTED ON THE *EFFECTS* OF NATURAL WEATHERING
  - Six month readings tells us very little about effects that are driven by discrete events (e.g., acid etch) or continuous weathering effects.
  - A good example in the conference was the automation of reading shown by the Rohm and Haas presenter (Schmitt, Chapter 26).
  - This could be used for more frequent/appropriate readings.

✦ NIST SHOULD BRING THE BIOLOGICAL COMMUNITY INTO THE DISCUSSION to deal with the bio-fouling issue.

✦ NIST SHOULD COLLABORATE WITH INDUSTRY TO DEVELOP SIMPLIFIED INDIVIDUAL TESTS TO REFLECT CURRENT KNOWLEDGE AS EACH STAGE PERMITS. This should be the number one goal for all levels of the food chain for the near term.

✦ ALL PARTIES SHOULD PURSUE AUTOMATION and high throughput techniques to substantially improve the productivity of the whole service life prediction effort.

✦ **Coatings Company R&D Labs Must Learn How to Reproducibly Prepare Paint Samples in the Laboratory.** Presently, have few controls, and poor reproducibility. Our plants produce products more reproducibly than our labs can (paint and panel preps). This gap needs to be filled promptly. Researchers know that they need to make single large batches of paints that they wish to test under different conditions to eliminate the preparation error, which is large. There are also shelf life effects, after a paint has been made. We have not talked about how to detect them, or how to deal with them. (This is probably the number one need for coatings companies.)

✦ **NIST Needs to:**
  - Teach industry how to create and poll very large databases.
  - Develop models to identify actual temperature, humidity, irradiance dose *inside* a given film as a function of depth in the film.
  - Move to engineering, quality, test methods, automation.
  - Translate learning into correlations. Too many publications stop short of doing this.

APPENDIX — MINER'S PLOT

Bathtub-Shaped Curve of
Hazard Rate vs. Total Lifetime of Population

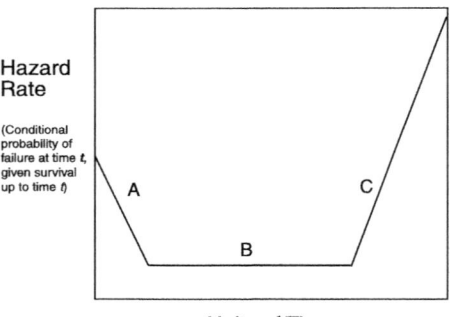

Units of Time

Region A: Early failures can be due to design deficiencies or manufacturing errors, but are more frequently related to improper installation. This can also be seen in some cases as a lack of *robustness* on the part of the product. Type A failures are commonly referred to as infant mortality or burn-in (electronics) failures. Most coatings service life test protocols are designed to ignore these kinds of failures as being unrepresentative.

Region B: Failures are random, and caused by "freak" events such as storms or accidents, which are unrelated to the normal survivability of the coating.

Region C: Failures represent the normal "wearing-out" process, and are indicative of end-of-life. These typically are referred to as the *capability* of the product. Most coatings service life test protocols are designed to detect only these kinds of failures.

Product service life and product liability are quite different issues. Most product failure claims originate from deficiencies in installation or robustness (type A, aka "infant-mortality failures"), rather than from less-than-expected long-term lifetimes (type C, aka "capability"). Some wide-ranging examples of abnormal early failures include early infant deaths (crib death, premature birth problems, immune deficiencies), burn-in failures of electronic circuits, early failures in mechanical devices due to installation errors, deficient training or capabilities of installers, and using product for inappropriate purposes.

Early failures usually have different causes than normal lifetime failures,[1] relating to errors in installation of the product or to insufficient tolerance of the range of adverse conditions that are routinely experienced during installation. An example of the latter would be attempting to establish adhesion while the substrate was wet. Thus, early failures occur and are observed very early in the life of a product, but are not related to the wear-out processes of longer lifetime failures. It is important for the reader to note that early failures are not predictors of wearing-out processes; they are different in origin, and therefore require different solutions than wear-out failures do.

A product is said to be robust if it has no early failures. In the case of coatings, both installation deficiencies and lack of tolerance of field conditions by the coating play significant roles in early failures. Unfortunately, coatings manufacturers rarely have control over the installation process, particularly for field-applied coatings.

Unfortunately, standard testing protocols tend to filter out these early failures as being unrepresentative or irrelevant.[2] It is common practice in the coatings industry to design experiments to avoid variables of type A, since they are viewed as non-representative of the long-term capability of the coating product. Herein lies the problem.

Author's comment: this plot and commentary were added to this chapter at the request of attendees who had not previously seen them. This appendix should be considered in the context of the chapter itself.

References

(1)   Floyd, F.L., "Risk Management: The Real Reason for Long Product Development Cycle Times." In *Service Life Prediction of Organic Coatings: A Systems Approach*, ACS Symposium Series # 722, Oxford University Press, p. 21+, 1999.

(2)   Floyd, F.L., "Reducing Product Development Cycle Times Without Increasing Risk," *J. Coat. Technol.*, *70*, No. 876, p. 70-81 (1998).

Index

Printed by Printforce, the Netherlands